Faunal Heritage of Rajasthan, India

General Background and Ecology of Vertebrates

B.K. Sharma • Seema Kulshreshtha
Asad R. Rahmani
Editors

Faunal Heritage of Rajasthan, India

General Background and Ecology of Vertebrates

Springer

Editors
Dr. B.K. Sharma
Associate Professor & Head
Department of Zoology
RL Saharia Government PG College
Kaladera (Jaipur), Rajasthan, India

Dr. Seema Kulshreshtha
Associate Professor & Head
Department of Zoology
Government Shakambhar PG College
Sambhar Lake (Jaipur), Rajasthan, India

Dr. Asad R. Rahmani
Director
Bombay Natural History Society
Mumbai, Maharashtra, India

ISBN 978-1-4614-0799-7 ISBN 978-1-4614-0800-0 (eBook)
DOI 10.1007/978-1-4614-0800-0
Springer New York Heidelberg Dordrecht London

Library of Congress Control Number: 2012954287

Printed on acid-free paper

Springer is part of Springer Science+Business Media (www.springer.com)

यज्ञवत् भूमंडलम् धत्त समृगवनकाननम् ।
तावत् तिष्ठति मेदिन्याम् संततिः पुत्रपोत्रिकी ।।

So long as the earth bears woods and forests full of wild animals
Will it continue to support the progeny of man.

- Atharva Veda

The ***Atharvaveda*** *(Sanskrit: अथर्ववेदः, Atharvaveda), a tatpurusha compound of atharvan, an ancient Rishi, and veda (meaning "knowledge") is a sacred text of Hinduism and one of the four Vedas, often called the "fourth Veda". The other three Vedas are the Rigveda, the Yajurveda and the Samaveda. The Atharvaveda is a collection of spells and incantations, apotropaic charms and speculative hymns.*

To my father (Late) Prof. J.K. Sharma, an epitome of valor, self esteem and versatility whose untimely and hapless demise left me speechless but his presence still dwells in my convictions and strength and to my mother, Mrs. Prabha Shalini, whose blessings continue to support my endeavors.

–Dr. B.K. Sharma
Chief Editor

Foreword

The contents of this diligently edited work enrapture me. They have an intimate link not only with my core subject of Zoology but also other areas of my interest as Secretary of the Zoological Society of London (ZSL) responsible for London and Whipsnade Zoos, the Institute of Zoology and our worldwide conservation programs. Fauna have always attracted me as an avid observer, interpreter, and reader; the present volume wonderfully describes and analyzes the vertebrate faunal abundance of Rajasthan, currently the largest state of the Indian republic.

Being home to the most exotic biological diversity, splendid ecosystems and colorful cultural heritage, Rajasthan has fascinated researchers, conservationists, academics, travelers, and tourists from around the globe. I am particularly impressed with the fact that through this well-researched work, the editors have achieved an extraordinary accomplishment not only in further unveiling the well-known Thar or Great Indian Desert but also putting in the spotlight the much lesser known yet ravishing wilderness, communities, lush green landscapes, and wetlands of Rajasthan. More than 600 illustrations are a direct testimony to this. These two volumes are an assemblage of what is bound to become some of the most sought after chapters and brilliantly synthesized scientific information available. The content of this monumental yet modern faunal treatise will surely make it a distinguished contribution to knowledge in the area of faunal ecology and conservation. The first book (Volume-1) entitled "*Faunal Heritage of Rajasthan, India: General Background and Ecology of Vertebrates*" in its 24 chapters covers a spectrum of vertebrate fauna of the region. Individual chapters dedicated to threatened faunal species are of special significance in the contemporary setting. The second book (Volume-2) entitled "*Faunal Heritage of Rajasthan, India: Conservation and Management of Vertebrates*" aptly describes the conservation- and management-related aspects spread over 20 chapters.

This publication will be highly appreciated since there is no comparable account currently available. I am delighted to find that the physiographic and biodiversity profile of Rajasthan, conservation strategies covering a vision on the future of the fauna of Rajasthan, and information that fills significant gaps in research each find a bold presence in these superbly edited volumes. In addition, separate chapters on ecotourism, community conservation, and wildlife trade covered in Volume-2 will be useful resources, introducing concurrent themes for researchers interested in this part of the world. The editors have effectively revised the image of the Thar from that of merely a desert to a more vivid landscape housing some of the most resplendent and majestically unique fauna and flora.

The opening chapters of the first volume provide a well-focused introduction to Rajasthan as a vivacious state of India. The historical, sociocultural, mythological, and anthropological aspects of faunal conservation and the tribes of Rajasthan together with the fossil records set the scene for the book. I am certain that this work will serve not only zoologists, wildlife biologists, conservationists and natural scientists, and social scientists but also the general reader. Students, teachers, and active researchers on wildlife and conservation biology will find these volumes particularly valuable as an important reference and textbook. Although, there are many lacunae in our knowledge about faunal ecology and its conservation, this formidable twin volume set will surely help in bridging the gaps, while enabling conservationists and policy makers to arrive at a consensus regarding future strategies in Rajasthan. The editors have superbly compiled the latest information on both the ecology of Rajasthan and the conservation of the region's myriad vertebrates.

Today, when mankind has encroached, exploited, and decimated the natural habitats of our planet, and we find ourselves in the midst of impending environmental calamities, these volumes will spur a sense of responsibility towards nature; they deserve to create mass awareness about sustainable development, conservation, and management of our forests, wildlife, and natural resources.

I extend my hearty congratulations on the publication of *Faunal Heritage of Rajasthan, India* as two separate yet closely linked volumes. I further take this opportunity to congratulate wholeheartedly the dynamic chief editor Dr. B.K. Sharma, for it was he who conceived, carried forward, and delivered this massive work. I wish him further good fortune in his academic and professional endeavors.

March 21, 2011

Paul H. Harvey CBE, FRS
Professor & Head
Department of Zoology, University of Oxford
Tinbergen Building, South Parks Road
Oxford OX1 3PS, United Kingdom

Preface

The Indian landmass, a cache of the top 12 mega-biodiversity regions of the world, proudly owns the famous ***desert state*** of Rajputana (old name) or Rajasthan with a human population of 68,621,012 (Census of India—2011). Endowed with diverse physiographic features, Rajasthan is often addressed as the state of ***Maru*** (desert), ***Meru*** (mountain), and ***Mal*** (plains). Best known as an exotic state of India where tradition and glory meet in a riot of colors against the backdrop of desert and greenery, Rajasthan has an unusual diversity of people, customs, cultures, costumes, music, manners, dialects, cuisine, and physiography. There is a haunting air of palpable romance about this famous abode of kings that never ceases to intrigue and enchant. The state owes much of its charisma to its enduring traditional way of life. So rich is the history of the land that every other village has its own tales of valor and sacrifice, the winds sing them and the sand shift to spread them further beyond. The panoramic outlook of the state is simply amazing with lofty hills of Aravallis and the golden sand dunes of the Great Indian or Thar Desert. In fact, no other region in the country is a conglomeration of so many paradoxes. Rajasthan is a land of superlatives where everything is breathtakingly fascinating. The state harbors a wealth of mesmerizing palaces and invincible forts of the erstwhile rulers and Maharajas, magnificent heritage ***havelies*** (palatial houses of olden times) and monuments, beauty of sorts, and natural resources. The vibrant and colorful attires, traditions, fairs, festivals, and pilgrimage sites attract a large number of national and international tourists. Rajasthan leaves a person truly spellbound with its glorious past, mysterious desert, art and craft, and rich cultural heritage. Located in the western part of the country, the state has been the hub of historical, cultural, archeological, and social activity. The triumph of Rajasthan's Rajput military architecture shone recently on the global stage with UNESCO granting six hill forts (Amber, Chittorgarh, Gangron, Jaisalmer, Kumbhalgarh, and Ranthambore) from the state World Heritage status in an unprecedented acknowledgement at the 37th meeting of the World Heritage Committee held at Phnom Penh, Cambodia on June 21, 2013.

Embracing 5.50 % of the total population of India, the state of Rajasthan presents an irregular rhomboid or kite shape in India's political map at 23°30" North latitude and 78°11" East longitude. Divided into 33 districts, the state covers an area of

342,239 km^2 (826 km × 869 km) which is equivalent to 10.41 % of the total area of the country. The 1,070 km long boundary forms the international border between India and Pakistan and touches four districts of Rajasthan, namely, Barmer, Jaisalmer, Bikaner, and Sri Ganganagar. Other states around Rajasthan are Punjab and Haryana in the north, Uttar Pradesh in the southeast and Gujarat in northwest. It is interesting to note that the size of Rajasthan is more than double of that of England. Physiographically, the state is bordered by the plains of the Sutlej–Beas Rivers in the northeast, the Ganga–Yamuna Rivers in the east, the Malwa Plateau in the southeast and the Gujarat Plain in the south. Out of the seven mountain ranges of the country, Aravalli and Vindhyas run across Rajasthan. In addition, Rajasthan also has the Indo-gangetic Plains. In spite of considerable aridity, unfavorable climatic conditions, and a limited forest cover, Rajasthan has an unmatched broad ecological spectrum with many biodiversity-rich areas. A wide variety of ecosystems, climatic and topographical conditions, and an intricate network of biological diversity and the tribes depending on them make Rajasthan a spectacular geographical region of the world.

Besides, there are a number of potential ecotourism sites boasting dense forests and scenic waterfalls still unexploited by the rapid pace of civilization. Hadoti region and Abu Hills of southwestern Rajasthan, for example, attract thousands of native and international tourists, naturalists, and researchers. Rajasthan has extensive wetlands too. In addition, the traditionally conserved ***Orans*** and ***Gauchars*** or Common Property Reserves and the age-old water conservation structures are also indicative of the wise use of the limited natural resources since ancient times.

This largest state of India can be divided into four major physiographic regions: (a) 640 km long Great Indian or Thar Desert with barren hills and rocky and sandy plains extending over an area of 175,000 km^2 in the west, (b) the forested Aravalli Hills covering the central districts while running from Khetri in the northeast to Khedbramha in the southwest over a length of 550 km and dividing the state obliquely into arid western part and semiarid semihumid eastern part, (c) The Eastern Plains covering the northeastern districts with rich alluvial soil and further named as the lowlands of the Chambal Plain, the Banas Plain, and the middle Mahi or Chhappan Plain, and (d) the southeastern Hadoti Plateau covering the southeastern districts which forms the Archaean Shield of the Deccan Peninsula and is divided into Vindhyan Scarpments and Deccan Lava Plateau. The desert occupies 61.1 % of geographical area followed by 23.3 % of the area covered by the Eastern Plains, 9.56 % by the forests (mainly confined to Aravalli and Vindhyan Hills) and 7.1 % by the vegetation. The average rainfall in Rajasthan is 54.78 mm. The climatic profile of Rajasthan can be seen as a transition between two major climatic regions of India, the semiarid east and the arid west. Water resources of the state include Chambal, Banas, and Mahi Rivers, their tributaries, and the fresh water and salt lakes, while Shekhawati, Marwar, Mewar, Wangar, Hadoti, Bangad, and Mewat present the cultural divisions. Forests of the state are divided into four broad types, namely, tropical thorn forests, tropical dry deciduous forests, central Indian subtropical forests, and mixed miscellaneous forests. The habitat variation in Rajasthan can be witnessed not only in the desert but also in the grasslands, dense forests,

Malwa Plateau, wetlands (both fluvial and nonfluvial), agricultural fields, ravines of Chambal River System and even the rocks, ruins, and civil structures. In the state of Rajasthan, aridity increases from east to west and south to north. The southern and southeastern parts of the state are richer in forest and biodiversity. A layer of soil of varying depth is present on the slopes of mountains of southern Aravallis, which makes it suitable for growing many mesophytic, tuberous climbers and lianas species. The area to its east is well-drained by several integrated drainage systems, while the area to the west has only one integrated drainage system, namely, the Luni drainage system situated in the southeastern part of the desert. However, Indira Gandhi Nahar Pariyojana (Indira Gandhi Canal Project or IGNP) has changed not only the fate of the Thar Desert due to the upsurge of a large irrigated area but also the vegetation and the habitat characteristics of the faunal elements.

The world has two types of desert climates, namely, warm and cold. The Thar Desert of Rajasthan falls under the former category with some extremely sensitive and fragile ecosystems surviving under severe and hostile conditions. The Thar Desert in India extends from the Sutlej River, surrounded by the Aravalli Range on the east, on the south by the salt marsh known as the Rann of Kutch (parts of which are sometimes included in the Thar) and on the west by the Indus River. The Indian Thar Desert joins the Iranian Desert and through it, the desert of the Middle East and finally the Sahara. The Thar Desert covers about three fifths of the total geographical area of Rajasthan and 9 % of the total geographic area of India. Of the total desert in India, 61 % falls in Rajasthan, 20 % in Gujarat, and 9 % in the Punjab and Haryana states. In Rajasthan, the Thar Desert covers 12 districts, namely, Bikaner, Churu, Hanumangarh, Sriganganagar, Pali, Jalore, Barmer, Sikar, Jhunjhunu, Nagour, Jaisalmer, and Jodhpur.

The Indian subcontinent is a major ecological region of the Indomalaya ecozone, one of the eight ecozones dividing earth's land surface, and the state of Rajasthan broadly falls under the Indomalaya ecozone. According to the Wildlife Institute of India (WII), India has a total of ten biogeographic zones and 26 biotic provinces. Rajasthan falls under the biogeographic zones 3 and 4, namely, the desert, covering the Thar and semiarid area covering the rest of the Rajputana or Rajasthan. The biological diversity of India is one of the most significant in the world since it has only 2.4 % of the total landmass of the globe but is home to over 7% of its animal species. Out of the total 1,196,903 animal species recorded from the world, India has 86,874 species, which means India is home to a little more than 7 % of the total animal species of the world.

The fauna of Rajasthan actually resembles that of West Asia and North Africa more than it resembles the fauna of most of the rest of India. The faunal richness of Rajasthan encompasses 140 species of fish, 14 amphibians, 67 reptiles (including eight endangered reptiles and five falling under Schedule I of the Indian Wildlife (Protection) Amendment Act, 2006, 477 birds (including 6 Critically Endangered, 5 Endangered, 12 Vulnerable, 19 Near-Threatened, 14 Red Data species, and one Conservation Dependent species) and 87 species of mammals (including seven of endangered mammals and ten species falling under Schedule I of the Act. Himalayan Tree Frog (*Polypedates maculatus*) has been recorded from Bansi (near Sitamata

Sanctuary), Banswara, and Jhalawar districts The reptilian species of significance include Isabelline Vine Snake (*Ahaetulla nasuta isabellinus*), Foresten's Cat Snake (*Boiga forsteni*) at Aravallis and a species of Lizard, *Phrynocephalus laungwalansis* endemic to Jaisalmer. Isabelline Vine Snake was recently seen in Phulwari and Kumbhalgarh Sanctuaries also. Indian Starred Tortoise (*Geochelone elegans*) was once common in some districts of Rajasthan but its population has severely declined recently. A significant population of Gharial *Gavialis gangeticus* is seen in the river Chambal even today.

The grasslands of eastern Rajasthan and parts of western Rajasthan hold significant number of the Lesser Florican *Sypheotides indicus* during monsoon. Unfortunately, the population of Indian White-rumped Vulture *Gyps bengalensis* and Indian Vulture (former name Long-billed Vulture) *G. indicus* has drastically reduced in the whole of Rajasthan by almost 99 % due to the killer drug diclofenac. The eastern plains are famous for Keoladeo National Park (KNP)—the only World Heritage Site in Rajasthan (so far as protected areas are concerned) which is home to thousands of migratory avifauna and other rare and endangered animals. The Southern Rajasthan harbors mammalian fauna such as the Mouse Deer or White-spotted Chevrotain (*Tragulus meminna*), Common Palm Squirrel (*Funambulus palmarum*), and Elliot's Giant Flying Squirrel or Large Brown Flying Squirrel (*Petaurista philippensis*) which are not found anywhere else in Rajasthan. The River Chambal has Gharial (*Gavialis gangeticus*), Marsh Crocodile or Mugger (*Crocodylus palustris*), and Gangetic River Dolphin (*Platanista gangetica*) apart from a variety of fishes. The presence of Wild Dog or the *Dhole* (*Cuon alpinus*), continues to be doubtful in Rajasthan.

The area south of 24°30 N latitude, Mt. Abu, Phulwari, Sitamata and Pratapgarh, Kumbhalgarh, and Shahabad (Baran district) is rich in biodiversity. Many species found in the Western Ghats and Deccan trap are also present in this zone and have small distribution range in Rajasthan. Western and northern distribution limits of many peninsular species end in this zone. Many of these species are generally not present in distribution north of 24°30 N latitude. Scimitar Babbler, *Pomatorhinus schisticeps obscurus*, Orange-headed Thrush (old name, White-throated Ground Thrush), *Zoothera citrina cyanotis*, Green Avadavat (*Amandava formosa*), Rajasthan Red-whiskered Bulbul (*Pycnonotus jocosus abunesis*) and Aravalli Red Spurfowl *Galloperdix spadicea caurina* are endemic to Mt. Abu and the surrounding area. In addition, both Jungle Babbler and Large Grey Babbler are found in the thorny deciduous forests/scrub jungles, whereas some species of creepers are also present in the open deciduous forests of Rajasthan. Only one species, namely Stoliczka's or White-browed Bushchat *Sexicola macrorhynca*, is confined to the Thar Desert. The Orange-headed Thrush is also seen in wet, cool, and shady pockets of Phulwari and Sitamata Sanctuaries. The White-naped Tit *Parus nuchalis* has been confined to arid zone of Rajasthan, Gujarat, and few pockets of South India. Leaf birds and fairy birds belonging to the family Irenidae, namely, Common Iora *Aeginnnthina tiphia* (Lineaus) and Marshall's Iora *A. nigeolutea* (Marshall) are also found at Mount Abu in Rajasthan while Pittas are winter visitors. Vindhyan Gorges (scrapland) has edaphic climax of *Anogeissus pendula* on its vast tract and comprises of Rock Bee

(*Apis dorsata*), Egyptian Vulture (*Neophron perctopterus*), Indian Vulture (old name Long-billed Vulture), *Gyps indicus*, Oriental Honey-buzzard (*Pernis ptilorhynchus*), Indian Pitta (*Pitta brachyura*), Tickell's Blue Flycatcher (*Cyornis tickelliae*), Indian Chat (old name, Brown Rock-chat), *Cercomela fusca,* and Eurasian Eagle-owl (*Bubo bubo*). Besides these species, the state holds Indian Skimmer *Rynchops albicollis*, congregations of Flamingo *Phoenicopterus roseus* at Sambhar Lake and Demoiselle Crane *Anthropoides virgo* at Khichan in Jodhpur district.

The mammalian fauna of the Thar Desert is diverse with nearly 68 species, which constitute about 18 % of total Indian mammals. The Wild Ass was not seen in Rajasthan for the last decade until a few were sighted in 2003 in areas adjoining Gujarat from where these animals had actually moved or rather sneaked. Except for Chinkara *Gazella bennettii* and in some areas Blackbuck *Antilope cervicapra*, the status of all the larger mammals is unsatisfactory and a few such as Caracal *Caracal caracal* is threatened. Of the 68 species, 29 species are listed in the Indian Wildlife (Protection) Amendment Act, 2006 and hence, need protection, though to a varying degree. Chinkara and Blackbuck are considered sacred by the *Bishnoi* community and are present in large numbers around the *Bishnoi* villages. The Nilgai *Boselephus tragocamelus* has a wide distribution in Rajasthan, but in the Thar Desert, unlike the Chinkara, it is not seen in the extreme arid areas where surface water is not available for most part of the year. Among bats, 12 species and subspecies are found in the Thar Desert. There is a great need to conserve the threatened wildlife of the Thar including some resident and migrant birds for which Rajasthan is globally important for conservation such as the Great Indian Bustard *Ardeotis nigriceps* (Critically Endangered), Lesser Florican *Sypheotides indica* (Endangered), Houbara or Macqueen's Bustard *Chlamydotis undulata* (Near Threatened), Stoliczka's or White-browed Bushchat *Saxicola macrorhyncha* (Vulnerable), Pied Tit or White-naped Tit *Parus nuchalis* (Vulnerable), Green Avadavat *Amandava Formosa* (Vulnerable), Demoiselle Crane *Anthropoides virgo*, Imperial or Black-bellied Sandgrouse *Pterocles orientalis* and vultures. Khichan in Jodhpur district is known for a large congregation of the winter visitor Demoiselle Crane *Anthropoides virgo*; a separate chapter has been dedicated to this species. Since effective ecological barriers are absent in Rajasthan, isolation is not effective and endemism is not so prominently seen.

Threatened mammals for which Rajasthan is globally important are Tiger *Panthera tigris*, Chinkara *Gazella Bennettii*, Nilgai *Antilope cervicapra*, Grey Wolf *Canis lupus,* and Caracal *Caracal caracal*. Presence of gorges (locally called *Khoh*) is a typical feature of Vindhyas, and they make an important tiger habitat. The population of Sloth Bear *Melursus ursinus*, an important species of the state confined to southern Aravallis and southeastern parts, is also decreasing. Large Brown Flying Squirrel (old name, Elliot's Giant Flying Squirrel) *Petaurista philippensis* is a characteristic fauna of the southern region. Likewise, the number of Golden Jackal, Gray Wolf, Bengal Fox, Red Fox, and Striped Hyaena is also declining in the agricultural zones. Mammals such as Gangetic River Dolphin *Platanista gangetica* and Smooth-coated Otter *Lutrogale perspicillata* are also present in the River Chambal. Two major carnivores, the Asiatic Lion *Panthera leo persica* and the Asiatic Cheetah *Acinonyx jubatus venaticus* became extinct during the last 65–100 years, and the Wild Ass *Equus hemionus khur* has become extinct in Rajasthan four decades ago. In addition, the traditionally conserved *Orans* and *Gauchars* or Common Property Reserves and the age-old water conservation structures are also indicative of the wise use of the limited natural resources since ancient times. The religious beliefs and sociocultural traditions of the people of Rajasthan have contributed a great deal to the preservation of wildlife. Temples dedicated to various animals are a strong testimony to this, indicating the faunal linkages of the people, whereas birds such as *Kurjan* (Demoiselle Crane), Parakeet, Indian Peafowl, and House Crow are favorite themes of the Rajasthani folk music since time immemorial. *Bishnois* of Rajasthan stand apart from countless other sects and communities in India for their commitment to protect wild plants and animals. Amrita Devi, a *Bishnoi* lady who along with 363 villagers was martyred in the year 1730 while trying to stop tree-cutting by men of the then-ruler at the Khejadi village near Jodhpur district, is a burning example of the passion of *Bishnois* toward biodiversity conservation. Saako-363 Amrita Ki Khejadi (Hindi: साको - ३६३ अमृता की खेजडी) is an upcoming Hindi movie produced by Suraj Bishnoi and directed and written by Kalyan Seervi under the banner of Shri Maruddhara Films Pvt. Ltd. This film is based on the true story of Amrita Devi – a Bishnoi woman who fought with and revolted against the Deewan (Chief Minister of the Ruler) of the then *Jodhana* realm and his men to save Mother Nature and to particularly protect the ambient flora and fauna in her locale, *Khejarli* Village near Jodhpur. Planned to be shot in Rajasthan, the movie has the famous Bollywood actress Gracy Singh as the main lead and is expected to release in December 2013. The story is about a fearless woman's trials and tribulations to save the environment, a topic so relevant in the present times. The makers of the movie who belong to the Bishnoi Community have added a special clause in the film agreement whereby the cast and crew have been asked to abstain from non-vegetarian food and alcohol till the shooting is completed. The ethics of conservation nurtured by saints and spiritual teachers such as Guru Jambheshwarji, the great environmentalist of the fifteenth century, are deep-rooted in the religions and culture of Rajasthan. Unfortunately, the current generation seems to have been distancing from religious ethos and values regarding zoolatry.

Tribes constitute 12 % of the total population of Rajasthan, among which 39 % are *Bhils*. *Meena* is the second largest tribe. In fact, the southern belt of the state including whole of the *Mewar* and partly *Marwar* (Sirohi district) is together known as the tribal belt. The other well-known tribes and nomads of Rajasthan include *Gadiya Lohar, Garasia, Saharia, Damor, Bawaria, Mogiya, Meo, Banjara* (traveling tribes), *Kathodi, Rebari* (cattle breeders of *Mewar* region), *Sansi,* and *Kanjar*. Unfortunately, hunting continues to be an integral part of the socioeconomic life of tribes such as *Mogiya, Bawaria,* and *Pardhi*. They were recently held responsible for the killing of hundreds of tigers and panthers in the protected areas (PAs) and wilderness of Rajasthan, Gujarat, and Madhya Pradesh states during the current decade. In addition, *Pardhi* tribe hailing from Gujarat is the most skilled of poachers and frequently operates in the neighboring states of Madhya Pradesh and Rajasthan and even in the far eastern states. Tribal rehabilitation programmes (especially for

Mogiya) have been quite successful in providing them alternative sources of livelihood, child education, and female empowerment. They have also introduced young men to tourism, antipoaching activities, and cultivation of medicinal plants.

The state animal is the *Chinkara or Indian Gazelle*, the state bird is the Great Indian Bustard *Ardeotis nigricepes* (*Godavan* in Hindi and Rajasthani dialect), the state tree is *Khejri Prosopis cineraria*, and the state flower is *Rohida Tecomella undulata*, popularly known as the Desert or Marwar Teak. Oil India Limited's discovery of natural gas in the year 1988 in Jaisalmer basin has made Rajasthan into one of India's major revenue-producing states. Recent discovery of the presence of oil and natural gas in Deengod area of Hadoti region by the Oil and Natural Gas Commission (ONGC) presents another success story. On 14th March, 2013, a memorandum of understanding (MoU) was signed between the Government of Rajasthan and Hindustan Petroleum Corporation Limited (HPCL) for setting up a refinery-cum-petrochemical complex in the Barmer region. The *Mangla* oil field of Barmer would be the largest hydrocarbon zone in India and an industrial hub for India's petrochemical sector. Commercial production via this project is likely to begin from January, 2017. Currently, the income from crude oil is 50,000 million rupees, which is likely to increase by 3,500,000 million rupees per annum after the refinery is established. The tentative name given to this mega project is "Rajasthan Refinery Pariyojna." At present, in Rajasthan, *Cairn* is producing 175,000 barrels of crude oil per day. There are a total of 25 oil refineries in the country, which have a total processing capacity of 200 million tons of crude oil. The 9 million ton refinery, to come up in the Barmer-Sanchore basin, has been a long-pending demand of Rajasthan. About half of the crude oil processed at the refinery would come from Barmer's Mangala, Bhagyam, and Aishwarya oilfields operated by *Cairn India*, while the balance would be imported. At present, *Cairn* produces 175,000 barrels of oil per day (8.75 million tons a year) from these oilfields, with a potential to produce as much as 300,000 barrels (15 million tons) per day. The project would pave the way for setting up other ancillary industries, generating employment opportunities for about 170,000 people. The proposed complex would be the first such one specifically designed to produce petrochemicals from indigenous crude oil. On June 24, 2013 the Government of Rajasthan has finalized Pachpadra in Barmer district of western Rajasthan as the location for setting up a refinery worth 37,00,000 million Indian rupees. A new company called *HPCL Rajasthan Refinery Limited* (HRRL) came into force following a joint venture agreement between the state Government and HPCL on July 11, 2013. The project is likely to begin this year itself in November.

Economy of the state rests largely on monsoon-dependent agriculture, mining, stone-cutting and polishing, cement, zinc, textiles, and tourism. Livestock of the state includes the highest population of camels in India. Apart from threats to biodiversity conservation, frequent drought, illiteracy, female foeticide, child marriage, unemployment, population pressure, water scarcity, and poverty are key issues to be seriously handled if Rajasthan has to become one of the leading states of India. Conservation landmarks of the state cover 2.80 % (9,121.61 km^2) of the total area with three national parks namely, Keoladeo and Ranthambhore and the recently notified Mukundara Hills; 26 wildlife sanctuaries (WLS); four conservation reserves; two eco-sensitive zones; two Ramsar Sites (Keoladeo National Park [KNP] and Sambhar Lake); KNP is also a World Heritage site; one proposed biosphere reserve [Desert National Park (DNP)]; two tiger reserves (Sariska Tiger Reserve and Ranthambhore National Park); five zoos/zoological gardens at Jaipur, Udaipur, Bikaner, Kota, and Jodhpur; one private zoo at Panchwati, Pilani; two biological parks; ten safari parks/deer parks; and 24 Important Bird Areas (IBAs) as identified by Bombay Natural History Society, Mumbai, India. KNP and Sajjangarh Wildlife Sanctuaries are walled protected areas (PAs) while the National Chambal Water Sanctuary is a ravine system. On the other hand, Taal Chhapar and Gajner Wildlife Sanctuaries in Churu and Bikaner districts are the PAs of arid zone. On May 17, 2013, the State Wildlife Board, Department of Forests, Government of Rajasthan has declared Jeenmata (Sikar District), Mansamata (Jhunjhunu District), Grass-farm Nursery (Jaipur District) and Mokhla (Jaisalmer District) as the new conservation reserves. The board also decided to increase the forest area of Sawai Mansingh WLS (situated near Ranthambhore Tiger Reserve) to 4,137.40 ha; of Kailadevi WLS to 9,624 ha; of Sariska Tiger Reserve to 39,816.98 ha; and of Bassi WLS to 5,396 ha. Likewise, 86.26 km^2 revenue area of Todgarh-Raoli WLS and some area of Ramgarh WLS which falls under the Bundi city will be excluded. It is interesting to note that majority of PAs of Rajasthan initially came into existence as hunting reserves and private zoos of former kings and royals. *Shikar* (hunting) was a favorite sport of the erstwhile rulers which always found a place in the itinerary of visiting viceroys and British officers in the Pre-Independence era. Royal families in Rajasthan also owned private zoos, most of which were taken up by the government following Independence and later developed as wildlife sanctuaries and national parks. Governed by the National Tiger Conservation Authority (NTCA, formerly "Project Tiger"), the tiger reserves of Rajasthan are of global significance. Following a ruling by the Supreme Court of India and subsequent orders issued by the Central Government, tourism activities will now be shifted from core areas of National Parks to buffer areas. To this end, a tiger safari will be created at the Olwari-Niwari forest area of Ranthambhore Tiger Reserve and Nahargarh Biological Park.

To keep the readers abreast with the overall view of the subject, relevant appendices have also been included. One of the appendices describes the names of different faunal species in local (*Rajasthani*) dialect. I would like to reiterate here that the desert is only a small part of the state of Rajasthan. Contrary to popular belief, Rajasthan has lush green fields, grasslands and cultivated lands, rocky protrusions, hilly terrains, jungles, and extensive wetlands too, strewn along and around the desert. Interestingly, the desert makes the state both famous and infamous due to its extremity of harsh climates and hardships faced by the human and animal inhabitants through various seasons round the year. It is also true that some of the oldest civilizations have emerged in these areas which are by definition called as the zones of scarcity and hardships. On the whole, the state of Rajasthan has remained an amphitheater of zoogeography. The present and past distribution of animals indicates that the state has witnessed many climatic upheavals. Two chapters have been wholly dedicated to understand the retrospective picture of the faunal diversity, geological scenario over a longstanding past (including the Akal

Wood Fossil Park at Jaisalmer) to present some interesting workable knowledge of the geography of Rajasthan. Through these introductory chapters, the authors have tried to convey a galaxy of important facts and information. For example, 40 % of Indian livestock, 80 % of the best camels, rich mineral and stone deposits like zinc, copper, marble (the Makrana marble was used in the construction of the world famous *Tajmahal* at Agra), granite, and 1 % of the total water of India (both ground and surface) is present in Rajasthan. It is not an exaggeration to mention here that there is no such publication currently available either in the Indian or the international market which exclusively deals with the scientific account of the vertebrate faunal diversity. I had been very keen to make it a unique and monumental work on Rajasthan's faunal wealth with well-elaborated and relevant contents and assemblage of varied titles. I have tried hard to give this book a strong scaffolding to carry on the commensurate weight of scientific contents, without losing elegance. The present edited volume has attempted to cover the pattern of distribution of vertebrate inhabitants and wildlife of Rajasthan—many of them finding a mention in the IUCN Red List of threatened species.

The untiring efforts in the direction of compiling this information brought an altogether fresh vision and insight into the present status of vertebrates in Rajasthan. One of the opening chapters deals with an interesting account of the fauna in retrospect, especially large mammals which gradually vanished over the 200 years. The killing of the last Asiatic Lion in 1876, the disappearance of the last Cheetah, the last sighting of the Siberian Crane (migratory) at Keoladeo National Park (KNP) in 2003, the White-naped Tit which vanished in the arid zone after 2005, many vanished species of vultures, the extinction of the Wild Ass, and the doubtful presence of *Dhole* (Wild Dog) indicate the gradually changing environment and anthropogenic pressures in the state. The underlying theme of this mega volume along an evolutionary gradient is self-evident while simultaneously indicating that vertebrate ecology and conservation are two inseparable aspects closely linked in the patterns that have been determined by the course of events in the remote past. In this volume, we have managed to select topics that will serve as a guide and stimulus for synthesis of knowledge that ought to flow from this work on the faunal diversity of Rajasthan mainly focusing the ecology of vertebrates. Contributors of this volume include both seasoned and young scholars, experienced ecologists, forest officials, teachers, social scientists, and life scientists. Please also see *Faunal Heritage of Rajasthan, India: Conservation and Management of Vertebrates*. Sharma B.K. et al. (eds.), Vol. 2, 2013, Springer (978-3-319-01344-2) for conservation management-related aspects.

It may seem illogical that a book on faunal diversity does not include invertebrates, but their inclusion would have made the book unwieldy and hefty. I am fully aware that a few faunal groups, especially the invertebrates, were left to be covered in the present endeavor since there are significant gaps in research and also the information received did not appear scientifically up to date despite collection of over 11 chapters. If feasible, attempts will be made to include these thoroughly revised and updated manuscripts and additional chapters on invertebrate faunal diversity of Rajasthan in a subsequent volume titled *Faunal Heritage of Rajasthan, India: Ecology and Conservation of Invertebrates*.

While exploring around to conceive this book during the early winters of 2006, the I was unpleasantly surprised to witness that though a good deal of research work was carried out on the flora and fauna of the state over the past decades, the available information was largely scattered, fragmented, and patchy. Though the Zoological Survey of India (ZSI), Kolkata; Bombay Natural History Society (BNHS), Mumbai; Wildlife Institute of India (WII), Dehradun; Salim Ali Centre for Ornithology and Natural History (SACON), Coimbatore; World Wide Fund for Nature-India (WWF-India), New Delhi; forest departments and other premier Governmental and non-governmental organizations (NGOs) of the country have done pioneering work on Rajasthan's fauna, such vital information and data have hardly been updated and are available only in discrete journals, separate volumes, monographs, conference proceedings, and small books dedicated to particular faunal groups and/or single species. The need of an efficient database center for providing updates on the current status of existing faunal species, their population and distribution has long been felt. In a recent development on May 12, 2013, the forest department, Government of Rajasthan has proposed to set up a training institute in the state where appropriate training in connection with wildlife, forests, and related aspects would be given by experts from India and abroad. A branch of this "institute of excellence" would be opened at Ranthambhore National Park in Sawai Madhopur. It is worthwhile to mention here that the Wildlife Institute of India (WII) situated at Dehra Dun (Uttrakhand) is currently the only such institute in India. A near total absence of relevant scientific information about the present status of vertebrate fauna and the poor state of efforts towards conservation and management of biodiversity in this part of the world has in fact propelled the Chief Editor to compile this edited volume. A few important papers invited as proceedings of the *National Conference on Conservation and Management of Faunal Diversity of Rajasthan* (*NCCMFDR*), organized by me at Jaipur, India from August 11–13, 2006 have also been included in this book.

It was a Herculean task to present the vertebrate fauna of Rajasthan in a systematic yet scientifically designed and tightly edited volume. A serious effort has been made to structure the manuscripts starting from lower to higher forms while placing them in a format comprising various faunal forms along with chapters focusing on the general biodiversity. For the ease of understanding by the reader, the entire manuscript has been split into four major sections. It was indeed tricky to provide appropriate headings to cover a wide variety of chapters under these heads. Last but not the least, the present edited volume is an earnest attempt towards the scientific documentation of existing vertebrate fauna of Rajasthan. It is hoped that these volumes will be useful for wildlife specialists, conservationists, environmentalists, zoologists, ecologists, researchers, students, policy-makers, and education administrators not only in Rajasthan and India but throughout the globe.

At this crucial juncture when the planet's natural resources are depleting rapidly, the animal life is being driven to its ultimate retreat in the fast diminishing ecosystems, wild creatures are annihilated, the insensitivity of humans towards fellow creatures is increasing and when man's outlook upon the domain of nature has drastically changed—the teachings of Indian philosophy, theology, moral and social sciences can help us to relive the times when the human race had comfortably flourished by affectionately mingling with nature. A serious approach towards wildlife and forests is still lacking in India, the need for which is paramount. In fact, the callous attitude of policy-makers, administrators, politicians, and the intelligentsia coupled with greedy businessmen have badly affected the pace of welfare efforts and implementation of laws. The forgotten concepts of social sciences and the concepts of animal liberation and animal rights also seem pertinent in the present milieu, if India has to survive as a country which always commanded respect of the rest of the world on account of the culture and traditional values. The biggest testimony to this is our honest consideration of the protective umbrella or the environment around us as Mother Nature. A few quotes relevant to the present context and worth mentioning here are: *Man is the only creature that consumes without producing. He does not give milk, he does not lay eggs, he is too weak to pull the plough, and he cannot run fast enough to catch rabbits, Yet, he is the lord of all the animals*—George Orwell; *Life is life—whether in a cat, dog or man. There is no\difference there between a cat or a man. The idea of difference is a human conception for man's own advantage*—Saint Sri Aurobindo; *We can judge the heart of a man by his treatment of animals*—Immanuel Kant; *The Greatness of a nation and its moral progress can be judged by the way its animals are treated*—Mahatma Gandhi.

With its ancient culture, and rich traditional heritage India will surely act as the *Vishwaguru* (world teacher) in the times to come. In the present scenario of terrible unrest, biodiversity conservation is something pragmatic that must be directly linked with education and incorporated in the curricula at schools, colleges, and universities not only in India but the whole world. In a country of rich traditional heritage where *ahimsa parmodharma* (a phrase in Sanskrit language which means that "non-violence is the topmost duty to the extent that it supersedes all other duties") and *Vasudhaiva Kutumbakam* (a phrase in Sanskrit language which means that "the whole world is one single family") are the guiding principles, destruction should have no place. It is high time that we come together to living with nature, commiserate with the harmless animals and join hands to create a symphony of peaceful coexistence. Nature conservation is the key to this concept.

It is a pleasant coincidence that the book was accepted for publication in 2010—the international Year of Biodiversity and was being worked upon through 2011 which marked the beginning of a crucial decade in the International calendar for biodiversity. This was the start of the United Nations "Decade on Biodiversity" and was declared "International Year of Forests." It was a great delight to see that the final proof reading of the manuscript was completed towards the end of 2012—which is marked as the "International Year of Sustainable Energy" and when India hosted the XI Conference of Parties (CoP) on Convention on Biological Diversity (CBD) at Hyderabad.

Jaipur, Rajasthan, India
July 07, 2013

Dr. B.K. Sharma
Chief Editor

Acknowledgements

At the outset, I would like to thank the residual faunal wealth of Rajasthan and the Bishnoi community that had inspired me to create this book. This volume is also dedicated to all life forms (both companion and free-living animals) flourishing in the lap of Mother Nature. In addition, the handful of people I wanted to thank but the list swelled as it generally happens.

Writing acknowledgement for something one has been doing for years can be both easy and tricky. Easy because the person is quite intimate with the innumerable events that have quickly passed by and quietly slipping down the memory lane is the only way to gather them back, this is generally a pleasure. However, it becomes tricky because one cannot always lay down the pain and unpleasant facts and circumstances associated with such a task.

The book actually got conceived during my involvement as Organizing Secretary of the "National Conference on Conservation and Management of Faunal Diversity of Rajasthan" (NCCMFDR) held during August 11–13, 2006. Sponsored by the University Grants Commission (UGC)—the apex body governing the higher education sector in India and the Department of Science and Technology (DST),

Government of Rajasthan, deliberations of this meeting formed the basis of the need to compile such a publication. The idea of compiling the faunal abundance of Rajasthan and aspects of its conservation management in one place arose from the fact that when I desperately searched for a book describing the fauna of Rajasthan, I was wonderstruck not to find one. I wish to humbly acknowledge Dr. A.K. Mathur, the then Principal, R.L. Saharia Government PG College, Kaladera (Jaipur), Rajasthan, India, for rendering his support while I organized the above conference.

Thanks are also due to the Eco-Ethics International Union (EEIU), Germany and colleagues and members, organizing committee of the NCCMFDR-2006 especially Drs. Medhatithi Joshi, Rakesh Lata, (Govt. College, Kaladera) and Abhimanyu Singh Rathod (B.N. College, Udaipur).

I would like to express my gratitude towards my teachers Professors A.L. Bhatia (Late) and Reena Mathur, Former Heads and Prof. Shekhar Verma, Department of Zoology, University of Rajasthan, Jaipur, for their affection, encouragement, and blessings.

I am indebted to Mr. V.D. Sharma, Former Principal Chief Conservator of Forests, Government of Rajasthan; Ms. Geetanjali Bhattacharya, Zoological Society of London (ZSL); Ms. Sally Walker of Zoo Outreach; Dr. Gobind Sagar Bhardwaj, IFS, Department of Forests, Government of Rajasthan; and Mr. Bittu Sahgal of Sanctuary Asia (Mumbai) for their kind support.

I would like to especially mention Dr. G.V. Reddy, I.F.S., Department of Forests, Government of Rajasthan; Dr. S.M. Mohnot, Emeritus Scientist & Director, School of Desert Sciences and Chairman, Primate Research Centre, Jodhpur; and Dr. B.K. Tyagi, Officer-in-Charge, Centre for Research in Medical Entomology (ICMR,) Madurai, Tamil Nadu for their numerous e-mails and valuable suggestions.

I am personally grateful and owe special thanks to Dr. Divyabhanusinh Chavda, Naturalist, Conservation Expert, President, WWF-India and Member, Cat Specialist Group, who despite being extremely occupied spared time for personal meetings and supported this project with his creative ingenuity and invaluable advice. He also made me aware of the sources of relevant information including his own books, other published works, and archival material which otherwise I would not have known. Often, I had to take leave to meet and visit related people and places in Rajasthan and elsewhere in India in order to procure additional materials for this book.

Apart from agreeing to write a chapter at a short notice, Dr. Jeffrey Snodgrass of Colorado University, USA rendered fruitful suggestions which helped in improving the manuscript; I gratefully acknowledge his inputs.

I feel greatly privileged to have received valuable contributions from eminent authors without whom this volume would not have seen the light of the day. I am fortunate that apart from experienced teachers and colleagues, I enjoyed the counsel of some very competent wildlife experts and conservationists during the preparation of this volume.

I take this opportunity to profoundly thank Professor Paul Harvey, FRS, Head, Department of Zoology, University of Oxford, for he penned the foreword despite other potential academic and professional commitments.

Without expressing deep gratitude and thanks towards my scholarly co-editors, the thanks giving will not culminate whose active cooperation actually helped me to complete the book. Were it not for the encyclopedic knowledge of Dr. Rahmani and the hard work of Dr. Kulshreshtha, this volume would have lacked in substance.

My heartfelt thanks to Mr. Sunil Singhal, Mr. Devendra Bhardwaj, Dr. Tejveer Singh, Mr. Niranjan Sant, Mr. Aditya Roy, Mr. Jaysukh Parikh, Dr. Ashish Kothari, Mr. Arfin Zukof (Listening Post), Dr. Rakesh Vyas, Ms. Sonali Singh, Dr. Anil Chhangani, Mr. J.K. Tiwari, Prof. K. Sankar, Ms. Urva Sharma, and Ms. Babette de Jonge (Wild Cats Magazine/Wild Cats World, Masai Mara 2009), Sanctuary Asia Photo Library and the Victoria & Albert Museum, London for providing rare and beautiful pictures from their personal collection. Thanks Mr. Rajiv Lochan, Drs. Narendra Jain and R.S. Bhatnagar, Mr. Anshul Sinha, Ms. Vartika Sinha Pooja & Vijay Bishnoi, Rakesh & Daisy and Mr. Umakant Baluni (UGC) for all your concern and support.

I am overwhelmed with gratitude, pride, and indebtedness towards my siblings—Veena, Anant, Anand, and Ashutosh for they remained a constant source of inspiration and whose everlasting affection helped a great deal while working on this project; thank you for your moral support and good wishes.

It is worthwhile to mention here that right from the call for papers and collection of manuscripts to the interactions with the publishers, editing and proof reading the entire work stretched over five long years. I owe much to my charming daughter Anushka, and son Divyam, who actually grew up with the preparation of this volume and most of the time witnessed me working in my study, enriched the result and remain a great source of comfort for me.

While writing the acknowledgements, my heart goes out to *Jugal Bhawan*—the 65-year-old ancestral house in Jaipur built by my grandfather (Late) Pt. Jai Nath Sharma where I was born with four siblings and lived as a joint family until the end of 2011. The building has been demolished only to be reborn as an apartment called *Jugal Enclave*. The place not only witnessed my upbringing but also my struggles and survival in general and with this writing project in particular. In fact, the majority of the work related to these volumes was completed in a quiet corner of this palatial bungalow.

I immensely thank my publisher Springer for entrusting upon me and making the delivery on time. Janet Slobodien, Publishing Editor; Mellisa Higgs, Assistant Editor; Felix Portnoy, former Production Editor, Jeffrey Taub, the new Production Editor; and Manoranjan Mishra, Project Manager at SPi Content Solutions—SPi Global and his entire team deserve special thanks for doing all possible efforts to bring this peer-reviewed volume in such a refined form. My sincere thanks are also due to the anonymous reviewers whose thoughtful suggestions have greatly helped in improving the original manuscript. I also owe a debt of gratitude to many teachers, friends, colleagues, and students and to all those who have contributed variously in bringing out this volume in its present shape whom I might have missed mentioning here.

For the one person who supported me selflessly and out of the way while I was putting all my heart and soul into this book, working long hours and late nights despite occasional illness and loads of other commitments, she cheerfully stood beside me, I am short of words to wholeheartedly thank my spouse without whom, this mammoth task would not have reached fruition.

I must share with the readers that I had originally planned an almost 1,200 pages-long single volume entitled "*Faunal Heritage of Rajasthan, India: Ecology and Conservation of Vertebrates*" covering both ecology and conservation management under one cover. In order to make a hefty volume handy, Janet Slobodien suggested me to split this book into two volumes, rather two separate books titled "*Faunal Heritage of Rajasthan, India: General Background and Ecology of Vertebrates*" and "*Faunal Heritage of Rajasthan, India: Conservation and Management of Vertebrates*." The splitting may has caused marginal delay in the publication of this work but I am sure, the readers and contributors would appreciate the need to do so.

Above all, I profusely thank God Almighty for giving me enough courage during the gestation and prolonged labor and as always for everything. I would be grateful to the esteemed readers for their comments on the newborn twins. I hope the editors justify their aspirations; however, constructive criticism and suggestions are invited to further improve this volume in its future revisions. The only thing now left to be added is that responsibility for imperfections and failings, if any, are mine alone.

Jaipur, Rajasthan, India
July 07, 2013

Dr. B.K. Sharma
Chief Editor

Contents

Contributors

Harendra Singh Bargali The Corbett Foundation, Nainital, Uttarakhand, India

Devendra Bhardwaj Department of Forests and Wildlife, Government of Rajasthan, Jaipur, Rajasthan, India

Gobind Sagar Bhardwaj Department of Forests and Wildlife, Government of Rajasthan, Jaipur, Rajasthan, India

Chhaya Bhatnagar Department of Zoology, Mohanlal Sukhadia University, Udaipur, Rajasthan, India

Subramaniam Bhupathy Sálim Ali Centre for Ornithology and Natural History, Coimbatore, Tamil Nadu, India

A.K. Chhangani Department of Environmental Science, Maharaja Ganga Singh University, Bikaner, Rajasthan, India

Sumit Dookia School of Environment Management, Guru Gobind Singh Indraprastha University, Dwarka, New Delhi, India

Shalini Gaur Desert Regional Station, Zoological Survey of India, Jodhpur, Rajasthan, India

Shilpi Gupta Wildlife Institute of India, Dehradun, Uttarakhand, India

G.R. Jakher Maharaja Ganga Singh University, Bikaner, Rajasthan, India

Bela Kothari Department of Anthropology, University of Rajasthan, Jaipur, Rajasthan, India

Seema Kulshreshtha Department of Zoology, Shakambhar Government P.G. College, Sambhar Lake (Jaipur), Rajasthan, India

Sanjeev Kumar Desert Regional Station, Zoological Survey of India, Jodhpur, Rajasthan, India

R.M. Lodha Department of Geography, Environment Studies Centre, Mohanlal Sukhadia University, Udaipur, Rajasthan, India

Meenakshi Mathur Department of Zoology, Mohanlal Sukhadia University, Udaipur, Rajasthan, India

Rajan Mathur Department of Forests and Wildlife, Government of Rajasthan, Jaipur, Rajasthan, India

Reena Mathur Formerly, Department of Zoology, University of Rajasthan, Jaipur, Rajasthan, India

U.B. Mathur Formerly, Geological Survey of India, Jaipur (Rajasthan, India)

B-198, University Marg, Jaipur, India

Devendra Mohan Department of Zoology, Jai Naraian Vyas University, Jodhpur, Rajasthan, India

S.M. Mohnot The School of Desert Sciences, Jodhpur, Rajasthan, India

Krishnendu Mondal Wildlife Institute of India, Dehradun, Uttarakhand, India

Shomita Mukherjee Sálim Ali Centre for Ornithology and Natural History (SACON), Coimbatore, Tamil Nadu, India

National Centre for Biological Sciences, Bangaluru, Karnataka, India

Rishad Naoroji Godrej Bhavan, Mumbai, Maharashtra, India

Qamar Qureshi Wildlife Institute of India, Dehradun, Uttarakhand, India

L.S. Rajpurohit Department of Zoology, Jai Narain Vyas University, Jodhpur, Rajasthan, India

Madhuri Ramesh Group for Nature Preservation and Education, Chennai, Tamil Nadu, India

Ashoka Trust for Research in Ecology and the Environment, Bengaluru, Karnataka, India

Ramkishor Department of Zoology, Jai Naraian Vyas University, Jodhpur, Rajasthan, India

N.S. Rathore Desert Regional Station, Zoological Survey of India, Jodhpur, Rajasthan, India

Mamta Rawat Ecology and Rural Development Society, Jodhpur, Rajasthan, India

Harkirat Singh Sangha Gautam Marg, Jaipur, Rajasthan, India

K. Sankar Wildlife Institute of India, Dehradun, Uttarakhand, India

Ravi Sankaran Formerly, Salim Ali Centre for Ornithology and Natural History (SACON), Coimbatore, Tamil Nadu, India

B.K. Sharma Department of Zoology, R.L. Saharia Government P.G. College, Kaladera (Jaipur), Rajasthan, India

G. Sharma Department of Zoology, Jai Narain Vyas University, Jodhpur, Rajasthan, India

Niranjan Sharma Formerly, Department of Botany, Government College, Kota, Rajasthan, India

Satish Kumar Sharma Sajjangarh Wildlife Sanctuary, Department of Forest and Wildlife, Government of Rajasthan, Udaipur, Rajasthan, India

Shailja Sharma Department of International Business and Management, Manchester Business School, The University of Manchester, West Manchester, MI, UK

Madhu Singh Department of Geography, Government College, Ajmer, Rajasthan, India

Partap Singh Department of Zoology, Dungar College, Bikaner, Rajasthan, India

Sunita Singh Department of Geography, M.V.S. College, J.R.N. Rajasthan Vidyapeeth University, Udaipur, Rajasthan, India

Y.P. Sinha Gangetic Plains Research Station, Zoological Survey of India, Patna, Bihar, India

Bhargavi Srinivasulu Wildlife Biology Section, Department of Zoology, University College of Science, Osmania University, Hyderabad, Andhra Pradesh, India

C. Srinivasulu Wildlife Biology Section, Department of Zoology, University College of Science, Osmania University, Hyderabad, Andhra Pradesh, India

J.K. Tiwari Centre for Desert and Ocean Village, Kutch, Gujarat, India

R.S. Tripathi Central Arid Zone Research Institute (CAZRI), Jodhpur, Rajasthan, India

M.K. Vijayvargiya Department of Forests and Wildlife, Government of Rajasthan, Jaipur, Rajasthan, India

Rakesh Vyas Hadouti Naturalists Society, Kota, India

Part I
Unveiling the Vivacious Rajasthan

Chapter 1
The Majestic Rajasthan: An Introduction

Seema Kulshreshtha, Shailja Sharma, and B.K. Sharma

Abstract This introductory chapter briefly displays an overall picture of the vivacious state of Rajasthan. Economy of the state is mainly based on natural resources while agriculture, mining, industry, and tourism are the major pillars of revenue. Despite the arid environment and scarcity of water, this desert state is endowed with a glorious historical past, forts and palaces and royal and cultural heritage. Rajasthan's cultural assets range from a jewelry, colorful attires and turbans, fairs and festivals to music and dances. Yet, the state is lagging behind in rural connectivity in terms of transport and communication, education, women empowerment, health and nutrition, and value addition to its natural resources.

Introduction

Rajasthan has a tremendous wealth of raw material and human resource which it needs to utilize as a mission with a vision to develop in a true sense. Unveiling the untapped resources and providing them a sound platform for better flowering is essential to surge ahead. The charm of vibrant Rajasthan (old name—Rajputana

S. Kulshreshtha (✉)
Department of Zoology, Shakambhar Government P.G. College,
Sambhar Lake (Jaipur), Rajasthan, India
e-mail: seema.zoology@gmail.com

S. Sharma
Department of International Business and Management, Manchester Business School,
The University of Manchester, Booth Street, West Manchester, M15 6PB, UK
e-mail: sharmashailja1988@gmail.com

B.K. Sharma
Department of Zoology, R.L. Saharia Government P.G. College,
Kaladera (Jaipur), Rajasthan, India
e-mail: drbksharma@hotmail.com

B.K. Sharma et al. (eds.), *Faunal Heritage of Rajasthan, India: General Background and Ecology of Vertebrates*, DOI 10.1007/978-1-4614-0800-0_1,

State), its royal heritage, colorful culture, mysterious sand-dunes, and Aravalli Mountain Ranges with varied wildlife make it a unique land of diversities and a true ambassador of India. Rajasthani folklores, brave warriors, fairy tales of romance and tragedies of princes, glory of forts and beautiful monuments tell the saga of bygone era. This desert land has simple people in colorful bright attires, impressive turbans, and medieval culture intermingled with modernization. Before unfolding the mysteries of the wonderful biological wealth of this very special state of India, let us flow with its magical charm as a lay man and the Jekyll and Hyde image of development.

Geographical Location

Situated in the northwest (23°3′ to 30°12′ North latitude and 69°30′ to 78°17′ East longitude) arid region, this largest state of India covers a total of 10.4% of the country with an area of 342,239 km^2 (869 km east to west and 826 km north to south). The state boundary runs for 5,920 km including 1,070 km, called "Redcliff Line" along the Pakistan border. This kite-shaped state as shown in the map, adjoins Punjab and Haryana in the north and northeast, Uttar Pradesh in the east, Madhya Pradesh in the southeast, and Gujarat state in the southwest. The Tropic of Cancer passes through 23 1/2° North latitude from the southern border of Dungarpur district and Kushalgarh *tehsil* of Banswara district (Fig. 1.1).

In fact, the Indian Subcontinent is a major ecological region of the Indomalaya ecozone, one of the eight ecozones dividing earth's land surface, and the state of Rajasthan broadly falls under the Indomalaya ecozone. According to the Wildlife Institute of India (WII), India has a total of ten biogeographic zones and 26 biotic provinces and Rajasthan falls under the biogeographic zones 3 and 4, namely, the desert, covering the Thar or the Great Indian Desert which is a mix of rocky and sandy desert and has a unique biodiversity, and semi-arid area covering the rest of the Rajputana or Rajasthan. A sizeable portion of Rajasthan state and its vast Thar Desert lie well within the Palaearctic and the Indomalaya Ecozones—two out of the eight ecozones dividing earth's surface. Physically, the Palaearctic is the largest ecozone which includes the terrestrial ecoregions of Europe, Asia, north of the Himalaya foothills, northern Africa, northern and central parts of the Arabian Peninsula. On the other hand, Rajasthan also seems to lie within the Indomalaya ecozone which extends across most of south and southeast Asia, and into the southern parts of the east Asia. Also called the Oriental Realm by some biogeographers, the Indomalaya Ecozone extends from Afghanistan and Pakistan through the Indian Subcontinent and southeast Asia to lowland southern China. Most of the Indomalaya was originally covered by forests, mostly tropical and subtropical broad leaf forests with tropical and subtropical dry broad leaf forests predominant in much of India and southeast Asia. Hence, it is clearly evident that the Indian Subcontinent is the major ecological region of the Indomalaya Ecozone and so is Rajasthan. Among the typical fauna characteristic of the Indo-malayan Ecozone, Rajasthan has tigers and leopards. Indomalaya is the most species-rich biogeographic ecozone with respect

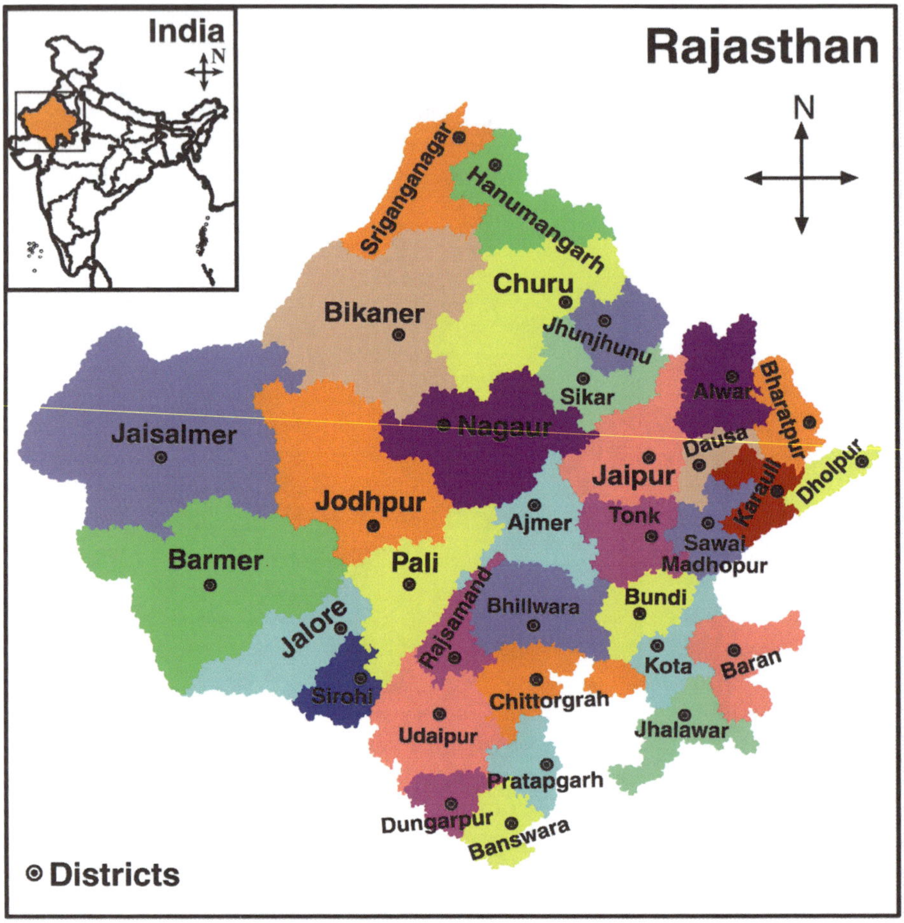

Fig. 1.1 Political map of Rajasthan (*Courtesy: Dr. B.K. Sharma*)

to the diversity of threatened reptiles. The Thar Desert of Rajasthan has been clearly placed under deserts and xeric shrublands as one of the Indomalayan-terrestrial ecoregions. It would not be out of place to mention here that the World Wide Fund for Nature, India (WWF-India) divides the Indomalaya Ecozone into three bioregions, which it defines as "geographic clusters of ecoregions that may span several habitat types but have strong biogeographic affinities, particularly at taxonomic levels higher than the species level (genus, family)". From a geographical point of view, the Hindukush, the Karakoram and the Himalayas are a major biogeographic boundry between the tropical and subtropical flora and fauna of the Indian Subcontinent and the temperate-climate Palearctic Ecozone.

Administrative Structure

The state of Rajasthan was formed in seven steps from 1949 to 1956 by integrating the erstwhile princely states of Rajputana and the centrally administered territory of Ajmer-Merwara. The foundation day of the state is called "Rajasthan Day" which falls on March 30 every year and is celebrated with gaiety. Presently, Rajasthan comprises of 33 districts clubbed into the following seven administrative divisions:

1. Ajmer Division consisting of Ajmer, Bhilwara, Nagour, and Tonk districts
2. Bharatpur Division comprising of Bharatpur, Dholpur, Karouli, and Sawai Madhopur districts
3. Bikaner Division comprising Bikaner, Churu, Hanumangarh, and Sri Ganganagar
4. Jaipur Division consisting of Jaipur, Alwar, Jhunjhunu, Sikar, and Dausa
5. Jodhpur Division includes Barmer, Jaisalmer, Jalore, Jodhpur, Pali, and Sirohi districts
6. Kota Division comprising Baran, Bundi, Jhalawar, and Kota
7. Udaipur Division consisting of Banswara, Chittourgarh, Dungarpur, Pratapgarh, Udaipur, and Rajsamand

Within the State, Ajmer and Jaipur divisions are relatively more developed than Kota, Udaipur, Bikaner, and Jodhpur. District Magistrate wields the specter of authority in the districts. Several civil servants under his/her aegis carry out the day-to-day governance. The districts are further divided into 241 *tehsils*. More than almost half of all the inhabited villages in the state, numbering 41,353, have fewer than 500 inhabitants each. The state has a total of 39,787 villages and 222 towns and cities. Rajasthan was the first state in India to start the *Panchayati Raj* in Oct 1959. At present, *Panchayati Raj* has a well-knit system of 32 *Jila Prishads* (District Administrative Blocks), 237 *Panchayat Samitis* (Development Blocks), and 9,189 *Gram Panchayats* (Village Development Blocks). On April 24, 1992, this system was provided constitutional status.

Demography

Population-wise Rajasthan stands at the eighth rank on the national ladder. According to the Census of India, 2011 [1] (March 31, 2011), the country has a total population of 1,210,193,422 having 17.64% population growth when compared to 2001 census. Rajasthan has a total population of 68,621,012 which is 5.67% of the total population of India. Population density in Rajasthan in the year 2001 was 165 people per km^2, which has now increased to 201 persons per km^2. Decadal growth rate of population in Rajasthan is 21.44. The latest census (2011) report displays that child sex-ratio (0–6 years) of 883 is much worse (926) than that of 2001. Female feticide is a rampant practice especially in western Rajasthan where the girl child is still considered unwanted. The districts of Jaisalmer, Barmer, Pali, Chittourgarh, Ganganagar, and Jhunjhunu are well-known for female infanticide. According to the census of 2001, the rural population of Rajasthan was estimated to be 43,267,678

Table 1.1 General statistics of Rajasthan

S. No.	Items		Particulars
1	Total geographic area		342,240 km^2
2	Reporting area for land utilization		342,650 km^2
3	Forest area		26,060 km^2
4	Area not available for cultivation		43,060 km^2
5	Permanent pastures and other grazing lands Area		17,070 km^2
6	Land area under misc. tree crops and groves		140 km^2
7	Cultivable wasteland Area		49,080 km^2
8	Fallow lands area other than current fallows		24,440 km^2
9	Current fallow area		24,150 km^2
10	Net sown area		158,650 km^2
11	Human population (Census of India, 2011)	Total	68,621,012
		Male	35,620,086
		Female	33,000,926
12	Density of population (Year 2011)		201 per person/km^2
13	Females per 1,000 males		926
14	Percent urban population (Year 2001)		23.4%
15	Literacy rate (Census of India, 2011)	Total	67.06%
		Male	80.51%
		Female	52.66%
16	Cities and towns		222
17	Villages	Inhabited	3,975,300
		Uninhabited	1,600
18	Livestock population		0.491 million

Source: Land use statistics, Ministries of Agriculture and Human Resource Development, Government of India, 2011 and Census Directorate Report, April, 2011

(76.62%), whereas the urban population was 13,205,444 (23.28%) [2]. This indicates that, even today, the people of the state are mostly rural pursuing primary activities like agriculture, horticulture, forestry, and animal husbandry. While literacy rate in India is 74.6%, it is 67.06% in Rajasthan. Census of India, 2011 points towards the unfair state of male to female ratio of children between 0 to 6 years of age in 15 districts of the state, which is surprising and directly indicates female feticide. Unfortunately, in the 2011 census, 533 villages did not show any girl child in the age group of 0–6 years and 74 villages had no women at all, which is a shame for the state. The rapidly growing population trends completely match with the projection and predictions by Kothari [3]. Interim data based on the latest population survey (Census of India, 2011) have been shown in Table 1.1.

Economy of the State

Agriculture

Rajasthan is the eight largest economy of the country which is largely agrarian based. The state accounts for 22.5% of the total national economy [4]. Rajasthan has 20 million hectares cultivated land but only 20% is irrigated. Three main seasonal crops, namely, *Rabi* crop during winter season, *Zayad* crop during summer season, and *Kharif* crop during rainy season are cultivated on a commercial scale. The food crops of the winter season include the cereals, mainly *Triticum aestivum* and *Hordeum vulgare* and few pulses like *Cicer arientium* and *Pisum sativum*. Important vegetables of winter season are *Brassica campestris var. raps*, different varieties of *Brassica oleracea, Daucus carota, Lycopersicon esculentum, Solarium tuberosum, Trigonella foenum-graecum*, etc. The condimental crops include *Coriandrum sativum* and *Foeniculum vulgare*. The oil-yielding crops that include *Brassica campestris var. sarson* and *B. nigra, Saccharum officinarum, Nicotiana tobacum*, and *Papaver sommiferum* are the cash crops of winter season. In the desert zones, the cultivation of *Rabi* and *Zayad* crops is limited to a very small area where irrigation facilities are available. The human population of these areas mainly depends on rainy season crops. During summers, when the soil and climate are dry, a few cucurbitaceous plants are cultivated for fruits and vegetables in the fields and on the sandy river-beds. The most common ones are the species of *Citrullus, Cucumis, Cucurbita, Lageneria Luffa, Momordica*, etc. The food crops of the rainy season include cereals like *Echinochloa frumentacea, Oryza sativa, Pennisetum typhoides, Setaria italica, Sorghum saccharatum,* Zea mays, etc. and pulses like *Cajanus cajan* and several species of Vigna. The important vegetables of the rainy season are *Abelmoschus esculentus, Capsicum annuum, Lablab purpureus, Cucumis melo var. culta*, and *Solanum melongena. Cyamopsis tetragonoloba*; besides providing edible pods and fodder, it is the main source of *Guar* gum particularly in desert zones (Table 1.2). Oil-yielding crops include *Arachis hypogaea, Gossypium species*, and *Sesamum indicum. Crotalaria juncea, Gossypium* species, and *Hibiscus cannabinus* constitute the fiber-yielding crops. The common fruit-yielding plants are *Aegle marmelos, Annona reticulata, A. squamosa, Carica papaya, Citrus* spp., *Mangifera indica, Psidium guajava, Punica granatum, Syzygium cumini, Ziziphus mauritiana*, etc.

Livestock

Livestock may be considered a natural gift to Rajasthan due to the arid environment of the state which makes the farmer dependent on animals. Animal husbandry is the biggest sector, next to agriculture. Rajasthan stands first in wool production and third in milk production in India. The overall population of livestock in Rajasthan was 57,899,870 in 2007 [4]. Actually, during the present times, the human density has outstripped the livestock density but on an average the ratio between the livestock and human population is 1:1. In Dungarpur, Banswara, Pali, Rajsamand, Udaipur, and Alwar districts, livestock densities are higher because of open forests and grasslands. In the western arid districts where grasslands and forest products are very scanty, the number of animals too is quite less. In the arid districts, goat, sheep, and camel (Fig. 1.2) constitute the main livestock, which can sustain on fewer fodder resources and little water. A large number of sheep and goat breeders from western Rajasthan migrate to the adjoining states of Punjab, Haryana, Madhya Pradesh, and Gujarat during summer, where fodder and water are still available in fields.

Table 1.2 Characteristics of agroclimatic zones of Rajasthan

Zone	Area sown (million ha) Total	Net sown	Districts covered	Range rainfall (mm)	Temperaure range Max	Min	Crops Kharif	Rabi	Soils
Ia	4.74	2.34	Barmer and Part of Jodhpur	200–370	40.0	8.0	Pearl Millet, Moth bean, Seasame	Wheat, Mustard, Cumin	Desert Soil, sand-dunes, coarse sand in texture, some places calcareous
Ib	2.10	1.50	Sriganganagar, Hanumangarh	100–350	42	4.7	Cotton, Cluster bean	Wheat, Mustard, Gram	Alluvial deposits calcareous, high soluble salt and exchangeable sodium
Ic	7.70	2.44	Bikaner, Jaisalmer, Churu	100–350	48.0	3.0	Pearl Millet, Moth bean, Cluster bean,	Wheat, Mustard, Gram	Desert soil and sand-dunes, Aeolian soils, loamy coarse soil
IIa	3.69	2.68	Nagore, sikar, Jhunjhunu, part of Churu	300–500	39.7	5.3	Pearl Millet, Cluster bean, Pulses	Mustard, Gram	Sandy Loam, Shallow depth Red soil
IIb	3.00	1.93	Jalore, pali, Part of Sirohi, Jodhpur	300–500	38.0	4.9	Pearl Millet, Cluster bean, Sesame	Wheat, Mustard	Red Desert Soil in Jodhpur, Jalore and Pali, sierozens in Pali and Sirohi
IIIa	2.96	1.77	Jaipur, Ajmer, Dausa, Tonk	500–700	40.6	8.3	Pearl Millet, Cluster bean, Sorghum	Wheat, Mustard, Gram	Sierozens, eastern part alluvial, northwest lithosols, Foothills brown soils
IIIb	2.77	1.41	Alwar, Dhoulpur, S. Madhopur, Bharatpur, Karouli	500–700	40.0	8.2	Pearl Millet, Cluster bean, Groundnut	Barley, Wheat, Mustard, Gram	Alluvial soil prone to water logging, recently calcareous nature has been observed
IVa	3.35	9.92	Bhilwara, Sirohi, Udaipur, Chiitorgarh	500–900	38.6	8.1	Maize, Pulses, Sorghum	Wheat, Gram	Soil is lithosols in foothills and alluvial in plains
IVb	1.72	0.57	Dungarpur, Udaipur, Banswara, W. Pratapgarh	500–1,100	39.0	7.2	Maize, Paddy, Sorghum, Black Gram	Wheat, Gram	Predominantly reddish medium texture, well drained calcareous shallow on hill, deep soil in valleys
V	2.70	1.27	Kota, Jhalawar, Bundi, Baran	650–1,000	42.6	10.6	Sorghum, Soybean	Wheat, Mustard	Black of alluvial origin and clay loam, ground water salinity

Source: Agriculture Department, State Government of Rajasthan, India

Fig. 1.2 Popularly called as the "Ship of Desert" the Dromedary or Arabian Camel (*Camelus dromedarius*) is the symbol of arid Rajasthan. The animal is known to survive on fewer fodder resources and little water. In this picture, a typical village boy is seen riding a camel (*Courtesy: Devendra Bhardwaj*)

Mineral Resources

Mining is the second largest sector after agriculture. In India, Rajasthan is the greatest producer of nonferric metals such as copper and zinc and accounts for 40% of the country's copper production and 100% of zinc production [5]. The province contains the world class Rampura-Agucha lead–zinc deposit along with several large deposits of lead and zinc in Rajpura-Dariba and Zawar belts. There are large copper mines at Khetri of Jhunjhunu district. The state also accounts for 85% of lead production, 94% of gypsum, 76% of silver ore, 68% of feldspar, 84% of asbestos, and 12% of mica. Makrana in the Nagour district of Rajasthan is a famous site where white marble is heavily mined and exported. In addition, there are number of deposits in Deri-Basantgarh belt. Since the last decade, several gold deposits have also been discovered in the southern part of the province in Banswara district. There are about 42 major and 28 minor minerals present in Rajasthan. This sector provides employment to about two million mine workers throughout the state. The huge sandy expanse of Rajasthan abounds in phosphate fluoride, phosphorite, rock-phosphate, clay, granite, dolomite, calcite, emerald, and garnet. Rajasthan is the sole producer of garnet (gem variety), jasper, and wollastonite. It shares 24% of the aggregate national production of nonmetallic minerals. The dry terrain of Rajasthan is a repository of colossal slabs of stone and accounts for 65% of India's total stone production. The state accounts for 90% of the marble, slate, and sandstone production (Fig. 1.3a, b). Rajasthan also houses rich salt deposits at Sambhar (Jaipur district) and stands as the third largest salt producing state. The share of Rajasthan in terms of reserve and production of principal minerals is shown in Table 1.3.

Fig. 1.3 (**a**) Marble mining in progress: mining is one of the major industries in Rajasthan. (**b**) A slate stone mine (*Courtesy: Sonali Singh*)

Table 1.3 Mineral reserves and production in Rajasthan

Minerals	Reserve (million tons per annum)	Production (million tons per annum)
Copper Ore	35	0.98
Lead and Zinc Ore	75	2.64
Gypsum	70	2.86
Limestone	1,990	21.19
Rock phosphate	60	1.18

Industry

Rajasthan began the journey of industrial development between 1950 and 1960 in Kota, Jaipur, Udaipur, Bhilwara, and other industrial estates. The main industries of Rajasthan include textile, rugs, woolen goods, vegetable oil, and dyes [6]. Textile is the chief industry flourishing mainly in Bhilwara district as India's second largest producer of polyester fiber. Beside these, Rajasthan is also involved as the fourth largest producer of cotton, spurn yarn, and wool in the country. In total, the production of textile accounts 21.96% in the state. Heavy industries consist of copper and zinc-smelting and manufacturing of railway rolling stock. The other industries of the private sector include steel, cement, ceramic and glassware, electronic, leather and footwear, stone and chemical industries. Rajasthan is also the major producer of cement and accounts for 15% of the total cement output of the country. The chemical industry produces calcium carbide, caustic soda, sulfuric acid, pesticides, insecticides, and fertilizers. The industrial sector of Rajasthan accounts for about 32.5% of the total share. The state has earned huge revenue from the petroleum and oil production by Cairn India situated at Mangala in Barmer-Sanchour Basin which started in August 2009. Other economic sources of the state include, transportation, energy, power and telecommunication, banking and nonbanking financial institutions.

Transport

Rajasthan is far behind in transport and communication due to the absence of waterways, proper expansion of train and road routes providing major infrastructural links. The total road and railway length in Rajasthan is 164,000 km and 5,895 km, respectively. Regular transport services connect cities by 19 National Highways (NH) of which NH8 being the busiest of all and connects Delhi–Jaipur–Ajmer–Udaipur–Ahmadabad–Mumbai.

Communication

Information Technology Enabled Service Policy (IT & ITES Policy), 2007, aimed at implementing e-Governance and capacity-building within Government domain; promoting investment in IT and ITES sector within the state; and skill development through enhancement of employment opportunities by developing the capability of the youth of the state and making them employable by the industry. Likewise, Rajasthan Cyber Cafe Rules, 2007 is a legal framework to prevent and curb cyber crimes, while Bharat Nirman Rajiv Gandhi Sewa Kendra is a facility to deliver G-2-C services to the citizens at block level in the rural areas. e-SANCHAR is integration of the IT with mobile telephony for providing information pertaining to citizen-centric services within remote rural areas of Rajasthan, helping the illiterate beneficiaries. The Project was awarded as the best project under the category of "Maximum Social Impact" by PC Quest's "Best IT Implementations Awards 2009."

Fig. 1.4 Lake Palace of Udaipur—a hotspot for royal style wedding especially liked by celebrities across the globe: Forts and Palaces of the erstwhile rulers with their medieval architecture make the state of Rajasthan one of the most sought after places in the world (*Courtesy: Sonali Singh*)

Touch Screen kiosks at district level have been installed in 33 district collectorates to facilitate efficient, timely, and cost-effective delivery of government information and services like land records, old-age pension, widow pension, individual beneficiary schemes, and so on to the citizens of the state [7]. Thirteen subdivisional headquarters have started e-SUGAM (single window system) for fast working, especially to serve the rural population of the state.

Tourism

Rajasthan is a major attraction for international tourists visiting India due mainly to its royal palaces (Fig. 1.4), forts, history, colorful culture, tiger reserves, and mysterious sand-dunes (Fig. 1.5). Tourist department of the Government of Rajasthan also runs the heritage train called "Palace on Wheels" in addition to heritage hotels which are nothing but palaces converted into hotels. Apart from this, the department also organizes festivals like *Kite* festival and *Teej* Festival at Jaipur and *Maru* festival at Jaisalmer and conducts tours to places of rare scenic beauty, raw forests, and wetlands (Fig. 1.6) which are quite popular amongst national and international tourists (for details please also see Faunal Heritage of Rajasthan: Conservation and Management of Vertebrates, Vol 2, Sharma B. K. et al. (eds.) 2013, Springer (978-3-319-01344-2), Chap. 18). Rajasthan tourism is regarded as the best tourism department in India. In the year 1989, tourism was actually bestowed the status of a fully fledged Industry [8].

Fig. 1.5 Sand-dunes near Bikaner (*Courtesy: Partap Singh*)

Fig. 1.6 Scenic beauty of Chandlai Wetland near Jaipur (*Courtesy: Devendra Bhardwaj*)

Sociocultural Composition: The Societies, Casts, and Tribes

The people of Rajasthan are divided into various casts, subcasts, societies, and sects. Hindus have Brahmins, Oswals, Mahajans, Jains, and Rajputs. Rajputs stand apart for their martial reputation and ancestry. Apart from this, every physiographic region has tribals depending upon the natural resources.

The principal tribes of Rajasthan are *Bhil* and *Mina,* while *Garasia, Sahariya, Damor, Gadia or Gaduliya lohar*, and *Raibaries* are amongst the smaller tribes. The tribes form approximately 12% of the total population of Rajasthan. The common traits shared by different tribes link their distant past. Each tribal community is distinguished by its costumes, festivals, ornaments, and rituals.

On one hand, there are staunch protectors like the *Bishnois* while on the other hand, some are indulged in hunting and poaching. Most of the tribal in Rajasthan are poor and earn their livelihood mainly from forest products and wildlife trade. The agriculture production by them fetches little that too is largely dependent on monsoon. They are generally illiterate and nonvegetarian who follow primitive life styles involving drinking and hunting. A few communities like *Natts* and *Banjaras* are nomads and eat all kinds of flesh except that of the jackal [9]. Meat-eating communities like *Rajputs* and *Muslims* have also promoted *Shikar* (hunting) in Rajasthan.

Likewise, *Bavarias* also have a well-organized system of committing crimes including illegal *Shikar*. Some of the tribals as *Madaries* exhibit monkey and sloth bear while *Kalbelias* exhibit snakes and mongoose [10]. *Kalbelias* and *Jogies* are also notorious cat-killers. Through generations of professional hunting traditions, they have mastered the skill [11]. Skin and fur of cats and other mammals are recovered from them during sudden raids. Tribal use hunting dogs and traditional weapons like arrows, nets, snare, etc. for their hunting operations whereas some well-off ones use guns, rifles, and motor vehicles for the purpose [12].

Tribal Settlements

The ancient history depicts that during the invasion of *Hunas, Aryans, Sakas,* and *Kushanas*, the tribal communities survived the attacks and maintained their culture and traditions. Most of them adopted Rajasthan and followed Hindu religion. The scheduled areas include Dungarpur, Banswara and Pratapgarh *tehsil* of Chittourgarh district and the foothills of Aravallis and Vindhyas. The heavily concentrated tribal area comprises Kherwara, Kotra, Gogunda, Phalasia *tehsils* of Udaipur district, the Abu road *tehsil* of Sirohi district, Achnera and Arnod *Panchayat Samitis* of Chittourgarh district, and Baran of Kota district. Nontribal population is generally predominant in sparsely populated tribal areas which include Tonk, Bhilwara, and Alwar districts. Tribal population in Rajasthan is mainly concentrated in the belts running from Sirohi through the Udaipur, Dungarpur, Chittourgarh, and Banswara districts to Bundi, Kota, Sawai, Madhopur, Tonk, and Jaipur districts (Fig. 1.7).

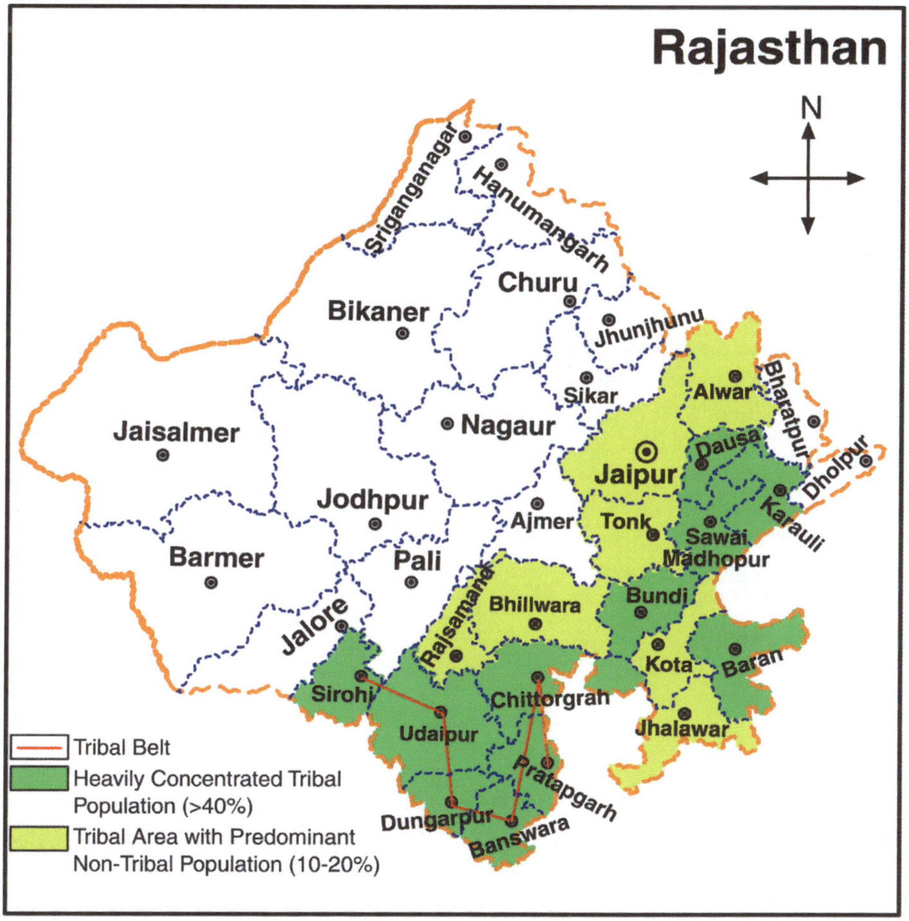

Fig. 1.7 Map of Rajasthan showing tribal areas (*Courtesy: Dr. B.K. Sharma*)

Bhils

Bhils comprise of 39% of the total tribal population inhabiting the mountainous southwestern part of Rajasthan embracing the wildest of the Aravallis. The *Bhils* live in *pals* or in clusters of detached huts amid the hills, each hut stands on a small mound in the midst of its path in a cultivated piece of land. The settlement or the *pal* is divided into a number of *paras* or *phalas* to afford cover and protection in case of attack. Such clusters of huts within a single enclosure form a typical *Bhil* habitation. They gather during the *Beneshwar* fair at Dungarpur district which is a Bhil dominated area. Banswara district is yet another stronghold of the *Bhils*. They are fine archers and the fact has been mentioned in the ancient Indian epics like *Ramayana* and *Mahabharata*. They were of great help to Rajput kings being warriors.

Witchcraft magic and superstitions are deeply rooted in them who use body parts of different birds and animals to cure ailments of both fellow humans and domestic cattle [13] (Fig. 1.8a–c).

Minas

The *Minas* constitute almost one-third to half of the tribal population of Rajasthan and are the second largest tribe who live on rocky elevations or in thick forests and their settlements are called *Mewasas*. The cluster of their houses is also called a *pal* and is named after the *gotra* to which most of the inhabitants belong. The *Minas* are largely settled in the villages of Jaipur, Sawai Madhopur, and Tonk districts of Rajasthan. *Minas* have two classes, the *Purana Basi Minas* who are mainly agriculturists and the *Naya Basi Minas* who belong to the light-fingered fraternity which prior to independence were subjected to daily attendance at the nearest police station under Criminal Tribes Act. The literacy rate among *Minas* is low. Initially they were spread all over Rajasthan but, later *Kacchawa Rajputs* forced them towards Aravallis. They also inhabit *Shekhawati* and parts of eastern Rajasthan. Originally, they were the ruling tribes but their downfall began with the upsurge of *Rajput* kings and *Minas* got sidelined when the British Government declared them a "Criminal Tribe" in 1924.

Garasias

A *Garasia* settlement is not a cluster of houses. The dwellings are scattered over slopes of Abu Road hills and mounds and the fields extending in the front. These solitary dwellings are made of bamboo and leaves and lightly plastered over with cow dung. Marriage invariably through "elopement" is an interesting custom.

Saharias

Saharias, the jungle-dwellers, are found in Kota, Sikar, and Sawai Madhopur districts. They mostly reside in Shahbad and Kishenganj *Panchayat Samitis* of Baran district and are considered the most backward tribe. Fishing, selling forest produce like honey, fruits, gum, forest-wood, *Tendu* or *Diospyros melonoxylon* leaves, and hunting are the main source of livelihood. Actually, "*Sehar*" means jungle (forest) in Urdu language and the erstwhile Muslim rulers gave them the name "*Saharia*" meaning the inhabitants of the jungle. It is far sure that they were among few of the early settlers of Rajasthan. Their clusters of huts are called *Saharana* (Fig. 1.9) and they speak local dialect influenced with Hadoti accent. They worship *Gogaji, Dhakar Baba, Lalbai*, and *Bejasan* as local deities. *Sawa* or *Echinocola* sp. grass seed is consumed by them along with a nonvegetarian diet.

Fig. 1.8 (**a**) *Bhils* of Rajasthan in a procession on the occasion of Rajasthan Day Celebrations (*Courtesy: Devendra Bhardwaj*). (**b**) *Bhil* females in a procession on the occasion of Rajasthan Day Celebrations (*Courtesy: Devendra Bhardwaj*). (**c**) A tribal girl with Four-horned Antelope (*Courtesy: Dr. Gobind Sagar Bhardwaj*)

Fig. 1.9 Tribal huts (*Courtesy: Devendra Bhardwaj*)

Mogiyas

Mogiyas are a seminomadic community [14] who usually live with big landowners. They earn their livelihoods by guarding vast agricultural lands from wild animals at night. This community has the skill of tracking various movements of the wild animals which got further refined as a result of their engagement with landowners. Later on this skill was used by modern hunters. Gradually with the declining tiger population in India and increasing demand for tiger skin, nail, and teeth, the demand of *Mogiya* hunters also increased. Around the same time the Indian Government too became serious about conserving tiger—the national animal of India. During this period, *Mogiyas* were considered to be a major threat and a lot of efforts were made to distance them from the forest and wildlife. Unfortunately, nothing much was really done to provide them with any concrete support which resulted in a precarious situation for *Mogiyas*. Despite being termed as a criminal community, *Mogiya* reform and rehabilitation program was initiated at Ranthambhore National Park following the proven involvement of this community behind majority of poaching cases. The idea was to discourage them from hunting by providing them with healthy alternative sources of livelihood and encouraging them to join mainstream of the society.

Fig. 1.10 Hamlet huts in a tribal village on the way to Phulwari Ki Naal Wildlife Sanctuary near Udaipur (*Courtesy: Sonali Singh*)

Gadia Luhars

Originally it is a martial tribe. They derive the name *Gadia Luhar* (*Gadia*-cart; *Luhar*-blacksmith) for they usually carry their entire home on a bullock-cart and are blacksmiths by occupation. They left their homeland along with *Maharana* Pratap, the legendary king who opposed Mughal Emperor Akbar and was ousted by him from Chittourgarh. This clan vowed to re-enter only after the victory of *Maharana* Pratap. Unfortunately, *Maharana* was killed in the battle field and hence, this tribe never ever settled and leading a nomadic lifestyle even today.

Damors

They are cultivators and manual laborers who migrated from the adjoining Gujarat state to Rajasthan to settle in Udaipur and Dungarpur districts. A typical *Damor* settlement near Udaipur is shown in Fig. 1.10.

The other known tribes of Rajasthan include *Meo, Bawaria, Banjara, Raibari, Kathodi, Kanjar, Saansi*, and *Natt. Bawarias* are one of the 200-odd notified tribes of India which are also considered as yet another criminal tribe. The *Meo* inhabit Alwar, Jaipur, Bharatpur, and Dholpur districts in large numbers. The *Banjara* is a nomadic community who travel in bullock-carts and earn their livelihood by manufacturing and repairing agricultural and household implements. The nomadic

Fig. 1.11 A typical *Raika* herder with his domesticated camels (*Courtesy: Ashish Kothari*)

Kathodi tribe lives in the Mewar region while the *Raibaris* of the Marwar region are cattle-breeders. The nomadic *Raibari* or *Raika* (Fig.1.11) are divided into two groups, namely, the *Marus* who breed camels and *Chalkias*, who breed sheep and goat. Camel is a means of transport in remote villages of the desert and the Border Security Forces (Fig. 1.12). *Kalbelia* (Fig. 1.13) or the snake charmers (known as *Saperas* in Hindi) is yet another well-known tribe of Rajasthan who earn their living exclusively by displaying snakes.

In the north and the west of Rajasthan, the *Jats* and *Gurjars* (Fig. 1.14a, b) are the largest agricultural communities. The *Gurjars,* mainly inhabiting the eastern Rajasthan are also traditionally involved with cattle keeping and milk supply. The Muslims constitute less than 10% of the total population of the state. Most of the Rajasthani Muslims are Sunnis. There also exits a small but affluent community of Shiaite Muslims known as *Bohras* occupying the southeastern Rajasthan.

The Religion

Hinduism is the religion of majority of the population generally practiced through the worship of *Brahma, Ganesha, Shiva, Shakti, Vishnu,* and other Hindu gods and goddesses. Nathdwara is an important religious center for the *Vallabhacharya* sect of Krishna followers. There are also followers of the *Arya Samaj*, a reforming sect of modern Hinduism. Jainism is also an important religion which has followers among the business community and wealthy section of society. Mahavirji, Ranakpur,

Fig. 1.12 Camels are frequently used by the Border Security Forces (BSF) on the border adjacent to Pakistan. The picture shows BSF sleuths in a procession on the occasion of Rajasthan Day Celebrations (*Courtesy: Devendra Bhardwaj*)

Fig. 1.13 Snake Charmer (*Sapera*) in Hindi displaying a snake (*Courtesy: Devendra Bhardwaj*)

Fig. 1.14 (**a**) A typical *Gurjar* man from Rajasthan. (**b**) *Gurjar* females standing in front of a hut (*Courtesy: Devendra Bhardwaj*)

Dhulev, and Karera are the chief centers of Jain pilgrimage in Rajasthan. The *Dadupanth* forms another important religious sect mainly comprising the followers of Saint Dadudayal, of the very beginning of 17th century, who preached equality of all men, vegetarianism, total abstinence from liquor, and lifelong celibacy. Islam is the religion of Muslims—state's second largest community who expanded in Rajasthan with the conquest of Ajmer by Muslim invaders in the late twelfth century when Muslim traders, craftsmen, and soldiers settled there. The Sufi Saint Khwaja Moinuddin Chishti (1141–1230 CE) popularly known as *Hazrat Khwaja Gharib*

Fig. 1.15 Victory Tower at Chittourgarh Fort is a unique example of the architecture of Rajasthan (*Courtesy: Sunil Singhal, Kota*)

Nawaj (benefactor of the poor) who taught peaceful coexistence was originally born in Afghanistan but later turned towards India reportedly after a dream in which Prophet Muhammad instructed him to do so. Following a brief stay at Lahore (Pakistan) he reached Ajmer in Rajasthan and settled here. The saint practiced the Sufi *Sulh-e-Kul* (meaning peace to all) concept to promote harmony and understanding between Muslims and non-Muslims. The world famous ancient shrine (also called *Dargah Sharif or Ajmer Sharif*) dedicated to the memory of the saint is situated in Ajmer city of Rajasthan where thousands of devotees belonging to all faiths and religions visit to pay their respects every day. It is said that, the Saint fulfills the wishes of people who pray before him. The population of Christians and Sikhs is small in the state.

The Architecture

Majestic forts and intricately carved temples, palaces and havelis (mansion), and even step-wells make Rajasthan a paradise for architecture buffs. Some of the prominent structures that represent the architectural heritage of Rajasthan are Jantar-Mantar, Hawa Mahal, City Palace, Sarga-Suli, Albert Museum and Amber Fort (Jaipur), Dilwara temples (Mount Abu), Victory Tower (Chittourgarh) (Fig. 1.15), City Palace and Lake Palace (Udaipur) and the Sonar fort at Jaisalmer (Fig. 1.16).

Fig. 1.16 Sonar (meaning golden) Fort of Jaisalmer (*Courtesy: Devendra Bhardwaj*)

For further details, please refer Chap. 18 from Faunal Heritage of Rajasthan: Conservation and Management of Vertebrates Vol. 2, Sharma B. K. *et al.* (eds.) 2013, Springer (978-3-319-01344-2).

The People

Unlike the olden days, the profession of the people no more decides their caste, and they are free to opt any profession. People of different castes and subcastes peacefully coexist in Rajasthan. Rajasthanis are sturdy and simple which makes Rajasthan one of the safest destinations worldwide.

Colorful Costumes

The bright costumes and attires of the Rajasthanis present a vivid contrast to the barren, colorless landscape and the monotony of its cloudless sky. A sparkling gold or silver *zari* highlight lends a perfect finish. The women wear *Lahnga* and *odhni,* popular as bridal dress all over north India along with silver jewelry. In fact, the age-old designs characteristic of a specific tribe reflect in their ornaments. Ornaments of neck, nose, ear, and hand are worn by Rajasthani women on daily basis. Exquisite designs of the jewelry of Rajasthan have made it popular not only amongst the women of India but also in foreign countries.

Fig. 1.17 *Rajput* (the warrior clan) women in the Thar Desert (*Courtesy: Devendra Bhardwaj*)

Impressive Turbans

Turban popularly known as *pagdi,* vary with the religion and region. The Rajput turbans are completely different from turbans of other casts. A typical Gujar turban is shown in Fig. 1.14a. Not only cuisine, water, and dialect but turbans also change within every 12 miles in Rajasthan. As a matter of fact, there are almost 100 different styles of wearing turbans highlighting the region, caste, and even the socioeconomic class of the individual. Brightly colored turbans are largely preferred while more special turbans in terms of color, fabric, and design are worn during family functions like wedding and other festive occasions.

Brave Patriotic Warriors and Graceful Women

The glorified history of chivalry and bravery of the great *Rajput* clan is reflected in the ancient monuments. The *Rajput* females (Fig. 1.17) of Rajasthan are identified for their beauty, grace, and bravery. Maharani Padmini Devi of Chittourgarh was one of them. She chose *Johar* (entered alive into the funeral pyre of her husband by literally taking his head in her lap) along with other royal widows to save her grace after her husband, the King of Chittourgarh was killed in the battlefield while fighting against the Mughals since, Allauddin Khilji, the then Mughal Emperor was infatuated by her beauty. *Johars* were performed 12 times by Rajput women to protect themselves from Mughal armies belonging to Emperors Akbar, Khilji, and

Tuglaq following many defeats of *Rajput* kings in the wars during 1301–1567 BC. Famous wars were fought between Mohammad Gouri and Prithviraj Chauhan (1191 and 1192 BC), Maharana Kumbha, Rana Sanga, Maharana Pratap, and Mughul warriors Khiljis, Baber, Shershah Suri, and Akbar. The fort of Chittourgarh is a living testimony which is spread over a 13 km long Aravalli hill top. It was the dream of every Mughal Ruler to win over the fort, hence received maximum attacks. Since, the *Rajput* kings initially were ignorant of the technology of tanks, gun powder, and canons, they were badly defeated most of the times. Panna Dhai, a sixteenth century governess of Udai Singh—the fourth son (Prince) of Maharana Sanga was another *Rajput* lady with great courage who sacrificed her own son to save the little king Udai Singh from his barbarian uncle Hammir, the then care-taker of the throne. *Rajputs* are famous for their bravery and it was said that even if the enemy cut their throats, their hands used to continue attacking the enemy with swords. Yet another legendry beauty, (Late) *Maharani* Gayatri Devi, the former "*Rajmata*" (King's mother) of Jaipur was once listed amongst the ten most beautiful women of the world. After independence, the democratic India saw the abolition of Princely states and *Jagirs* (Estates), land reforms and finally the Privy Purse withdrawal made the former kings lose their empire (1948–1956). In order to tide over the financial scarcity, they turned their palaces and other establishments into five star hotels, called "Heritage Hotels."

The Festivals

Apart from the major festivals like *Holi* (festival of colors), *Deepawali* (festival of lights), *Rakshbandhan* (symbolizing the bond of brother and sister), *Dussehra* (symbolizing the victory of good over evil), and *Navaratri* (a 9 day long worship of goddess Durga) celebrated all over India, the festivals characteristic of Rajasthan state include the following:

1. *Festival of Kites*: Celebrated as *Makar Sakranti* on January 14 every year, during which a variety of kites are flown in every household of the capital Jaipur. This is also celebrated in other parts of the country especially in the Indian capital New Delhi on August 15 to mark the Indian Independence day.
2. *Gangour*: Celebrated in the month of April by worshiping *Gangour*, a symbol of goddess *Parvati*. A special dance form called *Ghumar* is performed by the women on the occasion and a procession of *Gangour* is taken on the main streets of Jaipur—the famous pink city and the capital of Rajasthan.
3. *Akha Teej*: On the third day of the lunar month in April, being an auspicious day, large number of marriages, especially child marriages are solemnized where apart from adults even 6-month to 10-year-old children are also married off, however, they are only allowed to live together when they attain adulthood. Although, the Government has stringent rules against child marriage but it still prevails in the remote villages of Rajasthan.

Table 1.4 Fairs of Rajasthan

Name of district	Fair
Ajmer	Pushkar fair, Donkey fair, Khwaja Moinuddin Chishti's Urss (the biggest Muslim fair)
Beawer	Badshah fair
Baran	Sitabadi fair (biggest Saharia Tribal fair)
Bikaner	Karnimata fair (Deshnok) and Kapil Muni fair (Kolayat)
Bundi	Kajli Teej fair
Hanumangarh	Gogaji fair (Gogamedi)
Sikar	Jeen mata fair, Khatu Shyamji fair
Jhunjhunu	Rani Sati fair
Jaipur	Gangour fair, Teej Fair,Shakambhari Mata fair,
Alwar	Bharathari fair, Narayni Mata fair
Tonk	Shri Kalyanji fair (Diggi-Malpura)
Kota	Dushhera fair
Chittourgarh	Mata Kalika fair, Mata Kundalini fair, Sawariaji fair
Banswara	Ghotia Amba fair, Tripura Sundri fair
Dungarpur	Beneshwar Mahadeo fair (biggest tribal fair), Galiakot Urss

4. *Hariyali Teej*: Celebrated in the month of August during rains when women especially the newlyweds go to their parents' home and enjoy swings put up on the heavy twigs of big trees. A procession of *Teej* at Jaipur is famous the world over.
5. *Bachh-Baras*: Calf is worshiped as a symbol of son in the month of August.
6. *Goga Navami*: Birth day of the local deity, the king *Veer Gogaji*, is celebrated every year in August largely by the tribes at Gogamedhi of Churu district.

The Fairs

A large number of fairs are celebrated in every district of Rajasthan (Table 1.4).

The Cuisine

Rajasthani cooking has its roots in the lifestyle of the medieval Rajasthan when the chieftains were mainly at war and the focus was on edible items that could last for several days and could also be eaten without heating. Furthermore, the scarcity of water as well as fresh green vegetables had some impact on the art of cooking. In the desert belt of Jaisalmer, Barmer, and Bikaner, cooks use minimum of water and prefer, instead, to use more milk, buttermilk, and clarified butter. Generally, Rajasthani curries are brilliant red but not as spicy as they appear. Perhaps the best known Rajasthani food is the combination of *Dal, Bati, Churma*, and *kair sangri*. In addition, each region is distinguished by its own popular sweet—*Mawa Kachori* from Jodhpur, *Mawa* (Alwar), *Malpuas* of Pushkar, *Rasgullas* of Bikaner, and the

Ghevar of Jaipur just to name a few. Contrary to popular belief, people of Rajasthan are not all vegetarians. One of the unique nonvegetarian Rajasthani dishes is the *Junglee maans* and *Lal maans* (the red meat)—a specialty of Jodhpur. Some of the Maharajas apart from being great hunters relished the passion of cooking the "Shikars" (hunted animal) themselves for their chosen guests and the trend continues among the successive generations even today.

The Paintings

Traditional Rajasthani painters belonging to various regions have their own characteristic styles of paintings which include the following:

1. Jaipur and Alwar styles have green as the main color and *Peepal* (*Ficus religiosa*) trees are invariably seen in paintings. Miniature paintings on rice are also famous.
2. Jodhpur and Bikaner styles have yellow as the main color and mango (*Mangifera indica)* trees are integral to the paintings.
3. Udaipur style has red as the main color and *Kadamb* (*Neolamarckia cadamba*) trees are painted as a rule.
4. Kota and Bundi style have blue and a golden color, respectively, with Date Palm (*Phoenix dactylifera*) as the background.
5. The famous Kishangarh style has white and pink color and banana (*Musa paradisiacal)* tree is usually painted along. The world famous *Bani-thani* painting is considered as a symbol of Rajasthan.

The Music

Rajasthani music has a strong religious fervor and songs are dedicated to various saints and deities like Surdas, Kabirdas, Meerabai, Baba Ramdev, Tejaji, Gogaji, and others. *Maand* style songs of Rajasthan are surprisingly classical despite being folk (Fig. 1.18). Entertainers like the *Langas, Bhopas, Manganniyars, Mirasis*, and *Dholis* keep the musical traditions alive, whose education in classical music begins early and is known to be passed on to generations. They sing about folk heroes like Tejaji, Gogaji, and Ramdev and narrate heroic tales of battles, God, Goddesses, and even of legendary lovers and their tragedies. Distinctive traditions falling into this class are the "*Phad*" and "puppetry." The accompanying instruments are *sarangi, kamaycha, satara, nad, chang, dhol, bhapang, nagara,* and *morchang* (Jewish Harp) and even items of common use such as bells, thalis (metal dishes), and earthen pots. A variety of wind instruments include, *Algoja, Satara, Murla, Nad, Poongi* and *Shehnai.*, *Kartaal* and *Morchang* that are unique additions to the great repository of folk instruments. Music sung by women is mostly about water and wells and the style is called *panihari* and songs more and often mention of *moriya* (peacock), *koyaldi* (cuckoo), *sugga* (parakeet), *kaga* (crow), *Naag* (snake), and even *Bichhuda* (scorpion).

Fig. 1.18 *Maand* singers of Rajasthan (*Courtesy: Devendra Bhardwaj*)

Table 1.5 Dances of Rajasthan

District/area	Name of dance
Bikaner	Fire dance
Bharatpur	Bamm dance, Nautanki, Charkula dance
Baran	Shikari dance
Ajmer	Chari dance
Karouli	Languria
Kota	Chakri dance
Jodhpur	Ghudla dance and Thali dance
Jalore	Dhol dance
Rajsamand	Dang dance
Pali	Terah-Tali
Udaipur	Gawri dance, Ghumar dance
Jaisalmer	Kalbeliya dance
Jaipur	Famous Jaipur style of Katthak dance (Hindu style)
Shekhawati	Gindar dance, Chang dance

The Dances

There are wonderful varieties of folk dances in different regions of Rajasthan (Table 1.5 and Fig. 1.19).

Fig. 1.19 *Kalbelia* dancers of Rajasthan (*Courtesy: Devendra Bhardwaj*)

Table 1.6 Handicrafts of Rajasthan

Bikaner	Copy making, Matherna Art, Ustkala
Sikar	Batik and Khandela art
Jaipur	Hand-made paper, Blue pottery, Pomecha, Jaipuri quilt, Laharia, Sanganeri Prints of clothes, Enamel work on silver, Kundan Jewelry, Semiprecious stones
Alwar	Paper mache pottery
Tonk	Namda
Kota	Kota Doriya Saree, Moothda and Masooriya Sarees, Black Pottery
Chittourgarh	Woodwork, Gangour, Kavad, Bevan
Rajsamand	Terracotta, Pichhwai, Silver enamel work
Pratapgarh	Thewa art of jewelry, special curry spices
Jodhpur	Badla, Mojdi, Bandhej Prints and Bridal Chuda
Bikaner	Wool made of sheep and camel hair
Udaipur	Wooden toys

The Handicrafts

Rajasthan has rich art and handicraft specialties of every region which lure the visitors by their uniqueness (Table 1.6). Typical handicraft items from Rajasthan namely, blue pottery, artistically crafted bullock cart, and colorful Persian style wheel (called *Rahant*) used for irrigation are shown in Fig. 1.20a–c.

Fig. 1.20 (**a**) A vase made at Jaipur during the late nineteenth century—an example of glazed and painted frit-ware popularly known as "Blue Pottery." Museum number: IS.21-1891. Gallery location: Ceramics Study Galleries, Asia & Europe, room 137, case 16, shelf 4. *Photo ©Victoria and Albert Museum, London.* (**b**) A traditional and artistic bullock-cart from Rajasthan (*Courtesy: Devendra Bhardwaj*). (**c**) Colorful Persian wheel locally called *Rahant* is still used for lifting water from traditional wells for irrigation (*Courtesy: Dr. Satish Kumar Sharma*)

The State Symbols

The Government of Rajasthan has declared the following as state symbols with an aim to conserve them:

1. The State Bird—Great Indian Bustard *Ardeotis nigriceps* locally called *Godavan* (Fig. 1.21)
2. The State Animal—Chinkara *Gazella bennettii* (Fig. 1.22)

Fig. 1.21 Great Indian Bustard *Ardeotis nigriceps*—The state bird (*Courtesy: Jaysukh Parekh*)

Fig. 1.22 Chinkara *Gazella bennettii*—The state animal (*Courtesy: Dr. Anil Kumar Chhangani*)

3. The State Tree—*Khejadi or Khejri* or Khejri *Prosopis cineraria* (Fig. 1.23)
4. The State Flower—*Rohida Tecomella undulata* (Fig. 1.24)
5. The State Song—*Kesaria Balam aawo nee Padharo mhare Des*
6. The State Dance—*Ghumar*

Fig. 1.23 *Khejri or Prosopis cineraria*—The state Tree. Red clothes tied around the trunk depict that, the tree is worshiped in the villages of Rajasthan and elsewhere in India on the occasion of *Dushehara* festival. It is also popularly known as the *Shami* tree and has astrological significance too (*Courtesy: Tejveer Singh*)

Fig. 1.24 *Tecomella undulata*—The state flower, locally called *Rohida* (*Courtesy: Devendra Bhardwaj*)

A Glimpse of the Major Issues

The state of Rajasthan lags behind in certain aspects when compared to other developed states of India. With the rapid population growth, scarcity of water is one of the biggest issues since the state has only 1% of the total water of the country. Being the largest state of India from the standpoint of geographical area, Rajasthan has extreme climatic conditions. In addition, 66% of the land area is desert and the long international border with Pakistan in the western Rajasthan poses various problems and threats. An estimated 32,000 villages of the total 122,000 villages of Rajasthan still suffer from severe drinking water crisis and 90% of these are falling under the desert zone. Also, 10,000 villages are fighting with fluoride contamination while more than 21,000 villages struggle with the problem of salty water. Unfortunately, even today, 570 villages having a population of 4,000 or more are not yet connected with drinking water supply lines. Sikar, Jhunjhunu, Churu, Pali, Nagour, Alwar, Karouli, Dausa, and Sawai Madhopur districts of Rajasthan are severely affected with the problems related to drinking water. The Government requires a huge sum of 9,500 million rupees to handle these issues. Out of 239 water blocks, 198 have already gone under dark zone. In order to get rid of the serious water crisis, the state Government has time and again requested the central Government of India to bestow the status of "Special State" with a special financial package to Rajasthan. Other key issues include poverty, women and child health, illiteracy, unemployment, unsustainable development activities, pollution, desertification, floods and famine, electricity and roads to remote villages, irrigation facilities to poor farmers, applications of modern agricultural tools, crop protection, and the quality of education [15]. Female feticide and even killing the newborn girl continues to be a common practice in western Rajasthan which needs urgent attention and action.

Biotic pressure continues to be one of the most important conservation problems in Rajasthan. Other prominent conservation issues include, poaching and illegal trade; encroachment and mosaic human settlement inside PAs; scarcity of surface water, exotics and weeds; destruction of grasslands; drought; pollution; myths and zoophobia; mining; habitat loss, habitat alteration, habitat fragmentation and loss of eco-corridors; presence of canals, roads, railway track and electricity lines inside of PAs; *Aida* (Tribal treat)—a communal *shikar* (hunting) operation in Udaipur–Rajsamand–Bhim area; wells and water-storing tanks without parapet wall inviting accidents of wild animals; and wrong rescue practices, for example, Starred Tortoises thrown in water in the name of rescue, ultimately killing these non-aquatic reptiles. Poaching records of Wildlife Flying Squad (eastern zone) reveal 383 cases during 1974–1975 and 1997–1998. In addition, as many as 51 tribes/communities were found indulged in these heinous offences. For more details, please see (Chap. 4).

The Eastern plains are famous for Keoladeo National Park (KNP)—the only World Heritage Site in Rajasthan which is home to thousands of migratory avifauna and other rare and endangered animals. Unfortunately, the Siberian Cranes have stopped visiting it since 2003. The Southern Rajasthan harbors mammalian fauna such as the Mouse Deer or White-spotted Chevrotain (*Tragulus meminna*), Common Palm Squirrel (*Funambulus palmarum*) and Elliot's Giant Flying Squirrel or Large Brown Flying Squirrel (*Petaurista philippensis*) which are not found anywhere else in Rajasthan. Despite being rich in biodiversity, this part is heavily fighting to save its large number of wetlands and dense forests due mainly to unsustainable development and excessive human encroachment. The River Chambal has Gharial (*Gavialis gangeticus*), Marsh Crocodile or Mugger (*Crocodylus palustris*) and Gangetic River Dolphin (*Platanista gangetica*) apart from a variety of fishes. Mass mortality of *Gharial* in Chambal has recently shocked conservationists and the common men. The presence of Wild Dog or the *Dhole* (*Cuon alpinus*), also known as the Asiatic Wild Dog, Indian Wild Dog, or Red Dog and Wild Ass (*Equus hemionus khur*) also called Khur continues to be doubtful in Rajasthan. Moreover, escalating human population causing vast expansion of rain-fed cropping has resulted into shrinking of grazing area for wild herbivores, while overgrazing by livestock (their number being much above the carrying capacity of arid land) has almost reversed the natural vegetational succession leaving very little edible plants for the fauna in wilderness. As a consequence, the overall number of larger species is fast declining and that of smaller ones such as destructive rodents and insect pests is increasing. These facts point towards a dire state of affairs as far as conservational planning of the state is concerned.

On the other hand, continuous mining activities (since, Rajasthan has near monopoly with respect to mineral wealth and large deposits of stone) is posing a serious threat to the Aravalli Ranges and thus, to the faunal species present in this area. The Aravalli Ranges and its dry deciduous forests provide shelter to the threatened fauna such as the Sloth Bear (*Melursus ursinus*), Grey Wolf (*Canis lupus*), Striped Hyaena (*Hyaena hyaena*), Leopard or Panther (*Panthera pardus*) and a variety of cats. The destruction of the Aravalli Ranges due to mining is having a negative impact on the ecosystem and wildlife. We are also not able to save wetlands especially the lakes of the state which are either dried or vanished and are never filled in their full capacity. The Sambhar Lake, a Ramsar site of the arid zone, is continuously deteriorating and is being visited by only a few hundreds of wintering migratory birds such as the Greater and Lesser Flamingos (*Phoenicopterus roseus* and *Phoeniconaias minor*) since last decade while some of the other rare birds have stopped visiting the lake altogether due to drastic shrinkage in the water spread. Likewise, the White-naped Tit also known as the White-winged Tit (*Parus nuchalis*) has completely disappeared from some pockets of the state, Sambhar being one of the known areas. This has happened due to the callous attitude of administration towards the existing threats to its conservation. If proper conservation measures are taken, other wetlands of the state can greatly benefit both humans and wildlife. Lack of awareness is yet another issue directly linked with man-made threats to biodiversity conservation (Please also see Chaps. 2 and 3 from this Volume and Chaps. 2, 8, 18, 19 and 20 from Faunal Heritage of Rajasthan: Conservation and Management of Vertebrates Vol. 2 Sharma, B.K.; Kulshreshtha, Seema; Rahmani, Asad R. (eds.) for more pictures and details of issues pertaining to the wonderful faunal diversity and its conservation and management in Rajasthan).

Acknowledgments Authors are thankful to Mr. Devendra Bhardwaj, Department of Forests & Wildlife, Government of Rajasthan, Mr. Jaysukh Parekh, Dr. Ashish Kothari, Sonali Singh, Dr. Gobind Sagar Bhardwaj, Sunil Singhal, Kota, Tejveer Singh, Dr. AK Chhangani, Dr. Satish Kumar Sharma, and the Victoria and Albert Museum, London, for providing some rare pictures.

References

1. http://www.censusindia.gov.in/2011-provresults/DATA_FILES/Final%20PPT%202011_11.PDF
2. http://www.dmg-raj.org/why_rajasthan.aspx
3. Kothari D (2002) Population projections for Rajasthan and districts: 2002–2011 IIHMR Occasional Paper No.3, Institute of Health Management Research, Jaipur, India and Forum for Population Action
4. http://www.mapsofindia.com/maps/rajasthan/economy/
5. http://www.mapsofindia.com/maps/rajasthan/rajasthanminerals.htm
6. http://www.mapsofindia.com/maps/rajasthan/rajasthanindustry.htm
7. http://www.rajasthantourism.gov.in/Rajasthan/About-Rajasthan.aspx
8. http://esanchar.rajasthan.gov.in
9. Singh V, Pandey RP (1998) Ethnobotany of Rajasthan, India. Scientific Publisher, Jodhpur, India, pp 367
10. Deoras PJ (1981) Snakes of India. NBTI
11. Sankhla K, Sharma VD (1984) Vanishing cats of Rajasthan. Cheetal 26(1):5–23
12. Soni RG, Sharma SK (2000) Wildlife offences by communities in the eastern Rajasthan. Zoos Print J 15(8):321–325
13. Sharma SK (2002) A study on ethnozoology of southern Rajasthan. In: Trivedi PC (ed) Ethnobotany. Aavishkar Publisher, Jaipur, India, pp 239–253
14. Bhatia S (2007) Mogya reform project. Project report, Jul 2007
15. Kothari D (1990) Growing population in Rajasthan: Some emerging issues, IIHMR Occasional Paper No.1, Institute of Health Management Research, Jaipur, India, pp 48

Chapter 2
Physiography and Biological Diversity of Rajasthan

B.K. Sharma, Seema Kulshreshtha, Satish Kumar Sharma, R.M. Lodha, Sunita Singh, Madhu Singh, and Niranjan Sharma

Abstract This introductory chapter presents an overview of the physiography, zoogeography and biodiversity of the largest state of India, Rajasthan, popularly known as the desert state. Physiographically speaking, Rajasthan is divided into four major regions, namely, the Western Thar Desert, the Aravalli Range with Vindhyan Mountains, the Eastern Plains and the south-eastern Hadoti Plateau. About 640 km long western desert region, extending over an area of 175,000 km^2 is covered with stable and shifting sand dunes, barren hills, rocky structural plains,

B.K. Sharma (✉)
Department of Zoology, R.L. Saharia Government P.G. College,
Kaladera (Jaipur), Rajasthan, India
e-mail: seema.zoology@gmail.com

S. Kulshreshtha
Department of Zoology, Government Shakambhar P.G. College,
Sambhar Lake (Jaipur), Rajasthan, India
e-mail: drbksharma@hotmail.com

S.K. Sharma
Sajjangarh Wildlife Sanctuary, Department of Forest and Wildlife,
Government of Rajasthan, Udaipur, Rajasthan, India
e-mail: sksharma56@gmail.com

R.M. Lodha
Department of Geography, Environment Studies Centre, Mohanlal Sukhadia University,
Udaipur, Rajasthan, India

S. Singh
Department of Geography, M.V.S. College, J.R.N. Rajasthan Vidyapeeth University,
Udaipur, Rajasthan, India
e-mail: sunita_geog@yahoo.co.in

M. Singh
Department of Geography, Government College, Ajmer, Rajasthan, India

N. Sharma
Department of Botany, Government College, Kota, Rajasthan, India

B.K. Sharma et al. (eds.), *Faunal Heritage of Rajasthan, India: General Background and Ecology of Vertebrates*, DOI 10.1007/978-1-4614-0800-0_2,

sandy hammocks and low sand streaks. The entire Aravalli Range and hill tracts can be further divided into the north-eastern hill tracts or the Alwar Hills, the central Aravalli Range, the Mewar Hills and the Bhorat Plateau, the Abu Block and the Vindhyan Scarps. The Eastern Plains of Rajasthan include the lowlands of the Chambal Plain, the Banas Plain and the middle Mahi or Chhappan Plain. The Hadoti Plateau forms a part of the Archaean Shield of the Deccan Peninsula and is divided into Vindhyan Scarpments and Deccan Lava Plateau. Water resources of the state which include Chambal, Banas and Mahi Rivers, their tributaries, the fresh water and salt lakes, have also been emphasised briefly in this chapter. The climatic profile of Rajasthan has been presented as a transition between two major climatic regions of India, the semi-arid east and the arid west. The unique but diverse biotic richness of the state which supports its mesmerising landscape, especially the mysterious desert having its own common, threatened and endemic flora and fauna, reserve areas and wetlands, has been thoroughly described in the chapter.

Introduction

Biodiversity is the natural biological capital on the Earth. We the inhabitants of the planet Earth agree that, life here is now faced with the most severe extinction episode. Unfortunately, the destroying force this time is the mankind, not the physical environment which would ultimately lead to self-destruction. India, being the seventh largest country and one of the 12 mega biodiversity regions of the world, is blessed with an unmatched broad spectrum of biological and ecological heritage. Biogeographically, India is situated at the tri-junction of three realms, namely, Afro-tropical, Indo-malaya and Palearctic, and therefore, has characteristic elements from each of them and this makes the country rich and unique in biodiversity. The Wildlife Institute of India (WII) has identified ten bio-geographic zones in India, namely, Trans-Himalaya, Himalaya, Semi-arid, Desert, Gangetic Plain, Deccan, Western Ghats, North-east Coasts and Islands (Fig. 2.1). Diverse habitats falling in these zones harbour about 115,000 species of plants and animals. Amongst flora, the country can boast 49,000 species which accounts for 15% of the plants known the world over. The faunal wealth has nearly 81,251 species consisting of 80% insects. India's contribution to the world's biodiversity stands at 7%. The country has 12% of world's flora and 6.67% of fauna.

A sizable portion of Rajasthan broadly falls under the Indo-malaya ecozone—one of the eight ecozones dividing Earth's land surface. Indo-malaya has three major biomes in Rajasthan, namely, Deserts and Xeric Shrublands, Tropical and Sub-tropical Dry Broadleaf Forests and Tropical and Sub-tropical Moist Broadleaf Forests. The surface of the state of Rajasthan is as complex as its making which is evident from internal geological and external climatic forces prevailing since age old geological history. One of the major physiographic elements, the south-west and north-east Aravalli Hills, is a relic of the world's oldest fold mountains. Aravalli Hills bifurcate the state obliquely into western arid Rajasthan, consisting 3/5th part and eastern semi-arid Rajasthan covering 2/5th part of the state. The WII places Rajasthan under the bio-geographic zones 3 and 4, namely, the desert covering the Thar or the

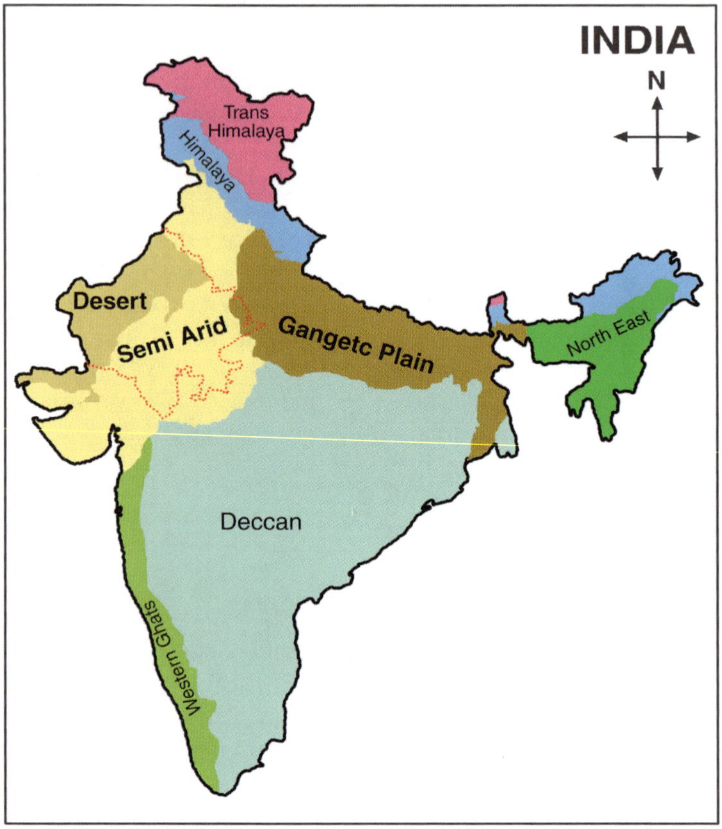

Fig. 2.1 Bio-geographic zones of India depicting location of the state of Rajasthan as "desert" and "semi arid" with a tiny portion in Gangetic Plains (*Courtesy: Dr. B.K. Sharma*)

Great Indian Desert which is a blend of rocky and sandy desert and has a unique biodiversity and the semi-arid area covering the rest of Rajputana or Rajasthan.

Physiography

Rajasthan is endowed with diverse physiographic features, particularly the Great Indian Thar Desert in the west, the Aravalli Mountains in the middle, the plains in the east and Hadoti Plateau in the south-east. Hence, it is generally addressed as the state of *Maru* (desert), *Meru* (mountain) and *Mal* (plains). Physically, the state is surrounded by the Ganga–Yamuna Plain in the east, Malwa Plateau in the south and the Sutlaj and Beas river-plains in the north and north-west. The Tropic of Cancer passes through south of Banswara district in the state (Fig. 2.2). Physiographically, Rajasthan may be divided into the following regions (Table 2.1):

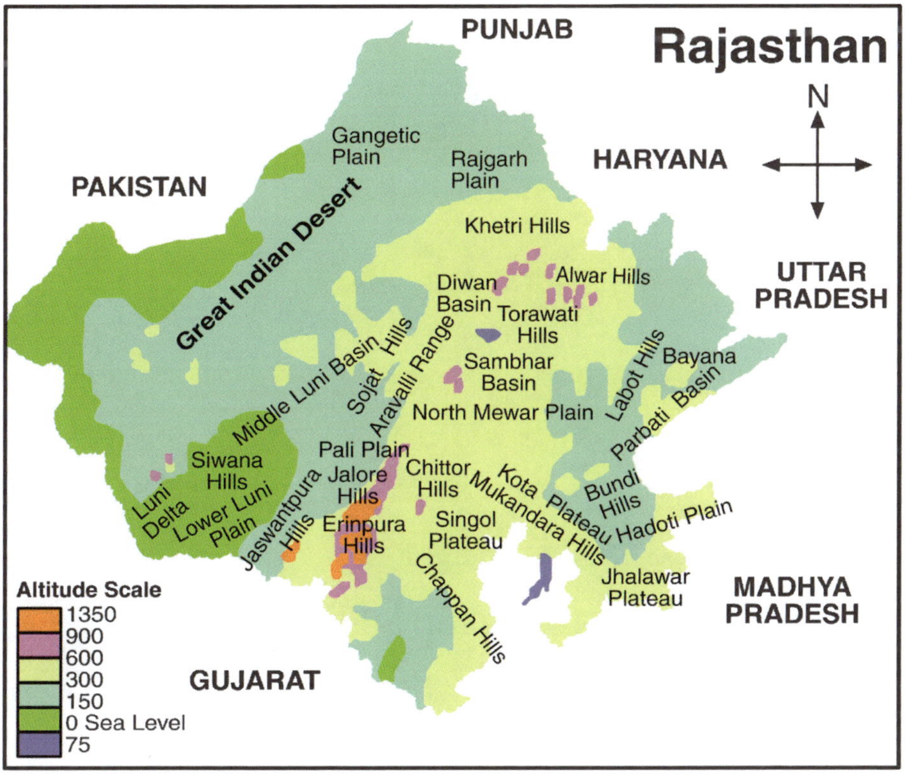

Fig. 2.2 Physical map of Rajasthan (*Courtesy: Dr. B.K. Sharma*)

1. The Western Desert Region

The region extends 640 km in the north-west to the south-east and 300 km in the west-east direction, covering an area of 175,000 km^2 being 61% of the state. It consists of about 40% human population and includes 12 districts of the state, namely, Shri Ganganagar, Hanumangarh, Bikaner, Churu, Nagour, Jodhpur, Jaisalmer, Barmer, Pali, Jalore, Sikar and Jhunjhunu. Its western boundary is known as Redcliff Line which constitutes the international boundary between India and Pakistan. The region is broadly divided into the Marusthali or desert region and semidesert or Bagar region (Fig. 2.3).

A. *Marusthali Region*: This is characterised by high intensity of sand-dunes, below 25 cm of annual rainfall, very sparse population, widely scattered rural settlements, high percentage of fallow land, very low percentage of net sown area and absence of industrial units due to poor resources and infrastructural base.

Table 2.1 Physiographic regions of Rajasthan

Physiographic Regions of Rajasthan

- 1 Western Desert Region
  - A Marusthali
    - a. Jaisalmer
    - b. Barmer-Phalodi
    - c. Bikaner-Churu
  - B Bagar or Semi desert plain
    - a. The Luni Basin or Godwar Region
    - b. The Shekhawati Region
    - c. Nagour Highlands
    - d. The Ghagghar Plain
- 2 The Aravalli Range & Hill Tracts
  - A. The Northern and North-Eastern Aravalli or Alwar hills
  - B. The Central Aravalli or Jaipur hills
  - C. The Southern Aravallis
  - D. Mewar and Bhorat plateau or Udaipur hills
  - E. Abu Block
- 3 The Eastern Plains
  - A. The Chambal Basin
  - B. The Banas Basin
  - C. The Chhappan Basin
- 4 The South-Eastern Hadauti Plateau
  - A. The Vindhyan Scarps
  - B. The Deccan Lawa Region

Fig. 2.3 Sand-dunes in the Thar Desert: a picture taken by (Late) Kailash Sankhla—a well-known conservationist from Rajasthan popularly known as the "Tiger Man," who pioneered the movement leading to the establishment of "Project Tiger" in India in 1972. A Chinkara can be seen amid the desert (*Courtesy: Bittu Sehgal/Sanctuary Asia Photo Library*)

This region has considerable variation in geographical elements which is evident in the distinguishing characteristics of its sub-regions as given below:

a. *Jaisalmer–Marusthali Region*: This region is the driest part of the region having most adverse environmental conditions resulting in the lowest population density (only 4 persons/km^2). Major towns of this region are Jaisalmer and Pokharan. The westernmost plain area of this region is the driest part, having no town while the eastern plain area is rich in gypsum deposit. Irrigation facilities are available at some places due to the Indira Gandhi Canal.

b. *Barmer–Phalodi Marusthali Region*: This region falls under dry region. The density of population is recorded to be only 10–15 persons/km^2. The exploitation of petroleum and natural gas in this region has given a boom to the economic development of the state. Barmer and Phalodi are the two major towns.

c. *Bikaner–Churu Marusthali Region*: Bhakhra dam located in the state of Punjab provides electricity to this region. It is further divided into Bikaner and Churu Plains.

B. *The Bagar or Semidesert Region*: Located between the Aravalli Range in the east and desert region in the west, the 75,000 km^2 Bagar region is relatively free from sand dunes. Small hills make a circumference of 65 km around Jaisalmer town. It has higher rainfall of 25–50 cm than the Marusthali region. Due to the higher population density, the settlement density also increases sharply. It is bounded by Ghagghar River in the north, Sikar-Shekhawati area in the north-east, Luni River basin in the south-east including Jodhpur–Barmer districts and western parts of Jalore, Sirohi and Pali districts. The average length and width of sand dunes sparsely found in this region is about 700 and 150 m, respectively. The Bagar Region is further divided into the following four sub-regions:

a. *The Luni Basin or Godwar Region*: This region is the most humid drainage area of the Luni, the biggest river of southern Rajasthan and its tributaries which rise from the southern hilly region of Ajmer and merges in the Rann of Kutch after flowing 480 km and passing through southern Jodhpur, Pali, Jalore and western Sirohi districts. Luni and its tributaries have built up alluvial plains which are covered by sand at some places brought by the south-western winds. The region is further divided into six sub-regions, namely, south-eastern Jodhpur plain, the Pali–Sojat plain, Luni–Sikri interfluve, the Jalore–Bhinmal plain, the Luni–Runn saline region and the south-eastern arid Barmer plain.

b. *The Shekhawati Region or Area of Internal Drainage*: Shekhawati Region falls in the north of the Luni basin. It includes Churu, Jhunjhunu, Sikar and northern part of Nagour districts. The region is surrounded by Aravalli Range in the east, Ghagghar Plain in the north, 25 cm isohyet in the west and 50 cm isohyet in the east. The whole region is formed by Barkhan type of sand-dunes. This is an area of internal drainage due to the sandy soil, meagre rainfall and walls of the dunes. Sambhar Lake lies within this drainage area where Mendha, Rupangarh and Khari Rivers coming from different directions merge (Fig. 2.4).

Fig. 2.4 Sambhar Salt Lake—a Ramsar Site and Wetland of International Importance still surviving in the arid zone of Rajasthan despite serious conservation threats (*Courtesy: Devendra Bhardwaj*)

c. *Nagour Highland Region*: Situated at an elevation of 300–500 m from the sea level, the Nagour Highland region is distinguished by an inland drainage system, salt lakes and rocky and hilly surface areas. The region is transitional between the more humid Luni basin on the south-east and the comparatively arid north-east Bagar. The eastern part has a few salty lakes, namely, Sambhar, Deedwana (now vanished), Nawa and Kuchaman. The region is famous for salt industries due to the abundant sodium chloride in these lakes.
d. *The Ghagghar Plain Region*: The Ghagghar Plain region covers almost 75% area of Ganganagar and Hanumangarh districts. This plain is formed by alluvial soil brought by Ghagghar, Sutlej, Saraswati (now eloped) and Chautang Rivers, which arise from the Himalayas. In fact, the ancient Vedic Saraswati River itself is termed as Ghagghar River which is called the "dead river" as no natural water flows in it. During the rainy season, the water mass from Punjab and Haryana brings floods to this river. Locally, the river is named as "Nali." Being the area of black soil, this plain is a fertile agricultural region of the state.

Fig. 2.5 Aravalli Hills during rainy season near Jaipur (*Courtesy: Devendra Bhardwaj*)

2. *The Aravalli Range and Hilly Tracts*

Aravalli Range is the principal dominant landform running across the state like a curved scimitar from south-west to north-east (Figs. 2.5 and 2.6). This range extends for about 692 km from Palanpur in Gujarat to Delhi. The loftiest and the most clearly defined section of the Aravalli Range is in the Mewar and Marwar Hills where it forms an unbroken range. Beyond Ajmer, it breaks up into discontinuous hills. The successive hills and rocky eminences stretch north-eastwards of Sambhar Lake through the districts of Jaipur, Sikar and Alwar up to Khetri (Jhunjhunu district). Here it terminates, though; detached hills can be seen as far as near Delhi. The average height of the Aravalli is 900 m. The south-western Aravalli Hills are much higher than the north ones. Here, the width of these hills is 100 m and average height is 1,000 m. The highest peak of Aravallis is Guru Shikhar in Mount Abu which is 1,722 m high. This region is characterised by a number of passes, namely, Sambhar Pass, Bar, Parwaria, Shivpura, Suraghat and Peepli. There exist irregular Torawati Hills in the east of Malkhet hilly area. Likewise, about 300–600 m high eroded hills and peaks are found around Alwar. The whole Aravalli Range and hill tracts can be sub-divided into the following physiographic units:

A. *The Northern Aravallis*: These include north-eastern hill tracts or the Alwar Hills covering Alwar and north-western Jaipur district. To the east and north, it merges with Ganga–Yamuna Plains.
B. *The Central Aravallis*: It extends from Devgarh in the south to Sambhar basin in the north Jaipur. This region includes the whole of Ajmer district having

Fig. 2.6 Aravalli hills with slopes and peaks during early summer (*Courtesy: Tejveer Singh*)

Merwara Hills, southern part of western Tonk district and northern part of Rajsamand district.

C. *The Southern Aravallis*: These include ten mountain ranges extending further into Sirohi, Udaipur, Rajsamand and Chittourgarh districts. It has two sub-units—the Mewar Hills and Bhorat Plateau and Abu Block.

D. *Mewar Hills and Bhorat Plateau*: It is the highest portion of the Aravalli Range and lies north-west of Udaipur between the fort of Kumbhalgarh and Gogunda, in the form of a Bhorat Plateau, having an altitude of nearly 1,225 m, with the highest peak rising to 1,431 m called Jarga Hill. To the south-west-south and south-east, the curved ridges reach up to eastern Sirohi, Dungarpur and northern Sabarkantha in Gujarat. Further to the south-west, the range is reduced to a few scattered butts hardly of 60 m height around Palanpur and Khed Brahma. The highest point is 1,290 m near Jharol, west to Bagpura Plateau. The western most ridge in the east Sirohi is locally known as "*Bhakar*" while towards east from Jaisamand lies a dissected plateau known as "*Lasadia*" (325–650 m). Towards east of the Bhorat Plateau, the southern most spur (500–600 m) is the most important one as it not only separates Udaipur basin from Jaisamand basin but also acts as a major "watershed" between the Bay of Bengal and the Arabian Sea drainage systems. Some spurs present a girdle of hills around Udaipur basin, locally known as "*Girwa*."

E. *The Abu Block*: It is an almost isolated hill mass known as Mount Abu which has been detached in the south-western parts and spreads in the form of cluster of hills in Sirohi and northern Banaskantha district of Gujarat. It has been separated from the main Aravalli Range by the wide valley of the west Banas and by a narrow pass near the village of Isra from Abu-Sirohi range in the west. The whole Abu Block is called a Great Inselberg (19 km × 8 km). The northern and western slopes are the gigantic blocks of syenitic rocks, e.g. Tod Rock or Horn Rock. The highest peak of Abu Block is "*Guru Shikhar*" having a towering elevation of 1,722 m from the sea level. Other prominent peaks adjoining Guru Shikhar are—Ser (1,597 m), Achalgarh (1,380 m), Jaswantpura (Dora Hill, 869 m), Siwana (Chhappan or Nakoda Hills), Jalore (Roja Bhakhar, 730 m; Israna Bhakhar, 839 m; and Jharola Hill, 588 m).

3. *The Eastern Plains*

The eastern part of Rajasthan is formed as a plain by Ganga and Yamuna Rivers and is bifurcated by 50 cm isohyet from west to east. It is bounded by the state of Uttar Pradesh in the north, Madhya Pradesh in the east, eastern edge of Aravallis in the west and Vindhyan Plateau in the south and south-east directions. The low hills having an elevation of 400–500 m are found in Bharatpur, Sawai Madhopur, Tonk, Dholpur and Bhilwara districts. The Eastern Plain of Rajasthan can be divided into three subunits, namely, the Banas Plain, the Chambal Plain and the middle Mahi or Chhappan Plain.

A. *The Banas Plain*: The literal meaning of "*Banas*" is "*Van ki Asha*" or "*Hope of the Forest.*" The Banas Plain is limited to the west by 50 cm isohyet in the eastern slopes of Aravallis, the great Indian watershed to the south, Banganga basin and Alwar Hill tracts to the north and the Chambal basin and Vindhyan Scarp to the east. The plain is drained by the River Banas and its tributaries, namely, Khari, Sodra, Bhosi and Morel on the left bank and Berach, Bejain and Golwa on the right bank. The southern part of the plain is called Mewar Plain while the northern part is known as Malpura-Karouli Plain.

B. *The Chambal Plain*: The Chambal River forming the Chambal Plain, rises from the northern edge of Vindhyan Scarps of Janapav Hill (616 m) near Manpur lying south of Mahu in Madhya Pradesh state and enters Rajasthan while passing through a gorge near Chaurasigarh. The average width of the ravine belt in the state is approximately 4,500 km^2 in the districts of Kota, Bundi, Tonk, Sawai Madhopur and Bharatpur. The most dominant is the Chambal belt extending over a length of 480 km from Kota to the confluence of the Yamuna River. On the north from Kota to Dholpur, an irregular and high wall of the upper Vindhyan Scarplands (Fig. 2.7) is marked by the watershed between the Banganga, a tributary of the Yamuna and by the Yamuna itself. The famous Keoladeo marshes are an integral part of this region (Fig. 2.8). The water is fed into these marshes via inundation from Rivers Gambhiri and Banganga. The southern boundary is demarcated by the tributaries, namely, Kalisindh, Parbati and Sip. Further, it is well marked by the ravines of the Kunwari River throughout the south-easterly course of the Chambal up to its confluence with the Yamuna.

Fig. 2.7 Vindhyan Gorges at Ranthambhore National Park, Sawai Madhopur (*Courtesy: Devendra Bhardwaj*)

Fig. 2.8 Marshes at the lake inside Keoladeo National Park, Bharatpur (*Courtesy: Devendra Bhardwaj*)

C. *The Middle Mahi or Chhappan Plain*: This plain area is formed by the Mahi River and its tributaries. The dissected region along with the hill tracts of Dungarpur and Banswara are locally known as *Bagar*. This plain is also named as the *Chappan Plain* where about 56 villages lie in Pratapgarh and Banswara districts. Presently this plain area includes five districts of southern Rajasthan, namely, south-eastern part of Udaipur, southern part of Banswara and Chittourgarh, Pratapgarh and Dungarpur.

4. The South-Eastern Plateau Region

Topographically, this region forms the northern part of the Malwa plateau, locally known as "*Hadoti Plateau region*." This region extends in the districts of Baran, Bundi, Jhalawar and Kota, and is limited by Madhya Pradesh state in the south-east and south-west direction, by Sawai Madhopur district in the north and by Tonk, Bhilwara and Chittourgarh districts in the north-west. The region is drained by the Chambal and its tributaries, namely, Kalisindh and Parbati. The Hadoti Plateau is characterised by crescent shaped hill ranges known as the Bundi and Mukundra Ranges. The eastern part of Hadoti Plateau is called "*Shahabad Highlands*" and towards south-west of this plateau region lays the "*Dug–Gangdhar Highlands*." Physiographically, the plateau is divided into two subregions:

A. *The Vindhyan Scarpland*: This region is formed by grey coloured sandstones which mark the topography between the Banas and Chambal River basins towards south-east direction. The region is bounded by Bundelkhand in the east, by steep slopes of scarps in the north-west on the left bank of the Chambal River.

B. *The Deccan Lawa Plateau*: The western part of Vindhyan Plateau of Madhya Pradesh lies in the form of three concentric scarp including Plateau region of Kota–Bundi, known as the "*Uparmal*." The deposit of Deccan Trap lawa appears at the bottom areas of the Vindhyan Scarps. The Chambal and its tributaries, Kalisindh and Parbati have built a triangular shaped alluvial basin with an elevation of 212 m. Bundi and Mukundra Ranges run across this region which have been eroded by rivers resulting in the poor accessibility because of the difficult terrain.

Agro-climatic Zones

Rajasthan state can broadly be divided into arid, semi-arid and sub-humid regions. The 143,842 km^2 of western Rajasthan is under the influence of arid climate where rainfall is low and highly variable and inhospitable living conditions prevail in an area of 9,290 km^2 towards the extreme. However, there is an improvement in rainfall pattern from the west towards the east of Rajasthan creating semi-arid conditions in

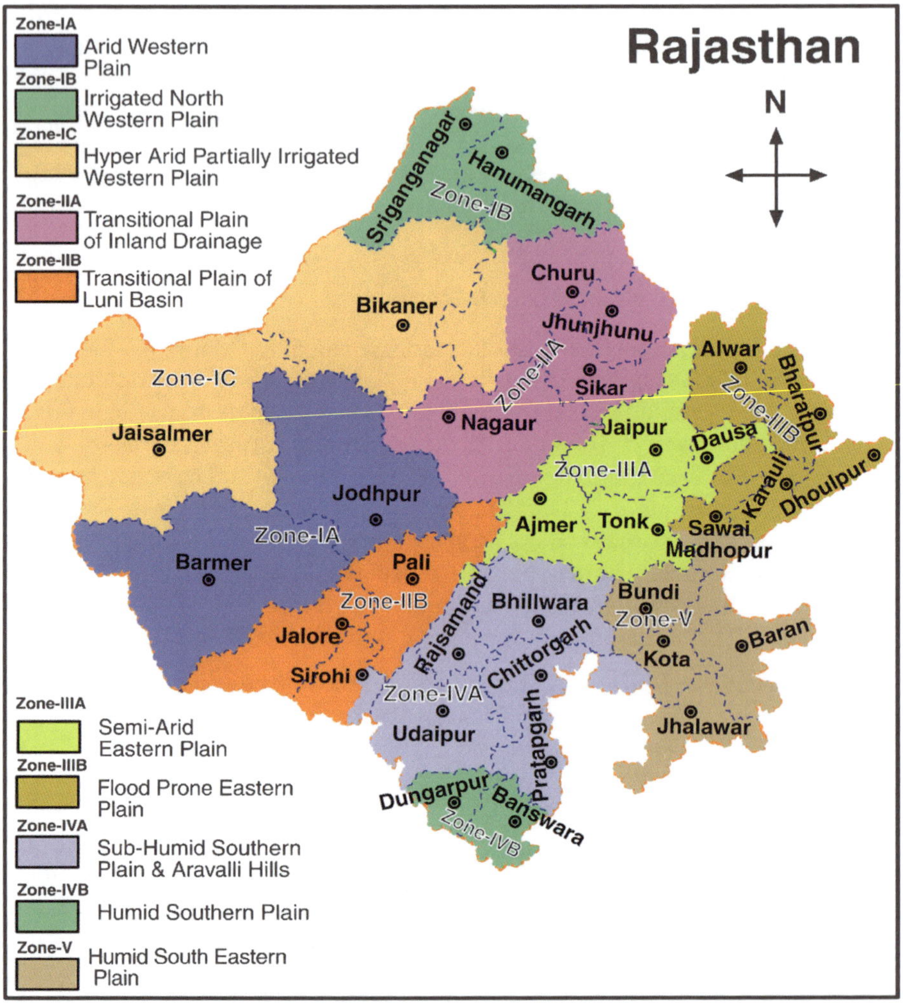

Fig. 2.9 Agro-climatic zones of Rajasthan (*Courtesy: Dr. B.K. Sharma*)

an area of 66,830 km² and sub-humid conditions in 3,210 km², covering parts of Jhalawar and Banswara districts in the extreme eastern part of the state. The state is divided into the following agro-climatic zones (Fig. 2.9).

IA. Arid western plain	IIIA. Semi-arid eastern plain
IB. Irrigated north-western plains	IIIB. Flood prone eastern plains
IC. Hyper arid partially irrigated western plains	IVA. Sub-humid southern plains and Aravalli
IIA. Transitional plain of inland drainage	IVB. Humid southern plains
IIB. Transitional plain of Luni basin	V. Humid south-eastern plains

Climate

Rainfall: The mean annual rainfall in Rajasthan varies from <100 mm in the extreme western parts of Jaisalmer district to >900 mm in the eastern parts of Jhalawar and Banswara. About 85% of the total rainfall is received during the southwest monsoon period extending from June to September. Remaining portion of rainfall is received in the cold weather period (January and February) in association with the local convective activity. The lowest recorded annual rainfall in the past 100 years was 24 mm in the western Rajasthan, whereas it was never below 120 mm in the eastern Rajasthan. The recorded highest rainfall was between 454 and 1,176 mm in the western Rajasthan, whereas it was between 1,223 and 1,698 mm in the eastern Rajasthan.

Wind Regime: Winds are light and variable during winter, but strong winds prevail during May to July in most parts of Rajasthan. Jaisalmer region experiences strongest winds with mean speeds of 23.4 km per hour during summer. Winds gradually get reduced towards the northern and eastern parts of the state. The wind directions are northeast to north in winter, while during the rest of the year, they are mostly south-westerlies or west south-westerlies. The frequency of dust storm during summer period is highest in the Ganganagar region. Bikaner, Jodhpur and Jaipur areas experience moderate number of dust storms, whereas the east and north-eastern part of the state experience lowest number of dust storms. Jaisalmer region is exposed to high winds, but it experiences lowest number of dust storms due to its geographic setup and soil conditions.

Air Temperature and Relative Humidity Conditions: The mean maximum air temperature in Rajasthan varies from 38.6 to 42.6°C in May and during this period, western part of the state forms the hottest place in the country. However, due to lack of proper negative cover, the sandy soils of this region cannot retain the heat resulting in moderate temperature during night even in summers. The winters are pleasant in the state with mean night temperature above 4.7–10.6°C. The recorded extreme temperature tours in the state were −4.4 to 50.0°C in the western Rajasthan and −2.8 to 47.8°C in the eastern Rajasthan.

Relative humidity in the state is 50–65% at 8.30 h and 30–5% at 17.50 h. The relative humidity in the arid region of western Rajasthan is also quite high compared to other semi-arid and sub-humid regions indicating that only because of unfavourable circulations of the atmosphere, low precipitation occurs in the arid region.

Soil

The soil in the south and south-eastern part of the state has remained fairly stable and has been formatted due to prolonged effects of climate and vegetation. But, in the northern and western part, large amount of variation in the climate has dominated the aeolian and alluvial processes influencing the soil formation processes.

In the northern region, the soils have developed from the sediments deposited by the Ghaggar River and its tributaries. Thus, the state is endowed with a variety of soil ranging from desert sand, basaltic silt, clay, loam, etc. The distribution of different soil groups is as follows.

(a) *Dune and Interdune soils*: These occur in continuous chain in two zones, the extreme west in the Barmer, Jaisalmer and Bikaner districts and the eastern part of Bikaner and Churu districts. A discontinuous zone of sand-dunes runs from Sanchor, Saila, Hinch, Didwana and Lakshmangarh. There are six types of sand-dunes, namely, obstacle, parabolic, longitudinal, transverse, barchans and shrub coppice. The soils associated with the dunes are very deep, fine, sandy and light yellowish brown to pale brown. Brown sandy soils occur on the sandy plains in Jodhpur, Barmer, Nagour, Bikaner and Jaisalmer districts.
(b) *Soils of Younger Alluvial Plain*: These are the soils of vast alluvial plain in the districts of Bharatpur, Alwar and parts of Jaipur. Soils of these regions are very deep, pale brown and sandy. The moisture retention capacity ranges from 200 to 215 mm/m. Brown alluvial soils occur in Bhilwara, Tonk and Ajmer districts and also scattered in the Udaipur, Chittourgarh and Jaipur districts while red loam soils are found in the southern part of Rajasthan covering Dungarpur, Udaipur and Chittourgarh districts. At places, these soils are associated with rocky undulating uplands. In such situations, soil erosion is very active. Black soils occur in the south and south-eastern part of Rajasthan state. In Kota, Bundi and Sawai Madhopur districts, the black soils have developed from the Vindhyan rock and Deccan Traps sediment deposited by the River Chambal and its tributaries, the Klaisindh and the Parvati. In Banswara, Jhalawar and Pratapgarh–Dhariyavad area of Udaipur–Chittourgarh districts, the black soils have developed from the Deccan traps.
(c) *Soils of Hilly Region*: The southern part of the state in Udaipur, Dungarpur, Banswara and Sirohi districts is occupied by the Aravalli, similarly the south-eastern part of Rajasthan covering Sawai Madhopur and Bundi districts is also hilly. The landscape of this region is very uneven, slopy, covered with stones and rocks. The soils, which occur in pockets, are very shallow, gravely, coarse, sandy, reddish brown to grayish brown and non-calcareous.

Water Resources of Rajasthan

Rajasthan is the driest state of India having only 1% of total water of the country. The water resources are scanty and unevenly distributed in relation to its area. In addition, water availability in general, is highly variable. Rainfall, the primary source of surface and underground water varies considerably from place to place and from year to year. Evaporation is also high throughout Rajasthan and on an average consumes 78% of rainfall that reaches the ground. High evaporation coupled with the variability of water availability make conservation and development of surface water resources more expensive and difficult. Every year water level is going down by 2 m and as much as 135% of the existing water is being extracted. Out of 239 blocks, 198 blocks have been declared as dark zones in the state.

Precipitation is recorded in Rajasthan by about 517 rain gauges setup, mainly, by the Revenue and Irrigation Departments of the state. Within an area of 342,232 km^2, the average annual rainfall is estimated as 47 cm. The mean annual rainfall of 47 cm over Rajasthan is equivalent to about 161,000 million m^3 (MCM) volume of water. On an average, there are 25 rainy days in a year in Rajasthan and for about eight days the rainfall is <2.5 mm. Taking this factor into account, it is estimated that, out of the average annual rainfall, 161,000 MCM is lost to the atmosphere. Of the remaining 160,300 MCM, 19,585 MCM flows as surface runoff and the rest 140,715 MCM seep into the ground. This shows that in Rajasthan, the average percentage of runoff to rainfall is only 12. Although, the average percentage of runoff to rainfall is low, its variability across Rajasthan is considerable.

According to the simulation studies carried out for each basin by Rajasthan State Water Resource Department, the total internal surface water resources in the State have been estimated as 21.71 BCM (17.6 MAF) at 50% dependability as against 19.56 BCM (billion m^3) (15.86 MAF—million acre feet) estimated earlier. The water availability at 75% dependability is only 14.12 BCM (11.45 MAF). However, the economically utilisable surface water in the present situation at 50% dependability is estimated as 16.05 BCM (13.02 MAF). Apart from this, the total external surface water resources from other States, under various interstate agreements, are 17.88 BCM (14.5 MAF). It has been assessed that mean annual natural replenishable ground water is 7.413 BCM (6.01 MAF) and total 10.09 BCM (8.18 MAF) of ground water including return flows from irrigated areas, urban and other water utilisation sectors is available in the State.

The Government of Rajasthan has recently adopted the State Water Policy, 2010 keeping in view the provisions contained in the National Water Policy and the specific conditions and problems of the state. The policy addresses all the issues for maximum development and optimum utilisation of scarce water resources in the state. The current problems being faced by the state and the future scenario with a long term vision up to the year 2045 emphasises the need of time-bound action plan for successful implementation of the State Water Policy.

Surface Runoff: The annual surface runoff in rivers of Rajasthan from rainfall is estimated as 19,585 MCM which equivalents to a mean runoff depth of 5.7 cm giving the surface runoff for different river basins in Rajasthan (Table 2.2). The table shows that about 75% of the surface runoff is concentrated in the three perennial rivers, namely, Banas, Mahi and Chambal, flowing in the southern and eastern part of the state (Figs. 2.10 and 2.11).

Owing to the limitations of topography, physiography and soils, only about 12,300 MCM of surface water can be utilised. In addition to this, about 15,000 MCM is brought in by the rivers (Bhakra Dam and Gang Canal) outside Rajasthan. Thus, as of today, the total usable flow in Rajasthan is about 27,300 MCM giving per capita amount of annual usable flow of the order of 680 m^3. On the other hand, from India as a whole, the per capita amount of annual usable flow is about 840 m^3. The surface water is disposed off in three ways. A part of it is stored in reservoirs, a part disappears

Table 2.2 Surface runoff in different river basins

S. No.	River basin	Area (km^2)	Annual flow (MCM)	Utilisable flow (MCM)
1.	West Banas	1,999	800	90
2.	Mahi	16,891	3,570	3,350
3.	Sabarmati	2,797	890	840
4.	Banas	45,543	4,360	3,100
5.	Chambal	31,393	6,130	2,980
6.	Luni	34,569	1,630	530
7.	Sahibi	4,532	280	150
8.	Banganaga	6,746	440	350
9.	Baran	3,225	230	120
10.	Gambhir	4,812	450	400
11.	Parvati	1,968	200	180
12.	Sukri	945	230	50
13.	Other	10,175	375	15
	Total	–	19,585	12,295

Fig. 2.10 Drainage map of Rajasthan (*Courtesy: Dr. B.K. Sharma*)

Fig. 2.11 River Chambal: a view from Gardiya village at Kota (*Courtesy: Sunil Singhal, Kota*)

as percolation from streams where ground water table is below the stream surface and rest finds its way to the sea. Yet another part of water stored in reservoirs is lost through evaporation and seepage and the rest is utilised for various purposes mainly irrigation. Of the 19,585 MCM of water that flows in the rivers annually at present about 8,650 MCM is stored in various reservoirs and tanks. There is evaporation loss of the order of 30% from reservoirs and more from tanks. The percentage loss is greater in tanks because they are relatively shallow. The remaining flow of about 10,935 MCM goes to the sea and adjoining states. Ground water is basically required for irrigation, domestic and industrial purposes and can be developed at a small capital cost in least possible time. The bulk of ground water is used for irrigation only. Intensive irrigation can be practised with double and triple cropping.

At present, 104 major and medium irrigation projects and 4,786 minor irrigation projects are being worked upon and irrigation potential thus created has increased to 2.81 million ha.

Biodiversity in Rajasthan

The spectacular state of Rajasthan exhibits a great variety of fauna and flora, ranging from the desert to grasslands and dense forests to wetlands, rivers and hilly areas. Agriculture fields, rocks, ruins and civil structures also are home to a great variety of biological diversity in the state.

Forests

The forests of the state can be divided into four broad types, namely, tropical thorn forests, tropical dry deciduous forests, central Indian sub-tropical hill forests and mixed miscellaneous forests (Table 2.3).

1. *Tropical Thorn Forests*: They are found in arid and semi-arid regions of western Rajasthan belonging to Jodhpur, Pali, Jalore, Barmer, Nagour, Churu and Bikaner districts extending from western Indo-Park border and gradually merging with the dry deciduous mixed forests of the Aravalli Hills and the south-eastern plateau. The main floral species found here include *Acacia nilotica*, *A. senegal*, *Calotropis procera*, *Capparis decidua*, *Commiphora weighttii*, *Opuntia dillenii*, *Prosopis cineraria*, *Capparis aphylla*, *Zizyphus* spp., *Flacourtia* spp., *Tecomella undulata*, *Tribulus terrestris*, *Zizyphus mumularia* and *Argemone mexicana*.
2. *Tropical Dry Deciduous Forests*: They are found in small patches in the northern and eastern slopes of Aravalli Ranges covering Alwar, Bharatpur and Dholpur districts. Sporadic growth of certain species is found along the dry river beds of Jalore, Nagour, Ganaganagar and Bikaner districts. The main species found in this forest are *Anogeissus pendula*, *Anogeissus latifolia*, *Acacia catechu*, *Terminalia tomentosa*, *T. balerica*, *T. arjuna*, *Boswellia serrata*, *Dendrocalamus strictus* and *Lanea grandis*.
3. *Central Indian Sub-tropical Hill Forests*: They are mainly abundant in central India particularly in Madhya Pradesh, parts of Gujarat and Maharashtra states. In Rajasthan, they are found in Sirohi district mostly on the hills girding Mt. Abu. These forests have semi-evergreen and some evergreen species of trees. The vegetation of Mt. Abu consists of plants similar to the sub-tropical region of Himalayas where they are well represented between 700 and 800 m altitudes.
4. *Mixed Miscellaneous Forests*: They are found in south-eastern and eastern parts of Rajasthan mainly Chittourgarh, Kota, Udaipur, Sirohi, Banswara, Dungarpur, Baran and Jhalawar districts. These Forests have *Anogeissus pendula*, *A. latifolia*, *Terminalia tomentosa*, *T. arjuna*, *T. chebula*, *Albizia lebbeck* and *Dalbergia paniculata*.

Table 2.3 Forest types in the state

S. No.	Type	Forest Area (ha)	Percentage of total forests area
1.	Dry teak forests	224,787	7.05
2.	Subsidary edaphic type of dry tropical *Anogeissus pendula* forests	1,902,775	59.65
3.	Northern tropical dry deciduous mixed forests	864,322	27.09
4.	Tropical thorn forests	185,452	5.81
5.	Sub-tropical evergreen forests	12,664	0.40

Source: Forests and Wildlife Department, State Government of Rajasthan, Jaipur

Biotic Richness of Rajasthan

The biological diversity of India is one of the most significant in the world since it has only 2% of the total landmass of the world containing about 6% of the world's known wildlife. Out of the total 1,196,903 animal species recorded from the world, India has 86,874 species, which means India is home to a little more than 7% of the total animal species of the world. Interestingly, out of a total of 86,874 animal species, insects alone comprise 68.52% and Chordates 5.70%. Among the large vertebrates, 173 species of mammals, 101 of birds, 15 of reptiles, three of amphibians and two of fishes are considered threatened out of 614 species of amphibians and reptiles; 1,225 of birds; and 350 of mammals. The International Union for Conservation of Nature (IUCN) has recently notified 132 species of plants and animals from India as "Critically Endangered" and falling in the Red List. Plants seemed to be the most threatened life form with 60 species being listed as "Critically

Table 2.4 Biotic richness of Rajasthan

Taxa	No. of species present in the state
Flora	
Angiosperms	
• Wild species	1,714
• Agro-horticultural species	320
Gymnosperm (wild species)	1
Cryptogamous	
• Liverworts and Hornworts	37
• Mosses	42
• Ferns	60
Red data plant species	7
Terrestrial orchids	8
Epiphytic orchids	6
Parasitic species	13
Carnivorous species	5
Fauna	
• Fish	140
• Amphibians	14
• Reptiles	67 (492)
• Birds	510
• Mammals	(7) 87
• Threatened mammals	16
• Threatened reptiles	8
• Critically Endangered bird species	6
• Endangered bird species	5
• Vulnerable bird-species	12
• Conservation dependent bird species	1
• Near Threatened species	19
Animals of Schedule I [Wildlife (Protection) Act, Amendment Act, 2006]	
• Mammalian species	10
• Reptile species	5

Endangered" and 141 as Endangered. The "Critically Endangered" list included 18 species of amphibians, 14 fishes and 10 mammals. There are also 15 bird species in the category. The agency listed 310 species as Endangered, including 69 fishes, 38 mammals and 32 amphibians. Two plant species were reported to be Extinct in the Wild. A Leaf-frog species and six plants were recorded as Extinct according to the latest assessment. Of the total 63,837 species globally assessed, the IUCN classified 3,947 as Critically Endangered, 81 as Extinct and 63 as Extinct in the Wild. In the lower-risk categories, there were 5,766 species in Endangered, 10,104 in Vulnerable and 4,467 in Near-Threatened categories. Scientific data regarding 10,497 species was not available and hence classified as data deficient. The threat level of as many as seven Indian bird species had increased in the last one year, say experts. According to the latest figure, 15 species of Indian birds including the Great Indian Bustard, Siberian Crane and Sociable Lapwing are there in the list of Critically Endangered birds. Rajasthan has a fauna differing from other parts of India and resembling those found in West Asia and North Africa.

The macro and micro habitat variations are remarkable in the state, making it rich from the biota point of view. The varied ecosystems of the state encompass about 2,208 species of plants including 2,034 angiosperms, 1 gymnosperm and 173 cryptogamous species. On the other hand, 792 faunal species in the diverse habitats consist of 140 fish species, 14 amphibian species, 67 reptile species, 477 avian and 87 mammalian species (Table 2.4). A number of researchers have contributed to enrich our knowledge about the biotic wealth of the state [1–19].

Flora of Rajasthan

According to the Botanical Survey of India [20–22], floral richness of the state can be conveniently studied in the following physiographic regions, namely, western sandy desert, the eastern gravelly/rocky plateau with outliers of Aravallis and the main Aravalli Range itself.

1. ***Flora of the Western Sandy Desert***

   The vegetation in this arid region is sparse consisting mainly of stunted, thorny or prickly shrubs and perennial herbs capable of drought resistance. Trees are a few and scattered. The ephemerals come up during the rainy season, complete their life cycle before the advent of summer and the bulk of the area is once more transformed into open sandy plain, desolate and barren. Since, the climate is more or less homogeneous, the vegetation can better be said to be edaphic controlled. The desert vegetation can be divided under the following heads:

   a. *Sand-Dunes and Interdunal Areas*: This is the most common habitat of this region. Sand-dunes of different types, magnitude and orientation are encountered. They may be stabilised, partially stabilised or unstabilised and barren. The common trees and shrubs found on stabilised and unstabilised sand-dunes are *Callrgonum polygonoides*, *Clerodendrum phlomidis*, *Haloxylon*

Fig. 2.12 *Acacia senegal* (locally called *Kumta*) (*Courtesy: Devendra Bhardwaj*)

salicornicum, *Lycium barbanum*, etc. On some of the dunes, trees like *Acacia senegal* (Fig. 2.12), *Prosopis cineraria* (Fig. 2.13), *Salvadora oleoides*, etc., are also common. The other common inhabitants of sand dunes include *Aerva javanica*, *Citrullus colocynthis*, *Crotalaria burhia*, *Dipterygium glaucum*, *Farsetia hamiltonii*, *Indigofera argentea*, *I. cordifolia*, *I. linifolia*, *Leptadenia pyrotechnica*, *Melhania denhamii*, *Sericostemma pauciflorum*, *Tephrosia falciformis*, *Tribulus longipetalus*, etc. The common grasses and sedges which play a vital role in soil conservation are *Aristida adscensionis*, *A. funiculata*, *Cenchrus biflorus*, *C. ciliaris*, *C. prieurii*, *C. setigerus*, *Cymbopogon jwarancusa*, *Dactyloctenium aegyptium*, *D. scindicum*, *Desmostachya bipinnata*, *Eragrostis species*, *Lasiurus sindicus*, *Latipes senegalensis*, *Ochthochloa compressa*, *Panicum antidotale*, *P. turgidum*, *Sporobolus helvolus*, *Stipagrostis hirtigluma*, *Cyperus atkinsonii*, *C. conglomeratus*, etc.

The interdunal gaps support more luxuriant vegetation due to greater availability of moisture. The common trees and shrubs are *Acacia jacquemontii*, *A. senegal*, *Prosopis cineraria*, *Salvadora oleoides*, *Tecomella undulata* (Fig. 2.14), *Calotropis procera* (Fig. 2.15), *Capparis decidua* (Fig. 2.16), *Ziziphus nummularia* (Fig. 2.17), etc. In addition, almost all the plants found on sand dunes grow more profusely in the interdunal area.

Fig. 2.13 *Prosopis cineraria* (popularly known as *Khejadi*) (*Courtesy: Dr. Seema Kulshreshtha*)

Fig. 2.14 *Teccomella undulata* (locally called *Rohida*) (*Courtesy: Devendra Bhardwaj*)

Fig. 2.15 *Calotropis procera* (*Aak* in Hindi) (*Courtesy: Devendra Bhardwaj*)

Fig. 2.16 *Capparis decidua* (*Kair* in Hindi) (*Courtesy: Devendra Bhardwaj*)

At the foot of the isolated hills and rock outcrops, the windward sides of which obstruct the movement of sand, obstacle sand-dunes are often formed. On these dunes a mixture of characteristic elements of sand-dunes and rock outcrops like *Acacia senegal*, *Anogeissus pendula*, *Commiphora wightii* (Fig. 2.18) and *Maytenus emarginatus* may be seen.

The unstabilised sand-dunes are least covered with plants. However, *Cyperus arenarius*, *Crotalaria burhia*, *Aerva javanica*, *Leptadenia pyrotechnica*, etc., are some of the pioneer species to colonise these dunes. The very crest of such sand dunes is often colonised by varying frequencies of *Lasiurus sindicus* and *Citrullus colocynthis*.

Fig. 2.17 *Ziziphus nummularia* (*Jhad-beri* in Hindi) (*Courtesy: Devendra Bhardwaj*)

Fig. 2.18 *Commiphora weighttii* (*guggul* in Hindi) (*Courtesy: Devendra Bhardwaj*)

b. *Sandy and Hummocky Plains*: These plains constitute the major part of the desert and plants growing in such habitats form some of the characteristic associations of the region. The common trees and shrubs found in these areas are *Acacia senegal*, *Calligonum polygonoides*, *Calotropis procera*, *Capparis decidua*, *Maytenus emarginatus*, *Prosopis cineraria*, *Salvadora oleoides*, *S. persica*, *Tecomella undulata* and *Ziziphus nummularia*. In addition, undershrubs and herbs like *Aerva javanica*, *Arnebia hispidissima*, *Boerhavia diffusa*, *B. elegans*, *Convolvulus microphyllus*, *Crotalaria burhia*, *Farsetia hamiltonii*, *Heliotropium strigosum*, *Indigofera cordifolia*, *Leptadenia pyrotechnica* and *Tephrosia purpurea* are also common. The most common creepers are *Citrullus colocynthis*, *C. lanatus*, *Cucumis melo* and *C. prophetarum*, and climbers are *Cocciria grandis*, *Momordica dioica*, *Mukia maderaspatana* and *Pergularia daemia*. Most of these plants possess well-developed root-system and occur in open clump formations with plenty of vacant spaces between them which are occupied by several ephemerals and grasses like the species of *Aristida* and *Cenchrus*, and *Dactyloctenium scindicum*, *Lasiurus sindicus*, *Ochthochloa compressa*, *Panicum antidotale* and *Stipagrostis hirtigluma* along with sedges like *Cyperus atkinsonii* and *C. conglomeratus*. The shifting dunes are often successful in overrunning the low vegetation.

c. *Gravelly/Rocky Plains*: Gravel, which is formed by the action of wind, covers fairly large areas of the region. Some of the common plants of these habitats are *Cleome vahliana*, *C. gracilis*, *Fagonia indica*, *Dactyloctenium aristatum*, *Indigofera linnaei*, *Heliotropium rariflorunm*, *Leptadenia pyrotechnica*, *Blepharis sindica*, *Sericostemma pauciflorum*, *Bouchnera marubifolia*, *Salvia aegyptia*, etc. Certain plants of gravel are prostrate and star like with the branches remaining appressed to the ground, e.g. *Euphorbia clarkeana*, *E. granulata*, *Indigofera cordifolia*, *I. hochstetteri*, *Mollugo cerviana*, *M. nudicaulis* and *Tribulus terrestris* (Fig. 2.19). The common trees and shrubs of this habitat are *Calotropis procera*, *Capparis decidua*, *Euphorbia caducifolia*, *Maytenus emarginatus*, *Prosopis cineraria*, *Salvadora oleoides*, *Ziziphus nummularia*, etc. At certain spots, the gravel plains maintain characteristic grass-legume associations. The common grasses are *Enneapogon brachystachyus*, *E. schimperanus*, *Melanocenchris abyssinica*, *M. jacquemontii*, *Oropetium thomaeuin*, *Tragus roxburghii*, etc.

   The gravel plains are marked by deep valleys or shallow depressions, particularly near the foot of hills. The typical rock plants which become prominent along with some rambers in such habitats are *Ephedra foliata*, *Asparagus racemosus*, etc. The rocks in the area represent various geological formations. The rocky plains maintain sparse vegetation without any true forests. *Anogeissus pendula*, *Asparagus racemosus*, *Balanites aegyptiacea*, *Corallo carpus epigaeus*, *Rivea hypocrateriformis*, etc. are common plants of these habitats.

d. *Isolated Hills and Rock Outcrops*: The region in the west of Aravalli consists of isolated hills of low elevation and of various origins. These hills are usually bare at the top, occasionally, however, *Boswellia serrata*, *Rhus mysurensis* among woody plants and *Convolvulus stocksii* and *Viola cinerea var. stocksii* among herbaceous ones are found. The hill-slopes maintain better vegetation due to the accumulation of sand and better water holding capacity. The common plants of these habitats are *Acacia senegal*, *Anogeissus pendula*, *Capparis decidua*, *Commiphora wightii*, *Euphorbia caducifolia*, *Grewia tenax*, *Maytenus emarginatus*, *Ziziphus nummularia*, etc.

Fig. 2.19 *Tribulus terrestris* (*Gokhru* in Hindi) (*Courtesy: Devendra Bhardwaj*)

Among climbers, *Abrus precatorius*, *Asparagus racemosus*, *Boerhavia verticillata*, *Rhynchosia minima*, *Rivea hypocrateriformis* are common. The undergrowth is characterised by *Anticharis senegalensis*, *Barleria acanthoides*, *B. prionitis*, *Blepharis linaraefolia*, *Boerhavia diffusa*, *Cleome brachycarpa*, *Fagonia indica*, *Heliotropium bacciferum*, *Seetzenia lanata*, *Sida cordata*, etc.

The vegetation is comparatively dense at the foot of hills and nearby areas. *Acacia senegal* is the main jungle forming tree in such habitats. Other characteristic elements include *Blepharis sindica*, *Lipidagathis trinervis*, *Melhania denhamii*, *Monsonia senegalensis*, *Pavonia zeylanica*, *Schweinfurthia papilionacea*, *Seddera latifolia*, *Tephrosia purpurea*, *Tridax procumbens*, *Aristida adscensionis*, *Enneapogon brachystachyus*, *Melanocenchris jacquemontii*, *Oropetium thomaeum*, *Stipagrostis hirtigluma*, *S. pogonoptila*, *Tragus roxburghii*, etc. *Lipidagathis bandraensis* and *Corbichonia decumbens*, the typical lethophytes of this region, are found in crevices of rocks. *Commelina albescens* and *Lindenbergia indica* are some of the other species growing from the crevices of rocks.

e. *Saline Tracts*: There are large saline tracts spread throughout the desert. The common plants of this habitat are *Cressy cretica*, *Haloxylon recurvum*, *H. salicornicum*, *Portulaca oleracea*, *Salsola baryosma*, *Sesuvium sesuvioides*, *Suaeda fruticosa*, *Tamarix indica*, *Trianthema triquetra*, *Zaleya govindea*, *Zygophyllum simplex*, etc.

Fig. 2.20 Typical grassland in the Thar Desert of Rajasthan—preferred habitat of Great Indian Bustard, variety of birds and Wild Ungulates (*Courtesy: Dr. Asad R. Rahmani*)

f. *Aquatic and Marshland Habitats*: The area maintains considerable number of fresh water, artificial, salt lakes and tanks with rocky substratum. These support a reasonable number of aquatic species like *Hydrilla verticillata*, *Lemna paucicostata*, *Najas graminca*, *N. welwitschii*, *Potamogeton crispus*, *P. pectinatus*, *Spirodela polyrhiza*, *Vallisneria spiralis* and *Wolffia* species. *Eichhornia cressipes* is also encountered in certain ponds and lakes.
g. *Miscellaneous Habitats*: The area has a number of old and historical buildings, forts, palaces, etc. On the walls of such establishments, plants like *Lindenbergia indica*, *Ficus religiosa*, *Salvadora oleoides*, *Tephrosia strigosa*, *Vernonia cinerea* are frequent. *Cistanche tubulosa* is a common root-parasite particularly on *Capparis*, *Calotropis*, *Salvadora* and *Prosopis* sp. *Ephedra foliata* is the only living gymnosperm found in this region.

Certain species escape from cultivation and imperfectly naturalise near habitations, e.g. *Citrullus lanatus* var. *fistulosus*, *Lycopersicum esculentum* and *Momordica charantia*. The area includes few biologically significant habitats like the Desert National Park (DNP) and the Akal Wood Fossil Park. The DNP situated in Barmer and Jaisalmer districts, occupies an area of about 3,000 km^2 and represents all the characteristics of the desert ecosystem. It has been established to preserve the flora and fauna of the Thar Desert and to restore the degraded natural ecosystem. It has a good concentration of Chinkara, Blackbuck and Great Indian Bustard (Fig. 2.20), the last being an endangered species. Grassland ecosystem belonging to a part of the DNP region is also an important habitat for a variety of some common and endangered faunal and floral species.

The Akal Wood Fossil Park, located at Hirchindani in Jaisalmer district, has further enhanced the importance of the Thar Desert by the discovery of fossil wood logs of the lower Jurassic Age. Saphari National Park, located at Kailana in Jodhpur district, is a recent effort being taken for the preservation of desert flora and fauna, including some of the rare and threatened plants and animals of western Rajasthan.

Endemic and Rare Taxa: The desertic zone due to its characteristic topography, geology, edaphic and climatic factors maintains a peculiar type of vegetation, not found elsewhere in India. About 16 taxa of angiosperms are endemic to the desert. They are *Cenchrus rajasthanensis*, *Convolvulus blatteri*, *Farestia macrantha*, *Pulicaria rajputanae*, *Ziziphus truncata*, *Abutilon bidentatum var. major*, *A. fruticosum var. chrysocarpa*, *Alysicarpus monilifer var. venosa*, *Barlaria prioritis var. diacantha*, *Cenchrus prieurii var. scabra*, *Cleome gynandra var. nana*, *Convolvulus auricomus var. ferruginosus*, *Ipomoea carica var. semine-glabra*, *Pavonia arabica var. glutinosa* and *P. arabica var. massuriensis*. Most of these endemic taxa have very restricted distribution probably due to the greatly disturbed ecosystems in the desert and due to the limited range of adaptability of these taxa. Besides these, the over exploitation of certain species for various uses in the desert pose a serious threat to them, e.g. *Citrullus colocynthis*, *Commiphora wightii*, *Ephedra foliate* and *Tecomella undulata*.

The other rare taxa found in Rajasthan are *Ammannia desertorum*, *Glossonema varians*, *Heliotropium rariflorum*, *Lineum indicum*, *Moringa concanensis*, *Seddera latifolia*, *Sesuvium sesuvioides*, *Tephrosia falciformis* and *Tribulus rajasthanensis*.

2. ***The Flora of Aravalli Range***

The second major physiographic region of interest is the Aravalli Range which runs diagonally across the state extending from Champaner in Gujarat in the south-west to near Delhi in the north-east for a distance of about 692 km. Within Rajasthan, the range runs from Khed Bramha in the south-west to Khetri in the north-east for a length of about 550 km.

The elevation of the Aravalli gradually rises in the south-west direction and so also the vegetation pattern and floral composition change due to changes in climatic and edaphic factors. On Khetri Hills (792 m), the vegetation is scruby and in degraded stage. The top of the hills is practically barren while on the slopes, where some sand and moisture accumulate, thick growth of plants such as *Acacia leucophloea*, *A. senegal*, *Balanites aegyptiaca*, *Capparis decidua*, *Euphorbia nivulia*, *Grewia tenax*, *Justicia adhatoda* and *Securinega leucopyrus* can be seen. On Harshnath Hills (913 m), the floral composition up to 600 m is similar to that of Khetri Hills; but above that, plant species like *Calotropis*, *Clerodendrum* and *Justicia* disappear and they are taken over by elements like *Dichrostachys cinerea*, *Euphorbia neriifolia* and *Triumfetta rhomboidea*. Trees like *Anogeissus latifolia*, *A. pendula* (Fig. 2.21), *Balanites aegyptiaca*, *Prosopis cinerarea* and *Wrightia arborea*, which are stunted at lower elevations, become more and more prominent.

Further south-westwards at Kho (920 m), Raghunathgarh (1,055 m) and Todgarh in Ajmer district, the scrub vegetation merges to some extent with the

Fig. 2.21 *Anogeissus pendula* (*Dhok* in Hindi) (*Courtesy: Devendra Bhardwaj*)

Fig. 2.22 *Boswellia serrata* (*Palash* in Hindi) (*Courtesy: Devendra Bhardwaj*)

deciduous type. The floral composition of these hills include *Anogeissus pendula*, *Acacia leucophloea*, *Bauhinia racemosa*, *Boswellia serrata* (Fig. 2.22) *Commiphora wightii*, *Dichrostachys cinerea*, *Mimosa hamata*, *Prosopis cineraria*, *Rhus mysorensis*, *Securinega leucopyrus*, *Sterculia urens*, etc.

From Bijapur forest range (1,100 m) in Pali district to further south-westwards, the hills are covered with mixed deciduous type of forests dominated by *Anogeissus pendula*. The other common associates at Bijapur are *Aegle marmelos*, *Anogeissus latifolia*, *Bauhinia racemosa*, *Boswellia serrata*, *Butea monosperma* (Fig. 2.23), *Cassia fistula* (Fig. 2.24), *Diospyros melanoxylon* (Fig. 2.25), *Mitragyna parvifolia*, *Wrightia tinctoria*, etc.

Fig. 2.23 *Butea monosperma* (*Dhak* in Hindi) (*Courtesy: Devendra Bhardwaj*)

Fig. 2.24 *Casia fistula* (*Amaltas* in Hindi) (*Courtesy: Devendra Bhardwaj*)

Fig. 2.25 *Diospyros melanoxylon* (*Tendu* in Hindi) (*Courtesy: Devendra Bhardwaj*)

Mt. Abu (1,727 m) at the south-western border of the state is the highest peak not only of the Aravallis but also between western Himalayas and Nilgiri Hills. The vegetation here falls into fairly distinct elevational zones, though they intermingle to certain extent. The chief components up to 1,300 m are the same between Bijapur and Mt. Abu with only a little change in the frequency and abundance. But above 1,300 m, the vegetation gradually changes to sub-tropical evergreen type with species like *Boswellia serrata*, *Carvia callosa*, *Crateva nurvala*, *Flacourtia indica*, *Girardinia zeylenica*, *Jasminum humile*, *Lannea coromandelica*, *Mallotus philippensis*, *Mangifera indica*, *Rosa brunoni*, *R. involucrata*, *Sterculia urens* and *Syzygium cumini*. At certain places *Albizia* spp., *Erythrina* spp., *Emblica officinalis*, *Kydia calycina*, *Trema orientalis*, etc., are also found at higher elevations. Several species of ferns and fern-allies also occur at Mt. Abu.

The ground cover comprises of grasses like *Acanthospermum hispidum*, *Blainvillea acmella*, *Sclerocarpus africanus*, and species of *Alysicarpus*, *Cassia* and *Desmodium* while *Borreria articularis*, *B. pusilla*, etc., become very dense at lower elevations at Mt. Abu. In the north-eastern direction from Bijapur, right up to Khetri Hills, not only the density of the above mentioned taxa decreases but species like *Boerhavia diffusa*, *Borreria articularis*, *Dactyloctenium aegyptium*, *Evolvulus alsinoides*, *Glossocardia bosvallea* and species of some ferns like *Actinopteris* and *Adiantum* become more common.

Endemic and Threatened Plants: Mt. Abu maintains characteristic vegetation due to a relatively high altitude coupled with the climatic and edaphic factors. The endemic taxa of Mt. Abu are *Bonnaya bracteoides*, *Dicliptera abuensis*, *Oldenlandia clausa*, *Strobilanthes hallbergii* and *Veronica anagallis var. bracteosa*. These taxa were described by Blatter and Hallberg between 1918 and 1931 with the specimens deposited in Blatter Herbarium, Bombay (BLAT).

Fig. 2.26 *Aerides maculosum* (*Courtesy: Dr. Satish Kumar Sharma*)

Fig. 2.27 *Acampe praemorsa* (*Courtesy: Dr. Satish Kumar Sharma*)

Since then, many botanists have explored Mt. Abu, but none could collect these taxa; probably they have become extinct.

Intrestingly, many varieties of Orchids *Aerides maculosum* (Fig. 2.26), *Acampe praemorsa* (Fig. 2.27), *Eulophia ochriata* (Fig. 2.28), *Perystylus constrictus* (Fig. 2.29) and *Nervilia aragoana* (Fig. 2.30) also occur in this region.

Fig. 2.28 *Eulophia ochriata* (*Courtesy: Dr. Satish Kumar Sharma*)

Fig. 2.29 *Perystylus constrictus* (*Courtesy: Dr. Satish Kumar Sharma*)

Fig. 2.30 *Nervilia aragoana* (*Courtesy: Dr. Satish Kumar Sharma*)

3. ***The Flora in the East of Aravalli***
   This area may be sub-divided into the following physiographic units for a better understanding of the floral composition.

   a. *Bhorat Plateau*: The vegetation on the hills is of mixed deciduous type, showing three altitudinal zones with the higher elevations dominated by *Boswellia serrata*, associated with *Anogeissus latijolia*, *Lannea coromandelica*, *Sterculia urens*, etc. In the middle zone, *Anogeissus pendula* is dominant and is associated with *Albizia odoratissima*, *Diospyros melanoxylon*, *Holoptelea integrifolia*, *Wrightia tinctoria*, etc. The lower elevations are dominated by shrubs like *Cassia auriculata*, associated with *Annona squamosa*, *Butea monosperma*, *Dichrostachys cinerea*, *Diospyros cordifolia*, etc. Undershrubs are usually not affected by elevation, the common ones being *Capparis sepiaria*, *Dyerophytum indicum*, *Grewia flavescens*, *Justicia adhatoda*, *Spermadictyon suaveolens*, *Woodfordia fruticosa*, etc. Bamboo species are also seen here (Fig. 2.31).
   b. *Banas Basin*: The forests located in the Banas basin are of the mixed deciduous type dominated by *Anogeissus pendula*, associated with *Acacia senegal*, *Bauhinia racemosa*, *Boswellia serrata*, *Capparis sepiaria*, *Cassia fistula*,

Fig. 2.31 *Bamboo* sp. (*Courtesy: Dr. Satish Kumar Sharma*)

Dichrestachys cinerea, *Diospyros melanoxylon*, *Lannea coromandelica*, *Wrightia tinctoria*, etc. There is no significant elevational effect on the vegetation of Banas basin except that *Boswellia serrata* becomes more abundant at the summit of the hills and *Butea monosperma* and *Cassia auriculata* on the outskirts. *Sterculia urens* is usually found at the top or on higher slopes. The vegetation is comparatively denser in the valleys of the hills.

c. *Chappan Plateau*: The forests are of deciduous type dominated mainly by *Tectona grandis*. The altitude does not have much effect on the zonation of vegetation except that on higher slopes teak is replaced by species like *Bauhinia racemosa*, *Boswellia serrata*, *Dalbergia latifolia*, *Emblica officinalis*, *Lannea coroman delica* and *Sterculia urens*. The vegetation is richer on the gentle slopes, where additional associates of teak like *Adina cordtfolia*, *Aegle marmelos*, *Albizia odoratissima*, *Diospyros melanoxylon*, *Hymenodictyon excelsum*, *Lagerstroemia parvifiora*, *Madhuca longifolia*, *Mitragyna pairvifolia*, *Terminalia arjuna* and *Wrightia tinctoria* occur. *Dendrocaiamus strictus* grows in isolated patches. Steep slopes with big boulders of rocks support *Euphorbia neriifolia*, *E. nivulia*, etc. At the foot of the hills *Nyctanthes arbortristis* is abundant. On the outskirts, dry teak forests are reduced to scrubs with many stemmed, crooked branchy teak stocks amidst *Acacia chundra*, *A. leucophloea*, *Butea monosperma*, *Holarrhena pubescens*, etc.

d. *Deccan Plateau*: The western part of Deccan Plateau extends to the south-eastern part of Rajasthan over Kota, Bundi and Jhalawar districts with the Vindhyan and Aravalli outliers. The forests here are of mixed deciduous type, showing altitudinal zonation of vegetation to the extent that *Adina*, *Aegle*, *Boswellia*, *Buchanania lanzen*, *Cassia fistula*, *Dendrocalcmus*, *Diospyros*, *Lagerstroemia*, *Lannea*, *Sterculia*, etc., are more abundant in the middle zone, while *A leucophloea*, *Butea monosperma*, etc., are abundant at the foothill or outskirts. *Anogeissus pendula* is the dominant species of these forests extending right from the base to the top of the hills.

   A natural belt of teak is found in the flat terrain along Parbati River from Atru to Kishanganj in Kota district. In the east of Kishanganj, the terrain becomes hilly and the forests, mixed deciduous of *Anogeissus pendula* type. There are three peripheral zones of vegetation in this flat terrain. The inner part is of pure *Tectona grandis*, with isolated trees of *Diospyros melanoxylon* and *Hardwickia binata* and naked ground floor. In the peripheral region, the destruction of the forest has resulted in scrub woodlands consisting of *Acacia catechu*, *A. nilotica* subsp. *indica*, *A. leucophloea*, *Balanites aegyptiaca*, *Butea monosperma*, *Diospyros melanoxylon*, etc. Shrubs and undershrubs are also abundant in the marginal zone. Separating the two, there is a thick belt of *Acacia catechu*, *A. nilotica* and *Ziziphus mauritiana* associated with teak.

   A sub-tropical evergreen forest named "Sitabari" situated near Kelwara village in Kota district, and considered sacred, supports species like *Ampelocissus latifoia*, *Bombax ceiba*, *Carissa congesta*, *Cayratia trifolia*, *Cissus repanda*, *Cordia dichotoma*, *Ficus virens*, *Hiptage benghalensis*, *Ixora arborea*, *Mangifera indica*, *Pueraria tuberosa*, *Schleichera oleosa*, *Syzygium cumini*, *Terminalia arjuna* and *Venda tassellata*. Many ferns, aquatic and shade-loving spermatophytes grow in this forest fed by a perennial stream.

e. *Vindhyan Scarpland*: In Dholpur and Bharatpur districts, most of the area is either under cultivation or is in the form of ravines. The hillocks are more or less naked with degraded and stunted trees and shrubs. In the protected forests, however, the trees attain considerable height, the common ones being *Acacia leucophloea*, *A. nilotica*, *Anogeissus pendula*, *Balanites aegyptiaca*, *Kirganelia reticulata*, *Maytenus emarginatus*, *Prosopis juliflora*, *Salvadora oleoides*, *S. persica*, etc. In addition, *Dichrostachys cinerea*, *Sterculia wrens*, etc., are also found.

   Keoladeo National Park occupying an area of about 29 km^2 and located near Bharatpur is also a protected forest which is comparatively denser than elsewhere. The area is an extensive marshland and it was once the hunting preserve of the princely family of Bharatpur. Besides the taxa already noted above, *Adina cordifolia*, *Azadirachta indica*, *Diospyros montana*, *Mitragyna parvifolia*, *Pithecellobium dulce*, *Phoenix sylvestris*, *Ziziphus mauritiana*, etc., also occur in this park where thousands of migratory birds from Afghanistan, Central Asia, Tibet, Siberia, China, etc., visit every year and make their nests. *Clerodendrum phlomidis*, *Justicia adhatoda*, *Ziziphus nummulaira*, etc., together with many climbers further make the habitat suitable for wild animals like python, porcupine, nilgai, spotted deer, hyaena, sambar and wild boar. The vegetation of wastelands is identical to other regions except that *Tamarix aphylla* and *Alhagi maurorum* are very common throughout the area. It is interesting to note that thick populations of *Carissa spinarum* are encountered near Sikandra village along the foothills of Bayana–Bharatpur. The area also maintains many aquatic and marshland plants, including algal flora.

f. *North-eastern Hilly Region*: North eastern hilly region (Alwar district) is traversed in the north and north east to south and south east by a number of parallel outliers of Aravallis rising up to 800 m. This region is open towards north-west and south-west with flat plains at 300–400 m above mean sea level (MSL). The hill slopes are excessively stony and the vegetation shows three distinct elevation zones, dominated by *Boswellia serrata*, associated with *Crateva nurvala*, *Euphorbia neriifolia*, *Holoptelea integrifolia*, *Mallotus philippensis*, *Terminalia bellerica*, etc., at higher elevations. The middle zone is dominated by *Anogeissus pendula*, associated with *Commiphora wightii*, *Lannea coromandelica*, *Wrightia tinctoria*, etc. In the basal zone, *Acacia nilotica* (Fig. 2.32), *Dichrostachys cinerea*, *Butea monosperma*, *Tecomella undulata*, etc., grow almost in equal proportions. In the cool and shady valley *Colebrookea oppositifolia*, *Dendrocalamus strictus*, *Mitragyna parvifolia*, *Wrightia arborea*, etc., grow in abundance. The ruthless destruction of vegetation in various ways has resulted in the naked hills with big boulders of rocks throughout eastern Rajasthan. The vegetation in such habitats is very sparse, stunted and bushy with plants like spiny *Euphorbias* and *Acacias*, D*iospyros cordifolia* and *D. montane*.

g. *Wasteland Vegetation in the East of Aravalli*: The vegetation of wastelands is almost identical in all the physiographic divisions in the east of Aravalli. It is semi xerophytic with sparse tree layer. The common trees are *Acacia leucophloea*, *A. nilotica*, *Aegle marmelos*, *Azadirachta indica*, *Balanites aegyptiaca*, *Butea monosperma*, *Cordia dichotoma*, *Dolichandrone falcata*, *Ficus benghalensis*, *F. religiosa*, *Mangifera indica*, *Phoenix sylvestris*, *Pithecellobium dulce*, *Prosopis cineraria*, *Salvadora persica*, *Ziziphus mauritiana*, etc. The progressive regeneration of most of the trees, both by seeds and coppice, is negligible due to great biotic pressure on them. *Prosopis juliflora* is the only species which shows highest survival percentage under adverse climatic and biotic conditions.

The wasteland habitats are more favourable for shrubs and undershrubs which form a comparatively dense layer. The most common shrubs are *Abutilon indicum*, *Acacia farnesiana*, *A. jacquemontii*, *Calotropis procera*, *Capparis decidua*, *C sepiaria*, *C zeylanica*, *Clerodendrum phlomidis*, *Dichrostachys cinerea*, *Ficus palmata*, *Kirganelia reticulata*, *Lantana camera*, *Leptadenia pyrotechnica*, *Mimosa hamata*, *Opuntia elatior*, *O. dilleni* (Fig. 2.33) *Solanum incanum*, *Ziziphus nummularia*, etc. A large number of

Fig. 2.32 *Acacia nilotica* (*Babool* in Hindi) (*Courtesy: Devendra Bhardwaj*)

Fig. 2.33 *Opuntia dillenii* (*Cactus*) (*Courtesy: Devendra Bhardwaj*)

Fig. 2.34 *Argemone mexicana* (*Satyanasi* in Hindi) (*Courtesy: Devendra Bhardwaj*)

undershrubs also grow in the wastelands, the most common ones are *Cassia auriculata*, *C. occidentalis*, *Desmodium gangeticum*, *Malvastrum coromadelianum*, *Pavonia zeylanica*, *Pupalia lappacea*, *Sida alba*, *S. cordifolia*, *S. ovata*, *Triumfetta pentandra*, *T. rotundifolia*, *Urena lobata*, *Xanthium strumarium*, etc.

The most common climbers are *Abrus precatorius*, *Argyreia sericea*, *Cissampelos pareira*, *Cocculus hirsutus*, *Cryptostegia grandiflora*, *Leptadenia reticulata*, *Marsdenia tenacissima*, *Mucuna pruriens*, *Pergularia daemia*, *Rivea hypocrateriformis*, *Teramnus labialis*, *Wattakaka volubilis*, etc. They usually occur among the trees and shrubs, particularly on the boundaries of fields and gardens. Besides these, a large number of annual and perennial herbs are encountered in the wastelands. With the advent of the winter season, a considerable number of compositaceous meadow herbs, in association with *acalypha ciliaia*, *Achyranthes aspera*, *Argemone mexicana* (Fig. 2.34), *Leucas cephalotes*, *Nepeta hindostana*, *Solanum nigrum*, etc., come into bloom. The most common compositaceous herbs are *Ageratum conyzoides*, *Echinops echinatus*, *Eclipta alba* and *Gnaphalium polycaulon. Lagascea mollis*, *Launaea procumbens*, *Pulicaria augustifolia*, *Sonchus oleraceus* and *Sphaeranthus indicus*; the last mentioned species often forms dense mats at certain places.

Decreasing moisture content, increasing temperature and high velocity winds during March and April result in the disappearance of most meadow herbs excepting those growing in somewhat moist habitats. The dry meadow herbs appearing during summer months are very few, e.g. *Alhagi maurorum*, *Chrozophora rottleri*, *Solanum surrattense*, *Tridax procumbens* and *Dipcadi serotinum* which possesses underground bulbs and become visible above the ground only during the summers.

Fig. 2.35 *Dalbergia volubilis* (*Courtesy: Dr. Satish Kumar Sharma*)

During the rainy season, the area presents a glorious appearance of a green carpet of semi xerophytic and meadow herbs which help in increasing the humus content of the soil and extend the vegetation to the barren areas. The life-cycle of most of the herbs comes to close before the winter sets in. The rainy season herbs are dominated by leguminous plants like species of *Alysicarpus*, *Cassia*, *Crotalaria*, *Indigofera*, *Psoralea*, *Tephrosia* and *Vigna*. The common associates of the above are *Aristolochia bracteolata*, *Borreria articularis*, *Cieome gynandra*, *C. viscosa Convolvulus prostrates*, *Evolvulus alsinoides*, *Glossocardia bosvallea*, *Merremia tridentate* and *Polygala erioptera* while *Tridax procumbens* and several grasses including *Aristida*, *Cenchrus*, *Chloris*, *Eragrostis*, *Tragus*, *Urochloa*, etc., are more common. On the boundaries of fields, gardens and orchards, *Antigonon leptopus*, *Canavalia ensiformis*, *Cardiospermum halicacabum*, *Coccinia grandis*, *Dioscorea bulbifera*, *Dalbergia volubilis* (Fig. 2.35), *Momordica dioica*, *Mukia maderaspatana*, etc., besides reducing wind velocity, also act as an obstacle in the path of moving sand particles. A few sciophytic humus and shade-loving herbs fairly distributed in the gardens, orchards and other similar habitats are *Baliospermum montanum*, *Biophytum sensitivum*, *Chenopodium album*, *Corchorus olitorius*, *Costus speciosus* (Fig. 2.36), *Ensete superbum* (Fig. 2.37), *Euphorbia hirta*, *Hybanthus enneaspermus*, *Lindernia crustacea*, *Oldenlandia corymbosa*, *Oxalis corniculata*, *Phyllanthus fraternus*, *Physalis minima*, *Solanum nigrum*, etc.

Fig. 2.36 *Costus speciosus* (*Courtesy: Dr. Satish Kumar Sharma*)

Fig. 2.37 *Ensete superbum* (*Courtesy: Dr. Satish Kumar Sharma*)

The neglected corners of fields and gardens which escape ploughing and have lower percentage of nitrates, humus and low water holding capacity are most suitable for the growth of lime-loving species like *Chenopodium album*, *Desmostachya bipinnata*, *Echinops echinatus*, *Euphorbia hirta*, *Perotis indica*, *Saccharum spontaneum* and *Vetiveria zizanioides*. Parks and play grounds are most suitable localities for many grasses which form dense mental and act as an effecient soil-binder, e.g. *Bothriochloa pertusa*, *Cynodon dactylon*, *Dichanthium annulatum*, *Imperata cylindrica*, *Iseilema laxum* and *Tragus roxburghii*. Inhabiting semi-demolished walls of ancient buildings and monuments, there are well recognisable associations of meadow herbs like *Bidens biternata*, *Euphorbia hirta*, *Glossocardia bosvallea*, *Haplanthodes verticillata*, *Indoneesiella echioides*, *Lindenbergia indicea* and *Tridax procumbens*. The seedlings of *Ficus benghalensis* and *F. religiosa* cause damage to the buildings by sending their roots deep down into the walls. *Utricularia exoleta* and *U. stellaris* are the insectivorous representatives of the area. *Aerides crispum* and *Vanda tassallata* are the common epiphytes inhabiting the region.

The floristic composition of the wastelands, i.e. the presence of sparse ground cover during winter and summer, abundant ground cover during rainy season, distantly scattered deciduous trees and comparatively dense growth of xerophytic shrubs suggest semi-arid climate in the area. Date Palm *Phoenix sylvestris* is found in almost every zone of the state as well as here (Fig. 2.38).

h. *Grassland in the East of Aravalli*: The grasslands, managed for grass production, subsequent grazing and to check the exposure of soil to the action of rain and high velocity winds, are a few (Fig. 2.39). They are usually managed on the outskirts of the forests, on naked hillocks, in the degraded forests, protected forests, wastelands, etc. The grasslands of such habitats are dominated by tall grasses like *Aristida adscensionis*, *Bothriochloa pertusa*, *Cenchrus ciliaris*, *Chloris barbata*, *Cymbopogon martinii*, *Dichanthium caricosum*, *Digitaria adscendens*, *Dinebra retroflexa*, *Eragrostis unioioides*, *Heteropogon contortus*, *Iseilema laxum*, *Pennisetum hordeoides*, *Sehima nervosum*, *Themeda quadrivalvis* and *Tripogon jacquemontii*. Besides, many stemmed, crooked, branchy *Bauhinia recemosa*, *Dichrostachys cinerea*, *Mimosa hamata*, *Tectona grandis*, *Ziziphus mauritiana*, etc., which are sparsely distributed in the grasslands, some weed species like *Alysicarpus tetragonolobus*, *Celosia argentea*, *Cleome simplicifolia*, *Crotalaria hirsuta*, *Ipomoea sindica* and *Cyperus rotundus* also compete with grasses in several ways and are harmful to grasslands.

Another category of grasslands occupy large, open undulating rocky terrain. These grasslands are deteriorated due to grazing, spread of undesirable thorny bushes and a number of weeds. The most common grasses of these habitats are *Alloteropsis cimicina*, species of *Aristida*, *Cenchrus ciliaris*, *Chloris montana*, *Cynodon dactylon*, *Dactyloctenium aegyptium*, *species of Eragrostis*, *Hackelochloa granularis*, *Melanocenchris jacquemontii*, *Setaria tomentosa*, *Sporobolus tenuissimus*, *Tetrapogon villosus*, *Tragus roxburghii*, *Tripogon purpurascens*, *Urochloa panicoides*, etc., which grow on rather thin layer of soil. Besides these, many characteristic long-stemmed grass species of the first category also form distant patches. The scope for the development

Fig. 2.38 A branched Date Palm *Phoenix sylvestris* (*Courtesy: Dr. Satish Kumar Sharma*)

Fig. 2.39 Grassland amid Aravallis (*Courtesy: Dr. Satish Kumar Sharma*)

of good pasture lands along suitable gentle slopes of hills, on the outskirts of forests and wastelands is very bright, if the work is done under proper scientific management.

i. *Aquatic and Marshland Vegetation in the East of Aravalli*: The area in the east of Aravalli receives enough rainfall, the water table is normal, a number of rivers pass through the terrain and maintains a large number of temporary and permanent lakes, tanks, ponds, puddles, etc. These habitats provide variable emporia for the growth of aquatic and marshland plants. It has been observed that, the deep water habitats are mostly used for irrigation, etc., the vegetation is poor in comparison to shallow tanks and ponds.

   On the basis of their contact with soil, water and air, the hydrophytes of the area may be broadly classified into (1) free floating species like *Pistia stratiotes* and *Utricularia stellaris*. (2) attached with floating leaves and/or shoots, species like *Ipomoea aquatica*, *Nymphaea nouchali*, *N. pubescens*, *Nymphoides cristata* and *Potamogeton nodosus*. (3) suspended submerged species like *Ceratophyllum demersum*, *Hydrilla verticillata*, *Najas minor*, *Nechamandra alternifolia*, *Potamogeton pectinatus* and *Zannichallia palustris*. (4) attached submerged species like *Ottelia alismoides*, *Potamogeton crispus* and *Vallisneria spiralis*. (5) aquatic and/or amphibious emerged plants like *Aeschynomene indica*, *Hydrolea zeylanica*, *Limmophila indica*, *Polygonum glabrum*, *Sagittaria sagittifolia* and *Typha angustata* constitute this group of hydrophytes. (6) marshland hydrophytes—a large number of plants grow in marshland habitats, particularly in low lands, rice fields and road-side puddles. The most common ones are *Ammannia baccifera*, *Hygrophila auriculata*, *Phyla nodiflora*, most of the sedges and few grasses like *Coix lacryma-jobi*, *Hemarthria compressa*, species of *Paspalidium*, *Paspalum*, etc.

   The area apparently is unsuitable for the growth of luxuriant vegetation but is rich in marshland species. The paucity of aquatic species may be due to rocky bottom and wide amplitude of water level in reservoirs. Most of the aquatic and marshland species grow in a number of associations except a few like *Pistia* and *Typha* which often form pure stands. *Eicchornia crassipes* is a troublesome American weed widely distributed throughout the area in different habitats. Most often, it completely covers the reservoirs and gives a beautiful appearance both in vegetative and blooming state.

Weeds and Alliens

A large number of weeds grow with the crops. Most of them are well-equipped for dissemination by wind, water, man and animals. In the winter season the highest weed density may be noted during the months of January and February. The typical weeds of the winter crops are *Ageratum conyzoides*, *Anagallis arvensis*, *Asphodelus tenuifolius*, *Chenopcdium album*, *Cynodon dactylon*, *Euphorbia dracunculoides*, *Fumaria indica*, *Lepidium sativum*, *Lathyrus aphaca*, *Melilotus alba*, *M. indicus*, *Oxalis corniculata*, *Polypogon monspeliensis*, *Striga angustifolia*, etc.

The typical weeds associated with the summer crops are *Alhagi maurorum Gomphrena celosioides*, *Solanum nigrum*, *S. surrattense*, *Tribulus terrestris*, *Ziziphus nummularia*, etc. The paucity in the number of summer weeds is due to unfavourable climatic and soil conditions. Some winter season weeds like *Euphorbia dracunculoides*, *Cynodon dactylon* and *Cyperus rotundus* which have lower susceptibility to receding soil moisture and rise of temperature give considerable cover during summers.

The density and frequency of rainy season weeds is higher due to high moisture content during this period. The common weeds of *Kharif* crop are *Aerva lanata*, *Alysicarpus longifolius*, *Ammannia baccifera*, *Caesulia axillaris*, *Celosia argentea*, *Commelina benghalensis*, *Corchorus aestuans*, *C. olitorius*, *C. trilocularis*, *Cyperus rotundus*, *Digera muricata*, *Eclipta alba*, *Euphorbia hirta*, *Launaea procumbens*, *Leucas cephalotes*, *L. nutans*, *Oldenlandia corymbosa*, *Oxalis corniculata*, *Trianthema portulacastrum*, *Vernonia cinerea*, etc.

It is interesting to note that, the majority of weeds are annual and therophytes, except *Cynodon dactylon*, *Cyperus rotundus*, *Oxalis corniculata* and *Ziziphus nummularia* which propagate by means of seeds as well as root-stocks. Some weeds like *Ageratum*, *Eclipta and Oxalis* show wide range of adaptability.

Further, with the advent of Indira Gandhi Canal and increasing canal irrigation facilities, a large number of weeds have migrated from Punjab to the canal command areas in western Rajasthan. These species are *Antirrhinum orontium*, *Arenaria serpyllifolia*, *Astragalus tribuloides*, *Centaurium centaurioides*, *Gastrocotyle hispida*, *Hypecoum procumbens*, *Kochia indica*, *Lophochloa pumila*, *Malcolmia africana*, *Malva sylvestris*, *Oenanthe javanica*, *Phalaris minor*, *Plantago amplexicaulis*, *Polygonum lanigerum, Psammogeton canescens*, etc.

Faunal Abundance of Rajasthan

The faunal richness of Rajasthan encompasses 140 species of fishes, 14 amphibians, 67 reptiles (including eight endangered reptiles and five falling under Schedule I of the Indian Wildlife (Protection) Amendment Act, 2006, 477 birds (including 6 Critically Endangered, 5 Endangered, 12 Vulnerable, 19 Near-Threatened, 14 Red Data species and one Conservation Dependent species) and 87 species of mammals (including 7 of endangered mammals and 10 species falling under Schedule I of the Act). For the better convenience, on the basis of physiography of the state, we have divided the faunal diversity of Rajasthan in following heads: Faunal richness south of 24°30 Latitude; Fauna of Mount Abu; Fauna of Vindhyan Gorges; Fauna of Arid Zone and Fauna of the Rest of Aravallis and Area East of Aravallis.

Faunal Richness South of 24°30′ Latitude

Among these areas, Mt. Abu, Phulwari ki Nal, Sitamata Wildlife Sanctuary (WLS), Pratapgarh, Kumbhalgarh and Shahabad (Baran district) are rich in faunal diversity.

Fig. 2.40 Himalayan Tree Frog *Polypedates maculatus* (*Courtesy: Dr. Satish Kumar Sharma*)

Non-chordates Among invertebrates, South Indian Praying Mantis (*Gongylus gongyloides*), Hooded Grasshopper (*Teratodes monticollis*), Red Weaver Ant (*Oecophylla smaragdina*), Tasar Silkmoth (*Bombyx mandarina*), Moon Moth (*Actias selene*), Teak Defoliater Moth (*Hyblaea puera*), Teak Leaf Skeletonizer (*Pyrausta machaeralis*) and *Julus* sp. are common. Presence of Carpenter Ant *Camponotus* spp. observed near Sannoti village in Pratapgarh tehsil of Chittourgarh district is a new record in Rajasthan. This species is fairly common in Ratan Mahal Wildlife Sanctuary, Dahod in the state of Gujarat adjoining Banswara district of Rajasthan.

Amphibians Himalayan Tree Frog (*Polypedates maculatus*) has been recorded from Bansi (near Sitamata Sanctuary), Banswara and Jhalawar districts (Fig. 2.40). This species is not seen anywhere else in the 600–750 mm rainfall zone. Probable distribution range of this species is above 750 mm rainfall zone in Rajasthan. Beside human settlement, it is equally at home in the dense forests. This species is abundant in dense forests of Satkosia Wildlife Sanctuary in Orissa State as well as the Nandan Kanan Zoo of Bhubaneswar, where it is seen breeding in the animal enclosures.

Reptiles In this group, Beaked Worm Snake (*Rhynotyphlops acutus*), Dumeril's Black-headed Snake (*Sibynophis subpunctatus*), Striped Keelback (*Amphiesma stolatum*), Green Keelback (*Macropisthodon plumbicolor*), Common Bronzeback Tree Snake (*Dendrelephis tristis*), Common Vine Snake (*Ahaetulla nasuta*), Isabelline Vine Snake (*A. nasuta* var. *isabellinus*), Slender Racer (*Coluer gracilis*), Banded Racer (*Argyrogena fasciolata*) and Forsten's Catsnake (*Boiga forsteni*) are generally found in this area. Probably Bamboo Pit Viper (*Trimeresurus gramineus*) is also present in this zone (please refer to Chap. 11 for details).

Fig. 2.41 Alexandrine Parakeet *Psittacula eupatria* Pair: male feeding a female (*Courtesy: Sunil Singhal, Kota*)

Birds: The major avians present in the area are Painted Francolin (*Francolinus picta*), Great Thick-knee (*Esacus recurvirostra*), Alexandrine Parakeet (*Psittacula eupatria*) (Fig. 2.41), Common Flameback *Dinopium javanense* (Fig. 2.42), Grey Junglefowl (*Gallus sonneratii*) (Fig. 2.43), Crested Treeswift (*Hemiprocne coronata*), Oriental Dwarf Kingfisher (*Ceyx erithacus*), Brown-headed Barbet (*Megalaima zeylanica*), Asian Brown Flycatcher or Red-breasted Flycatcher (*Muscicapa dauurica*) (Fig. 2.44), Black-hooded Oriole (*Oriolus xanthornus*) (Fig. 2.45), Indian Scimitar Babbler (*Pomatorhinus horsfieildii obscurus*) (Fig. 2.46), White-throated Ground Thrush or Orange-headed Thrush (*Zoothera citrina cyanotis*), Black-lored Tit (*Parus xanthogenys*), Purple Sunbird (*Nectarinia asiatica*) (Fig. 2.47), Scaly-breasted Munia (*Lonchura punctulata*), Red Avadavat (*Amandava amandava*) (Fig. 2.48), Green Avadavat (*A. formosa*) (Fig. 2.49), Mt. Abu White-throated Babbler or Tawny-bellied Babbler (*Dumetia, hyperythra abuensis*), Rajasthan Red-whiskered Bulbul (*Pycnonotus jocosus abuensis*) (Fig. 2.50) and Aravalli Red Spurfowl (*Galloperdix spadicea caurina*). Due to habitat availability, Alexandrine Parakeet is now extending towards north and has reached up to Behror (Alwar) area and IGNP area in the Thar [13].

Mammals: Apart from the common mammals, Mouse Deer or White-spotted Chevrotain (*Moschiola meminna*), Common Palm Squirrel (*Funambulus palmarum*), Elliot's Giant Flying Squirrel or Large Brown Flying Squirrel (*Petaurista philippensis*) are characteristic fauna of the southern region.

Fig. 2.42 Common Flameback *Dinopium javanense* (*Courtesy: Devendra Bhardwaj*)

Fig. 2.43 Grey Junglefowl *Gallus sonneratii* (*Courtesy: Dr. Anil Kumar Chhangani*)

Fig. 2.44 Asian Brown Fly-catcher *Muscicapa dauurica* female (*Courtesy: Sunil Singhal, Kota*)

Fig. 2.45 Black-hooded Oriole *Oriolus xanthornus* (*Courtesy: Sunil Singhal, Kota*)

Fig. 2.46 Indian Scimitar Babbler *Pomatorhinus horsfieldii obscurus* (*Courtesy: Devendra Bhardwaj*)

Fig. 2.47 Purple Sunbird *Nectarinia asiatica* feeding chicks (*Courtesy: Sunil Singhal, Kota*)

Fig. 2.48 Two breeding males of Red Avadavat *Amandava amandava* endemic to Indian Subcontinent (*Courtesy: Sunil Singhal, Kota*)

Fig. 2.49 Green Munia or Green Avadavt *Amandava formosa* endemic to Indian Subcontinent (*Courtesy: Sunil Singhal, Kota*)

Fig. 2.50 Red-whiskered Bulbul *Pycnonotus jocosus abuensis*—An endemic subspecies (*Courtesy: Devendra Bhardwaj*)

Fauna of Mt. Abu

The abundant faunal diversity of Mt. Abu has Mt. Abu Scimitar Babbler (*Pomatorhinus schisticeps obscurus*), White-throated Ground Thrush or Orange-headed Thrush (*Zoothera citrina cyanotis*), Green Avadavat or Green Munia (*Amandava formosa*), Rajasthan Red-whiskered Bulbul (*Pycnonotus jocosus abunesis*), Isabelline Vine Snake (*Ahaetulla nasuta isabellinus*) and Foresten's Cat Snake or Reddish Peninsular Cat Snake (*Boiga forsteni*). White-throated Ground Thrush is also seen in wet, cool and shady pockets of Phulwari and Sitamata Sanctuaries. Likewise, Isabelline Vine Snake was recently sighted in Phulwari and Kumbhalgarh Sanctuaries too. Mt. Abu has the biggest population of Green Munia in Rajasthan. A small population of this species is also present in Kumbhalgarh area. The grasslands found near the waterbodies in the eastern half of the Aravallis provide habitat to the Sarus Crane *Grus antigone* [23] (Fig. 2.51a–c).

Fig. 2.51 (**a**) Sarus Crane *Grus antigone*. (**b**) At the nest with eggs. (**c**) In flight (*Courtesy: Sunil Singhal, Kota*)

Fauna of Vindhyan Gorges

Presence of gorges (locally called *Khoh*) is a typical feature of Vidhyas. Many gorges like Menal, Kundakhoh (Shahabad), Gautmeshwar Mahadeo (Arnod), Yogeshwar Mahadeo, Liliya Mahadeo (Kherot), Padajhar and Kalsiya Mahadeo (both located in Bainsroadgarh Sanctuary) are typical examples of "Khoh" habitat. Seasonal waterfalls, stream-beds, water-pits (locally called *darrah* or *dah*) and vertical banks are characteristic features of the Khoh habitat. Water drips, seeps and flows through rocks during rainy season in the bank zone of gorges, making them fluvial while some are perennial, too.

Rock Bee (*Apis dorsata*), Egyptian Vulture (*Neophron percnopterus*) (Fig. 2.52a, b), Long-billed or Indian Vulture (*Gyps indicus*) (Fig. 2.53), Oriental Honey-buzzard (*Pernis ptilorhynchus*) (Fig. 2.54a, b), Indian Pitta (*Pitta brachyura*) (Fig. 2.55), Tickell's Blue Flycatcher (*Cyornis tickelliae*), Indian Chat

Fig. 2.52 Egyptian Vulture *Neophron percnopterus*. (**a**) An adult. (**b**) Juvenile (*Courtesy: Sunil Singhal, Kota*)

Fig. 2.53 Indian Vulture *Gyps indicus* (*Courtesy: Sunil Singhal, Kota*)

Fig. 2.54 (**a**) Oriental Honey-buzzard *Pernis ptilorhyncus* (*Courtesy: Sunil Singhal, Kota*). (**b**) Oriental Honey-buzzard at the muddy water spot (*Courtesy: Anish Andheria/Sanctuary Asia Photo Library*)

Fig. 2.55 Indian Pitta *Pitta brachyura* (*Courtesy: Sunil Singhal, Kota*)

Fig. 2.56 Brown Rockchat or Indian Chat *Cercomela fusca* (*Courtesy: JK Tiwari*)

or Brown Rock chat (*Cercomela fusca*) (Fig. 2.56), Eurasian Eagle-owl (*Bubo bubo*) (Fig. 2.57a, b) etc. constitute the important fauna of the Vindhyan gorges.

Fauna of Arid Zone

Amphibia: Marbled Balloon Frog (*Uperodon systoma*) (Fig. 2.58) found in most parts of Rajasthan is more common in many parts of Sikar (especially abundant in Neem-Ka-Thana area) and Jhunjhunu districts.

Reptiles: *Stenodactylus orientalis* (Sind Gecko), *Cyrtodactylus scaber* (Roughtail Gecko), *Phrynocephalus laungwalensis* (Laungawala Long-headed Lizard), *Uromastix hardwicki* (Spiny-tailed Lizard) (Fig. 2.59), *Ophiomorus tridactylus* (Three-toed Snake Skink), *Acanthodactylus arabicus* (Arabian Fringe-fingered Lizard), *Ophisops jerdoni* (Punjab Snake-eyed Lacertid), *Varanus griseus* (Desert Monitor) (Fig. 2.60), *Naja oxiana* (Central Asian Cobra) (Fig. 2.61), *Spalerosophis arenarius* (Red-spotted Royal Snake), *Lytorhynchus paradoxus* (Sind Awl-headed or Long-nosed Snake), and *Psammophis schokari* (Schokari Sand Racer also called Afro-Asian or Forskal Sand Snake) (*Psammophis schokari*) are the common reptiles of the Thar. Spectacled Cobra *Naja naja* (Fig. 2.62) earlier thought to have been vanished from Rajasthan was recently sighted near Jaipur.

Fig. 2.57 Eurasian Eagle-owl *Bubo bubo*. (**a**) Adult (*Courtesy: Jaysukh Parekh*). (**b**) Juvenile (*Courtesy: Sunil Singhal, Kota*)

Fig. 2.58 Marbled Balloon Frog *Uperodon systoma* (*Courtesy: Dr. Satish Kumar Sharma*)

Fig. 2.59 Spiny-tailed Lizard *Uromastix hardwickii* (*Courtesy: Clement Francis/Sanctuary Asia Photo Library*)

Fig. 2.60 Desert Monitor of the Thar *Varanus griseus koniecznyi* (*Courtesy: Dr. Gobind Sagar Bhardwaj*)

Birds: Varieties of lark, sandgrouse, stone curlew, courser, chat, wheatear, etc., are characteristic birds of the desert. Eurasian Thick-knee (*Burhinus oedicnemus*), Indian Courser (*Cursorius coromandelicus*) (Fig. 2.63), Cream-coloured Courser(*C. cursor*), Greater Hoopoe Lark (*Alaemon alaudipes doriae*) (Fig. 2.64), Crested Lark (*Galerida cristata chendoola*) (Fig. 2.65), Black-crowned Sparrow Lark (*Eremopterix nigriceps*) (Fig. 2.66), Variable Wheatear (*Oenanthe picata*) (Fig. 2.67), Rufous-tailed or Kurdish Wheatear (*O. xanthoprymna*), Desert Wheatear

Fig. 2.61 Central Asian Cobra *Naja oxiana* (*Courtesy: Dr. Anil Kumar Chhangani*)

Fig. 2.62 Spectacled Cobra *Naja naja* (*Courtesy: Dr. Anil Kumar Chhangani*)

Fig. 2.63 Indian Courser *Cursorius coramandelicus* (*Courtesy: Sunil Singhal, Kota*)

Fig. 2.64 Greater Hoopoe Lark (*Alaemon alaudipes doriae*) (*Courtesy: JK Tiwari*)

Fig. 2.65 Crested Lark (*Galerida cristata chendoola*) (*Courtesy: Devendra Bhardwaj*)

Fig. 2.66 Black-crowned Sparrow Lark (*Eremopterix nigriceps*) (*Courtesy: JK Tiwari*)

(*O. deserti*), Isabelline Wheatear (*O. isabellina*) (Fig. 2.68), Ashy-crowned Sparrow Lark *Eremopterix griseus* (Fig. 2.69a, b), Chestnut-bellied Sandgrouse *Pterocles exustus* (Fig. 2.70a, b, c), Spotted Sandgrouse (*Pterocles senegallus*), Black-bellied Sandgrouse (*P. orientalis*) Macqueen's Bustard or Houbara Bustard (*Chlamydotis undulata*) (Fig. 2.71), Lesser Florican *Sypheotides indicus* (Fig. 2.72) and Great

Fig. 2.67 Variable Wheatear (*Oenanthe picata*) (*Courtesy: JK Tiwari*)

Fig. 2.68 Isabelline Wheatear (*O. isabellina*) (*Courtesy: Devendra Bhardwaj*)

Fig. 2.69 Ashy-crowned Sparrow Lark *Eremopterix griseus*. (**a**) Male. (**b**) Female (*Courtesy: Sunil Singhal, Kota*)

Indian Bustard (*Ardeotis nigriceps*) (Fig. 2.73a, b) are commonly seen in the desert zone. Studies on the bustards estimated that more than half of the Great Indian Bustards (GIB) population in India is present in Rajasthan, mainly in the Thar Desert. Of the eleven arid and semi-arid districts of Rajasthan, the GIB is found in six, namely Bikaner, Jodhpur, Jaisalmer, Barmer, Pali and Jalore. In addition, the grasslands of eastern Rajasthan and parts of western Rajasthan hold significant numbers of Lesser Florican during monsoon. Stoliczka's or White-browed Bushchat *Sexicola macrorhyncus* is endemic to the Thar Desert of Rajasthan (Fig. 2.74). Unfortuntely, numbers of White-rumped Vulture *Gypus bengalensis* and Indian Vulture *G. indicus* have drastically come down by almost 99% due to the killer-drug diclofenac. Khichan village in Jodhpur district is particularly known for a large congregation of the winter visitor Demoiselle crane.

Mammals: Desert or Long-eared Hedgehog (*Periechinus auritus*), Indian Desert Cat or Wild Cat (*Felis silvestris*) (Fig. 2.75), Red Fox (*Vulpes vulpes*) (Fig. 2.76), Desert Fox *Vulpes vulpes pusilla* (Fig. 2.77), Indian Desert Gerbil (*Meriones hurrianae*) (Fig. 2.78), Mole Rat or Short-tailed Bandicoot Rat (*Nesokia indica*), Desert or Indian Hare (*Lepus nigricollis dayanus*) (Fig. 2.79), Little Hairy-footed Gerbil (*Gerbillus gleadowi*), Indian Gazelle or Chinkara (*Gazella bennetti*) (Fig. 2.80a, b) and Blackbuck (*Antilope cervicapra*) (Fig. 2.81a, b) are the important mammals present in the west of Aravallis. Tall shady trees and surface water is now freely available in the command area of IGNP. Due to habitat availability, Flying Fox *Pteropus gigantius* and Short-nosed Fruit Bat *Cynopterus sphinx* (Fig. 2.82) are also spreading their range in the Thar.

Fig. 2.70 (**a**) Chestnut-bellied Sandgrouse *Pterocles exustus*—a male (*Courtesy: Dr. Gobind Sagar Bhardwaj*). (**b**) A female (*Courtesy: Sunil Singhal, Kota*). (**c**) A pair amid forest (*Courtesy: Baiju Patil/Sanctuary Asia Photo Library*)

Fig. 2.71 Houbara Bustard *Chlamydotis undulata* (*Courtesy: Aditya Roy*)

Fig. 2.72 Lesser Florican *Sypheotides indicus* performing an aerial display to achieve conjugal bliss by attracting its future mate (*Courtesy: Dr. Gobind Sagar Bhardwaj*)

Fig. 2.73 Great Indian Bustard (GIB) *Ardeotis nigriceps*: (**a**) Male. (**b**) Male with Gular's Pouch—a typical display during courtship (*Courtesy: Dr. Asad R. Rahmani*)

Fig. 2.74 Stoliczka's or White-browed Bushchat *Sexicola macrorrhyncus* is confined to the Thar Desert of Rajasthan (*Courtesy: JK Tiwari*)

Fig. 2.75 Indian Desert Cat or Wild Cat *Felis silvestris ornata* (*Courtesy: Devendra Bhardwaj*)

Fig. 2.76 Red Fox *Vulpes vulpes* (*Courtesy: Devendra Bhardwaj*)

With the incoming of IGNP, batracofauna will improve qualitatively as well as quantitatively in the Thar area. Few frog breeding centres can be started in this zone, since water is no more a limiting factor. Since, man-made irrigation facilities have been developed during last 60–70 years by Gang Canal and Indira Gandhi Canal in Bikaner, Jaisalmer and Jodhpur districts, heavy afforestation activities after Independance have grossly altered the habitat and crop pattern as well as the avian spectrum in many areas of the Thar.

Fig. 2.77 Desert Fox or Whitefooted Fox *Vulpes vulpes pusilla* peeping out of the den (*Courtesy: Nikhil Devasar/Sanctuary Asia Photo Library*)

Fig. 2.78 Indian Desert Gerbil *Meriones hurrianae* peeping out of its burrow in the Thar (*Courtesy: Devendra Bhardwaj*)

Fauna of the Rest of Aravallis and Area East of Aravallis

Reptiles: Testudine population is on the decline in this zone whereas Indian Flap-shelled Turtle (*Lissemys punctata*) (Fig. 2.83) was common until 1970s. Turtle is still commonly seen in village ponds. Nearly every open well had at least one or a

Fig. 2.79 Desert or Indian Hare *Lepus nigricollis* (*Courtesy: Arfin Zukof* www.listeningpostindia.com)

Fig. 2.80 (**a**) A male Chinkara or Indian Gazelle *Gazella bennetii* (*Courtesy: Sunil Singhal, Kota*) (**b**) A female Chinkara (*Courtesy: Devendra Bhardwaj*)

Fig. 2.81 (a) A pair of Blackbuck *Antilope cervicapra* (*Courtesy: Devendra Bhardwaj*). (b) A male Blackbuck drinking water (*Courtesy: Bernard Castelien/Sanctuary Asia Photo Library*)

Fig. 2.82 Greater Short-nosed Fruit Bat *Cynopterus sphinx* is spreading it's range to the Thar (*Courtesy: Sunil Singhal, Kota*)

Fig. 2.83 Indian Flap-shelled Turtle (*Lissemys punctata*) (*Courtesy: Devendra Bhardwaj*)

Fig. 2.84 Indian Star Tortoise *Geochelone elegans* with young one (*Courtesy: Dr. Satish Kumar Sharma*)

few *L. punctata*. The population present in the wells could not breed, as sand bed is needed for egg-laying which was not available due to cleaning of the wells. This situation took a heavy toll on this species.

Indian Star Tortoise (*Geochelone elegans*) (Fig. 2.84) was once common in Ajmer, Beawar, Sojat, Bhim, Kareda, Mandal, Bhilwara, Mandalgarh, Mavli, Nathdwara, Udaipur, Sadri, Ghanerao and surrounding zone. A big chunk of its population has reached the elite households for keeping as pets or an ornamental animal. This is unnatural and should be ceased because captive population of *G. elegans* is deprived from breeding, while its number in the wild is going down at an alarming rate. Construction of *pucca* (permanent) fence wall and ditch fences around agricultural fields is restricting its entry in its feeding area whereas rain water kills them. Flood also decimates a large population of this species when they reach the low-lying areas.

Fig. 2.85 Common Garden Lizard *Calotes versicolor* is a common arboreal lizard of Rajasthan (*Courtesy: Dr. Anil Kumar Chhangani*)

Calotes versicolor is a common arboreal lizard of Rajasthan distributed almost in the whole state (Fig. 2.85). Three distinct breeding colour patterns, namely, Udaipur pattern, Jaipur pattern and Alwar pattern have been recorded in southern, central and northern Aravallis, respectively. The Chamaeleon *Chamaeleon zeylanicus* (Fig. 2.86) is another arboreal lizard present in majority of the forests of western Aravallis as well as in Kumbhalgarh, Todgarh-Raoli and Jamwa Ramgarh Wildlife Sanctuaries.

Rajasthan Luminous Gecko (*Eublepharis macularius*), a small glowing gecko was first recorded by Indian herpetologist Dr. R.C. Sharma near Madar in Ajmer district. Turkish Gecko (*Hemidactylus turcicus*) is also commonly distributed in Madar and surrounding area [16].

Birds: Indian Peafowl *Pavo cristatus* (Fig. 2.87), Painted Francolin *Francolinus pictus* (Fig. 2.88) and Black Francolin *Francolinus francolinus* (Fig. 2.89) are abundant in Sariska WLS. The population of Siberian Crane *Leucogeranus leucogeranus* that used to winter in the marshes of the only World Heritage site in Rajasthan—Keoladeo National Park (KNP), Bharatpur has now become extinct. Some other aquatic avians of KNP are: Common Coot *Fulica atra* (Fig. 2.90) Pheasant-tailed Jacana *Hydrophasianus chirurgus* (Fig. 2.91), Lesser Whisteling-duck *Dendrocygna javanica* (Fig. 2.92), Dalmatian Pelican *Pelecanus crispus* (Fig. 2.93), Spot-billed Pelican *Pelecanus philippensis* (Fig. 2.94), Bar-headed Goose *Anser indicus* (Fig. 2.95),

Fig. 2.86 Chameleon *Chamaeleo zeylanicus* is also an arboreal lizard (*Courtesy: Dr. Anil Kumar Chhangani*)

Black-crowned Night Heron *Nycticorax nycticorax* (Fig. 2.96) and Painted Stork *Mycteria leucocephala* (Fig. 2.97). Sambhar Lake in the semi-arid region is a Ramsar site which holds significant congregations of Flamingo *Phoenicopterus roseus*. The surrounding environment of the lake used to support the vulnerable bird species of scrubland, namely, the Pied or White-naped Tit *Parus nuchalis* which has almost vanished. Chambal is the last remnant breeding ground for Small Indian Pratincole *Glareola lactea* (Fig. 2.98).

Mammals: The Bengal Tiger *Panthera tigris tigris* (Fig. 2.99a, b), Leopard *Panthera pardus* (Fig. 2.100), Small Indian Civet *Viverriculla indica* (Fig. 2.101), Common Palm Civet *Paradoxurus hermaphroditus* (Fig. 2.102), Thick-tailed Pangolin *Manis crassicaudata* (Fig. 2.103), Indian Crested Porcupine *Hystrix indica* (Fig. 2.104a, b), Sambar *Rusa unicolor* (Fig. 2.105a, b), Chital or Spotted Deer *Axis axis* (Fig. 2.106a–c) are present in the Aravallis from Mamer (near Gujarat border) to Sariska in Alwar district. The most beautiful microchiroptera of India, the Painted Bat (*Kerivoula picta*), has been reported only once in Rajasthan from a single locality—Tatarpur on Alwar-Behror road. Among ungulates, Nilgai or Bluebull *Boselaphus tragocamelus* (Fig. 2.107a, b) is a common animal of this zone while Four-horned Antelope (*Tetracerus quadricornis*) is confined to the dense hilly forests of Rajasthan (Fig. 2.108a, b). Wild Boar (Fig. 2.109) and Striped Hyaena *Hyaena hyaena*

Fig. 2.87 Indian Peafowl *Pavo cristatus* is one of the most beautiful birds (*Courtesy: Dr. Anil Kumar Chhangani*)

Fig. 2.88 Painted Francolin *Francolinus pictus* (*Courtesy: Dr. Gobind Sagar Bhardwaj*)

Fig. 2.89 Black Francolin *Francolinus francolinus* (*Courtesy: Dr. Gobind Sagar Bhardwaj*)

Fig. 2.90 Common Coot *Fulica atra* (*Courtesy: Nayan Khanolkar/Sanctuary Asia Photo Library*)

Fig. 2.91 Pheasant-tailed Jacana *Hydrophasianus chirurgus* (*Courtesy: Devendra Bhardwaj*)

Fig. 2.92 Lesser Whisteling-duck *Dendrocygna javanica* (*Courtesy: Dr. Gobind Sagar Bhardwaj*)

(Fig. 2.110a, b) are also found here. Two Tiger projects at Sariska (Alwar district) and Ranthambhore (Sawai Madhopur district) are confined to this region.

Nag Pahar of Ajmer, Jamwa–Ramgarh WLS and Todgarh-Raoli WLS are also rich in wild fauna. Once famous for Sambar *Rusa unicolor*, this area now has its dwindling population while nearly the whole populations of Indian Gazelle or Chinkara (*Gazella bennetti*) and Blackbuck (*Antilope cervicapra*) has been wiped out. A small population of Blackbuck is still seen in Jodia village of Alwar, Kushtala, Devpura, Dekna and Itawa up to Chouth-ka-Burwara in Tonk and Sawai Madhopur districts [12].

Until 1970s, many village ponds, rivers and streams were semi-perennial in this zone, but now they have become seasonal, causing water deprivation to wild animals. Flying foxes are rapidly displacing and even disappearing from the area owing to the non-availability of surface water round the year.

Fig. 2.93 Dalmatian Pelican *Pelecanus crispus* (*Courtesy: Devendra Bhardwaj*)

Fig. 2.94 Spot-billed Pelican *Pelecanus philippensis* (*Courtesy: Dr. Satish Kumar Sharma*)

Fig. 2.95 Bar-headed Goose *Anser indicus* (*Courtesy: Sunil Singhal, Kota*)

Fig. 2.96 Black-crowned Night Heron *Nycticorax nycticorax* (*Courtesy: Dr. Gobind Sagar Bhardwaj*)

Fig. 2.97 Painted Stork *Mycteria leucocephala* breeding (*Courtesy: Dr. Anil Kumar Chhangani*)

Sloth Bear *Melursus ursinus*, an important species, is only confined to the southern Aravallis and south-eastern parts of the state (Fig. 2.111a, b). Its presence in northern most distribution limit in Rajasthan ends near Beawar in Ajmer district. Golden Jackal *Canis aureus* (Fig. 2.112a, b), Grey Wolf *Canis lupus*, Red Fox *Vulpes vulpes* and Striped Hyaena *Hyaena hyaena* are also decreasing in the area. Habitat loss has made the wild animals homeless in this part of the world.

Fig. 2.98 Chambal is the last remnant breeding ground for Small Indian Pratincole *Glareola lactea* (*Courtesy: Sunil Singhal, Kota*)

Fig. 2.99 Bengal Tiger *Panthera tigris tigris* (**a**) A tigress caressing her young one at Ranthambhore National Park, Sawai Madhopur (*Courtesy: Dr. Gobind Sagar Bhardwaj*) (**b**) A Tiger in a rare pose at an unidentified location (*Courtesy: Aditya Singh/Sanctuary Asia Photo Library*)

Fig. 2.99 (continued)

Fig. 2.100 Leopard *Panthera pardus* (*Courtesy: Dhritiman Mukherjee/Sanctuary Asia Photo Library*)

Fig. 2.101 Small Indian Civet *Viverricula indica* (*Courtesy: Nayan Khanolkar/Sanctuary Asia Photo Library*)

Fig. 2.102 Common Palm Civet or Toddy Cat *Paradoxurus hermaphroditus* (locally called *Bijju*) (*Courtesy: Dr. Satish Kumar Sharma*)

Endemism in Rajasthan

Since effective ecological barriers are absent in Rajasthan, isolation is not effective. That is the reason why endemism is not prominent. Two main centres of endemism have developed in the state, namely, Mt. Abu and Thar Desert, having endemic species of Indian Subcontinent as given below:

A. *Endemism in Mt. Abu Area*: Mt. Abu Scimitar Babbler, *Pomatorhinus horsfieldii obscurus* and Rufous-bellied or Tawny-bellied Babbler *Dumetia hyperythra*

Fig. 2.103 Thick-tailed Pangolin *Manis crassicaudata* (*Courtesy: Dr. Gobind Sagar Bhardwaj*)

Fig. 2.104 (**a**) Indian Crested Porcupine *Hystrix indica* (*Courtesy: Dr. Anil Kumar Chhangani*). (**b**) Two Porcupines drinking at a water body (*Courtesy: Gertrud Helmut Denzau/Sanctuary Asia Photo Library*)

Fig. 2.105 (**a**) Male Sambar *Rusa unicolor* (*Courtesy: Sunil Singhal, Kota*). (**b**) Female Sambar (*Courtesy: Dr. Satish Kumar Sharma*)

abuensis, Rajasthan Red-whiskered Bulbul *Pycnonotus jocosus abuensis*, Aravalli Red Spurfowl *Galloperdix spadicea* and Yellow-legged Green-Pigeon *Treron phoenicoptera* are not found in other parts of Rajasthan except Mt. Abu and the surrounding area.

B. *The Thar and Other Areas*: White-naped Tit (*Paras nuchalis*) is seen in Rajasthan, Gujarat and a few pockets of South India. Stoliczka's or White-browed Bushchat (*Saxicola macrorhyncus*) is mainly confined to Rajasthan and border zone of Pakistan. A reptilian species, namely, *Phrynocephalus laungwalensis* is endemic to Jaisalmer.

Fig. 2.106 (**a**) Herd of Chital or Spotted Dear *Axis axis* drinking water in the wild (*Courtesy: Bernard Castelian/Sanctuary Asia Photo Library*). (**b**) A male Spotted Dear (*Courtesy: Dr. Anil Kumar Chhangani*). (**c**) Spotted Deer in a rare pose (*Courtesy: Dr. Gobind Sagar Bhardwaj*)

Indo-malayan, West Asian, Oriental, North-African and Western Ghat Elements of the Faunal Diversity of Rajasthan

Rajasthan has a fauna distinct from other parts of India and resembling those found in West Asia and North Africa. The Indian Subcontinent is one of the major ecological regions of the Indo-malaya Ecozone and so is Rajasthan. Among the typical faunal characteristic of the Indo-malayan ecozone, Rajasthan has animals like Tiger and Leopard. Leaf birds and fairy birds belonging to the family Irenidae, namely, Common Iora *Aeginnnthina tiphia* (Shaubigi in Hindi) (Fig. 2.113a, b), Marshall's Iora *A. nigrolutea* (Fig. 2.114) and minivets are found at Mount Abu in the Rajasthan

Fig. 2.107 (**a**) Nilgai or Bluebull *Boselaphus tragocamelus* female feeding its fawn (**b**) A herd of Bluebull (*Courtesy: Dr. Gobind Sagar Bhardwaj*)

while pittas are winter visitors. In addition, both Jungle Babbler *Turdoides striatus* (Fig. 2.115) and Large Grey Babbler *Terdoides malcolmi* are found in the thorny deciduous forests/scrub jungles, whereas some species of creepers are also present in the open deciduous forests of Rajasthan. As a matter of fact, Indo-malaya is the most species rich biogeographic ecozone with respect to the diversity of threatened reptiles, tigers and leopards.

Out of 65 mammal species, 32 mammals have palearctic (refers to eco-regions of north of Himalayas) affinities, 30 are oriental and three species *viz.*, *Mus platythrix*, *Mus phillipsi* and *Cremnomys cutchicus* are abundant [24] in the region. Most of chiropterans, rodents and all primates, pholidata and lagomorpha are oriental in

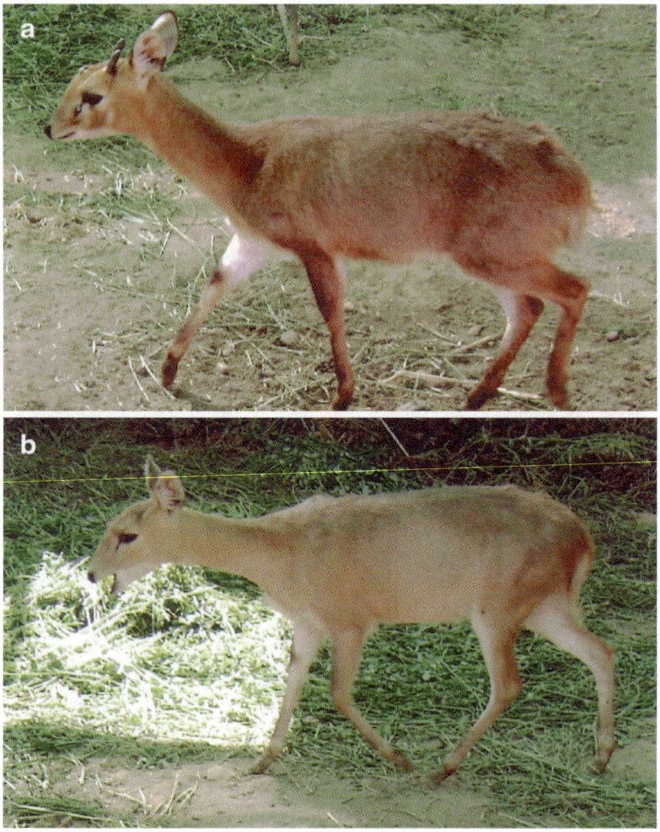

Fig. 2.108 (**a**) A male Four-horned Antelope (Chow-singha in Hindi) *Tetracerus quadricornis* (**b**) A female Four-horned Antelope (*Courtesy: Dr. Satish Kumar Sharma*)

Fig. 2.109 Wild Boar *Sus scrofa* (*Courtesy: Devendra Bhardwaj*)

Fig. 2.110 Stripped Hyaena *Hyaena hyaena*. (**a**) Puppies (*Courtesy: Devendra Bhardwaj*). (**b**) An adult (*Courtesy: Dr. Satish Kumar Sharma*)

distribution. Some species of bats (*Triaenops persicus* and *Rhinopoma muscatellum* of family *Rhinolophidae* and *Rhinopomatidae* respectively) and rodents, the birds of family Paridae, Great Tit *Parus major* (Fig. 2.116), *Parus nuchalis* or White-naped Tit and *Prinia burnesii* or Rufous-vented Prinia belonging to family Sylviidea are also seen in the desert [25].

As a consequence of its geomorphological configuration, the Thar Desert is a conglomeration of Saharan, Turanian, Oriental and Peninsular biological texa. The genera *Gazella* and *Meriones* are spread from the Saharan to the Thar and further to Gobi Desert in the north. Since, a fair numbers of species of these genera occur in

Fig. 2.111 (**a**, **b**) Sloth Bear *Melursus ursinus*—an important species confined to southern Aravallis and south-eastern parts of Rajasthan (*Courtesy a: Sunil Singhal, Kota*; *Courtesy b: Dhritiman Mukherjee/Sanctuary Asia Photo Library*)

the northern Africa, they are thought to have originated in the Sahara Desert. However, the species *Meriones hurrianae* has an Iran-Tharian distribution. It does not occur east of the Aravallis. A number of mammals are limited to an extent to the oriental region such as Blackbuck *Antilope cervicapra*, Nilgai *Boselaphus tragocamelus* and Four-horned Antelope *Tetracerus quadricornis* with their western limit in the Thar. A few texa, originated in the peninsula (Deccan), such as *Chameleo zeylanicus*, *Bandicota bengalesis* and *Golunda ellioti*, have invaded the Thar Desert during the recent past while about 4% of the vertebrate fauna is macro-endemic.

Fig. 2.112 (**a**) Golden Jackal *Canis aureus* (*Courtesy: Devendra Bhardwaj*). (**b**) Golden Jackal young ones (*Courtesy: Nayan Khanolkar/Sanctuary Asia Photo Library*)

Rajasthan has already lost large mammals like Indian Cheetah and Asiatic Lion long back while Wild Ass and Dhole were not seen in the recent past. There are certain species especially carnivores such as tiger, leopard, Caracal *Caracal caracal* (Fig. 2.117), Jungle Cat *Felis chaus* (Fig. 2.118), Rusty-spotted Cat *Prionalurus rubiginosus* (Fig. 2.119) and wolf are facing the risk of extinction in Rajasthan due to loss of their habitat.

Apart from the above mentioned species, a number of other amphibians and reptiles also occur in the state among which, Black-spectacled Toad formerly known as Common Asian Toad *Duttaphrynus melanostictus* (Fig. 2.120), Ornate Narrow-mouthed Frog or Ornamented Pygmy Frog *Microhyla ornata* (Fig. 2.121), Bullfrog *Hoplobatrachus tigerinus* (Daudin, 1803) synonym *Rana tigrina* (Fig. 2.122), varieties of lizards like Termite Gecko *Hemidactylus triedrus* (Fig. 2.123), Common

Fig. 2.113 (**a**) Common Iora *Aegithina tiphia* Female (*Courtesy: Sunil Singhal, Kota*). (**b**) Common Iora male in breeding plumage (*Courtesy: Aditya Roy*)

Fig. 2.114 Marshall's Iora *Aegithina nigrolutea* in non-breeding plumage (*Courtesy: Devendra Bhardwaj*)

Fig. 2.115 Jungle Babbler *Turdoides striatus* (*Courtesy: Dr. Anil Kumar Chhangani*)

Fig. 2.116 Great Tit *Parus major* (*Courtesy: Sunil Singhal, Kota*)

Fig. 2.117 Caracal *Caracal caracal* (*Courtesy: Dr. Gobind Sagar Bhardwaj*)

Fig. 2.118 Jungle Cat *Felis chaus* (*Courtesy: Dr. Gobind Sagar Bhardwaj*)

Fig. 2.119 The Vulnerable Rusty-spotted Cat *Prionailurus rubiginosus* (*Courtesy: Dr. Satish Kumar Sharma*)

Fig. 2.120 Common Asian Toad or Black-spectacled Toad *Duttaphrynus melanostictus* (*Courtesy: Dr. Satish Kumar Sharma*)

Indian Wall Lizard *Hemidactylus flaviviridis* (Fig. 2.124), Skink *Mabuya carinata* (Fig. 2.125), Glass snake *Ophiosaurus gracilis*—a limbless Lizard (Fig. 2.126), Common Indian Monitor *Varanus bengalensis* (Fig. 2.127) and Gharial *Gavialis gangeticus* along the River Chambal (Fig. 2.128). A small population of Mugger *Crocodylus palustris* (Fig. 2.129) strangely survives in the *nullahs* (drains) at Kota.

Fig. 2.121 Ornate Narrow-mouthed Frog or Ornamented Pygmy Frog *Microhyla ornata* (*Courtesy: Dr. Satish Kumar Sharma*)

Fig. 2.122 Bullfrog *Hoplobatrachus tigerinus* (Daudin, 1803). Synonym *Rana tigerina* (*Courtesy: Devendra Bhardwaj*)

Fig. 2.123 Termite Gecko *Hemidactylus triedrus* (*Courtesy: Dr. Satish Kumar Sharma*)

Fig. 2.124 Indian Wall Lizard *Hemidactylus flaviviridis* (*Courtesy: Devendra Bhardwaj*)

Since, there are more than 490 species of avians found in Rajasthan; we have tried to cover the pictures of a cross section of common as well as important avifaunal species. The common avifaunal species of Rajasthan are represented by raptors such as White-rumped Vulture *Gyps bengalensis* at Sariska tiger reserve (Fig. 2.130), Himalayan Vulture *Gyps himalayensis* (Fig. 2.131), Cinereous Vulture *Aegypus monachus* (Fig. 2.132), Red-headed Vulture *Sarcogyps calvus* (Fig. 2.133), Osprey

Fig. 2.125 Skink *Eutropis carinata* (*Courtesy: Devendra Bhardwaj*)

Fig. 2.126 Glass snake *Ophiosaurus gracilis*—a limbless Lizard (*Courtesy: Dr. Gobind Sagar Bhardwaj*)

Pandion haliaetus (Fig. 2.134), Black-winged Kite *Elanus caeruleus* (Fig. 2.135), Short-toed Snake Eagle *Circaetus gallicus* (Fig. 2.136), Crested Serpent Eagle *Spilornis cheela* (Fig. 2.137), Tawny Eagle *Aquila rapax* (Fig. 2.138), Changeable Hawk Eagle *Nisaetus cirrhatus* (Fig. 2.139), Shikra *Accipiter badius* (Fig. 2.140) and White-eyed Buzzard *Butastur teesa* (Fig. 2.141). Still other common avians include Lesser Goldenback *Dinopium benghalense* (Fig. 2.142), Common Myna

Fig. 2.127 A Common Indian Monitor *Varanus bengalensis* (*Courtesy: Dr. Anil Kumar Chhangani*)

Fig. 2.128 Gharial *Gavialis gangeticus* at the River Chambal (*Courtesy: Sunil Singhal, Kota*)

Fig. 2.129 A small population of Mugger *Crocodylus palustris* occurs at some streams (*nullahas*) in Kota (*Courtesy: Dr. Anil Kumar Chhangani*)

Fig. 2.130 Indian White-rumped Vulture *Gyps bengalensis* at Sariska WLS (*Courtesy: Dr. Ashish Kothari*)

Fig. 2.131 Himalayan Vulture *Gyps himalayensis* immature (*Courtesy; Devendra Bhardwaj*)

Fig. 2.132 Cinereous Vulture *Aegypus monachus* (*Courtesy; Devendra Bhardwaj*)

Acridotheres tristis (Fig. 2.143), Yellow-wattled Lapwing *Vanellus malabaricus* (Fig. 2.144), Red-wattled Lapwing *Vanellus indicus* (Fig. 2.145), Brahminy Starling *Sturnus pagodarum* (Fig. 2.146), Indian Robin *Sexicoloides fulicatus* (Fig. 2.147), Common Hoopoe *Upupa epops* (Fig. 2.148), Black Drongo *Dicrurus macrocercus* (Fig. 2.149), Greater Coucal *Centropus sinensis* (Fig. 2.150), Little Green Bee-eater *Merops orientalis* (Fig. 2.151), Indian Roller *Coracias benghalensis* (Fig. 2.152), Common Kingfisher *Alcedo atthis* (Fig. 2.153), Baya (the weaver bird) *Ploceus megarhynchus* (Fig. 2.154), European Roller *Coraceus garrulous* (Fig. 2.155), Grey Francolin *Francolinus pondicerianus* (Fig. 2.156), Barn Owl *Tyto alba* (Fig. 2.157)

Fig. 2.133 Red-headed Vulture *Sarcogyps calvus* (*Courtesy: Dr. Anil Kumar Chhangani*)

Fig. 2.134 Osprey *Pandion haliaetus* taking off (*Courtesy: Dr. Rakesh Vyas*)

Fig. 2.135 Black-winged Kite *Elanus caeruleus* (*Courtesy: Dr. Gobind Sagar Bhardwaj*)

Fig. 2.136 Short-toed Snake Eagle *Circaetus gallicus* pulling the snake from other parent (*Courtesy: Dr. Gobind Sagar Bhardwaj*)

Fig. 2.137 Crested Serpent Eagle *Spilornis cheela* (*Courtesy: Devendra Bhardwaj*)

Fig. 2.138 Tawny Eagle *Aquila rapax* (*Courtesy: Devendra Bhardwaj*)

Fig. 2.139 Changeable Hawk Eagle *Nisaetus cirrhatus* (*Courtesy: R.S. Tomar*)

Fig. 2.140 Shikra *Accipiter badius* male, spreading its tail and one wing (*Courtesy: Sunil Singhal, Kota*)

Fig. 2.141 White-eyed Buzzard *Butastur teesa* (*Courtesy: Dr. Gobind Sagar Bhardwaj*)

Fig. 2.142 Lesser Goldenback *Dinopium benghalense*—Male (*Courtesy: Sunil Singhal, Kota*)

Fig. 2.143 Common Myna *Acridotheres tristis* (*Courtesy: Sunil Singhal, Kota*)

Fig. 2.144 Yellow-wattled Lapwing *Vanellus malabaricus* (*Courtesy: Dr. Gobind Sagar Bhardwaj*)

Fig. 2.145 Red-wattled Lapwing *Vanellus indicus* (*Courtesy: Dr. Gobind Sagar Bhardwaj*)

Fig. 2.146 Brahminy Starling *Sturnus pagodarum* (*Courtesy: Sunil Singhal, Kota*)

Fig. 2.147 Indian Robin *Sexicoloides fulicatus* (*Courtesy: Dr. Gobind Sagar Bhardwaj*)

Fig. 2.148 Common Hoopoe *Upupa epops* (*Courtesy: Dr. Gobind Sagar Bhardwaj*)

Fig. 2.149 Black Drongo *Dicrurus macrocercus* (*Courtesy: Sunil Singhal, Kota*)

Fig. 2.150 Greater Coucal (Koyal in Hindi) *Centropus sinensis* (*Courtesy: Devendra Bhardwaj*)

Fig. 2.151 Little Green Bee-eater *Merops orientalis* (*Courtesy: Sunil Singhal, Kota*)

Fig. 2.152 Indian Roller *Coracias benghalensis* (*Courtesy: Sunil Singhal, Kota*)

Fig. 2.153 Common Kingfisher *Alcedo atthis* (*Courtesy: Sunil Singhal, Kota*)

Fig. 2.154 Baya weaver *Ploceus philippinus philippinus* nests are being mercilessly destroyed due to chopping of *Acacia* trees (*Courtesy: Sunil Singhal, Kota*)

Fig. 2.155 European Roller *Coraceus garrulous (Courtesy: Rajat Bhargava)*

Fig. 2.156 Grey Francolin *Francolinus pondicerianus* (*Courtesy: Sunil Singhal, Kota*)

and Indian Lark *Mirafra erythroptera* (Fig. 2.158). Sociable Lapwing *Vanellus gregarius* is a Critically Endangered species (Fig. 2.159) and Large-billed Crow *Corvus macrorhynchos* is sighted quite less specifically in local urban areas (Fig. 2.160).

The important avifauna of Rajasthan include Brown Fish-Owl *Ketupa zeylonensis* (Fig. 2.161), White-throated Kingfisher *Helcyon smyrnensis* (Fig. 2.162), Stork-billed Kingfisher *Pelargopsis capensis* (Fig. 2.163), Pied Kingfisher *Ceryle rudis* (Fig. 2.164), Sykes's or Tawny Lark *Galerida deva* (Fig. 2.165), Bay-backed Shrike *Lanius vittatus* (Fig. 2.166), Long-tailed Shrike *L. schach* (Fig. 2.167), Black-breasted Weaver *Ploceus benghalensis* (Fig. 2.168), Booted Warbler *Hippolias caligata* (Fig. 2.169), Kentish Plover *Charadrius alexanderinus*

Fig. 2.157 Barn Owl *Tyto alba* (*Courtesy: Devendra Bhardwaj*)

Fig. 2.158 Indian Lark *Mirafra erythroptera* (*Courtesy: Dr. Gobind Sagar Bhardwaj*)

Fig. 2.159 Sociable Lapwing *Vanellus gregarius* is a Critically Endangered bird (*Courtesy: Devendra Bhardwaj*)

Fig. 2.160 Large-billed Crow *Corvus macrorhynchos* (*Courtesy: Sunil Singhal, Kota*)

Fig. 2.161 Brown Fish-Owl *Ketupa zeylonensis* (*Courtesy: Anish Andheria/Sanctuary Asia Photo Library*)

Fig. 2.162 White-throated Kingfisher *Helcyon smyrnensis* (*Courtesy: Dr. Ashish Kothari*)

Fig. 2.163 Stork-billed Kingfisher *Pelargopsis capensis* (*Courtesy: Sunil Singhal, Kota*)

Fig. 2.164 Pied Kingfisher *Ceryle rudis* (*Courtesy: Sunil Singhal, Kota*)

Fig. 2.165 Sykes's or Tawny Lark *Galerida deva* (*Coutresy: Dr. Gobind Sagar Bhardwaj*)

Fig. 2.166 Bay-backed Shrike *Lanius vittatus* (*Courtesy: Sunil Singhal, Kota*)

Fig. 2.167 Long-tailed Shrike *Lanius schach* (*Courtesy: Sunil Singhal, Kota*)

Fig. 2.168 Black-breasted Weaver *Ploceus benghalensis* in non-breeding plumage (*Courtesy: Sunil Singhal, Kota*)

Fig. 2.169 Booted Warbler *Hippolias caligata* (*Courtesy: Sunil Singhal, Kota*)

Fig. 2.170 Kentish Plover *Charadrius alexanderinus* (*Courtesy: Sunil Singhal, Kota*)

(Fig. 2.170), Streak-throated Swallow *Hirundo fluvicola* (Fig. 2.171), Bluethroat *Luscinia svecica* (Fig. 2.172) and Crested Bunting *Melophus lathami* (Fig. 2.173).

Likewise, still other mammals found in Rajasthan are Indian Crested Porcupine *Hystrix indica*, Indian Grey Mongoose *Herpestes edwardsii* (Fig. 2.174), Indian Hedgehog *Paraechinus micropus* (Fig. 2.175), Indian Long-eared Hedgehog

Fig. 2.171 Streak-throated Swallow *Hirundo fluvicola* collecting mud for nesting (*Courtesy: Sunil Singhal, Kota*)

Fig. 2.172 Bluethroat *Luscinia svecica* (*Courtesy: Sunil Singhal, Kota*)

Fig. 2.173 Crested Bunting male *Melophus lathami* (*Courtesy: Sunil Singhal, Kota*)

Fig. 2.174 Indian Grey Mongoose (Newla in Hindi) *Herpestes edwardsii* (*Courtesy: Devendra Bhardwaj*)

Hemiechinus collaris (Fig. 2.176) and Five-stripped Palm Squirrel *Funambulus pennantii* (Fig. 2.177) while Dromedary or Arabian Camel (*Camelus dromedarius*) is a typical mammal of the arid zone popularly known as "Ship of the desert" (Fig. 2.178). In addition, a small population of Hog Deer *Axis porcinus* (Fig. 2.179) sighted at KNP, Bharatpur in 2010 and Barking Deer or the Southern Red Muntjac *Muntiacus muntjak* (Fig. 2.180) whose presence in Rajasthan is doubtful need a bold mention apart from Hanuman Langur or Northern Plains Gray Langur *Semnopithecus entellus* (Fig. 2.181a, b), Rhesus Monkey *Macaca mulatta*

Fig. 2.175 Indian Hedgehog *Paraechinus micropus* (*Courtesy: Devendra Bhardwaj*)

Fig. 2.176 Indian Long-eared Hedgehog *Hemiechinus collaris* (*Courtesy: Sachin Rai/Sanctuary Asia Photo Library*)

Fig. 2.177 Five-stripped Palm Squirrel *Funambulus pennantii* (*Courtesy: Devendra Bhardwaj*)

Fig. 2.178 Dromedary or Arabian Camel (*Camelus dromedarius*)—a typical mammal of the arid zone of Rajasthan (*Courtesy: Devendra Bhardwaj*)

Fig. 2.179 Hog Deer *Axis porcinus* was reported to have been seen in Keoladeo National Park, Bharatpur in 2010 and in July, 2011 also (*Courtesy: Dr. Gobind Sagar Bhardwaj*)

Fig. 2.180 Southern Red Muntjac or Barking Deer or the Muntjac *Muntiacus muntjak* is now extinct from Rajasthan (*Courtesy: Dr. Gobind Sagar Bhardwaj*)

Fig. 2.181 (**a**) Hanuman Langur or Northern Plains Gray Langur (Adult female) *Semnopithecus entellus* with a baby (*Courtesy: Goutam Sharma*). (**b**) Hanuman Langur infant (*Courtesy: Dr. Anil Kumar Ghhangani*)

Fig. 2.182 Rhesus Monkey *Macaca mulatta* pair (*Courtesy: Goutam Sharma*)

Fig. 2.183 Smooth-coated Otters *Lutrogale perspicillata* are found at River Chambal. A small population has been seen at KNP, Bharatpur. (**a**) Smooth-coated Otter resting on the rock (**b**) Smooth-coated Otters in their natural habitat of River Chambal (*Courtesy: RS Tomar, Kota*)

Fig. 2.184 Ganges River Dolphin *Platanista gangetica* (*Courtesy: Rakesh Vyas*)

(Fig. 2.182), Smooth-coated Otter *Lutrogale perspicillata* (Fig. 2.183a, b) found at River Chambal as well as at KNP, Bharatpur and the Ganges River Dolphin *Platanista gangetica* (Fig. 2.184) inhabiting the National Chambal Sanctuary.

Conservation Problems in Rajasthan

The varied ecosystems and habitats of the desert state of Rajasthan are facing multiple conservation problems due mainly to unsustainable natural and anthropogenic developmental activities. The major ones are detailed as under:

1. Climate Change

Pant and Maliekal [26] reported that the climate of Rajasthan and northwest India was subjected to large-scale fluctuations during the past 10,000 years while the recent arid-phase has a history of more than 3,000 years. When the civilisations of Harappa and Mohenjodaro in the Indus Valley flourished during 2500 and 1700 B.C., the mean annual rainfall was between 500 and 800 mm [27]. World Meteorological Organization (WMO) projected an increase of 0.1–0.3°C by the year 2000 in South Asia resulting in a decrease in annual crop production [28]. The impact of this projected climate change by the turn of 21st century [29] is more likely in the arid ecosystem than in semi-arid or sub-humid regions. In the context of global warming leading to climatic change, several studies made on long term climatic changes and its variability in different locations of Rajasthan, confirm an increase in annual rainfall at some places in the region.

2. Impact of Indira Gandhi Canal Project

The introduction of canal irrigation in the arid region has brought many changes, not only in vegetation and land use but also in rainfall patterns, particularly in Sri Ganganagar district which is under such irrigation since past several decades. Due to this increased water availability, xeric faunal species are being replaced by "mesic" species and some minor pests of the region are becoming a major problem [30, 31]. The Gang Canal and the Indira Gandhi Canal (IGNP) were initiated in the years 1927 and 1961 to divert water from the eastern rivers of the Indus system, namely, the Ravi and Beas Rivers to the arid western Rajasthan. The IGNP canal, utilising 7.59 MAF of Ravi–Beas water is planned to irrigate 11.43 lakh hectares in Sri Ganganagar, Bikaner and Jaisalmer districts of Rajasthan.

3. Mining Activities

Fauna is greatly affected by mining activities which are rampant in many parts of the state. The state has 21.7% share of India's total mines. Rajasthan is the major producer of zinc, lead, silver, marble and gypsum. These spoiled soils should be properly reclaimed to regenerate vegetation to support the shaking ecological balance.

4. Population Pressure

Besides urbanisation, industrialisation has rapidly taken place in the districts of Jaipur and Jodhpur while Kota, Pali, Udaipur and Ajmer districts are emerging as new industrial centres of the state. The pressure of increasing population too is leading to massive degradation of natural resources.

5. Oil and Natural Gas Exploration

Petroleum and natural gas exploration and related developmental activities have gained momentum during the past two decades after the largest oil discovery at an oil mine named Mangla in Barmer–Sanchore basin in the year 2004. A total of 480 million tons of oil in-place reserves (3.5 billion barrels) have been discovered in the 22 fields of Barmer–Sanchore Block. Out of these, five fields were used for commercial production since 2009. Oil exploration leads to *in situ* burning of waste gases and oil spill which badly affects the surrounding environment. Burning the oil generates a large amount of smoke, which contains particulates and toxic gases above the level of tolerance for the human population, birds and mammals for several miles downwind. Once coated by oil, neither birds nor mammals have responded well to rehabilitation efforts, the success rate of wildlife rehabilitation has only been moderate at best [32, 33].

6. Poaching

Poaching records of Wildlife Flying Squad (eastern zone) revealed 383 cases from 1974–1975 to 1997–1998. As many as 51 casts/tribes/communities were found indulged in offences related to animal killings and wildlife trade [34].

A few other prominent conservation problems in the state are scarcity of surface water, drought, rapidly increasing desertification due to deforestation, rapid urbanisation and overgrazing (please refer to Chap. 20 from Faunal Heritage of Rajasthan: Conservation and Management of Vertebrates Vol. 2 Sharma, B.K. *et al.* (eds.) 2013, Springer. ISBN 978-3-319-01344-2 for details).

Conservation Efforts

The Government of Rajasthan is now becoming particular about wildlife conservation. Many NGOs, nature lovers, Village Forest Protection and Management Committees and Eco-development Committees are dedicated to wildlife conservation at the moment. *Bishnois* of Rajasthan (and also from other states) are popularly known for their conservation values. KNP and Sajjangarh WLS have been walled around to ensure better protection. Desert Wildlife Sanctuary is proposed for Biosphere Reserve under MAB, the Man and Biosphere Programme. A Biosphere Reserve status is also needed for Aravallis to save the gene pool of this area. Likewise, southern Aravalli is rich in medicinal plants and need special attention, too. However, revival of tiger at Sariska Wildlife Sanctuary, problem solving approach to Keoladeo water crisis, formation of a new Forest and Ecotourism Policy and the ongoing efforts to bring Cheetah back to the wild are a few appreciable steps taken recently. A biodiversity rich area can only survive with the awareness of local people. In the fast changing climatic conditions of the state, joint effort of all the stakeholders is crucial for its conservation. Please also see Chap. 3, from this book and Chaps. 1, 8, 18 and 20 from Faunal Heritage of Rajasthan: Conservation and Management of Vertebrates, Vol. 2 Sharma, B.K. *et al.* (eds.) 2013, Springer. ISBN 978-3-319-01344-2 for more pictures and details.

Acknowledgements The Authors are extremely grateful to Mr. Sunil Singhal, R.S. Tomar (Kota), Dr. Gobind Sagar Bhardwaj (Department of Forests and Environment, Government of Rajasthan), Dr. Anil Kumar Chhangani (MGS University, Bikaner), Mr. Jaysukh Parekh, Mr. Aditya Roy, Dr. Ashish Kothari (Bengaluru), Mr. Bittu Sahgal (Sanctuary Asia Photo Library), Dr. Satish Sharma, Mr. Devendra Bhardwaj (both Department of Forests, Govt. of Rajasthan), Mr. J.K. Tiwari, Dr. Asad R. Rahmani (BNHS, Mumbai) and Mr. Arfin Zukof (Listening Post, New Delhi) for providing excellent pictures.

References

1. Adam A (1990) The Western Rajputana States. Junior Army and Navy Stores Ltd, London
2. Ali S, Ripley SD (1993) Handbook of the birds of India and Pakistan. Oxford University Press, Mumbai, pp 266
3. Bhandari MM (1990) Flora of the Indian desert. MPS Repros, Jodhpur, pp 435
4. Daniel JC (2002) The book of Indian reptiles and amphibians. BNHS/Oxford University Press, Mumbai, pp 238
5. Ghosh AK, Baquri QH, Prakash I (eds) (1996) Faunal diversity in the Thar Desert: gaps in research. Scientific Publishers, Jodhpur, pp 294
6. Grimmett R, Inskipp C, Inskipp T (1999) Pocket guide to the birds of the Indian subcontinent. Oxford University Press, New Delhi, pp 384
7. Islam MZ, Rahmani AR (2004) Important Bird Area in India: priority sites for conservation. India Bird Conservation Network: Bombay Natural History Society and Birdlife International, UK, pp 1133
8. Gupta R, Prakash I (eds) (1975) Environmental analysis of the Thar Desert. English Book Depot, Dehra Dun, pp 484

9. Prakash I (1994) Mammals of the Thar Desert. Scientific Publishers, Jodhpur, pp 114
10. Prater SH (1994) The book of Indian Animals. BNHS Publication, Bombay, pp 348
11. Rahmami AR (1997) Wildlife in the Thar. Worldwide Fund for Nature, New Delhi, pp 100
12. Reddy GV (2002) Management plan of Ranthambhor Tiger Reserve, 2002–03 to 2011–2012. Dept. of Forest, Rajasthan
13. Sharma SK (2001) Impact of Indira Gandhi Canal on the desert Avifauna of Rajasthan. A report submitted to Ministry of Environ & Forest, Govt. of India, New Delhi
14. Sharma SK, Rahmani AR (2000) Birding in Badopal wetland, District Hanumangarh, Rajasthan. Zoos' Print J 15(12):15–16
15. Sharma SK, Singh BP (2001) Birds of the gorges of Vindhyas in Rajasthan State. Zoos' Print J 19 and 21(2):167–169
16. Tikader BK, Sharma RC (1985) Handbook: Indian Testudines. ZSI, Calcutta, pp 156
17. Tikader BK, Sharma RC (1992) Handbook: Indian lizards. ZSI, Calcutta, pp 430
18. Whistler H (1938) The ornithological survey of Jodhpur State. J Bomb Nat Hist Soc 40:213–235
19. Whitaker R, Captain A (2004) Snakes of India: the Field Guide. Draco books, Chennai, pp 481
20. Shetty BV, Singh V (1987) Flora of Rajasthan, vol 1. Botanical Survey of India Publication, Calcutta, India
21. Shetty BV, Singh V (1991) Flora of Rajasthan, vol 2. Botanical Survey of India Publication, Calcutta, India
22. Shetty BV, Singh V (1993) Flora of Rajasthan, vol 3. Botanical Survey of India Publication, Calcutta, India
23. BirdLife International (2001) Threatened birds of Asia: the BirdLife. International Red Data Book, Cambridge, pp 3038
24. Chakraborthy S, Agrawal VC, Kazmi SL (2005) Mammalian diversity in the Thar desert of Rajasthan. In: Tyagi BK, Baqri QH (eds) Changing faunal ecology in the Thar Desert, vol 12. Scientific Publishers, Jodhpur, pp 231–266
25. Gosh PK, Goyal SP, Prakash I (1979) Metabolism and eco-physiology of Rajasthan desert rodents. Thermoregulation at a moderately low temperature (21°C) during winter. J Arid Environ 2:77–83
26. Pant GB, Maliekal JA (1987) Holocene climate changes over north-west India: an appraisal. Climatic Change 10:183–194
27. Singh G (1971) The Indus valley culture seen in the context of post-glacial climatic and ecological studies in north-west India. Archeaol Phys Anthropol Oeania 6:177–189
28. Houghten JT, Jenkins GJ, Ephraums JJ (1990) Climatic Change. The IPCC Scientific Assessment WMO/UNEP, Cambridge
29. IPCC (1990 and 2007) Climatic change: the IPCC scientific assessment: report prepared for IPCC by Working Group for World Meteorological Organization/United Nations Environment Program
30. Rao AS (2005) Impact of introduction of IGNP canal irrigation on Micro- and Secular changes in Climate of Thar desert region. In: Tyagi BK, Baqri QH (eds) Changing faunal ecology in the Thar Desert. Scientific Publishers, Jodhpur, pp 37–44
31. Rao AS (2009) Climate and microclimate changes influencing the fauna of the Hot Indian Arid Zone. In: Sivaperuman C, Baqri QH, Ramaswamy G, Naseema M (eds) Faunal ecology and conservation of the Great Indian Desert. Springer, Heidelberg, pp 13–24
32. Dockery DW, Schwartz J, Spengler D (1992) Air pollution and daily mortality: associations with particulates and acid aerosols. Environ Res 59:362–373
33. Pope CA III, Schwartz J, Ransom MR (1992) Daily mortality and PM-10 pollution in Utah valley. Arch Environ Health 47:211–217
34. Soni RG, Sharma SK (2000) Wildlife offences by communities in eastern Rajasthan. Zoos' Print J 15(8):321–325

Chapter 3
Historical, Sociocultural, and Mythological Aspects of Faunal Conservation in Rajasthan

B.K. Sharma, Seema Kulshreshtha, and Shailja Sharma

Abstract This chapter exhibits the unique history, religions, and sociocultural traditions of the people of Rajasthan which have contributed a great deal in the preservation of wildlife. The ethics of conservation are nurtured by saints and spiritual teachers like Guru Jambheshwarji, the great environmentalist of the fifteenth century and are directly linked with the religion. The temples of animals such as *Garuda* (Brahmini Kite) at Chittourgarh, *Karni mata* (the Goddess *Durga*) temple of Bikaner district, famous for its thousands of rats inhabiting the premises, temples of snake deities like *Gogaji* and *Tejaji* in Gogamerhi (Hanumangarh district), and the *Mahishasur* temple of a buffalo demon who situated at Oriya village near Mount Abu town and festivals for animals like snake worship on *Nagpanchmi* and calf and cow worship on *Bachh-baras* present the religious aspects of conservation. Likewise, the sacrifice of animals to please the deities is an age-old tradition among the Rajput community in Rajasthan and despite being banned, it still continues clandestinely even today. Cattle fairs like, Camel fair at Jaisalmer, Donkey fair at Pushkar, and the Elephant fair at Jaipur strongly indicate the sociocultural aspect of faunal linkages to the people of Rajasthan. Birds like *Kurjan* (Demoiselle Crane), parakeet, Indian Peafowl and House Crow have been favorite themes of the Rajasthan's folk music

B.K. Sharma (✉)
Department of Zoology, R.L. Saharia Government P.G. College,
Kaladera-303801 (Jaipur), Rajasthan, India
e-mail: drbksharma@hotmail.com

S. Kulshreshtha
Department of Zoology, Shakambhar Government P.G. College,
Sambhar Lake (Jaipur), Rajasthan, India
e-mail: seema.zoology@gmail.com

S. Sharma
Department of International Business and Management, Manchester Business School,
The University of Manchester, Booth Street, West Manchester, M15 6PB, UK
e-mail: sharmashailja1988@gmail.com

B.K. Sharma et al. (eds.), *Faunal Heritage of Rajasthan, India: General Background and Ecology of Vertebrates*, DOI 10.1007/978-1-4614-0800-0_3,

since ages. Interestingly, the ancient literature of Rajasthan mentions elephant, tiger, bear, horse, cat, and many beautiful birds in the context of war, hunting, weather and climate forecasting in addition to romance and agriculture. The chapter also presents a wonderful account of the fauna in retrospect. Asiatic Lion and Indian Cheetah were present in the eighteenth century but vanished due mainly to hunting. *Shikar* (hunting) was a favorite sport of the erstwhile rulers which always found a place in the itinerary of visiting Viceroys and British officers. Royal families also owned private hunting preserves, most of which were taken up by the government after India's Independence and developed as wildlife sanctuaries. The chapter mentions about Amrita Devi, a *Bishnoi* lady with great courage and conviction who along with 366 villagers was martyred in the year 1730 while trying to stop tree-cutting by men of the then ruler at the famous Khejadi village near Jodhpur district. It is for the commitment for protecting wild animals, especially Blackbuck and Chinkara, that the *Bishnoi* community stands apart from countless other sects and communities in India. On the other hand, hunting is an integral part of the socioeconomic life of most of the tribals like *Mogiya*, *Bawaria*, and *Pardhi* which have been held responsible for the killing of more than 500 tigers and an equal number of leopards in the national parks of Rajasthan and Madhya Pradesh states during the past two decades.

Introduction

India is a land of spiritual wisdom which has enriched the lives of people since time immemorial. Many religions based on this extraordinary heritage and strength have flourished here. Every religion has made commandments intermingled with love, compassion, and empathy towards all life forms and a sense of responsibility for ecological balance.

Indian *Vedas* are centered on nature worship and Vedic hymns symbolize nature's power. It is the simplest way of making the common people love and respect Mother Nature along with its biodiversity. *Vedas* also glorify the medicinal values of herbs and shrubs. Trees, such as, Peepal *Ficus religiosa*, Banyan *Ficus bengalensis*, Beel *Aegel marmelos*, Ashoka *Polyallhia longifolia*, Aak *Calotropis procera*, and *Tulsi* (Basil) *Ocimum sanctum* are considered sacred and worshipped in every Hindu household. Likewise, prehistoric rock paintings exhibiting animal figures have been found at Sitamata Wildlife Sanctuary, Udaipur (Fig. 3.1).

The intimate association of flora and fauna with humans has been depicted beautifully in epics like *Mahabharata* and *Ramayana*. The famous ancient Sanskrit poet and writer Kalidasa had mentioned climate, ecosystems, water cycle, human–nature relationship, ancient geography, and a divine feeling towards nature with a wide vision coupled with an in-depth scientific knowledge. He mentioned many terrestrial animals from the tropical and subtropical forests of western parts like Nilgai, Chital, Sambar and Blackbuck as the symbol of a couple's love, sensitivity, and unity in his famous epics *Ritusamharam*, *Meghadutam*, *Kumarsambhava*, and

Fig. 3.1 Prehistoric Rock painting at Sitamata Wildlife Sanctuary, Udaipur, Rajasthan depicting wild animals. (*Courtesy*: *Dr. Gobind Sagar Bhardwaj*)

Abhigyan-Shakuntalam [1] (Fig. 3.2). In addition, *Varah* (Wild Boar) groups have been shown defeating a lion in other epics. Likewise, *Vyaghra* (tiger) was shown as a beautiful but brutal creature of the temperate regions in India including Rajasthan. Birds like *Krounch* or *Kurj* (Demiossellle Crane) were mentioned as the migratory birds of western states of India especially Rajasthan. In the same manner, *Gridh* (vulture) as a raptor, *Chakrawak* (Ruddy Shelduck) as the symbol of melancholy in the memory of a departed partner, *Kridamayur* (Peacock) as an indicator of rainfall, and *Rajhans* (flamingo) as the symbol of togetherness are still other creatures mentioned by Kalidasa along with the fauna of northern and southern India. Creation of sacred groves, protected forests for *Ashramas* (hostels), and *Orans* (Common Property Reserve) was a way to conserve the biodiversity in this part of the world too like rest of the country [1].

Animals in Religions and Indian Mythology

Animals have always been an essential part of the Indian religion and mythology. Most of the Hindu Gods and Goddesses have some or the other animal or bird as their mount or *vahana* (vehicle). To name a few, Goddess Durga's lion, Goddess Lakshmi's owl, Lord Ganesha's rat, Vishnu's *Garuda*, Karthikeya's peacock, and Lord Shiva's *Nandi*, the bull (Fig. 3.3). On the other hand, many animals command respect and are worshipped due to the manifold benefits they bestow upon humans. The cow is considered as the holiest of all the animals living on earth. It is worshipped mainly on account of its nutritious milk, the cow dung (*gobar*) which is

Fig. 3.2 A work of art exhibiting the famous epic "Abhigyan-Shakuntalam." penned by Kalidasa—the well-known Indian Sanskrit poet of the medieval period showing that, wild animals and humans peacefully coexisted well in those times unlike today (*Courtesy*: *Devendra Bhardwaj*)

Fig. 3.3 A statue of *Nandi* the bull—mount (vehicle) of Hindu God Lord Shiva (*Courtesy*: *Devendra Bhardwaj*)

Fig. 3.4 Brahminy Kite (*Haliastur indus*) finds a bold mention as *Garuda* in both Hindu and Buddhist mythology (*Courtesy*: *Dr. Satish Kumar Sharma*)

considered antibacterial and used for coating mud houses and *Gau-Mutra* (cow urine) used in traditional and alternative medicine as a remedy to a number of common as well as serious ailments. The cow is the most important and sacred of all because it is considered the mother of all demigods and human beings. People in many parts of the world drink cow's milk and eat butter, ghee (clarified butter) or cheese made of the milk. Therefore, from a moral point of view the cow is considered as mother.

Linked to the holy cow is a festival called *Bachh-Baras* (*Bachh* means Calf and *Baras* means 12th day of the lunar month) celebrated all over Rajasthan by the Hindu women wherein calves are worshipped as the symbol of a son. Lord Krishna is a transcendental cowboy and his topmost spiritual kingdom is known as "*Goloka*" or the planet of cows.

The *Garuda* (Sanskrit: *garuda* = *an* eagle) is a large mythical bird that appears in both Hindu and Buddhist mythology. *Garuda* is the Hindu name for the constellation Aquila, and the Brahminy Kite *Haliaster indus* is considered to be the contemporary representation of *Garuda* [2] (Fig. 3.4). According to the Hindu mythology, *Garuda* is the mount (*vahana*) of Lord Vishnu and is depicted as having a golden body of a strong man with white face, red wings, an eagle's beak, and a crown on its head. Legend says that the span of wings of this ancient deity was massive enough to block out the sun. His stature in Hindu religion can be gauged by the fact that an independent *Upanishad*, namely the *Garudopanishad*, and a *Purana*, the *Garudpurana* is especially devoted to this bird. The Vedas provide the earliest reference of *Garuda*, though by the name of Śyena, where this mighty bird is said to have brought nectar to the Earth from heaven. Worship of *Garuda* is believed to remove the effects of poisons from one's body. The story of *Garuda's* birth and deeds is mentioned in the first book of the great epic *Mahabharat* [3]. G*aruda's* father was the creator-*rishi* Kasyapa while his mother Vinata's sister was Kadru, regarded as the mother of serpents. According to the Mahabharata, *Garuda* had six sons from

Fig. 3.5 The *Garuda* statue at *Garuda* temple in Chittourgarh Fort, Rajasthan (*Courtesy*: *Jai Prakash Bhatnagar*)

whom descended the race of birds and Lord Vishnu was their protector [4]. In *Bhagvad Gita* (Chap. 10, Verse 30), in the middle of the battlefield at *Kurukshetra*, Lord Krishna explained his omnipresence by calling himself as the son of *Vinata*. He further said, "I am in the form of *Garuda*, the king of the bird community." *Garuda* played an important role in *Krishnavatar* (reincarnation) in which Lord Krishna and Satyabhama, his wife rode on *Garuda* to kill *Narakasura*, a demon king. On another occasion, Lord Hari (Vishnu) rides on *Garuda* to save the devotee elephant named Gajendra. It is also said that *Garuda's* wings chanted the *Vedas* while flying. Furthermore, the elite bodyguards of the medieval Hoysala kings of Karnataka were called *Garudas*, because they served the king in the way *Garuda* served Vishnu. In addition, powerful warriors advancing rapidly on doomed foes are linked to *Garuda* swooping down on a serpent [5] and defeated warriors were like snakes beaten down by *Garuda* [6]. In the battle of *Mahabharata*, the field marshal Acharya Drona used a military formation named after *Garuda* [7] and Lord Krishna even carried the image of *Garuda* on his banner [8]. Garuda temple at Chittourgarh Fort is situated in front of the *Barah Mandir* (a grand temple devoted to the 12 incarnations of Lord *Vishnu*) where a huge statue of the holy bird *Garuda* can be witnessed beneath an umbrella even today (Fig. 3.5). This is a living testimony of *Garuda's* presence during the ancient period.

Hindu mythology fondly mentions Lord *Narasimha* (*Nar* meaning human and *Simha* meaning lion) as an incarnation of Lord Vishnu who appeared in the form of half man and half lion to save and bless his greatest follower *Prahlada*, a 12-year old prince and killed the demon king Hiranakashyapa who being an atheist, tortured Prahalada. *Narsimha* temples are located at Amer and Gudha in Jaipur district, Garhi in Hindon district, Hasampur in Kothputli, and Holidada in Ajmer district. Likewise, the fish form of God as *Matsayavtar*, turtle form as *Kurmavtar*, pig form as *Varahavtar* are also mentioned in ancient Hindu literature. *Ganesha* the elephant

Fig. 3.6 An idol of the Hindu God Lord *Hanumana* (*Courtesy: Dr. Seema Kulshreshtha*)

God and *Hanumana* the monkey God (Fig. 3.6) are well known and routinely worshiped all over India by the Hindus.

The snake ranks sacred second to the cow. Because of its swift and gliding movement, scaly skin, hypnotic eyes and poisonous bite, it is feared and, therefore, is a subject of myths and legends. Inevitably, it is worshipped in the hope that veneration would protect its devotees. Snakes or *nagas* were usually mentioned in ancient Hindu literature as gigantic cobras with several hoods, or with a human head and serpent body. They were considered the kings of all snakes, capable of assuming beguiling human forms. According to Hindu mythology, they lived in *Pataal loka* (the mysterious under-water world situated deep beneath the earth) and their capital city Bhogwati was considered the richest and the most beautiful in the entire universe.

There are various beliefs about the origin of snakes. The most popular belief, according to the *Puranas*, is that snakes are the progeny of Sage Kashyapa and Kadru. Besides, snakes are also commonly associated with Lord Shiva who wears a serpent around his neck. For this reason, snakes are widely worshipped, usually as a stone with a snake carved on it by *Shaivas*. *Nagapanchami*, celebrated on the fifth day of Craven is an important snake festival. On this day, snakes are worshipped and released into the forests with the hope to gain knowledge, wealth, and fame, and milk is traditionally offered to them (it is a belief that snakes drink milk). Certain

Fig. 3.7 A typical temple of the local deity *Veer Tejaji* near Sambhar town (Jaipur) who is known to save lives of people dying of fatal snake bites. Several such temples exist in every nook and corner in the villages of Rajasthan (*Courtesy: Dr. Seema Kulshreshtha*)

snakes called *ichhadhari sarp* (who can change its shape and form at will) are believed to have the power to fulfill human desires. It is also a myth that a female *ichhadhari* snake takes a definite revenge if anybody kills her partner. It is also a famous superstition that *Naga-mani*, the precious jewel on such snake's head, if procured can fulfill all wishes of a human being. Indian film industry has been making popular films exclusively based on snakes and the various myths and superstitions linked to them.

In Rajasthan, temples of snake deities like Gogaji and Tejaji are located in Gogamerhi in Hanumangarh district where snake-bite cases are treated by swaying peacock feather broom together with magic charms. Temples of Tejaji along with snakes carved on stones or huge wall paintings of snakes can be found in every village (Figs. 3.7 and 3.8). *Karni mata* (the Goddess Durga) temple at Deshnok in Bikaner district is famous for its thousands of rats inhabiting the premises. It is a popular belief that, sighting the only white rat present among them is highly auspicious and the person, who spots it, is sure to receive the divine blessings of the Goddess.

Unfortunately, in this technological age, the ancient custom of sacrificing male goat and male buffalo to please Hindu deities still continues among certain communities and sects especially the *Rajput* community (the warrior clan) of Rajasthan. Dried skins of goats thrown after sacrificing can be seen hanging on the trees outside the famous temple of Goddess Shakambhari located on the periphery of Sambhar salt lake (Fig. 3.9). On the other hand, by their teachings of *Ahimsa* or nonviolence, the three major religions of India namely, Hinduism, Jainism, and Buddhism stress on the conservation of our faunal diversity.

Fig. 3.8 Painting at Tejaji temple showing the deity with a snake (*Courtesy: Dr. Seema Kulshreshtha*)

Demoiselle Crane Anthropoides virgo *and Indian Peafowl* Pavo cristatus

Kurjan or Demoiselle Crane *Anthropides virgo* (Fig. 3.10a, b) is a favorite theme of the folk music and *Kurjan* songs are popular in social and family gatherings particularly in the villages of Rajasthan. It is a migratory bird of the temperate regions. During winter, a large number of Demoiselle Cranes come to Khichan village in Jodhpur district where they are fed by villagers. The *Kurjan* or *Kuraj* represent the far away lands of their origin for the native Rajasthanis. Due to the harsh weather conditions and uncultivable soils, many Rajasthani men leave for foreign lands in search of occupation and livelihood. They leave their wives, children, and families behind and often, the separation is a prolonged one. With the settlement of *Kurjan* in their locale, the heart of the Rajasthani woman is filled with longing for her departed spouse. The overwhelming emotions take recourse in music and, thus, are born the *Kurjan* songs of the region. In fact, the lonely woman entreats these charming birds to look for her beloved in the alien lands and to guide him safely home. Every August as these avian visitors fly in, the women await the homecoming of their spouses and break into deep melancholy when disappointed. At the onset of summer, the women bid farewell to their spouses and also expect the bird to remind them of their sadness

Fig. 3.9 Dried skin of a male goat hanging on a tree. Goats are regularly sacrificed to please the Goddess Shakambhari at her famous temple situated in Sambhar town of Jaipur district in Rajasthan (*Courtesy: Dr. Seema Kulshreshtha*)

and yearning. With the beginning of monsoons, the rustic damsels wait for the return of their partners with the hope that the coming of the *Kurjan* may finally bring back their beloveds. *Kurjan* songs are the evergreen essence of Rajasthani folk songs and are sung on almost all cultural occasions. Likewise, folk songs also mention other birds like peacock, popularly known as the "*Moriyo*" in local dialect (Fig. 3.10c), parrot, crow, and pigeon symbolizing a couple's emotions and relationships.

Guru Jambheshwarji: The Environmentalist Saint from Rajasthan

No one can imagine that more than 500 years ago, a simple man from a remote desert village, without any formal education, understood the importance and utilized his wisdom in influencing generations of people to conserve biodiversity by weaving it with religion. Guru Jambheshwar, also known as Jambhoji was the founder of the *Bishnoi* religion. He taught the locals to stop idol worship and instead preached them to respect and protect the environment especially the animals and plants by considering them as God's creations. He launched the well known eco-religious revolution at Samrathal Dhora on the eighth day of a black fortnight of the month (*Amavsaya*) of

Fig. 3.10 (**a** and **b**) Demoiselle Crane *Anthropoides virgo* and (**c**) Indian Peafowl *Pavo cristatus*. Locally called *Kurjan* and *Moryo* respectively, both are historically linked with romantic folklore. They are behind the theme of some well-known Rajasthani folk songs, versions of which have been frequently adapted in Bollywood—the massive Hindi film industry based at Mumbai, India [*Courtesy*: (**a**) *Sunil Singhal*, *Kota*, (**b**) *Sachin Rai/Sanctuary Asia Photo Library*, (**c**) *Anil Kumar Chhangani*]

Kartika in 1485 A.D. according to Indian Lunar Calendar later came to be known as *Bishnoism*. Jambhoji was born in a remote village Pipasar of Jodhpur district in 1451 A.D. He was the only child of Lohatji Panwar (father) and Hansa Devi (mother). For the first seven years, Jambhoji remained a silent and introvert child. He then spent about 27 years as a transcendental cowboy like Lord Krishna (incidentally both shared an identical birthday falling on the auspicious *Janmashtami* which is celebrated with gaiety as the birth anniversary of Lord Krishna even today). At the age of 34 Jambhoji founded the *Bishnoi* religion. Interestingly his teachings were presented in a poetic form, known as *Shabadwani*. Although, he preached for another 51 years traveling across the country and wrote many *Shabads*, i.e., verses of *Shabadwani* though hardly 120 *Shabads* are available at present [9].

Bishnoism, as mentioned earlier revolves around 29 commandments. Out of these, eight are prescribed to preserve biodiversity and encourage good animal husbandry. *Bishnoi* community observes socioreligious gatherings known as *Melas* twice a year at Mukam in the Bikaner district where Guru Jambhoji's mortal remains were consigned to earth in 1536 A.D.

The 29 Commandments of Bishnoi Community to Protect and Conserve Biodiversity

Rule No. 19. Not to fell green trees

"जीव दया पालणी रूख लीलो न घावै।'

Rule No. 22. Provide a common shelter (*Thhat*) for goat and sheep to avoid being slaughtered. No *Bishnoi* should sell a male goat or sheep and instead should be sent to *Thhat*. In later years, most of the *Bishnois* even left the business of rearing goats and sheep.

> "जीवां ऊपरि जोर करीजै अतिकाल हुयी भारी।!'
> (If you kill innocent animals, your end will be horrible!)
>
> "रे विनही गुन्हे जीव क्यू मारौ थे तकि जांनो तकि पीड़ न जांणों,
> विणि परचै वाद निवाज गुजारो चरि फिरि आवै सहजि दुहावै!
> तिंहका खीर हलाली।, तिंहकै गलै करद क्यों सारो।' [9]
> (You might not have imagined how much pain these dumb animals felt while getting killed by you. When you like the cows and also consume their products, then it is ignorant and cruel of you to cut its throat! Being a human, such inhumane behavior towards animals is not fair!)

Rule No. 23. Not to have bulls castrated. In the rural India, bulls are castrated before they are used as bullocks for agricultural purposes. Jambhoji prohibited this activity for his disciples and asked *Bishnois* to rear the bovines like their son/daughters.

Rule No. 28. Not to eat meat or other nonvegetarian food. The underlying rationale of this commandment is to protect the animals/birds from being slaughtered by creating a market barrier.

Jambhoji stipulated that no trees were to be cut, and hunting was also forbidden.

"वरजत मारै जीव तहां मर जाइयै" [9]
(If someone is not listening to your request to stop killing an animal then sacrifice your life to save it!).

सिर साँटे रूँख मिले, तो भी सस्ता जाण।
(It is still cheaper, if you save a tree even after sacrificing your head).

This, in fact, became the main commandment of every religious *Bishnoi* to follow.

His followers, some of whom may have thought of Jambhoji as an incarnation of Vishnu were also enjoined to have compassion for all living beings, gave up all intoxicants, swear by the tenets of *ahimsa* (nonviolence) and *satya* (truth), and adhered to a vegetarian diet. It is commonly seen that flora and fauna profusely flourish wherever *Bishnoi's* are found. Even in the times of severe hardship and drought, the Blackbuck, and the Chinkara could count upon them for food and water. In nutshell, Jambhoji was a great visionary, who had foreseen the consequences of man's actions by destroying Mother Nature for economic development and other selfish gains. He actually made the people realize the need for environmental protection and weaved his principles and teachings into the religious commandments put forth by him so as to internalize them by making them easy to follow.

Bishnois: The Great Conservationists and True Followers of Guru Jambheshwar

The *Bishnoi* belong to a community of nature worshipers prevailing in the western parts of Rajasthan. They are called *Bishnoi* ((बीस) *Bis* meaning 20 and (नौई) *Nau* meaning nine) for being the followers of the 29 commandments of Guru Jambheshwar. They also have a sizeable presence in the neighboring states of Gujarat, Haryana, and Delhi but the *Bishnoi* of Rajasthan are particularly known for their nature conservation related activities. *Bishnoi* consider the protection of wildlife and trees as a religious and moral obligation and are known to sacrifice even their lives to protect plants and animals. They particularly consider the state tree *Khejadi Prosopis cineraria* and the antelopes among animals as divine. They are so dedicated to this noble religion that poachers and hunters do not dare to touch any animal in *Bishnoi* villages. Flora and fauna enjoy heaven in these villages and hence, despite being the arid zone, these villages are filled with a rich biodiversity. Interestingly, herds of antelopes can be seen wandering fearlessly near their houses (Fig. 3.11). Blue colored clothes are a dogma for *Bishnois* since the vegetable dye used to color them is extracted by cutting large amounts of shrubs. They do not burn the dead bodies in order to save wood and therefore burial is an acceptable way. The tradition of sacrificing lives to protect trees is called *Khadana* or *Saka*. The fourth *Saka* of Amrita Devi was the biggest among all (for further details please also refer to Chap. 4). In memoriam of this *Saka*, "Khejadi day" is celebrated every year on

---

जम्भवाणी (शब्दवाणी/वेदवाणी) के हिन्दी में उद्रत पद "धर्म और पर्यावरण", भाग–1, सम्पादक डॉ0 किशनाराम बिश्नोई एवं डॉ0 नरसीराम बिश्नोई, कॉमनवेल्थ पब्लिशर्स, दरिया गंज, दिल्ली (भारत) द्वारा प्रकाशित पुस्तक से साभार।

Fig. 3.11 A Blackbuck freely wandering close to human settlements around Khejarli village in Jodhpur: the village is largely represented by the conservationist *Bishnoi* community of Rajasthan (*Courtesy*: *Dr. Anil Kumar Chhangani*)

Fig. 3.12 *Bishnoi* Community's temple at Khejarli village erected at the place of the famous massacre in the year 1730 (*Courtesy*: *Dr. Anil Kumar Chhangani*)

Fig. 3.13 A painting at the Khejarli temple clearly depicting the scene of Amrita Devi's martyrdom (*Courtesy*: *Dr. Anil Kumar Chhangani*)

September 12 at the Khejadi village, which is marked by plantation of 363 saplings of various species [10]. As a mark of respect, a temple of Amrita Devi has also been constructed in the village (Fig. 3.12) in which some paintings depict the scene of the *saka* by Amrita Devi and other villagers (Fig. 3.13) and a cenotaph stands tall, on which the names of all the 363 martyrs are boldly engraved (Fig. 3.14). Three most important *Sakas* of *Bishnoi* community [11] are mentioned in Table 3.1.

Love for animals among *Bishnoi* is evident from an incident when in Nadori village at the Hisar district in Hariyana state on May 10, 1978 Ramidevi, wife of Rameshwar Dhamia breast-fed the new born blackbuck fawn, while the herd was being chased by hunters and few other pregnant female blackbuck were struggling for life (please also see Fig. 3.6, Chap. 4).

As mentioned elsewhere, apart from rulers of the princely states, the tradition of hunting among tribes still prevails in Rajasthan (for details please also refer to Chaps. 1 and 4). These tribal poachers were regularly exploited by well-established wildlife traders and smugglers. This organized nexus actually led to the killing of a variety of wildlife including tiger, leopard, small cat, peacocks, etc. in majority of the protected areas of Rajasthan and subsequently led to the famous Sariska debacle in 2005 as a shocking news to the Indian Government, international and national tourists, wildlifers and the public, not a single tiger was left alive in this world famous tiger reserve. Perhaps, this was the most shameful period in the history of faunal conservation in India. As a patch up work and in order to replenish the tiger population at Sariska, tigers from Ranthambhore National Park were reintroduced here and the program is still going on. The recent death (suspected to have died of poisoning by the villagers) of one of the relocated big cats at Sariska has further proved that wildlife trafficking and other potential issues like village relocation, etc. are far stronger than the government efforts.

Fig. 3.14 The Cenotaph of 365 Bishnoi martyrs at Khejarli who laid down their lives in 1730 while protecting trees especially *Khejadi* or *Prosopis cineraria* (later designated as the State Tree) (*Courtesy*: *Dr. Anil Kumar Chhangani*)

Traditional Water Conservation in Rajasthan [12]

British chroniclers in 1856 found a rich and flourishing tank culture for potable water storage in India. Rajputana (now Rajasthan) in addition to harboring natural and man-made lakes, also has other strategic water storage structures. This method of harvesting run-off rain water has been extremely successful in low to moderate rainfall areas of Rajasthan. It is well known that, the then Prince Dheer Deo of Bundi got 23 huge water tanks excavated in and around Kota in the early fourteenth century. Various popular water conservation structures and methods were traditionally used in Rajasthan, the major ones are mentioned below.

Johad: These structures are built by constructing semicircular mud walls on small streams, the primary function of which is to recharge the ground water.

Tanka: These are 3–8 m deep tanks excavated to store run-off water from the roofs of large forts and palaces to store drinking water. The Taragarh fort of Bundi district has worth-praising *tankas* for their design and capacity. In the desert regions of Rajasthan *tankas* are usually kept covered.

Baori: *Baoris* (step-wells) have their own underground water source and are popular structures with halls, balconies, steps, and *jharokhas* (small windows). Abhaneri

Table 3.1 Important *Sakas* made by Bishnois for protecting trees, deer, and wildlife in Rajasthan

S. No.	Samwat	Year (A.D.)	Location of event	Present district of area	Special note
1.	1661	1604	Raivasadi village	Jodhpur	Two women named Karma and Gaura sacrificed their lives to save Khejri trees
2.		1604–1616	Tilvasani village	Bilada tahsil	Khivani Devi, NEetu Devi, Nain, Motaram Khokhar sacrificed their lives to save Khejadi tree
3.	1700	1643	Polawas village	Nagour	Buchoji scarified his life to save trees
4.	1783	1730	Khejadi village	Jodhpur	363 men and women scarified their lives to save Khejri trees
5.		1940	Bahavalpur Bishnoi villages	Pakistan	Army personnel by train passing from these villages killed one deer; villagers attacked them and tolerated the court proceedings
6.		1947	Baravan village	Badmer	Chimnaram and Pratapnarayan Bishnoi sacrificed lives for protecting deer from hunters
7.		1948	Rohichkal village and Bhagtasani	Jodhpur	Chunnilal Bishnoi and Arjunram Bishnoi sacrificed lives for saving wildlife
8.		1948	Rotu and Banada villages	Jodhpur	Dhonkalram Bishnoi and Lalaram Bishnoi along with two other Bishnois sacrificed lives for saving wild animals
9.	1600–1673	1543–1616	Bishnoi village	Jodhpur	Damodevi, Rudi devi and other Bishnois sacrificed lives to save He-goats
10.	1914	1857	Chindad village	Hariyana	Bishnois protected cows and bulls from muslims and sacrificed lives

Baori at Bandikui of Dausa district is famous for its architecture. An artistic *Baori* called *Raniji ki Baori* near Amer (Jaipur) at Bhavni village and *Sura ki Baori* near Jaipur amidst deep forest of Nahargarh Biological Park are now being revived (Fig. 3.15a, b).

Kund: These structures store run-off rain water and the stored water fulfills various domestic needs of the people.

Fig. 3.15 An age old traditional step-well locally called *Baori* frequently used for water conservation in Rajasthan in the past. These structures are now being revived to fight with the increasing water crisis. (**a**) *Raniji ki Baori* at village Bhavni on Jaipur–Delhi highway near Amer (Jaipur). (**b**) *Sura ki Baori* amidst deep forests at Nahargarh Wildlife Sanctuary near Jaipur (*Courtesy*: *Devendra Bhardwaj*)

It is worth mentioning here that many of these traditional structures are still being used for water conservation. Government and the people of Rajasthan have started reviving and restoring these ancient structures being fully aware of their suitability and utility in the arid environments of this desert state.

Gaumukh: (Gau = Cow, Mukh = Mouth): These are mysterious cow-faced drips usually made of carved marble or other stone found at many ancient temples in Rajasthan. Lord Shiva temple at the Chittourgarh fort and Galta Kund at Jaipur are key examples. No one till date knows about the source of the sweet, clean water continuously dripping out and falling on the *Shivlinga* situated below the same. A separate *kund* (water reservoir) invariably lies about 6–7 m underneath the temple where water gets collected to be later used for various purposes. It has been observed that, during peak summer the water stream of the *Gaumukh* is generally thin while in rains it is heavy.

The Fauna in Retrospect

Archaeological surveys across the Indian subcontinent especially with respect to faunal remains of the times following Holocene period have revealed the presence of animals like buffalo, Chital, Chinkara, Sambar, Wild Boar, cow and hare at Tilwara and Bagor in Rajasthan [13–15]. On the other hand, Asiatic Lion and Indian Cheetah—the most sought after among large mammals, comfortably inhabiting the open scrubs and massive grasslands of Rajasthan for centuries have got completely extinct in the recent past (75–150 years) due mainly to unchecked and indiscriminate hunting by the erstwhile kings and Royals from far and beyond. Unfortunately, the beautiful Wild Dog or *Dhole* and Wild Ass or *Khur* too met the same fate in this state but it is a sign of great relief that, these two species are still being reported from the adjoining state of Madhya Pradesh and Gujarat.

Records of the Asiatic Lion **Panthera leo persica** *from Rajasthan*

History reveals that lion did survive well in the Asian belt extending from Palestine in the west to Palamau in the eastern part of India but is now extinct in the whole of Asia barring the Gir forests of Gujarat (Fig. 3.16). Dr. Divyabhanusinh has given a detailed and well-researched account of the historical occurrence of lions and their extinction in India in his well-known books *The story of Asia's Lions* [13] and *The Lions of India* [14]. According to him, evidences of the presence of lion in India dates back as early as first century though in small numbers in restricted areas [15–19].

Fig. 3.16 An Asiatic Lion (*Panthera leo persica*) in its natural habitat at the famous Gir Forest National Park, Gujarat (*Courtesy*: *Dr. Divyabhanusinh*)

Records of the Mughal period throw enough light on the history of these large fury creatures. More than two dozen paintings from the Mughal period reveal that they inhabited scrubs and grasslands in Rajasthan (Fig. 3.17). In fact, Babar (1483–1526) who used to keep written records is largely credited for making a foundation for Mughals in India. In the year 1555, Humayun brought paper and his own painters [20]. Mughal Emperor Akbar (1542–1605)—the son of King Humayun went a little beyond and even maintained his own studio [20] during 1592 and 1594 with the assistance of 49 artists. This was a time when a large number of paintings depicting hunting of large carnivores like lion and tiger were probably made. *Akbarnama* was commissioned by the Mughal King Akbar as the official chronicle of his reign written in Persian language by his biographer and court historian Abu'l Fazl between 1590 and 1596 [21]. It is thought to have been illustrated between c. 1592 and 1594 by the artists from Akbar's studio. After Akbar's death in 1605, the manuscript remained in the library of his son, Jahangir (r. 1605–1627) from whom it was inherited by Shah Jahan (r. 1628–1658)—the son of Jahangir. In the year 1896, this manuscript was purchased by the Victoria and Albert Museum, London, from Mrs. Frances Clarke, the widow of Major General John Clarke, an official who bought it in India while serving as the Commissioner in Oudh province between 1858 and 1862. It is thought to be the first illustrated copy of the *Akbarnama*. It drew upon the expertise of some of the best royal painters of the time, many of whom receive special mention by Abu'l Fazl in the *A'in-i-Akbari* ("Institutes of Akbar"—a sixteenth

Fig. 3.17 Akbar slays a tigress which attacked the royal entourage. A painting from the "*Akbarnama*" (Book of Akbar) shows the Mughal Emperor Akbar (r. 1556–1605) slaying a tigress which attacked the royal entourage near Narwar, Gwalior (central) India, in 1561. The royal entourage disturbed a female tiger, which sprang out from the forest and lashed out to protect her five cubs. The emperor's companions were said to have frozen in terror, but the emperor reacted instantly, killing the tigress with one blow of his sword. His men then killed the five offspring. The event is depicted over two pages, the other page being Museum no. IS.2:18-1896. This is the right side of a double composition. The image is overlaid by a panel of Persian text (*four lines*) on the *left-hand side* of the page. Place of origin: possibly made in India or Pakistan; date made: 1590–1595; artist: Basawan and Tara the Elder; materials and techniques: opaque water color and gold on paper; museum number: IS.2:17-1896; gallery location: in store (*Photo ©Victoria and Albert Museum, London*)

century document and the third volume of the *Akbarnama* which recorded the administration of Emperor Akbar's empire [22]). The inscriptions in red ink on the bottom of the paintings refer to the artists and indicate that this was a royal copy made for Akbar himself. *Akbarnama* mentions that, Akbar did hunt lions in the Mewat region of Rajasthan, parts of Mathura in Uttar Pradesh and some border areas of Haryana state [21]. Abul Fazal has clearly mentioned in *Ain-i-Akbari* about a man-eater who attacked Akbar's elephant and had to be killed by his courtiers at Bari in Dholpur district of Rajasthan [23].

Akbars's son Jahangir too loved nature and was particularly observant and interested in the faunal and floral elements. His personal diary *Tuzuk-i-Jahangiri* written in Persian language has ample stories pertaining to the hunting of lion [24]. He himself hunted in open scrub land and even mentioned about prey species like Bluebull and Blackbuck. Hunting had become a pleasure sport for the army men during the three years period when Jahangir stayed in Ajmer until 1616. Rupbas, Bari, Jodhpur, and Merta in Rajasthan were among the 16 locations used as imperial hunting grounds. Jahangir wrote about his close encounters with lion at Bak-Bhal and Nag Tah at Bari near Dholpur, Palam near Delhi, Rupbas near Agra and in Ajmer and Mandu in 1610 killing seven lions. Interestingly, Jahangir even shared his hunting skills with the then Mewar prince who was visiting his court during 1615 [24–26]. Records of hunting at a rate of one lion per week in 1611 clearly indicate that, Rajasthan had lions in abundance. There are paintings of Jahangir's encounter with lion at Bari, Dholpur [27]. Paintings (circa 1693) also show Maharaja Anup Singh of Bikaner and his three brothers hunting lion with the help of nets [28].

In the late eighteenth century Central India was a strong hold of lions but 200 years after Jahangir (who only killed 86 lions in 39 years) they had almost vanished from India. Interestingly, during that period a British officer Colonel Akland Smith on the other hand, is said to have killed 300 lions while Colonel William Rice and Colonel D killed 14 [29] and 80 lions [30], respectively. Nine lions were reportedly killed in 1866 by an unknown hunting party in the vicinity of Kotah [31]. Raja Bishen Singh of Bundi in Rajasthan killed another 100 lions besides tigers in the year 1830 [32]. Paintings still preserved at a museum in Kotah show lion and tiger hunting from a machan. Mr. Blanford wrote in the *Journal of the Asiatic Society of Bengal*, 1867 that, in central and western India lions existed from Gwalior to Kotah, Mt. Abu and Deesa and southward near Allahabad and in parts of Kathiawad's famous Ghur jungles.

In addition, Gazetteers of Western *Rajputana* (now Rajasthan) and Jodhpur Residency mention that, a full grown female lion was killed in Abu by a *Bhil* Shikari in 1872 while the last four lions were shot near Jaswantpura (Jodhpur) in the same year [32]. These records are ample evidence to state that by the year 1872 the lion had become extinct in Rajasthan. Although, stray reports of lion sighting around Mt. Abu kept pouring-in even after 1872 [33], the animal had actually disappeared from rest of India. Since, there occurred no natural calamities during the said period, it is obvious that, the erstwhile Royals and British hunters contributed a great deal in wiping out the entire lion population from India except at the Gir Forests in Gujarat where the last two Nawabs of Junagarh during the nineteenth century made really serious efforts to protect them in the wild. History reveals that Maharaja Kishen Singh of Bharatpur (Rajasthan) had a tamed lion which was displayed on a special cart during the visit of the British King Edward VIII in 1921 [13] (Fig. 3.18). In 1942, the then Viceroy Lord Linlithgow, the Vicereine, their daughter, and the daughter of the former Viceroy Lord Irwin also hunted a lion (Fig. 3.19). Soon after Independence in 1947, kings and other Heads of princely states needed a formal permission of the Saurashtra *Rajpramukh* (the then Head of state) to hunt lion at the Gir Forest (Fig. 3.20).

Fig. 3.18 The state lion of Bharatpur with its caretaker standing in procession to welcome HRH the then Prince of Wales (later King Edward VIII) on his visit to the state in 1921 (*Courtesy*: *Dr. Divyabhanusinh*)

Fig. 3.19 The Last Viceregal Shikar (Hunting), 1942—The Viceroy Lord Linlithgow, the Vicereine, their daughter, and the daughter of the former Viceroy Lord Irwin stand in front of the viceregal lion trophy (*Courtesy*: *Dr. Divyabhanusinh*)

Fig. 3.20 Maharaja Sawai Mansingh-II of Jaipur, the then *Rajpramukh* (Head of the state) of Rajasthan with his lion trophy in Independent India (1948) at Gir Forests in the erstwhile Princely Saurashtra state. He took permission for this hunt from Saurashtra *Rajpramukh* of Gujarat (*Courtesy*: *Dr. Divyabhanusinh*)

The Lost Glory of the Gorgeous Asiatic Cheetah Acinonyx jubatus veneticus: *A Historical Perspective*

Historical accounts maintained by the Government and royals are a direct testimony to the fact that, the Asiatic Cheetah (*Acinonyx jubatus veneticus*) thrived well in the immediate past until before India became Independent in 1947. Although, there is evidence of a few cheetahs present in Iran still surviving, their number is rapidly falling. Africa is the only place in the whole world where cheetah can still be found in the wild in large numbers. Dr. Divyabhanusinh in his famous book *The End of a Trail: The Cheetah in India* presents a detailed, yet very interesting account of this magnificent wild cat [34]. Though, the issue of the taxonomic classification of cheetah remained a bit controversial for a long time however, it is now abundantly clear that the Asiatic Cheetah is (Fig. 3.21) *Acinonyx jubatus veneticus* while the African race is *Acinonyx jubatus jubatus* (Fig. 3.22). Cheetah—the swiftest animal on planet earth enriched semidesert areas, scrub jungles, and grasslands of India and so as Rajasthan, feeding on Blackbuck, Chinkara and hare. Unlike many other wild species of animals, breeding of cheetah in captivity has never been easy.

Fig. 3.21 An Asiatic Cheetah *Acinonyx jubatus veneticus* with a small girl (probably British) who is clearly seen holding the chain in her hand. *Hyderabad, c. 1900*

It is an interesting fact that, cheetah is the only feline which literally remained as a trusted royal pet and hardly attacked man or got agitated by the crowds. The abundance of cheetah in India and its friendliness with locales is clearly shown in Fig. 3.21. During the nineteenth century, the population of cheetah rapidly declined in India and by the middle of twentieth century not a single animal was left. Cheetah's presence in Rajasthan was well recorded in the Mughal period especially during King Aurangzeb's rule (1667). Detailed accounts written in *Mir-al-i-Ahmadi* in early eighteenth century mentions that, several areas were earmarked in India for procuring cheetahs for the imperial court namely, Pattan, Bhatner, Bhatinda, and Hissar in Punjab, Jodhpur, Nagour, Merta, Jhunjhunu, Amarsar and Dholpur in Rajasthan, Jamnagar and Sidhpur in Gujarat and Alapur near Gwalior in Central India [35]. This is direct evidence to prove that, cheetahs were available in large numbers in the grasslands, scrublands, and semiarid regions of the then western India [34].

In addition to the above facts, several paintings at Royal museums and palaces across India and the ones preserved at Victoria and Albert Museum, London, are still other potential evidence of the presence of cheetahs in the wild and as pets held

Fig. 3.22 African Cheetah *Acinonyx jubatus jubatus* in the wild somewhere in South Africa. The Indian race of Cheetah *Acinonyx jubatus venaticus* or the Asiatic Cheetah vanished from Rajasthan as well as other parts of India by the year 1948. Note the white tip of the tail; the Asiatic Cheetah has a black tip (*Courtesy*: *Babette de Jonge (Wild Cats Magazine/Wild Cats World), Masai Mara 2009*)

by the erstwhile royals and kings especially the Mughal emperors like Akbar. One such painting from the *Akbarnama* shows Emperor Akbar assisting in lifting a cheetah out of a pit (1590–1596). This was the first time that, Akbar had caught a cheetah using techniques devised to ensure that the animals were not harmed. Cheetahs were also largely tamed and used in hunting (Fig. 3.23). Another painting from the London museum prepared by the Mughal court artists La'l and Kesav Khord depicts the Mughal Emperor Akbar (r. 1556–1605) hunting a Blackbuck using his trained cheetahs (Fig. 3.24). To this end, a painting of c. 1725–1730 from Udaipur shows Maharana Sangram Singh hunting Blackbuck with the help of cheetah. Similarly, the City Palace Museum at Jaipur clearly details coursing with cheetah in a painting from the rare collection of Maharana Ramsing II (1850–1860) where he is shown hunting Wild Boar and Blackbuck. Cheetahs were regularly used by the royal hunting parties consisting, in addition, Caracal and hawk during the late eighteenth and early nineteenth century (Fig. 3.25). They were, in fact, given proper training and kept nicely by their keepers along with Caracal (Fig. 3.26). Huge bullock-carts were normally used to carry them to the forests and other hunting sites (Fig. 3.27).

The skins of dead Cheetahs displayed at Lalgarh Palace in Bikaner with black tipped tails—a typical characteristic of the Asiatic or Indian Cheetah has been historically linked with the animal being shot in Rewa in Madhya Pradesh in 1925 by Maharana Sardul Singh of Bikaner (Rajasthan) [34] (Fig. 3.28).

Fig. 3.23 Akbar assists in capturing a Cheetah. A painting from the "*Akbarnama*" (Book of Akbar) shows the Mughal Emperor Akbar (r. 1556–1605) assisting in lifting a cheetah out of a pit. This was the first time that, Akbar had caught a cheetah using techniques devised to ensure that the animals were not harmed. In those years cheetahs were tamed and used in hunting. Place of origin: India or Pakistan; date made: ca. 1590–1595; artist: Tulsi Narayan; materials and techniques: opaque water color and gold on paper; museum number: IS.2:2-1896; gallery location: South Asia, room 41, case T (*Photo ©Victoria and Albert Museum, London*)

The *Sa'idnama*—a Persian manuscript written around mid-nineteenth century available at Tonk in Rajasthan has about nine pages dedicated only to the cheetah giving their location at Pali in Rajasthan in addition to the ecology, behavior, capturing methods, breeding and caring for cubs, training, ailments, life history, and varieties [34]. This is perhaps one of the last few surviving repositories clearly mentioning the interesting man–cheetah interaction.

According to Dr. Divyabhanusinh, the three last surviving cheetahs were shot dead by Maharaja Ramanuj Pratap Singh of Madhya Pradesh in 1947—the year India became independent of the British rule. A report on this was published in *Journal of Bombay Natural History Society* with a comment by the editor that, "these were probably the very last remnants of a dying race." Surprisingly, reports of cheetah sighting [36] came in 1952 from areas lying between Bangalore and Andhra Pradesh. Later, officials of the Rajiv Gandhi Foundation also reported falling of a cheetah in a well near Hyderabad (Andhra Pradesh) in 1957 but the rescue attempt was unfortunately unsuccessful. This was probably the last authentic sighting of cheetah in Independent India. In addition, history reveals that, Maharaja Brij Singh of Bharatpur presented an Indian cheetah to his brother-in-law, Jaya Cham Rajendra Wadiar of

Fig. 3.24 Akbar hunts with trained cheetahs. A painting from the *Akbarnama* (1590–1595) depicts the Mughal Emperor Akbar (r. 1556–1605) hunting Blackbuck with his trained cheetahs. Place of origin: possibly made in India or Pakistan; date made: 1590–1595; artist: La'l and Kesav Khord; materials and techniques: opaque water color and gold on paper; museum number: IS.2:92-1896 (*Photo ©Victoria and Albert Museum, London*)

Mysore in the early 1950s which was later sent to Mysore zoo [34]. Hitherto, it is generally believed that, the last cheetah of Indian origin in the wild in India were definitely those shot in 1947 by Maharaja Ramanuj Pratap Singh of Madhya Pradesh. In 1956, eight cheetahs were present at the Kolhapur Chhatrapati Palace but out of them only a pair was left until 1958 and that too died in 1960 [37].

Cheetah Reintroduction Program

The issue of cheetah reintroduction came up in the meeting of experts at the Ministry of Environment and Forests (MoEF), Government of India, several times during the past decades. Dr. Asad R Rahmani, one of the editors of the present volume was initially commissioned in the mid-1980s to conduct surveys to find out suitable sites for its relocation in India. A report on this was submitted to the Government of India by the Wildlife Institute of India (WII) in 2010 [38] wherein ten sites were selected from seven landscapes located in the states of Rajasthan, Gujarat, Madhya Pradesh, Uttar Pradesh, and Chhattisgarh for their potential to harbor viable reintroduced

Fig. 3.25 A royal hunting party comprising of hawk, Caracal and cheetah; circa 1920. The animals seem quite used to of each other's presence and appear to live in close proximity when tamed

cheetah populations [39]. The Shahgarh landscape and Desert National Park of Rajasthan were found to be suitable for introducing cheetah when the reports last came in. Cheetah for this program will be sourced from Africa. Since, the villagers and Border Security Force have severely objected to the above plan, Mukundra Hills at Kota and Darra Wildlife Sanctuary are also being seriously considered for this purpose. A separate section on cheetah relocation in Rajasthan covering the recent efforts by a team of experts entrusted for the job by the Ministry of Environment and Forests (MoEF), Government of India, can be found in Chap. 44.

Wild Dog (Dhole) Cuon alpinus dukhunensis *(Pallas, 1811)*

Hunting with a special breed of domestic dogs (Fig. 3.29), was a common practice among the princes in pre-Independent India which continued until the first half of

Fig. 3.26 A cheetah (tied on the cot) and a Caracal seen (tied with a pole on the ground) in their trainer's household, circa 1890 (*Courtesy*: *Gobind Ram and Oodey Ram Firm, Jaipur, India*)

Fig. 3.27 A cheetah party returning from the hunt—The big cat is tied with a rope (chain) around the neck like pet dogs of today which is a testimony to the abundance of this magnificent animal in those years. A hunted antelope can be seen staked below the cheetah's cot. *Hyderabad State, c. 1900*

Fig. 3.28 Three cheetahs shot from the car in Rewa (Madhya Pradesh) in 1925 by Maharaja Sardulsingh of Bikaner (Rajasthan)

Fig. 3.29 Wild Dog or Dhole *Cuon alpinus dukhunensis* has now completely disappeared from Rajasthan but is still found in the nearby state of Madhya Pradesh (*Courtesy*: *Niranjan Sant*)

the twentieth century [13]. Small groups comprising 5–6 of such dogs were known to keep a tiger at bay and even kill it. Tigers and leopards were known to run down by these creatures to be later shot dead by the royal hunters towards the end of a chase. Until early 1950s, some 30–40 Wild Dogs were still kept in bastion of the fort in Karouli of Rajasthan where they were fed with pig meat [13]. By Independence,

Fig. 3.30 Caracal or *Siyahgosh* (in Persian) *Caracal caracal* was frequently used for shikar along with Cheetah (*Courtesy*: *Dr. Gobind Sagar Bhardwaj*)

status of the prey base was quite sufficient in the forests of the state and Wild Dogs were widely distributed across Rajasthan. With the passage of time, scarcity of wild herbivores turned them towards villages for food where they became goat-killers, especially in southern Rajasthan. Due to the increased antipathy of locales and tribes for Wild Dogs, they were systematically killed in large numbers by professional hunters upon requests by the villagers. Now they are not found in Rajasthan, though, in the bordering state of Madhya Pradesh, the species can still be sighted in small numbers. Hunting dogs—a domesticated variety of a carnivora of unknown origin (may be the Wolf *Canis lupus*, often confused by many as Wild Dogs or Dhole) are in fact, a breed of trained domestic dogs, with long legs and very thin body that were frequently used by the erstwhile kings and royals for hunting small game. Even these days, the Mogiya tribe uses them to hunt hare, antelopes etc. This must be made clear at this juncture that, these so called hunting dogs are not Wild Dogs or Dhole or Whistling Dogs *Cuon alpinis*, they are *Canis familiaris*.

Hunting with Caracal Caracal caracal [34]

Caracal (called *Siyahgosh* in Persian on account of its characteristic black ears) *Caracal caracal* (Fig. 3.30) is the only other feline apart from Cheetah which was used for hunting by the Mughal Emperors from Firoz Shah Tuglaq to Akbar the Great. Like Cheetah, Caracal too cannot breed freely in captivity. The animal was also found in Baluchistan and Sind (now in Pakistan). In India, Caracal has been reported from Agra and Allahabad in Uttar Pradesh, Ajmer, Sawai Madhopur and Alwar in Rajasthan, Malwa and Kutch in Gujarat, and some parts of Punjab

Fig. 3.31 Asiatic Wild Ass or *Equus hemionus khur* flourishes in the Little Rann of Gujarat and occasionally seen in the border areas of Rajasthan (*Courtesy*: *Aditya Roy*)

and Central India. It has been widely painted in Shikar pictures. Interestingly, the distribution range of Caracal coincided well with that of Cheetah. A versatile hunter with lightening speed and agility—Caracal may even kill larger preys like Chinkara, although, in practice they were employed for hunting small animals such as patridges and hares. It has been known to attack its keeper and hold his neck being more agile than cheetah. During state period, it also wore a waist band and collar-like cheetah [34]. Amritsar in Punjab was a famous market where a variety of animals were sold and Caracal was one of them [33]. In Rajasthan, Caracal can still be seen at Ranthambhore National Park and Sariska Tiger Reserve though their number is reported to be rapidly falling.

Rhinoceros in Rajasthan

Studies published in 1973 have mentioned about remains of the greater one-horned Rhinoceros, *Rhinoceros unicornis* from Rajasthan [40].

Indian Wild Ass or Khur Equus hemionus khur

The Indian Wild Ass or Khur *Equus hemionus khur* is usually sandy in color, but varies from reddish gray, fawn, to pale chestnut (Fig. 3.31). The animal possesses an erect, dark mane which runs from the back of the head, neck and along the back, to the root of the tail [41]. Wild Ass usually grazes between dawn and dusk feeding on grass, leaves and fruits, agricultural crop, *Prosopis* pods, and saline vegetation. Arid zones, grassland, and scrubs are its preferred habitat. It is one of the fastest of Indian animals, with speeds clocked at about 70–80 km/h and can easily outrun a jeep. Stallions live either solitarily, or in small groups of twos and threes while family herds remain large.

Once it was found in Jaisalmer and Bikaner of the Rajasthan state in western India, western Pakistan, Sindh, Baluchistan, Afghanistan, and south-eastern Iran. Today, its last refuge lies in the Indian Wild Ass Sanctuary, Little Rann of Kutch, and surrounding areas of Greater Rann of Kutch in the Gujarat State of India. It is also seen in the districts of Surendra nagar, Banskantha, Mehsana, and other Kutch districts. This side of the Rann, in Rajasthan, can be considered part of the larger ecotone, a transition area between marine and terrestrial ecosystem. The Wild Ass population which had got confined to the Little Rann some time in the past has now spread to the Greater Rann as well, bordering Rajasthan, Pakistan, and the Arabian Sea. This population of about 4,038 animals is perhaps the only gene pool of Indian Wild Ass (*Khur*) in the entire world and one of the six geographical varieties or subspecies surviving on the planet Earth. The residents of Kukaria, a village bordering the flat salt lands of the Rann, talk about frequent raids on their cropland by Wild Ass herds.

There is historical evidence to show that the beautiful animal known for its stamina and speed used to inhabit the desert areas of Rajasthan in the past. An enlarged image of a painting from *Akbarnama* at Victoria and Albert Museum, UK, has the illustration which depicts the Emperor Akbar falling into a mystical trance in 1571 while on an Indian Wild Ass shoot, with several of them having been shot by him [42] (Fig. 3.32). Emperor Jahangir in his book *Tuzk-e-Jahangiri* writes that India's Mughal Emperors and noblemen from the time took great pleasure in hunting Wild Asses [23] whose meat is reportedly good to eat as Emperor Jehangir testified in *Jehangirnama*. It is thoroughly unknown as to how the Indian Wild Ass disappeared from its former haunts in parts of western India and Pakistan since the animal was never a hunting target of Indian Maharajas and colonial British officials.

From 1958–1960, the Wild Ass became a victim of the disease called *surra*, caused by *Trypanosoma evansi* and transmitted by flies, which resulted in a dramatic decline of its population in India. In November and December 1961, the Wild Ass population was reduced to just 870 following the outbreak of South African Horse Sickness [43]. Besides disease, the Ass's other threats include habitat degradation due to salt activities, invasion of the *Prosopis juliflora* shrub, and encroachment and grazing. Conservation efforts since 1969 have helped boost the animal's population to 4,000 in Gujarat's Wild Ass Sanctuary [44].

Sighting of Asiatic Wild Ass in Rajasthan has been reported in 2002 [45]. It emerged in Rajasthan terrain in 2009 when Muslim herdsmen in Khejariali village (not to confuse with Khejrali village near Jodhpur, known for brave *Bishnois*) of Jalore district's Sanchore *tehsil* (where a 60 km [2] area was transferred to the Rajasthan Forest Department by the revenue authorities in 2007) complained of the invasion by a "donkey that looked like a horse." Later it was found that, the animal which attacked the herdsmen's horses was actually a Wild Ass [46, 47].

Fig. 3.32 Akbar lost in the desert while hunting Wild Asses. A painting from the "Akbarnama" (Book of Akbar) ca. 1590–1595 shows the Mughal Emperor Akbar (r. 1556–1605) lost in the desert while hunting Wild Ass. This illustration from the Akbarnama depicts the emperor Akbar falling into a mystical trance while on a desert hunt in 1571. Akbar is shown seated and withdrawn in a clearing. On the right, corpses of Wild Asses lie on the ground near a tree. Place of origin: possibly made in north India or Pakistan; date made: 1590–1595; artist: Mahesh and Kesav; materials and techniques: opaque water color and gold on paper; museum number: IS.2: 84-1896, gallery location: in store (*Photo ©Victoria and Albert Museum, London*)

The Gujarat Ecological Education and Research Foundation (GEER) report has recommended that the Thar Desert in Rajasthan should be developed as an alternative site for reestablishing the Indian Wild Ass population via reintroduction program [48].

The Migratory Bird: Siberian Crane Leucogeranus leucogeranus

Siberian Crane *Leucogeranus leucogeranus* (Fig. 3.33) used to be the most charismatic and rare bird at Ghana or the Keoladeo National Park of Bharatpur. At one time, hundreds of "Sibes" used to winter in the Ghana Bird Sanctuary. Like white ghosts in the mist, they were lured by other north Indian wetlands from far and near. The "Sibes" used to visit Ghana from their breeding grounds in Siberia in search of food owing to the nonavailability of summer supplies due to extreme cold. No Siberian Crane was sighted in Bharatpur since 2003.

Interestingly, the seventeenth century court painter Mansur painted a fairly accurate depiction of a Siberian Crane, suggesting that the birds visited these wetlands, but that they were nevertheless rare enough for an artist of his repute to single them out for special consideration.

Fig. 3.33 Siberian Crane *Leucogeranus leucogeranus* has stopped visiting Keoladeo National Park (KNP) since 2003 due probably to the continued water crisis which is now solved but the bird is yet to reappear (*Courtesy*: *Niranjan Sant*)

Historic Tiger Reintroduction and Affability to Tiger in Rajasthan

Rulers of the erstwhile Dungarpur state are known for their love and knowledge for the wildlife. The entire tiger population of Dungarpur state was wiped out during *Chhapania Kaal* (the terrible famine of 1956) but many tigers remained alive in the adjoining Mewar state. The rulers of Dungarpur successfully attracted tigers of nearby Mewar state towards their forest by introducing baiting on the Dungarpur—Mewar border. Once, tigers of Mewar started their movement towards Dungarpur border, the baits were gradually shifted deeper into the forests. Thus, a new population of tigers was pulled inside the forests of Dungarpur. Perhaps, this was the first tiger reintroduction case in the history of Rajasthan and the world.

Samadhi (grave) of a tiger named "*Bokha*" is still lying close to the Dungarpur city that died on December 16, 1934. By the order of Maharaval of Dungarpur, Bokha was buried towards southern end of Gap Sagar and a small grave was constructed in its memorial. This is a unique example of humanly treatment given to a tiger. The latest tiger introduction was done at the famous Sariska Tiger Reserve, Alwar during 2008. Two enclosures were made in Naya Pani zone having an area of 1.0 ha each. Wild tigers were lifted from Ranthambhore National Park to be later housed in Naya Pani Enclosure of Sariska. After a few days of acclimatization, they were finally released in the wild habitat of Sariska forests.

History of Faunal Conservation in Rajasthan

In the third century BC, Emperor Ashoka was perhaps the first ruler in the world to make rules for conservation of wild animals. His decrees, engraved on stone slabs are still available, however; information about laws or rules relating to wildlife hunting during the medieval period is not available. *Chankaya*, the great Indian politician of the medieval period has mentioned that a king shall protect forests, shall not hunt animals and birds, and his forests should be enriched with the flora and fauna [49].

On one hand, the religion in Rajasthan had a wide impact in the conservation of wildlife, on the other, hunting of wild animals continued as a favorite sport of rulers and their families which carried on uninterrupted during and after the British rule. Historical records of various palaces in Rajasthan are full of detailed stories of hunts and trophies. Hitherto unexplored and inaccessible heritage structures near Jaipur at Nahargarh Wildlife Sanctuary like Gopal Vilas *Haudi* (hunting tower) (Fig. 3.34) and Ramsagar *Shikar-Haudi* (Fig. 3.35) were made by former rulers amidst dense forests where royal guests also joined them. These structures are currently being revived for national and international tourists and other visiting dignitaries.

Before Independence, erstwhile princely states (*Riyasats*) had their own management of forests [10, 50]. Rulers of these princely states in British India were keen lovers of "*shikar*." Acts and Rules were notified by different states for hunting of game (wildlife) providing enough safeguards to ensure effective conservation of wildlife. Salient features of the hunting rules of the princely states were commonly displayed. Shooting of big game was completely prohibited except with the written

Fig. 3.34 Unexplored heritage structure Gopal Vilas *Haudi* (Hunting Tower) made by the erstwhile rulers amid the dense forests of Nahargarh Biological Park near Jaipur (*Courtesy*: *Devendra Bhardwaj*)

Fig. 3.35 Hitherto unexplored and inaccessible heritage structure Ram Sagar *Shikar-Haudi*, an airy palace made of Jharokhas (small windows and corridors) amidst deep forests, once a hunting lodge of the former rulers of Jaipur is being revived to facilitate visitor's halt at Nahargarh Biological Park near Jaipur (*Courtesy*: *Devendra Bhardwaj*)

permission of rulers. Trapping of small game was prohibited and heavy penalties were prescribed for breach of rulers. Special *Shikar-khana* departments were created for the purpose of managing *shikars* by the rulers or their guests. Most of the kings were interested in hunting and due to their autocratic attitude, public was not allowed to enter these forest freely for having direct benefits. The British officers also joined the ruling families in *shikars*. In 1887 the British enacted "1887 Wild Birds Protection Act No. 10" which was applied to very limited areas near Cantonments. It restricted hunting and/or trapping of some birds during their breeding seasons. There was no restriction on hunting in the nonbreeding season and so, the slaughter of wild animals continued. For the first time, in 1912, some restriction on killing of wild animals was imposed by enacting "1912 Wild Birds & Animals Protection Act No. 8." This was applicable all over British India and penalties were prescribed for offences. It was amended in 1935 to implement its applicability in different provinces. Notification of "Reserve Areas" by the provincial governments for banning hunting and trapping of wild animals in notified areas was also done. In 1935, the powers to make laws and rules for wild animals and birds were delegated to the Provincial governments (Table 3.2).

After India's Independence in 1947, Rajasthan government enacted "The Rajasthan Wild Animals and Birds Protection Act, 1951." It was the first generalized law for wildlife that extended to almost the whole of British India and specified the seasons when hunting was prohibited, listed some animals, whose hunting required a license, declared "Reserved Areas" where hunting of animals was

Table 3.2 Important decrees by the erstwhile kings and royals and the British Government for the protection of forests

Year of the order	Order/decree regarding forest and faunal conservation
Jaipur state	
1885	British Govt. deputed Mr. Mc Moyar for inspection of forests of Jaipur state to give suggestions for better management
1886	On the basis of Mc Moyar's report, forests of Jaipur state were divided into three categories and standard silviculture was recommended to manage them
1936	Bhai Sadhu Singh was appointed as the first forest superintendent of Jaipur state
1943	Forest and Shikar-khana Departments were separated from each other in Jaipur state
1947–1949	Three departments namely Shikar-khana, Forest and Grass Farm forest were combined as a single department at Jaipur
Jodhpur state	
1604	First Saka (Sacrification) happened in Samdari village of Jodhpur district
1707	Maharaja Ajit Singh ordered not to cut trees in *Bishnoi* villages.
1764	Maharaja Vijay Singh ordered for release of all the goats and sheep from butchers and also to purchase all *Khejadi* trees.
1794	Maharaja Bheemsingh ordered not to cut green *Khejadi* trees; not to impose any taxes and not to kill any animal.
1787	Not to castrate bulls in order to follow the 29 commandments of Bishnois.
1832, 1834, 1874	Maharaja Mansingh ordered to summon whoever cuts *Khejadi* and shoot or hunt animals.
1906	The first ever 40 year work plan prepared for the forest of Jodhpur state by Sh. Chaturbhuj Gahlot.
Ajmer state:	
1850	Forest demarcation started in Ajmer-Merwara territory by Mr. Dixan.
1869	Rights and privileges of the general public with respect to the forests were notified.
1872	Forest Department came into existence.
1893	First work plan was prepared for the management of forests.
1918	Merwara forest subdivision was constituted at Beawar for the Ajmer-Merwara territory.
1936	Forests of Ajmer were divided into two circles by Mr. Kumbus, namely, coppis with standard circle and grazing circle.
Udaipur state:	
1878	Gulab Bagh Zoo established in Udaipur city by the then Maharana of Mewar.
1930–1947	Until 1930, there was no forest management plan in existence in Udaipur state. During 1941, tentative work plan for the forests was prepared. During 1947 further amendments were made.
Tonk state:	
1901	The post of forest superintendent was created to manage the forests and Sh. Maula Bux was appointed the first superintendent.
Banswara state:	
1901	Forest Department constituted.

(continued)

Table 3.2 (continued)

Year of the order	Order/decree regarding forest and faunal conservation
Abu:	
1869	Forest demarcation started in the forests of Abu.
Alwar state:	
1900	First forest settlement started by M.O. Duryer.
1903	Forest department came into existence.
Dungarpur state:	
1911	Forests of Dungarpur state were divided into two categories (1) Shikar Reserve for private and (2) village forest for use by common public.
Kota state:	
1915	Sh. Dulichand was appointed as first forest settlement officer.
1938	Organized management of forest was started by Sh. Sitaram Puri.
1940	*Kattha* extraction from *Khair* (*Acacia catechu*) tree began for the first time via *Handi* system by Nandan Bhargawa.
Bikaner state:	
1929–1930	Demarcation of forest began under Bikaner Forest Act.
Bharatpur state:	
1898–1899	Creation of the wetlands of Keolaleo by the Maharajah of Bharatpur.
1937	Dr. Salim Ali reported sighting of 11 Siberian Cranes at the Keoladeo National Park (then a private hunting reserve of the erstwhile Maharajah of Bharatpur).

completely banned and mentioned penalties for offences committed under the Act. In 1957, the state government notified a revised list of animals protected under the Act and also notified the closed season for different wild animals. Most of the present wildlife sanctuaries of Rajasthan state were notified as Game Reserves under this Act. In 1958, the state government made rules to implement the 1951 Act and defined the Game Reserves as "Game Sanctuaries." It also notified "Rajasthan entrance to the Game Sanctuaries Rules 1958." Keeping in view the precarious situation of wildlife in the country, the Government of India finally enacted *Wildlife (Protection) Act, 1972* which was adopted by the state government of Rajasthan on September 1, 1973. This was further amended as *Wildlife (Protection) Amendment Act, 2002*. In 1976, through the 42nd amendment in Constitution (Article 49) that states "The State shall endeavor to protect and improve the environment and to safeguard the forests and wildlife of the country," the subjects of "Forests and Protection of Wild Animals & Birds" were transferred from *State list* to *Concurrent list* of the *Constitution of India*. This was a major step for the protection of Wildlife in India.

Every Protected Area of Rajasthan has its own interesting history on account of being private hunting reserves of the erstwhile maharajas [51]. Before Independence, forests were managed keeping in view three major objectives: (1) to manage them as *shikargarh*, (2) to extract forest produce, and (3) to regenerate them back. Co-operation of public on Joint Forest Management (JFM) basis was lacking. West Bengal was the first state of India to adopt JFM in 1989. In Rajasthan, it was adopted in 1991 when government issued certain guidelines. Important government orders and incidents after independence are presented in Table 3.3.

Table 3.3 An overview of the events pertaining to nature conservation in Rajasthan after Independence

Year	Important orders post-Independence
1953	Rajasthan Forest Act came into existence
1954	Rajasthan Forest Training center was established in Alwar to train lower field staff of the Forest Department. A zoo was also established in Kota city
1955	Ranthambhore, Sariska, Darrah, and Van Vihar declared as wildlife sanctuaries
1956	Silviculture Division was established to organize and promote research work on forestry aspects. Ajmer forest division unified with state Forest Department. Keoladeo Ghana, Bharatpur notified as a wildlife sanctuary
1957	First forest labor co-operative society named Ambavi-Daiya-Patiya society constituted
1960	Mt. Abu notified as wildlife sanctuary
1961	Sir H.G. Champion traveled across the forests of Rajasthan to reclassify the types
1969	State trading wing started. Dantiwara Project launched. Ticketing started in Jaipur Zoo and a price of 10 paisa per ticket was decided. Meeting of International Council for bird preservation held at Bharatpur
1970	Sitamata forests declared as sanctuary. Kadana Project started. Publication of *Van Sampada*—a biannual departmental magazine began in January
1971	Kumbhalgarh and Talchhapar declared as sanctuaries
1972	Dr. CM Mathur was the first forest officer of the state who got a Ph.D. degree in forestry
1973	Ranthambhore (Sawai Madhopur district) declared as Tiger Reserve under Project Tiger. An IUCN meeting was held at Bharatpur
1974	A portion of Chambal River and forest strips on its banks were declared as the National Chambal Sanctuary
1975	Jawahar Sagar notified as sanctuary
1976	(Late) Mrs. Indira Gandhi, the then Prime Minister of India visited Keoladeo Ghana, Bharatpur
1977	Sahibi River project started
1978	Sariska (Alwar district) notified as a Tiger Reserve. After Ranthombhore, it was the second Tiger Reserve of Rajasthan
1979	Sariska included in Project Tiger
1980	Ranthambhore declared as national park. Nahargarh notified as sanctuary
1981	The status of Keoladeo Ghana (Bharatpur) was raised as National Park. Desert National Park given the status of a wildlife Sanctuary
1982	Great Indian Bustard declared as the state bird of Rajasthan on May 21
1983	Rohida, Chinkara, and Khejadi (August 31) were declared as Rajasthan's state flower, animal, and tree, respectively. Kailadevi, Phulwari-ki-Nal, Todgarh-Raoli, Ramgarh Vishdhari, Bainsroadgarh, and Shergarh were declared as new wildlife sanctuaries
1984	Sawai Mansingh and Bandh Buretha declared as sanctuaries
1985	Social forestry project started
1986	Forestry Training Institute established at Jaipur. Indira Priyadarshni Vriksh Mitra Award started by Govt. of India. So far, four forest officers of the state namely, Mr. A.K. Updhayay, Dr. D.N. Pandey, Dr. S.K. Sharma, and Mr. B.L. Yadav have received it. Forest training school started at Jodhpur
1987	Sajjangarh notified as sanctuary
1988	*Salar (Boswellia serrata)* gum tapping stopped
1990	Bassi announced as sanctuary

(continued)

Table 3.3 (continued)

Year	Important orders post-Independence
1992	*Padam Shree* was awarded to the famous tiger man Kailash Sankhla, a forest officer from Rajasthan to recognize his efforts to save the tiger. Aravalli Afforestation Project started in the state
1993	Dhundh and Bandi River Projects were started in Jaipur district
1996	Rajasthan's State Forestry Action plan (year 1996–2016) published
1997	Amrita Devi Bishnoi memorial award started
1998	India Eco-development Project was launched at Ranthambhore National Park
2001	Forest development activities started under the Forest Development Authority (FDA) scheme
2005	An interpretation center was established at Keoladeo National Park, Bharatpur with the support of an Austrian company. Tigers exterminated from the Sariska Tiger Reserve. State Government constituted a "Task Force" to look into wildlife management issues. As many as 133 recommendations were given by the task force

Table 3.4 History of ex situ conservation in the state of Rajasthan

S. No.	Name of zoo	Year of establishment	Location
1.	Jaipur Zoo	1876	Ram Niwas Garden, Jaipur
2.	Udaipur Zoo	1878	Gulab Bagh, Udaipur
3.	Jodhpur Zoo	1936	Ummed Bagh, Jodhpur
4.	Bikaner Zoo	1922	Bikaner
5.	Kota Zoo	1954	Kota
6.	Deer Park, Shri Goverdhan Trust, Udaipur	1963	Goverdhan Vilas, Udaipur
7.	Mrigvan	1970	Chittourgarh
8.	Municipal Council Beawar Zoo	1954	Beawar
9.	Panchwati Deer Park	1967	Pilani
10.	Safari Park, Haridasji Ki Magri	1963	Udaipur

Ex Situ Conservation in Rajasthan

Zoological Gardens or Zoos are known for recreation, research, and rehabilitation activities. All the five big cities of the state have zoos. Except Kota zoo, all zoos of Rajasthan were established before Independence. Jaipur Zoo situated at Ram Niwas Bag is the oldest Zoo of Rajasthan and fifth organized zoos of India harboring tiger, leopard, Sloth Bear, Chinkara, Blackbuck, Lion-tailed Macaque and a variety of birds and reptiles. The Jaipur zoo is also known for successful breeding of Gharial and some wild ungulates. Table 3.4 depicts a list of state's zoos.

Increasing Involvement of Locals to Conserve the Biodiversity in Rajasthan

The locals, forest-dwellers and tribals largely depend upon the forest resources for their livelihood. A huge amount of minor forest produce particularly nontimber is also harvested by them. To seek cooperation of the locals and to manage the forest resources in a better way, the state government has now accepted Joint Forest Management (JFM) Policy. Few important landmarks of JFM are as follows:

1991	Government issued an order on March 15 to develop and manage the forest land with the help of local people. A similar order was later issued on April 26 to develop and manage other community lands on JFM basis
1995–1996	*Janta Van Yojna* was launched in the state. This plan promotes Village Forest Protection and Monitoring Committees (VFPMCs), Panchayats and related experienced organizations for afforestation on forest panchayat and other types of land
1997	Government accepted the procedure of constituting an Eco-Development Committee (EDC) in villages near Protected Areas (PAs) to seek their cooperation for eco-development activities in and around Sanctuaries and National Parks
2000	Government has issued a combined and amended order on October 17 to promote afforestation on forest and all other types of Government lands

Oran and *Gouchar*: The Traditional Common Property Reserves of Rajasthan

Oran (Fig. 3.36) and *Gouchar* are common properties of a village traditionally developed for the purpose of habitat conservation and for common use with respect to fodder, grazing of livestock and fuel routinely and more importantly during the times of environmental distress like famine or very low rain fall. It is well known that, a few *Orans* in Rajasthan were so green and healthy in terms of availability of food and water that even the tiger visited them for shelter during peak summer apart from the usual wildlife. These days, though, the government has taken legal rights of such lands and agriculture is prohibited. Also called as wasteland, these lands are now controlled by the local *Panchayat Samiti*.

Human race today has realized the impending global crisis owing to gradual destruction of the biodiversity and disruption of ecological balance. Ancient Indian literature provides a cache of primitive, yet, scientific knowledge in addition to the sustainable environmental ideology. Only scientific solutions are not sufficient for achieving the ideal environment but a blend of cultural guidance, a sense of divinity, worship of nature and self-discipline are strictly needed. To this end, ancient Indian knowledge contained in the Vedas and epics, saints like Guru Jambheshwarji and the *Bishnoi* Community of Rajasthan are the lanterns of wisdom to guide the human race towards nature conservation. Please see Chaps. 1, 2, 3 and 4 from this book and Chaps. 1, 8, 18 and 20 from Faunal Heritage of Rajasthan: Conservation and Management of Vertebrates Vol. 2, Sharma B. K. *et al.* (eds.) 2013, Springer (978-3-319-01344-2) for more details.

Fig. 3.36 *Oran*—A typical Common Property Reserve in a village inside the Sariska Tiger Reserve, Alwar (*Courtesy*: *Ashish Kothari*)

Box 3.1 Royal Conservation Order Through Epigraph

An epigraph is present near Bansi town on Udaipur–Pratapgarh road where an order is carved on a stone by Maharaja Hari Singh to conserve the waterfowl. The stone was erected on the bank of a village pond in the year 1937. The order is inscribed in Hindi language appealing not to kill the aquatic life of the pond.

Acknowledgments The authors are extremely grateful to Dr. Divyabhanusinh Chavda, President, WWF-India, Mr. Niranjan Sant, Mr. Aditya Roy, Ms. Babette de Jonge, and the Victoria and Albert Museum, London, for providing rarest of rare pictures and paintings.

References

1. Rastogi Vandana(Ed.) (2005) Appendix III- Animals mentioned by Kalidasa. Pp 374-385. In: Prachin Bharat mei Paryavaran Chintan (Environmental thoughts of Ancient India). Publication Scheme, Jaipur, India, pp 400.
2. Russel RV, Lal H (1916) The tribes and castes of the central provinces of India. Vol. I. Published Under the Orders of the Central Provinces Administration. Mc Millan and Co., Limited St. Martin's Street, London, pp 2231.

3. Adi Parva, Sections 23. Mahabharata, Book I.
4. Udyoga Parva, Section 101. Mahabharata, Book V.
5. Karna Parva, Section 77. Mahabharata, Book VIII.
6. Karna Parva, Section 59. Mahabharata, Book VIII.
7. Adi Parva, Section 140. Mahabharata, Book I.
8. Karna Parva, Section 85. Mahabharata, Book VIII.
9. Bishnoi Kishnaram, Bishnoi Narsiram (Eds.) (2000) Dharam aur Paryavaran (Religion and Environment). Commonwealth Publishers, Delhi, India, pp 285.
10. Sharma RK (2007) Paryawaran Prashasan Avam Manav Paristhitiki (Hindi). Rajasthan Hindi Granth Academy, Jaipur; India, pp 363.
11. Ghosh A. Virasat (Hindi) (2007) Forest Department, Rajasthan. India.
12. Anonymous (2007) Lakes and wetlands of Rajasthan. Department of Environment, Government of Rajasthan, Jaipur, India, Publication. Centre for Environment Education, Nehru Foundation for Development, Thaltej Tekra, Ahmedabad, India.
13. Divyabhanusinh (2005) The Story of Asia's Lions. Marg Publications, Mumbai, India, pp 259.
14. Divyabhanusinh (2008) The Lions of India. Black Kite, Ranikhet, India, pp 267.
15. Chattopadhyay UC (2002) Researches on Archaeology in Holocene period (including the Harrappan traditions in India and Pakistan). Archaeology and Interactive Disciplines, Vol. III, pp 364–422. Indian Council for Historical Research and Manohar Publication, Delhi, India.
16. Jarrige Jean Francis, Hassan Usman (1989) Funerary Complexes of Baluchistan at the end of the third millennium in the light of recent discoveries at Mehergarh and Quetta. In: Ferfelt Karan, Sorensen Per(eds). South Asian Archaeology. Curzon Press, London.
17. Misra VN (2000) Climate as factor in the rise and fall of Indus Civilization- Evidence from Rajasthan and beyond. In: Lahiri Nayanjot (Ed.). Decline and fall of Indus Civilization. Permanent Black, New Delhi, India, pp 239–51.
18. Mukherjee BN (1969) Nana on Lion: A study in Kushan Numismatic Art. The Asiatic Society Calcutta.
19. Kinnear NB (1920) The Past and Present Distribution of lions in South Eastern Asia. J Bomb Nat Hist Soc 27(1):34–39.
20. Beveridge Annette Susannah (tr.) (1979) Babur Namah (Memoirs of Babar), Vols. I & II. 1922. Oriental Books Reprint Corporation, New Delhi, India.
21. Beveridge H (tr.) (1979) The Akbarnama of Abul Fazal. Vol II.1904. ES Publication, New Delhi, India.
22. Majumdar RC (2007) The Mughul Empire, Mumbai: Bharatiya Vidya Bhavan, p 5.
23. Blochman H (tr.) (1973) The A'in-i-Akbari by Abul Fazal, Vol. I. 2nd ed., 1873. Revised and edited by Col. DC Phillot. Oriental Books Reprint Corporation, New Delhi, India.
24. Alexender R, Henery B (1980) The Tuzuk-i- Jahangiri or Memoirs of Jahangir (1909-14). Munshiram Manoharlal, New Delhi, India.
25. Das Ashok Kumar (1978) Mughal Painting during Jahangir's Time. The Asiatic Society, Calcutta, India, pp 306.
26. Divyabhanusinh (1999) Hunting in Mughal Painting. In: Varma Sonprakash(Ed.). Flora and Fauna in Mughal Art. Marg Publication, Mumbai, India, pp 94–108.
27. Thackston Wheeler M (1999) The Jahangirnama: The memoirs of Jahangir, The Emperor of India. Freer Gallery of Art/ Arthur M Sackler Gallery, Washington DC and Oxford University Press, New York.
28. Hermann G (1950) The Art and Architecture of Bikaner State. Bruno Cassirer, Oxford.
29. William R (1884) Indian Game (From Quail to Tiger). WH Allen & Co., London, pp 380.
30. Newall Maj. Gen. DJF (1984) The Highlands of India. Being a Chronicle of Field Sports and Travel in India. Vol II (1882–87). Logos Press, New Delhi, India.
31. Mahesh R (1996) Fencing the Forest: Conservation and Ecological change in India's Central Provinces 1860–1940. Oxford University Press, Delhi, pp 245.

32. Tod James (1972) Annals and Antiquities of Rajasthan. Vol II. 1832. KMN Publishers New Delhi, India.
33. Joslin P (1973) The Asiatic Lion: A study of Ecology and Behavior. Ph. D. Thesis, University of Edinburgh.
34. Divyabhanusinh (1995) The End of a Trail: The Cheetah in India. Banyan Books, New Delhi, pp 248.
35. Irfan H (1986) An Atlas of Mughal Empire: Political and Economic maps with detailed notes. Oxford University Press, Delhi, India.
36. Kirkpatrick KM (1952) A record of the Cheetah *Acinonyx jubatus* Erexleben in Chittor district Madras state. J Bomb Nat Hist Soc 50:931–32.
37. Gee EP (1964) The wildlife of India. Collins, London, pp 192.
38. Gowda CD Krishna. (1982) Endangered species of Human Beings. Gnu's letters, Mysore. 1(2): p3
39. Ranjitsinh MK, Jhala YV (2010) Assessing the potential for reintroducing the cheetah in India. Wildlife Trust of India, Noida, & Wildlife Institute of India, Dehra Dun, TR2010/001.
40. Banerji S, Chkaraborty S (1973) Remains of the Greater One-horned Rhinoceros, *Rhinoceros unicornis* Linn. from Rajasthan. Sci Cult, Calcutta. 30(10):430–431.
41. Menon V (2003) A field Guide to Indian Mammals. Penguin Book India (P) Ltd, pp 208.
42. Illustration of a Painting from 'The Akbarnama; of Abul Fazal. Vol II. Depicting Akbar on a hunt of wild ass in Deserts of Rajasthan. Victoria and Albert Museum .website: VAM.ac.uk
43. Bavadem Lyla (2000) Gujarat's thirst; Distress migration of people and large-scale death of livestock have peaked. And this time the urban segments are as badly hit as the rural areas. Frontline magazine, 17(10); May. 13–26.
44. Bavadem Lyla (2001) Dealing with drought - Drought stalks Gujarat once again but the government relies on short-term crisis management measures instead of evolving a long-term, region-specific strategy to deal with this recurring phenomenon. Frontline Magazine, 18(12). June 9–22.
45. Sangha HS (2002) Sighting of the Indian wild ass *Equus onager* in Rajasthan: a northward range extension. J Bomb Nat Hist Soc, 100 (2&3): 617–621, 42.
46. Sebastian Sunny (2009) Wild Ass sighted in Rajasthan villages along Gujarat. The Hindu, India's National Newspaper; Sep 13.
47. Gole Susan (1998) Maps of Mughal India drawn by Col. Jean-Baptist Joseph Gentil, Agent for the French Government to the Court of Shuja-ud-dullah at Faizabad in 1770 Manohar Publications, New Delhi.
48. Roberts TJ (1977) The Mammals of Pakistan. Earnest Benn Ltd., London, pp 148–149.
49. Koutilya. Koutilya Artha- Shashtra. Vyakhya- Vachaspati Garola. Chokhambha Vidhya Bhavan (1962) Varanasi, Uttar Pradesh, India, pp 1011.
50. Sharma RK (2002) Joint Forest Management (Hindi). Avishkar Publishers & distributors, Jaipur; India, pp 259.
51. Sharma Shruti, Praveen B (2006) Management Plan: Keoladeo National Park, Bharatpur. Dept. of Forest and Wildlife, Rajasthan; 2002–2006, pp 182.

Chapter 4
An Anthropological Account of Bonhomie and Opprobrium between Communities and Animals in Rajasthan

Bela Kothari and B.K. Sharma

Abstract The present chapter is based on the anthropological literature on the communities of Rajasthan which are directly associated with animals economically and culturally. In this discussion these communities have been grouped as: hunters, pastoralists, entertainers, and conservationists. The text is arranged to provide some understanding of (1) the condition and present status of the communities, (2) the animals or specific breeds in terms of economy and socio-cultural association, and (3) consequences of environmental degradation on the communities.

Introduction

Around 15,000 years ago, man began experimenting while domesticating plants and animals in some parts of the world depending less on big-game hunting and more on relatively stationary food resources like fish, small game and wild plants. Cultural anthropologists by convention, use specific technical terms to describe the different ways developed by man for sustenance. These include (1) foraging, (2) fishing, (3) hunting, (4) pastoralism, (5) extensive tillers, and (6) intensive tillers [1]. The factors crucial for conducting these activities are the social organization to produce, the technologies available, and the environment in which the people operate.

B. Kothari (✉)
Department of Anthropology, University of Rajasthan, Jaipur, India
e-mail: belaalok@yahoo.com

B.K. Sharma
Department of Zoology, R.L. Saharia Government P.G. College,
Kaladera–303 801 (Jaipur), Rajasthan, India
e-mail: drbksharma@hotmail.com

B.K. Sharma et al. (eds.), *Faunal Heritage of Rajasthan, India: General Background and Ecology of Vertebrates*, DOI 10.1007/978-1-4614-0800-0_4,

According to the Anthropological Survey of India report, the occupational pattern of the total 4,635 communities located all over the country is as follows: 2,943 are agriculturists, 999 practice animal husbandry, 387 are fishermen, 311 are weavers, 237 are hunters and gatherers, and 37 are pastoralists. 2,662 communities exist as unskilled workers, 1,621 are artisans, 1,487 communities are in the unorganized sector, 3,690 in the organized sector, and 2,653 in business, trade and industry. It is important to mention here that many of the communities are involved in more than one type of occupation [2].

The state of Rajasthan being the focus of the present chapter has a total of 228 communities. The occupational pattern of these 228 communities throughout India is of similar type. Out of these 228, some communities are economically and culturally associated with animals directly by means of the following activities:

1. Hunting
2. Animal husbandry/Pastoralism
3. Entertainment
4. Conservation

Most of these communities have developed the expertise to handle a particular type of animal. It is pertinent to mention here that with the growing modernization and the ensuing socio-cultural changes, many of the communities are undergoing transformation at a very fast pace. The communities discussed in this chapter are those which are sizeable in population and still bear the attributes, to a greater or lesser degree which they have been known for.

The Hunters

The hunting communities include trappers of birds and animals. They trap hare, rat, francolin, mongoose, porcupine, monkey, and wild boar. Hunting communities are those which catch animals for food or to operate illegal wildlife trade. The communities falling under this category are *Kathodi, Bawaria, Aheri, Tirgar, Dhimar,* and *Sahariya.*

Kathodi

This rather unique but now vanishing community is the "monkey eating" tribe of Rajasthan. It lives in Udaipur district in southern Rajasthan. The *Kathodis* are migrants from Maharashtra where their traditional occupation has been the extraction of catechu (*Kattha*) from trees [3]. Their total population [4] is 2,922. For food, they hunt animals like rats, hare, pigeons, wild boar, partridges, junglefowl, and monkeys. Traditional weapons like stick, spear, catapult, and axe are used in hunting. For hunting monkeys two methods are generally practiced. In one technique, loops made from iron are used which are fastened to the trees. As soon as any body part

Fig. 4.1 *Jhonga*: A fishing trap made of bamboo strips used by *Kathodis*

of monkey gets into the loop the rope from the other side is stretched and it gets trapped. The second technique is as follows: Bengal gram nuts are scattered at a particular place for a few days regularly which results in making the monkeys habitual of visiting the same place. Then a trap in the form of a very narrow mouthed earthen pot with nuts in it is half dug in the soil. As the monkey puts his hand in the pot his fist full of nuts gets fixed in it [5]. The *Kathodi* men come out from hiding behind the trees and kill the monkey.

When they go to the forest for collecting catechu and other forest produce, they catch fish with the help of an instrument called "*jhonga*" (Fig. 4.1) which is made from bamboo [6]. They do not domesticate cattle.

The community is very backward in terms of their economic and educational status, and many of the ill-practices like gender discrimination, child marriage, and *purdah* (veil) system are prevalent among the people.

Bawaria

They are also known as *Baheliya and Mogiya* [7]. They got their name from "*bawar*" or noose with which they snare wild animals [8]. They have been known as the greatest threat to the country's wildlife. They were branded as criminal tribes in 1871 by the British for hunting and poaching. In 1952 they were included in the de-notified and Nomadic Tribes category.

The noose which they use as a trap is made of leather. They set long lines of these loops with running knots in the grass across the jungles. They arrange two rows of scarecrows consisting of bits of rag tied on the trees and grass along this noose. Next, they drive deer and other animals by frightening them, around these lines of scarecrows. In this course, the feet of the prey get entangled and they are caught.

Very cruel animal capturing and killing practices are prevalent amongst them. This includes torturing the captured birds and keeping them alive until a purchaser is found. They insert a feather through the eyelids of peacocks by which they are effectually blinded; whereas, they disable smaller birds, by breaking both their legs and wings [8]. Recently, the community was found to be killing partridges. The community has been involved in 99 wildlife offences registered in eastern Rajasthan [9]. They were also held responsible for killing peafowls by scattering poisonous grains along with healthy ones, owing to which the bird dies. Their total population is 60,121 persons [4].

Aheri

They are also known as *Ahehiya* and *Heri*. According to the census [4], the total population is 2,748. The *Aheris* are hunters and fowlers by descent. Their name is said to signify cowherd from "*her*," a herd of cattle. They are a vagrant community by habits. Almost no animal is considered non-edible by them. Socially, they are alienated from the community and are made to live beyond the village. What beef is to Hindus and pork is to Muslims, horse flesh is to the *Aheri* [7]. They live in Kota and Jhalawar districts of Rajasthan. They also assisted the rich people in their hunting game during the feudal era. When hunting was banned in 1972, they were deprived of their major source of livelihood and most of them started working as laborers in the agricultural fields.

Tirgar

The *Tirgars* or arrow makers dwell in the villages of the districts of Barmer, Jalor, and Sirohi and also in southern parts of Rajasthan in the districts of Banswara and Dungarpur. Their population [4] is 3,824. The community is traditionally engaged in making and repairing bows and arrows. According to telltales, the *Tirgars* supplied bows and arrows to Indian epic *Ramayana* characters Lord Rama during day time and to Ravana during night [10]. However, they still continue hunting practices considered illegal according to Indian law. Owing to same, they have taken up other professions for their livelihood. And thus, a great transformation as a result of modernization has been observed in their community.

Dhimar

They are also known as *Dhinwar* derived from "*dhi*" meaning fish and "*mar*" means—killer. They dwell in the north-eastern part of the state in the districts of

Alwar, Bharatpur, and Dholpur. They use various kinds of nets for fishing. One of the traditional nets is triangular in shape about 150 ft wide base and 80 ft in height to the apex. The size of each mesh varies from an inch width at the top to 3 in. at the bottom. The ends of the base are weighted with stones and the net is, then sunk into a river, so that the base rests on its bed and the top is held by men in boats at the surface. Then other Dhimars beat the surface of the water for some distance with long bamboos on both sides of the net, driving the fish downwards [8]. The boat is made from hollowed trunk of a tree. They worship their fishing nets, and the reverence for the knitted thread is such that they do not touch or wear a shoe made of thread.

Sahariya

The word *Sahariya* is derived from the Persian word "*seher*" meaning jungle or those who live in wilderness. *Sahariyas* dwell in the Baran district of Rajasthan. Their population [4] is 76,237. They are classified under the Primitive Tribal Group category. Forest produce and hunting was their main occupation. During the times of feudalism they were forced to work as bonded laborers, and no *Sahariya* could think of escaping it. They were given a kilogram of wheat per day for subsistence. After the Independence of India in 1947 the *Sahariyas* were allotted land but this could not help in improving their quality of life. Quite a number of *Sahariyas* are facing scarcity of food and resultantly many starvation deaths have taken place. One reason behind this might be the fact that the *Sahariyas* work for 2–3 days and then they rest until the money and food with them is finished.

Pastoralists

Pastoralism is a term used to refer to a way of life which involves the herding and management of large number of domestic animals including cattle, sheep, camels, and horses. These animals are the basis of daily subsistence. The communities which belong to this category are the *Raika, Gujjar,* and *Rathi.*

Raika

This is a camel-breeding caste living in Bikaner, Jaisalmer, and Barmer, the western districts of Rajasthan. They are also known as *Rebari*. The camel known as the "ship of the desert" is an essential part of Rajasthan's biodiversity and the key for preserving the identity of the *Raika* community. The population of Dromedary camel of the *Raikas* is fast dwindling. India was once proud of having the third largest camel population in the world after Somalia and Sudan. By 1997 their population dropped to 911,000, i.e., 11.6% decrease over a five years period beginning from 1992. More alarming is the decrease in the number of young camels by 50% during this time period, indicating a drop in camel breeding activities [11].

The *Raikas* are probably the only camel pastoralists in the world who have developed a deeply ingrained prohibition against the use of the camels for meat. This is supported by their belief that they have been created by God. They assert that their ancestors were created by Mahadeo (Lord Shiva, the God of destruction in Hindu mythology) in order to take care of the first camel which was created by the Goddess Parvati (spouse of Lord Shiva) for her amusement.

The community outlawed the sale of female camels, as the female camels were part of the dowry and as a part of marriage ritual; the bridegroom had to ride the camel. Due to intimate association with camels the *Raikas* have accumulated a cache of indigenous knowledge related to camel management, breeding, disease treatment, and behavior. Cash cropping of former grazing tracts, along with the creation of natural reserves out of open pastures, has restricted the range of the wandering *Raikas* [12]. Due to a decline in their traditional occupation, they were forced to live in abject poverty and worked as laborers.

Gujjar

Being traditionally a pastoral community of cattle herders, the *Gujjars* or *Gurjars* possess a large population of livestock. The wealth of *Gujjars* comprises buffaloes, although they keep cows and goats too. In the districts of Ajmer and Tonk they also keep sheep. In the Himalayas, the *Gujjars* are a purely pastoral and almost nomadic people, taking their herds into the higher ranges in summer and descending with them into the valleys or plains in winter. The saying about them is "*Ahir, Gadaria, Gujjar; E tinon taaken ujar*" or "the *Ahir, Gadaria, Gujjar;* are ever, in the look out of wasteland." It is said that even if a *Gujjar* is a cultivator in plains he is a bad cultivator and more given to keeping cattle than to following the plow. During their journeys across the country, they also get the wool sheared and sell it to their permanent customers. They earn their living exclusively by the sale of milk, butter, ghee (that is clarified butter oil), and other produce of their cattle. The women went to the markets every morning with earthen pots filled with milk, butter milk and ghee but these days they have started supplying their milk to dairy cooperatives. According to one estimate the *Gujjars* own approximately 30% of the total cattle of Rajasthan and 77% of the *Gujjar* families are engaged in the traditional occupation of cattle herding even today. As high as 42 livestocks per family were found in *Nagour* district [13]. *Gujjar* women acted as wet mothers for the princely families. The reason behind this is uncertain, but its association with their dairying occupation could be an interesting clue.

In southern Rajasthan, a caste named *Gairy* is known for herding sheep and goats. *Gujjars* also have their herds along with cattle. It is believed that those *Gujjars* who remained shepherds became, in due course of time, a different caste of *Gadaria* or *Giary* while some *Gujjars* took to cultivation as well.

Table 4.1 Cattle fairs of Rajasthan

Fair	Place	Animal/Breed	Period Held
Ramdev fair	Manasar (Nagore)	Nagori bull	February–March
Mallinath fair	Tilwara (Barmer)	Tharparkar cow	March–April
Shribaldev fair	Merta City (Nagore)	Nagori bull	April
Tejaji fair	Parbatsar (Nagore)	Nagori bull	July–August
Gomati Sagar fair	Jhalrapatan (Jhalawar)	Haryanavi cow	April
Gogamedi fair	Nohar (Hanumangarh)	Haryanavi cow	July–August
Jaswant fair	Bharatpur	Haryanavi cow	September
Chandra Bhaga fair	Jhalrapatan (Jhalawar)	Malvi bull	October–November
Pushkar fair	Pushkar (Ajmer)	Gir cow	October
Shivratri fair	Karouli	Haryanavi cow	July–August

Note: The correspondence between the dates of the Indian Calendar and the dates of the Gregorian calendar is not consistent. The Indian Calendar is Luni-Solar Calendar wherein the year is Solar but the months are Lunar and are based on the cycles of Full Moon and New Moon days. Thus, the numbers of days in a particular month vary from the year to year

Rathi

This is a Muslim community whose name is associated with the breed of cattle they keep, locally called *Rathi*. The *rathi* cow is a hybrid breed which is popular for the quantity of milk it gives. The community is distributed in the districts of Bikaner, Barmer, and Jaisalmer. Besides cattle, the R*athi* also keep sheep, goats, and camels.

A cattle fair is held every year in Barmer during which a lot of trading of this particular breed takes place. There are about ten large animal fairs held in Rajasthan every year for the trade and exchange of animals. These animal fairs are organized according to the lunar calendar known as *Vikram Samvat* in India. Each fair is known to be associated with a folk deity or a local hero. In addition, each of these fairs has become popular for a particular breed of cattle. These fairs are not just occasions to commemorate the great deeds of saints and heroes but have metamorphosed as giant centers of cattle trading. The people attending these fairs initiate the trade by first paying respect to the deity and praying for the health and prosperity of their cattle.

The popular animal fairs of Rajasthan are enlisted in the Table 4.1.

The *Tharparkar* and the G*ir* breeds of cow are famous for their milk production. The *Nagori* bulls are known for their strength and capacity of carrying load and are used to plow the agricultural fields. The *Haryanavi* breed of cow has dual purpose of dairying and for carrying load [14].

Entertainers

Communities that belong to this category are *Kalbelia, Madari,* and *Shikari*.

Fig. 4.2 A *Kalbelia* tent-hut with modern amenities like TV and tape recorder seen in the backdrop

Kalbelia

The *Kalbelias* are nomadic snake-charmer and snake-catcher. They travel in nuclear families from place to place entertaining people with their performances in towns and villages and only 9,401 people out of the total population of 75,118 are literate [4]. Interestingly, it is a popular belief that their guru Kanipavji in Jodhpur had magical powers. Looking at the people afraid of snakes and scorpions he would catch them and would treat snake bites with the use of herbs and spiritual healing [10]. His followers continued this profession of catching snakes and entertaining people by making them dance on the tune of their musical instrument—*pungi*. They are also known to eat snakes.

The women dance with veiled faces, mimicking the movements of a snake. They wear black dress and have performed in some of the most famous centers of cultural performance globally, such as the Royal Albert Hall in London. Fig. 4.2 depicts the rapid changes in the material culture brought about by contact with the mainstream culture.

Madari

People of *Madari* community are known to apply ashes (*bhabut*) to their bodies, wear iron chains round their head and neck and carry a black flag and turban. They are shunned by other castes and live in isolated areas. Their total population is 32,625 and the total literate population [4] is 9,892.

Shikari

Hunting or *shikar* has been a favorite amusement of kings and prosperous communities and people. For a successful hunt parties were sent out in search of the prey. The *Shikari* community served the princely states and was among companions on the hunting expeditions. On their report and guidelines, hunting expeditions were organized. The people of the community decoyed the prey to a convenient spot from where the hunters could shoot the prey [15]. It was considered a point of honor to hunt a tiger or a boar face to face. Hunting was banned by the Government of Rajasthan in 1972 under the Wildlife (Protection) Act.

Animal fights between elephants, tigers, leopards, boars, bulls, and cocks were among the themes of entertainment during feudal era. On the occasion of a birthday and a religious festival elephant fights were organized for which an open ground was preferred. The fighting elephants would meet each other face to face on the opposite sides of a wall about 1.4 m wide and 2 m in height. When the fight grew fierce and the animals became deadly, they had to be separated by thorny spears [16].

The Conservationists

Bishnoi Community

The *Bishnoi*, a religious sect turned caste residing in western Rajasthan is a living apostle of eco-friendliness and conservation (Fig. 4.3, 4.4, and 4.5). The *Bishnois* have inherited their name from the 29 precepts (*bis*=twenty, *noi*=nine) laid down

Fig. 4.3 A *Bishnoi* woman breast-feeding an orphan fawn and her own baby simultaneously (*Courtesy: Himanshu Vyas*)

Fig. 4.4 Rana Ram Bishnoi has planted and raised 20,000 trees in last 38 years. Every morning and evening he feeds wild gazelle and birds near his house in a small village in the Thar Desert (*Courtesy: Franck Vogel*)

Fig. 4.5 Lalu Ram Bishnoi brings a gazelle orphan near the herd to find a fostermother (*Courtesy: Franck Vogel*)

by their founder Saint Jambheshwarji in the fifteenth century, and they follow these virtues of moral conduct. Out of the 29 precepts eight have been prescribed to preserve biodiversity and encourage good animal husbandry. The *Bishnois* believe in the sanctity of God's creatures. They manage sacred groves locally called *Orans*, which are small patches of vegetation traditionally protected and managed by local

Fig. 4.6 Khejarli temple's Head Priest—Swami Heeranand shows a painting depicting the scene when 363 *Bishnois* sacrificed their lives to protect trees in 1730 AD (*Courtesy: Franck Vogel*)

communities for food, water and as shelter to wild animals and birds. Some of these groves are more than 500 years old [17]. They consider it their religious duty to save the Blackbuck and Chinkara or the Indian Gazelle. The Blackbuck is under Schedule I of the Indian Wildlife (Protection) Act, 2006. Their regard and love for the wildlife is not only restricted to stop animal killing in their own community but is courageously shown by their efforts in preventing the others too from doing the same. Consequently, their villages are swarming with antelopes and other animals.

The conservation and eco-friendly efforts of the *Bishnois* received international attention in 1998 when they chased a popular Indian film actor Salman Khan and others for gunning down two Blackbucks and three Chinkara. A legal case was registered against the actor and he was arrested due to the agitation led by the *Bishnoi* community. A large number of incidents are recorded in history when *Bishnoi* men and women have laid down their lives for the protection of flora and fauna from the outsiders.

An incident took place in *Vikam Samvat 1778* (1730 AD), when King of Jodhpur Abhay Singh decided to construct a new palace. As wood in that desert area was scarce, it was decided to cut down *khejri* trees sacred for the *Bishnois* growing in abundance in *Bishnoi* inhabited areas. Khejarli or Khejadi is a village in Jodhpur district of Rajasthan, India, 26 km south-east of the city of Jodhpur. The name of the town is derived from Khejri (*Prosopis cineraria*) trees, which were in abundance in the village. In anticipation of resentment from the community the king sent a large army for cutting down the trees. In this village 363 *Bishnois* (69 women and 294 men), led by Amrita Devi sacrificed their lives in 1730 AD while protecting green Khejri trees considered sacred by the community, by hugging them, this incident is the first event of "Chipko Movement" in the recorded history (Fig. 4.6). This compelled the king to take back his orders [18].

Discussion

Rajasthan is divided into various eco-cultural zones. The demographic profile of many tribes and communities has geographical reasons that are responsible for different ways of making a living. The subsistence economy based hunting communities even today hunt wild animals, many of them are now included in the category of endangered species. Due to deforestation and shrinking cultivable land areas and a change in the forest rights many of the communities are being forced to live a life of penury in the absence of alternative means of livelihood.

These communities enjoy privileges under the Constitution of India. The *Kathodi* and the *Sahariya* are included in the Scheduled Tribe list; the *Aheri, Madari* in the Scheduled Caste category; and *Gujjar, Bishnoi, Bawaria, Raika, Gairi* in the Other Backward Classes (OBC) category.

In Rajasthan, largely in arid and semiarid regions, animal husbandry has not been optimally practiced. Whereas, countries like Denmark and Sweden have high profit earning dairy industry. Promotion of animal husbandry can serve two purposes, building dairy industry on one hand, and improving the condition of pastoralists on the other hand by preventing their switching over to other professions and migration to cities in the need of wages. Some of the high milking breeds of cow as those maintained by the *Rathi* community at the local level should be encouraged. One of the major threats to human survival is the deterioration of environment through resource depletion. The *Bishnoi* community teaches us how each individual can play a significant role in the conservation of flora and fauna by imposing limitations at a personal level and how the quality of living can be raised by being content.

Clarification

The caste-based census was last conducted by the Government of India in the year 1931.

Therefore, recent census details of various castes in India are not available. However, the census details regarding communities included in the Scheduled Lists (that is Scheduled Tribe and Scheduled Caste) are being released regularly by the Directorate of Census. Therefore, population details of the scheduled communities included in the discussion have been referred from the Census of India 2001.

Please see Chaps. 1 and 3 for relevant pictures and more details about communities and tribes.

Acknowledgments The authors are thankful to Mr. Himanshu Vyas and *Hindustan Times* for permitting the use of their photograph. Wildlife photographer Mr. Franck Vogel deserves special thanks for his photographs. Thanks are also due to the anonymous referees for valuable suggestions.

References

1. Williams TR (1987) Cultural anthropology. Prentice Hall, Upper Saddle River, NJ
2. Singh KS (1993) Anthropological atlas, anthropological survey of India. Oxford University Press, Daryaganj, New Delhi
3. Singh KS (1998) People of India. Rajasthan, Vol. XXXVIII, 1, 2nd ed. Popular Prakashan, Mumbai, India
4. Census of India (2001) Rajasthan district census handbook. General Administration Department, Government of Rajasthan, Jaipur
5. Sharma SK (1998) Ethno-Zoology. Himanshu Publications, Udaipur, India, pp 53–66
6. Swarankar D (2002) Kathodi—A Social Description. Unpublished field report, Department of Anthropology: University of Rajasthan, Jaipur, India
7. Denzil I (1883) Punjab castes. Low Price Publications, Delhi, India (reprinted in 2008)
8. Russell RV, Lal H (1916) The tribes and castes of the central provinces of India. Cosmo Publications, Delhi, India (reprinted in 2003)
9. Soni RG, Sharma SK (2000) Wildlife offences by communities in eastern Rajasthan. Zoos Print 15(8):321–325
10. Hardayal Singh, Mardumsumari Raj Marwar (1891) Marwar census report. Shri Jagdishsingh Gehlot Shodh Sansthan: Jodhpur (reprinted in 1997)
11. Kohler-Rollefson I (2002) Best of both worlds http.www//rolexawards.com/en/the.laureates/ilsekohlerrollefson-the-project.jsp
12. Agrawal A, Saberwal VK. Raika pastoralists in Rajasthan. http://www.india.seminar.com/2006/564
13. Government of Rajasthan (2007) Report of the High Powered Committee on Gujjar's Demand, Jaipur.
14. Singhal PK (2008) Rajasthan. S.V. Publication, Jaipur, India
15. Sharma GN (1970) Rajasthan studies. Laxmi Narain Agrawal, Agra, India, pp 104–107
16. Rajasthan State Gazetteer (1995) Vol 1& 2. Directorate, District gazetteer of Jaipur: Government of Rajasthan
17. Bishnoi KR, Bishnoi NR (2000) Religion and environment, vol 1. Commonwealth Publishers, Delhi, India
18. Choudhary K (2002) Bishnoi sect—An anthropological study of the religious and environmental aspects. Unpublished field report, Department of Anthropology, University of Rajasthan: Jaipur

Chapter 5
Fossil Records of Rajasthan

U.B. Mathur

Abstract This chapter focuses on the fossil fauna of Rajasthan depicting the evolutionary history of almost 3,800 million years. Evidences of life during the Proterozoic eon have been found in abundance in the form of stromatolites, formed by blue-green algae some 1,500 million years ago. Fossil records of many younger geological periods known from Rajasthan as the sediments of almost whole of the Paleozoic, the era of great evolutionary significance, are not present. The sediments and fauna of the Permian, Triassic, Jurassic, and Cretaceous periods of the Mesozoic era had left some incomplete fossils. There is only a single record of dinosaurian remains of the Jurassic period from Rajasthan, although the same makes an interesting history in the adjoining states of Gujarat and Madhya Pradesh. The marine Jurassic fauna of Rajasthan contains a fair representation of the rich diversity of invertebrate life such as the ammonites, the relatives of squids, nautilus, and octopi of today. The most common vertebrates recorded from the Eocene deposits of Srikolayatji (Bikaner district) and Kapurdi (Barmer district) are the marine Cartilagenous Fish teeth of Elasmobranchs (Sharks, Skates and Rays). Rajasthan has three National Fossil Parks already in existence, namely, Stromatolite Fossil Parks at Jhamarkotra and Bhojunda in Udaipur and Chittourgarh districts, respectively, and the Akal Fossil Wood Park in Jaisalmer district. The former two needs to be further developed and brought to the limelight to attract geo-tourism.

Introduction

The Earth is ~4,600 million years old. Throughout the Archean eon, there was little to no free oxygen in the atmosphere (<1% of present level). Whatever little oxygen was produced by cyanobacteria got probably consumed in the weathering process. In the

U.B. Mathur (✉)
Formerly, Geological Survey of India, B-198 University Marg,
Jaipur 302015 (Rajasthan, India), India
e-mail: ubmathur@gmail.com

B.K. Sharma et al. (eds.), *Faunal Heritage of Rajasthan, India: General Background and Ecology of Vertebrates*, DOI 10.1007/978-1-4614-0800-0_5,

reducing atmosphere composed mainly of methane, carbon dioxide and hydrogen, the life that first appeared was anaerobic, which spent its life-cycle in an oxygen-free environment. It appeared ~3,800–3,100 million years ago in the form of microscopic objects, closely resembling microfossils, reported from the carbonaceous sediments of Canada, Transvaal (South Africa) and Western Australia. They had simple morphologies without any structural complexity. However, they had superficial resemblance with organic spheroidal bodies. The flora and fauna of Rajasthan in stratigraphy sequence and their importance in Indian stratigraphy is described below.

Archean (~3,800–2,500 Million Years Ago)

About 3,800 million years of geological history is recorded in the rocks of Rajasthan (Table 5.1)

The oldest rocks are found in the southwestern Rajasthan. They were formed between 3,800 and 2,500 million years ago (Archean eon). There is no record of life in the Archean of Rajasthan, while several geological formation elsewhere in India (Schists, and Iron Ore Formation in Karnataka, Orissa and Madhya Pradesh) contain alga-like spheroids, bacteria-like organisms, etc.

Proterozoic (~2,500–550 Million Years Ago)

The Proterozoic eon (subdivided into Paleoproterozoic, Mesoproterozoic, and Neoproterozoic era) represents a period before the first abundant complex life on Earth appeared. In Paleoproterozoic, the level of atmospheric oxygen was probably only 1–2% of its current level, which increased slowly up to 10%. Most of this was released by cyanobacteria, which increased in the fossils records about 2,300 million years ago. Present level of atmospheric O_2 was probably not achieved until ~400 million years ago.

The atmosphere changed from "reducing environment" in the Paleoproterozoic to "oxygenated" condition by the end of the Mesoproterozoic. Life existed in the form of bacteria and single-celled organisms.

In Rajasthan the Proterozoic sediments are grouped lithostratigraphically under Aravalli and Delhi Supergroups. They were deposited mostly in shallow seas. These marine sediments were uplifted into great mountain chains, perhaps higher than the Himalayas. Those mountains were subsequently eroded away and reduced to small hills as we see them today.

In Rajasthan, the evidences of life during the Proterozoic eon are found in abundance in the form of the organo-sedimentary structures formed by blue-green algae (cyanobacteria), known as stromatolites [1–4] (Fig. 5.1). The oldest stromatolites (>1,500 million years old) are found in rock phosphate bearing sediments of Aravalli Supergroup. Prolific development of different types of stromatolites (*Collenia columnaris*, *C. kusiensis*, *Baicalica prima*, *Minjaria calceolate*, etc.) can be seen at many places in carbonate deposits encircling Udaipur valley, particularly around Jhamarkotra, Maton,

Table 5.1 Geological time scale showing major global bioevents, Stratigraphic succession in Rajasthan and its major fossil elements

Geological time scale (m.y. = Million years)			Important global bioevents	Geological succession in Rajasthan	Faunal elements found in Rajasthan
Cenozoic (65 m.y to Present)	Quaternary (1.8 m.y. to Present)	Holocene (10,000 years ago to Present)	Repeated cooling (Ice Ages) and warming	Shumar Fm	–
		Pleistocene (1.8 m.y. to 10,000 years ago)	Large terrestrial mammals mammoths, mastodon, First modern Man	Nagour Gypsum beds	Freshwater ostracodes, gastropods, alga (*Chara*), and reworked marine foraminifers, reptilian remains
	Neogene (25–1.8 m.y.)	Pliocene (5.3-1.8 m.y.)	First australopithecines; Tool-making neanderthals	Mar Fm.	Angiospermous plant fossils –
		Miocene (25-5.3 m.y.)	Large sharks, whales; First hominids; Formation of Himalayas	–	
	Paleogene (65–25 m.y.)	Oligocene (38-25 m.y.)	First grasses and whales	–	–
		Eocene (54-38 m.y.)	Ancestral horses appear; First marine and large terrestrial mammals	Bandah Fm./Jogira Fm. Khuiala Fm.	Larger and smaller foramini-fers, ostracodes, echinoids, crabs, lobsters, fishes etc.
		Paleocene (65-54 m.y.)	Diversification and spreading of mammals; First true birds appear. Climate warm	Palana Fm./Akli Fm.Sanu Fm.	microflora and a few leaves

(continued)

Table 5.1 (continued)

Geological time scale (m.y. = Million years)		Important global bioevents	Geological succession in Rajasthan	Faunal elements found in Rajasthan
Mesozoic (245–65 m.y.)	Cretaceous (140–65 m.y.)	First flowering plants; First primates; Extinction of dinosaurs and ammonites at the end of the Cretaceous.	Intertrappean Fm.	Gastropod (*Physa*) & freshwater ostracodes
			Habur Fm.	ammonites, bivalves, gastropods.
		First primates.	Barmer Sandstone Fm.	Oncolites.
		Climate very warm	Pariwar Fm.	Upper Gondwana flora; Reptilian bones
	Jurassic (210–140 m.y.)	First birds; Conifers flourish; First belemnites; Desert widespread	Bhadasar Fm.	–
			Baisakhi Fm.	–
			Jaisalmer Fm.	Ammonites, brachiopods corals, bivalves, microfossils etc.
			Lathi Fm.	Fossil wood
	Triassic (245–210 m.y.)	First cycades, lizards, dinosaurs, Mammals; Ferns flourish	–	–
Paleozoic (540–245 m.y.)	Permian (290–245 m.y.)	First mammal-like reptile	Badhaura Fm.	Brachiopods, bryozoans crinoids, conulariids
	Carboniferous (365–290 m.y.)	Coal Age; First conifers, reptiles	Bap Boulder Beds	–
	Devonian (410–365 m.y.)	Age of fishes, First jawless fishes, amphibians, ammonites	–	–
	Silurian (440–410 m.y.)	First land plants, ferns, sharks, Boney fishes	–	–
	Ordovician (500–440 m.y.)	First land plants, First fishes, corals starfishes, sea urchins, bryozoans	–	–
	Cambrian (540–500 m.y.)	Proliferation of life in ocean; First invertebrates (Trilobites, foraminifera, brachiopods, nautiloids, clams, crustaceans, crinoids etc.); Small Shelly Fauna (Hard bodied) appear	–	Trace fossils of Trilobites

Precambrian (3,800–540 m.y.)		Neoproterozoic (1,000–540 m.y.)	Ediacaran (630–540 m.y.)	Abundant soft bodied multicellular organisms representing precursor of sea anemones, jelly fishes, arthropods, trilobites, brachiopods, mollusks, and bryozoans; Small Shelly Fauna (SSFs)	Bandah Fm. Birmania Fm. Marwar Supergroup: Nagour Gr., Bilara Gr., Jodhpur Gr. Malani Rhyolite	Trace fossils of trilobites, stromatolites, “MISS”
			Cryogenian (850–630 m.y.) Tonian (1,000–850 m.y.)	Extensive and intensive glaciations	Vindhyan Supergroup	Stromatolites
	Proterozoic (2,500–540 m.y.)	Mesoproterozoic (1,600–1,000 m.y.)		Transition to an oxygenated atmosphere	Delhi Supergroup	Trace fossils
		Paleoproterozoic (2,500–1,600 m.y.)		First eucaryotes (multicellular organisms) & magnetofossils appeared; Oxygen-free environment	Aravalli Supergroup	Stromatolites
Hadean (4,500–3,800 m.y.)	Archean (3,800–2,500 m.y.)			First single celled organisms (Bacteria and blue-green algae)	Bhilwara Supergroup	No fossil known
	Pre-Archean (4,500–3,800 m.y. ago)			Rocks that were formed before the first appearance of life	Rocks of this period not known	–

Fig. 5.1 Stromatolite from Aravalli Supergroup (Proterozoic) of Jhamarkotra, Udaipur district

Box 5.1 Sromatolites

Stromatolites are commonly thought to have been formed by the trapping, binding, and cementation of sedimentary grains by micro-organisms, especially cyanobacteria and single-celled algae. The oldest stromatolites are known from ~3,000 million years old Pre-Cambrian rocks. They declined sharply in both diversity and number during the Late Proterozoic, although they are present, but not so common in the following Paleozoic era. Today, stromatolites are quite uncommon in marine environments. They are found only at a few places (such as Shark Bay in Australia, the Bahamas, the Indian Ocean and Yellowstone National Park etc.), and are regarded as "living fossils."

Sisarama, Neemach Mata, Bargaon, Kanpur, Kharbaria-Ka-Gurha, and Dakankotra [1, 4]. A variety of stromatolites are displayed in the Stromatolite Park at the Mining Block-H of Jhamarkotra, which is 25 km southeast on Udaipur–Salumbar road linking Jhamerashwara temple (Box 5.1).

Stromatolites are of great use in correlation of stratigraphic rock formations of the Paleoproterozoic–Mesoproterozoic era, which contain microfossil of correlative value in Rajasthan. The stromatolites of Aravalli Supergroup are pre-Riphean (Paleoproterozoic) in age and its stromatolites are quite different from the Riphean forms found in other basins of India.

The younger Proterozoic rocks known in the Indian Stratigraphy as Delhi Supergroup, contain the sediments of northeastern Rajasthan, evidences of multicellular animal life in the form of trace fossils, fusiform and spindle-shaped structures, and bioturbation signature, such as trails and tracks of organisms [5–8]. These are found in the quartzites exposed near Galtaji, Kanak Vrindawan, Amer and Nahargarh Fort hill (Jaipur district), Behrod (Alwar district), and Bhojgarh village (Jhunjhunu district) [8]. These structures are possibly very primitive locomotion or crawling marks left out on unconsolidated sediments by soft bodied organisms, which lived on shallow marine shelf. At this time the oxygen level was still very low in the atmosphere as a result of which the animals were not able to live in deep burrows, their skeleton had no hard parts, and their bodies were thin so that oxygen could be absorbed through the surface of the organism.

Meso-Neoproterozoic (~1,600–630 Million Years Ago) of Cis-*Aravalli Region*

A younger group of rock formations, very well exposed in eastern Rajasthan and Madhya Pradesh, are known as Vindhyan Supergroup. A Meso-Neoproterozoic age (~1,750–650 million years) is generally assigned to it. The upper parts of its sequence are of great significance in Indian stratigraphy, as they belong to the Late Proterozoic period when certain significant life patterns had evolved. In the classical literature, this time span is called Vendian or Ediacaran. Since this constitutes the time span ranging from 650 million years up to the base of the Early Cambrian (550 million years), its relevance in the evolution of early biosphere is very significant. The fossils of Mesoproterozoic era that have been found in the upper part of Vindhyan Supergroup (Dholpur Shale Formation—the youngest Bhander Group) are *Chuaria–Tawuia* (the carbonaceous compressions of spherical, spheroidal, or sausage-shaped structures) [9, 10]. The limestone deposits of Bhagwanpura near Chittourgarh, and Bundi contain many form genera of stromatolites [11], particularly of *Baicalia* Group. Straiform, columnar, and nodular stromatolites, which can be seen in the exposures in Stromatolite Park near Bhojunda village, about 6 km southwest of Chittourgarh city on either side of the Chittourgarh–Udaipur State Highway, and adjacent to the Sheep Rearing Farm. A great variety of trace fossils have been reported from shallow water marine Vindhyan rocks exposed in Mewar region of Rajasthan (Kota, Chittourgarh, and Bundi districts) [12].

So far no fossils of Neoproterozoic have been found in eastern Rajasthan.

Meso-Neoproterozoic (~1,600–630 Million Years Ago) of Trans-*Aravalli Region*

On the western side of the Aravalli Ranges in Rajasthan, the stratigraphic equivalents of the Vindhyans were earlier known as "*Trans*-Aravalli Vindhyans." These rock formations have been renamed as the Marwar Supergroup [13]. The exposed rocks of this Supergroup are Meso-Neoproterozoic in age, and are the platform sediments deposited in an ancient sea referred to as Prototethys [14]. They contain fossils of algal stromatolites that are frequently observed in the limestone exposed near Bilara in Jodhpur district [15–17].

In the sandstones exposed near Sursagar in Jodhpur district are seen a variety of sedimentary structures like ripple marks, rill marks, etc., which appear to have been microbially fixed. However, it still needs to be confirmed if the structures in sandstone are really a product of active sediment fixation by microbial mats or not.

The above structures are known as "MISS" (Microbially Induced Sedimentary Structure). They occur in many parts of the world in the sandstone deposits of Archean Protorozoic age, and as such form one of the oldest evidences of life on the Earth. The modern analogue of "MISS" occurs in shallow water marine sand deposits.

There is a report of a "brachiopod" *Orthis* from the Marwar Supergroup. However, on closer examination it was found to be a nodule, and not a fossil [18, 19]. There is yet another report of some casts (filled with calcareous white powdery matter) of "bivalves" on the hill on which the temple of *Mataji* (Goddess) is located, south of Bogoria nera Bhupalgarh (Balru) in Jodhpur district [15]. This report is yet to be confirmed and documented.

The Small Shelly Fauna (SSF) is the name given to an obscure collection of small hard-shelled fossils found worldwide in beds a bit older than the earliest trilobites and archeocyathids. The first appearance of small shelly fossils marks the start of Cambrian period. The diminutive shell-like forms are a millimeter or two in size. They are in all probabilities fragments of unknown creatures.

Neoproterozoic (Ediacaran/Vendian Period) (~630–540 Million Years)

Multicellular animals (metazoans) evolved at the end of the Proterozoic between the time span ranging from about 630 million years to 540 million years, which is known as Ediacaran or Vendian. The soft bodied metazoans fossils (none had any hard parts) have been found in 30 localities all over the world, except Antarctica. Originally, most of the fossils were interpreted to represent types of Cnidarians (anemones and jelly fishes). Other forms are thought to represent precursor of arthropod, trilobite, brachiopods, molluscs, and even bryozoans. The Ediacaran biota exhibits a vast range of morphological characteristics and size ranging from a few millimeters to meters.

Ediacaran biota has been found in Krol Group of Nainital district, UP, Uttarakhand [20], Bhima Group of Karnataka [21], and Upper Vindhyan of central India [22]. In Rajasthan medusoidal Ediacaran fossil, viz., *Marsonia artiyansis*, has been reported from the Jodhpur Group in association with trace fossils, such as *Planolites* sp., *Skolithos* sp., and algal mats from brownish shales exposed around Artiya Khan and Dhoru villages near Jodhpur [23].

Small shelly fauna (SSF) appeared just before the beginning of the Cambrian period, about 540 million years ago. Skeletal material consisted of calcium carbonate (aragonite or calcite) and varieties of calcium phosphate. The nature of the animals that produced them cannot be assigned confidently to any living phylum. Many of these fossils are dissociated sclerites that give few clues as to the nature of the animals that bore them.

SSF was reported from the Vindhyan sediments exposed near Maihar, Madhya Pradesh and near Rohtasgarh, Bihar [24]. This claim was disputed by several geologists and paleontologists who neither accepted the biogenic nature of these reported fossils [25, 26], nor as SSF [27]. SSF has been recorded from Nainital [25], Kumaun [28], and Kashmir Himalayas [29].

MISS occur in many parts of the world in the sandstone deposits of Archean age and as such form one of the oldest evidence of life on the Earth [30–32]. The modern analogue of "MISS" occurs in shallow water marine sand deposits.

In western Rajasthan, the Sonia Sandstone Formation, at the base of the Jodhpur Group is thought to be 600 million years old Ediacaran [33], on the basis of radiometric dating of acid volcanic rocks [34–36]. The sandstone exposed near Sursagar in Jodhpur district contains a spectacular range of MISS [31].

Some of the concretions found in the sandstone near Jodhpur represent doubtful ediacaran life [19]. In the same geological horizon and locality megaplant fossils have been recorded [37]. The fossils show well-developed thallus, branching pattern, development of possible oogonia and zoospores, and antheridia. Showing morphological similarity with the extant Vaucheriacean plant, they are associated with well-developed microbial mats and circular structures with medusoidal affinity.

The Nagour Formation—the youngest of the rock formation of Jodhpur Supergroup have yielded excellently preserved trace fossils produced by trilobites. They have been identified as *Cruzana* sp., *Dimorphichnus* sp. (Figs. 5.2 and 5.3), *Rusophycus* sp., and *Aulichnites* sp. The presence of trilobites in the Nagour Sandstone gives the trace fossil-bearing horizon a Lower Cambrian age [38]. Thus, the new finding supports a Lower Cambrian age to the upper part of the Marwar Supergroup [39], a view refuted by later workers [40].

The sediments of Marwar Supergroup from Bikaner – Nagour Basin, Rajasthan have also yielded Acritarchs [41]—the complex multicellular organisms that appeared in many parts of the world in late Proterozoic (about 600 million years ago), but suddenly became abundant at the beginning of early Paleozoic (Cambrian Period—about 540 million years ago). They are interpreted as the fossilized remains of resting cysts of green algae or other complex-celled organisms.

Paleozoic

The beginning of the Paleozoic is marked by a sudden appearance of abundant hard-shelled fossils of organisms such as trilobites, brachiopods, crustaceans, mollusks, cnidarians, annelids, etc. This was the era of great evolutionary significance. Though, fossils of some of these groups are well represented in the Himalayan sediments, but there is no record of Cambrian, Ordovician, Silurian and Devonian rocks in Rajasthan. Only a part of Upper Carboniferous and Lower Permian sediments are present in Rajasthan. The transitory period between Carboniferous and Permian known as Permo-Carboniferous (~300 million years ago) was a period of great

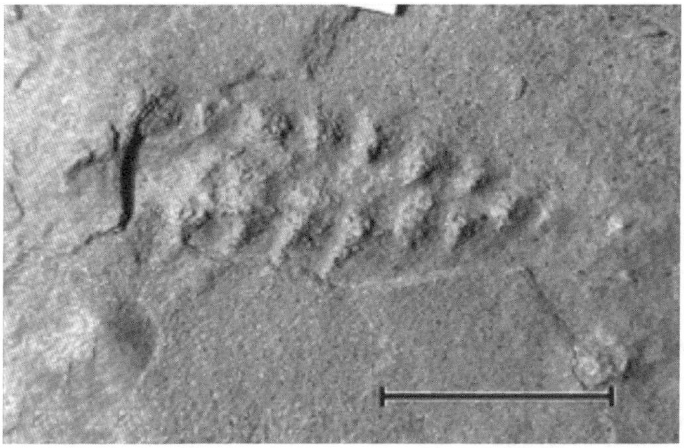

Fig. 5.2 Trace fossil of Trilobite (*Cruziana* sp.) from Nagour Sandstone, Nagour Group, Marwar Supergroup; Scale = 1 cm

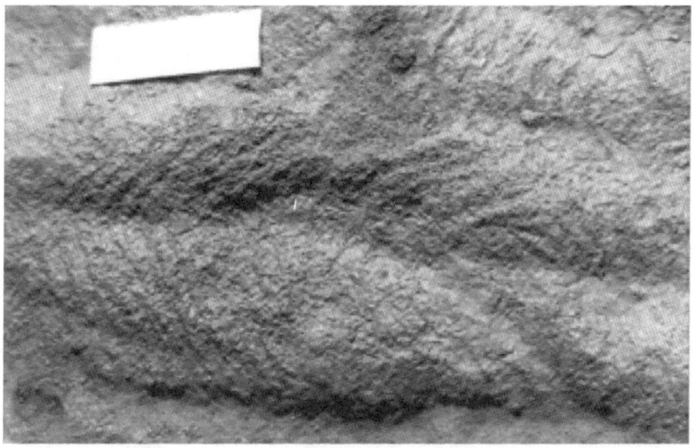

Fig. 5.3 Trace fossil of Trilobite (*Dimorphichnus* sp.) from Nagour Sandstone, Nagour Group, Marwar Supergroup; Scale = 1 cm

glaciation. At that time, a large part of the Indian Peninsular India was covered by an extensive mantle of ice. In the glaciated plains of western Rajasthan, boulder beds were deposited [42]. The boulder beds near Bap in Jodhpur district are the remnants of deposits laid down by glaciers of late Carboniferous period.

Permian Period (~300–250 Million Years Ago)

In the Permian period, the atmospheric oxygen reached its present levels, generating the ozone shield that screened out ultraviolet rays, and allowed complex life to live in the shallow seas, and finally on land. At that time, India was a part of Gondwanaland (India, Australia, South America, Africa and Madagascar), and was close to the South Pole.

Fig. 5.4 Brachiopod (*Spirifer*) from Badhaura, Jodhpur district

As the climatic conditions changed in the Permian, from glacial to warmer, swamps and forest developed in the valleys and depressions left by the Permo-Carboniferous glaciers of a large part of the Peninsular India. Testimony of such climatic conditions during later part of Permian is borne by the coal deposits with plant fossils of Permian age found in Madhya Pradesh, Bengal, Bihar, Jharkhand and Andhra Pradesh. At that time, a variety of marine animals lived in the warm waters of western Rajasthan Sea. They included brachiopods (Fig. 5.4), bivalves, gastropods, bryozoans, corals, crinoids, etc. [43–50]. The brachiopods from Badhaura Formation include *Spirifer, Notospirifer, Cyrtella, Aulosteges, Derbyia, Tibetospirifer, Semilingula, Crassispirifer, Brachythyrinella, Permasyrinx, Trigonotreta, Sulcicosta, Betaneospirifer* and *Elasmata retusus;* the gastropods include *Peruvispira*, *Bellerophon*, *Stachella*, and *Mourlonia*; some of the bivalves identified are *Nuculopsis, Phestia, Megadesmus, Merismopteria, Chaenomya, Aviculopecten*, and *Paleocorbula* (Fig. 5.4).

The species found in Rajasthan are remarkably akin to the Permian marine fauna of Salt Range (Pakistan), and Umaria -Manendragarh districts of Madhya Pradesh. This provides a proof that Rajasthan Sea of Permian period was connected with that of Salt Range of Pakistan in the west and Madhya Pradesh in the east. However, some of the workers are of the opinion that age of Bap-Badaura Formations is middle Asselian, and older than all other known Permian fauna found in Peninsular India [47].

Most of the Permian rocks have been either eroded away or probably destroyed during construction of National Highway No.15 and a branch canal of Indira Gandhi Nahar Project (IGNP) in Rajasthan. Their remnants are, however, found today near Bhimji ka Gaon, north-west of Badhaura, and between the west of Badhaura and the northeast of Bari Sird, in Jodhpur district. The area is bounded by latitudes 27° 16′–27° 38′ and longitudes 72° 06′–72° 40′ and falls in the Survey of India topographic sheets 45 A/3, 7, and 10 [51].

Mesozoic Era (~250–65 Million Years Ago)

The highly fossiliferous sediments of the era are exposed both in the Peninsular India as well in the Himalayas. The former were deposited in rivers and lakes, and are known for a rich flora, known as Gondwana Flora. The fossils of invertebrate marine fauna deposited in the Mesozoic sea (known as Tethys Sea), are found of the uplifted sediments in the Himalayas from Kashmir to Nepal and beyond. This biota is of great evolutionary significance. However, in comparison to the Himalayan fauna, Rajasthan Mesozoic fauna is too meager.

Triassic Period (~250–200 Million Years Ago)

There are no Triassic (~250–200 million years) deposits exposed in Rajasthan.

Jurassic Period (~200–145 Million Years Ago)

The Jurassic period—the second of the three divisions that make up the Mesozoic era, saw warm tropical greenhouse conditions worldwide, shallow continental seas, the breakup of Pangaea, cosmopolitan flora and fauna, and the triumph of the majestic dinosaurs and the great sea reptiles. During that time, there were no polar ice sheets, and no equatorial rainforests. The warm environmental conditions that had started in Permian persisted for about 200 million years, to at least the late Mesozoic. The effect of this warming was pronounced on the fauna and flora of the Jurassic as well the Cretaceous period.

Land plants abounded in the Jurassic, but gymnosperms were different from what we see today. Ferns, ginkgoes, bennettitaleans, and true cycads flourished in the Jurassic. This period is called "the Age of Cycads" as cycads were most abundant and diverse at that time. Conifers continued to be the most diverse large trees, including close relatives of living redwoods, cypresses, and pines.

The Gondwana rocks of the Peninsular India contain fossils of all the above plant groups in abundance. However, in Rajasthan rare leaf impressions of only a few species are preserved. Abundant remains of silicified tree trunks (Fig. 5.5) that are present in the Lathi Sandstone Formation of Jaisalmer suggest that this region also had forests but conditions for preservation of plant vegetation were not favorable [50].

Besides fossil wood, a rich palynological assemblage is known from the Lathi Formation of Jaisalmer. This assemblage comprises fern spores and abundant gymnosperms [51]. There were no angiosperms as they were yet to appear on the surface of the Earth.

Dinosaurs dominated the land fauna during Jurassic. Their fossils are found in the Rajasthan's neighboring Kutch district of Gujarat [52]. However, there is a single record of doubtful dinosaurian remains from Jurassic of Rajasthan (Fig. 5.6) [53]. The presence of large quantity of wood fossil in the Jurassic of Rajasthan is suggestive that a thick forest existed in the area at that time. It is very likely that sauropod dinosaurs had made the area their abode, but conditions conducive of their preservation did not exist.

The Jurassic marine rocks are best developed near Jaisalmer town, Hamira, Joyan, Jaisalmer Fort hill, Bada Bagh, Kuldhar, and Ludharwa. The last two sites are extremely rich in cephalopods—the relatives of the squids, nautili, and octopi of

Fig. 5.5 Gymnospermous wood fossil, Akal Fossil Park, Jaisalmer district

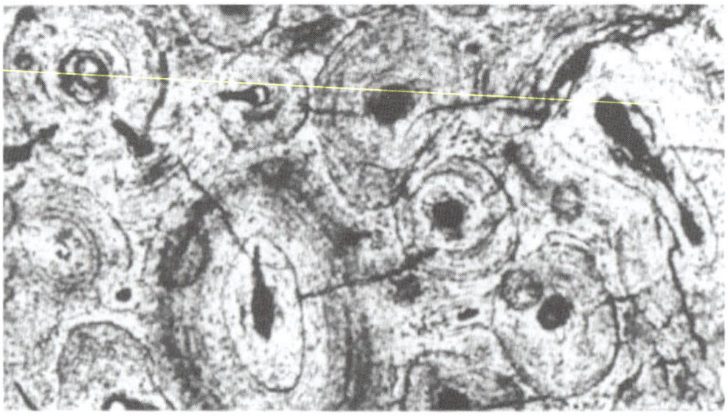

Fig. 5.6 Photomicrograph of the cortex of a long bone of a dinosaurian fossil, Jaisalmer

today. These include ammonoids (Fig. 5.7), and belemnoids (Fig. 5.8). The marine Jurassic fauna of Rajasthan contain a fair representative of the rich diversity of invertebrate life. Ammonoids dominated, followed by brachiopods, bivalves (Fig. 5.9), microforaminifers, ostracodes, and nanoplanktons [54–63]. Brachiopods are represented by the family Terebratulidae [64] and Rhynchonellidae (Fig. 5.10) that had survived the mass extinction of brachiopod at the end of Permian.

The Jurassic was a golden age for marine reptiles Plesiosaurs, and the fish-like Ichthyosaurs. They shared the oceans with large marine crocodiles, and with modern-looking sharks and rays. First bird appeared in Jurassic and so did a number of early mammals. For years, the mammals living in the Jurassic and the Cretaceous periods have been thought of as tiny shrew-like creatures scurrying through the underbrush. Now the discovery of a beaver-like mammal from the Jurassic of China has demolished that image [65]. There is no record of fossils of these animals so far from Rajasthan Jurassic sediments.

Cretaceous (~145–65 Million Years Ago)

The Cretaceous was the warmest period in the geological history. The sea levels were high, and the greenhouse conditions of the Cretaceous saw a diversification of the planktonic foraminifers. The major extinctions at the end of the Cretaceous included many planktonic foraminifers.

Box 5.2 Lathi Formation

Fossil wood of Lathi Formation has been preserved for the posterity at a site 17 km from the main Jaisalmer city near Barmer Road at Akal. The fossil trunks lie scattered in Akal National Fossil Wood Park. Fossilized tree trunks are of various sizes with the largest being 13 m in length and 1.5 m in width. Covering about 10 km^2 of bare hillside, the Fossil Park contains 25 petrified trunks, in all.

Fig. 5.7 Ammonite from Kuldhar, Jaisalmer district

Fig. 5.8 "Guards" of *Belemnites* from Kuldhar, Jaisalmer district

Life during this period was in a way not much different from the preceding Jurassic period. There was no burst of diversity in life; dinosaurs both great and small continued to roam the forests of ferns, cycads, and conifers. The marine life dominated by ammonoids, belemnoids, other molluscs, fishes, etc., were hunted by pterosaurs and birds who flapped and soared in the air above.

Cretaceous saw the first appearance of many life forms that evolved quickly in the coming Cenozoic period. Perhaps the most important was the first appearance of the flowering plants—the angiosperms. By the end of the Cretaceous, a number of angiosperms had evolved. Another event that took place was the mass extinction of

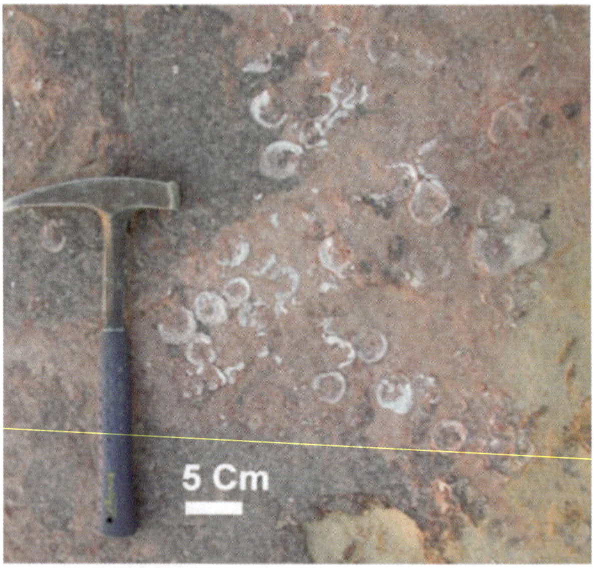

Fig. 5.9 Fossiliferous limestone from Jaisalmer

Fig. 5.10 Brachiopods (*Rhynchonella)* from Jaisalmer district

dinosaurs, the marine reptiles such as the ichthyosaurs and plesiosaurs, the flying pterosaurs, and the ammonoids. But many groups of organisms, such as flowering plants, gastropods and pelecypods (snails and clams), amphibians, lizards and snakes, crocodilians, and mammals continued to survive. During a part of the Cretaceous Period, sea continued to occupy a part of western Rajasthan.

In western Rajasthan, the marine Cretaceous fossiliferous limestone (Fig. 5.11) is exposed near Habur and Kuchri. Here a rich ammonoid fauna of epinereitic to infranereitic environment is known. The ammonoid fauna is highly cosmopolitan unlike the Jurassic fauna that is endemic [66].

The marine Cretaceous fauna comprising ammonoids, echinoids, bivalves, bryozoans, and foraminifers is very well known from parts of MP, Gujarat, Tamil Nadu, Khasi and Jaintia hills of Meghalaya.

Fig. 5.11 Fossiliferous limestone from Habur, Jaisalmer district

The non-marine Cretaceous sediments of Barmer basin contain oncolites. The source of these carbonate oncolites has been interpreted as biogenic under lacustrine condition [67].

In the Jaisalmer district, the freshwater Cretaceous deposits are known to contain the typical upper Gondwana fossil flora, together with fossil tree trunks and fragmentary bones of reptiles [68].

As the Cretaceous period was coming to its end, there was stupendous outburst of volcanic activity with short periods of quiescence. When volcanic activity stopped for some time, lakes developed in the depressions in volcanic rocks (known as Deccan Traps) in which sedimentation took place, and freshwater animals thrived. Today, the lake sediments of that time are found sandwiched between two succession volcanic flows (Deccan traps). They are known as "Intertrappeans." The Deccan Traps are found over a vast area in southern Jhalawar and in the eastern parts of Chittourgarh and Banswara districts, while the Intertrappean rocks are found in a restricted area in southern and south-eastern Rajasthan. The frequently found Intertrappean fossils are right-handed gastropod genus *Physa* (Fig. 5.12), micro-gastropods and ostracodes [69, 70].

Cenozoic Era

Paleocene (65–54 Million Years Ago)–Eocene (54–38 Million Years Ago)

The end of the Mesozoic was marked by one of the most significant periods of global climate change. A sudden change in oceanic and atmospheric circulation, led to the extinction of numerous deep-sea benthic foraminifers. A variety of larger foraminifera which included nummulids (Fig. 5.13a), discocylinids, alveolinids (Fig. 5.13b), etc., appeared in the early Eocene and quickly diversified.

The Eocene saw the appearance of a number of direct evolutionary ancestors of modern animals like hoofed mammals—perrisodactyls and artiodactyls, which

Fig. 5.12 A gastropod fossil (*Physa*) from Jhalawar district

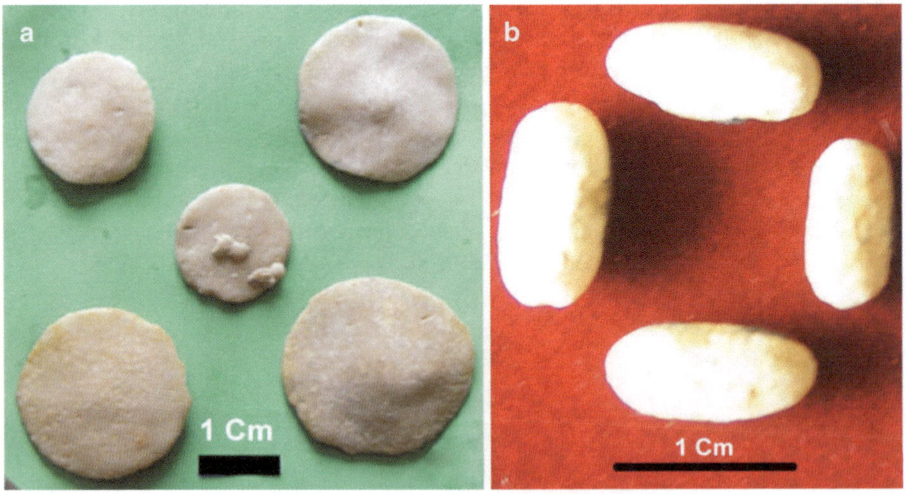

Fig. 5.13 (**a, b**) Larger foraminifers (*Nummulites* and *Alveolina*) from Bikaner and Jaisalmer districts, respectively

include proto-horses, tapirs, rhinoceroses, and camels. Many of the mammals quickly grew from small to huge size, before suddenly dying out in a mass-extinction at the end of the Eocene. The rodents replaced the multituberculates. Primates including forest-dwelling ancestors of today's lemurs and tapirs flourished. The Eocene saw the appearance of modern birds, as well as the flightless birds. Giant flightless birds

Fig. 5.14 Bony fish fossil from Kapurdi, Barmer district

are thought to be the descendants of the great theropod dinosaurs. The first aquatic mammals, whale and sea cow appeared in the oceans during Eocene.

None of the above vertebrates, other than fishes, are preserved in Paleocene–Eocene rocks of Rajasthan.

The most common vertebrates recorded from the Eocene deposits of Srikolayatji (Bikaner district) and Kapurdi (Barmer district) are the marine cartilagenous fish teeth of Elasmobranchs (sharks, skates, rays) [86]. Interestingly, sharks have few hard parts that can be preserved but they are well-represented in fossils records by ~2,500 species the world over. Their ancestry goes back more years than 200 million before the earliest known dinosaur. The skeletal remains of the bony fishes (Fig. 5.14) of families Clupeidae, Euzaphlegidae, Bothidae, Luvaridae, and Carangidae are also common in Eocene marine sediments of Rajasthan. The bony fishes Teleosts and Holostean that arose in Triassic were the most successful fishes as they are the largest class of vertebrates in existence today.

The Eocene sediments are exposed in three sedimentary basins in western Rajasthan viz., Bikaner (Palana, Srikolayatji), Jaisalmer (Khuiala–Bandah), and Barmer (Kapurdi and Nagurda area) areas. The exposed sedimentary sequences of the former two basins are largely shallow marine, while that of the Barmer basin are coastal to continental deposits.

Among marine invertebrates known from Rajasthan are foraminiferans (Fig. 5.13a, b) ostracodes, bivalves, gastropods (Fig. 5.15), echinoids, crabs (Fig. 5.16), shrimps (Fig. 5.17), and annelids (Fig. 5.18). At places foraminifera constitute bulk of the rock, and the fossiliferous rock is popularly known among the geological community as "foraminiferal limestone" or "Nummulitic limestone." Many of the larger foraminiferal species have restricted age and wide geographical distribution, and as such serve the purpose of age fixation and correlation. In the Jaisalmer and Bikaner fossil localities, the shallow water marine sediments are known for prolific development of microfauna, consisting predominantly of foraminiferans and ostracodes [71–86]. They are best exposed in area in northwestern extremity of the Jaisalmer district. From the Eocene of Jaisalmer, bivalves and gastropods are well known. They have also yielded nannoplanktons—the microscopic calcite skeletal elements produced largely by one-celled golden-brown algae [85].

The lignite deposits of Giral near Barmer, and Palana near Bikaner contain a rich and diversified palynoflora of fungal remains, pteridophytic spores, and angiospermic pollens, dinoflagellate cysts are also present. The palynological assemblage is dominated by angiospermic pollens of the families: Arecaceae, Liliaceae, Oleaceae,

Fig. 5.15 Bivalves and gastropod from Kapurdi, Bikaner district

Fig. 5.16 Fossil crabs and shrimp (Decapod Crustacean) from Karpurdi, Barmer district

Caeselpiniaceae, Rubiaceae, Meliaceae, and Myricaceae. The abundance of the palynotaxa having affinity with *Proxaperitites, Spinizonocolpites, Palmidites, Palmaepollenites, Kapurdipollenites,* and *Retiverrumonosulcites* indicate that the deposition of the sediments took place under coastal environment. The palynotaxa and the abundant fungal spores are also suggestive of tropical to subtropical conditions of deposition under warm and humid conditions [87–93].

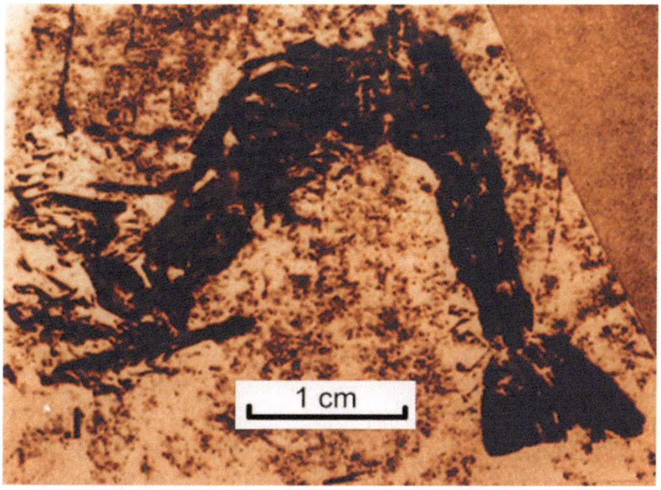

Fig. 5.17 Fossil crabs and shrimp (Decapod Crustacean) from Karpurdi, Barmer district

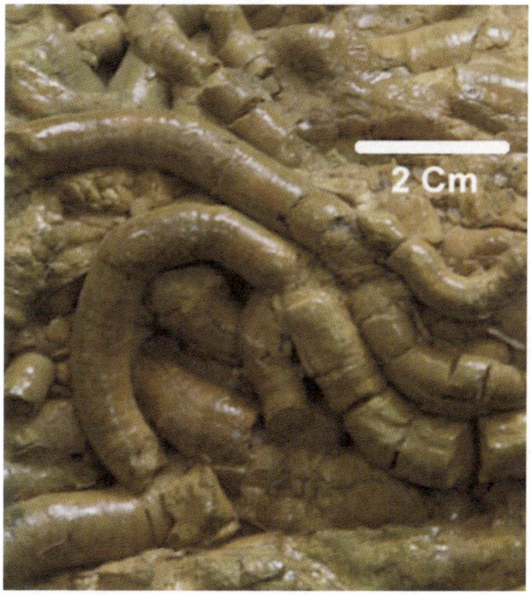

Fig. 5.18 Serpulid tubes (Annelida) from Sanu, Jaisalmer district

Among the floral elements, the fossil leaf impressions of two genera *Mesua* and *Garcinia* (Family Guttiferae), and fossil wood (two families of gymnosperms and three families of angiosperms) are known from the Middle Eocene of Kapurdi area, Barmer district [94–96].

An assemblage of lower vertebrates comprising fish, crocodilians, and snakes is recorded from subsurface beds of the Akli Formation of Giral lignite mine about 40 km NNW of Barmer [97].

Oligocene (~38–24 Million Years Ago)

Oligocene sediments are no where exposed in Rajasthan.

Miocene–Pliocene (~24–1.8 Million Years Ago)

Miocene (~23–5.3 million years ago) and Pliocene (~5.3–1.8 million years ago) are the two divisions of the Neogene. During these periods new forms of mammals and bird appeared whose close relatives are alive today. During the Pliocene marine fauna were essentially modern, although continental fauna were a bit more primitive than that of today. Primates continued to evolve. The first bipedal ancestors of humans are known to have evolved in Pliocene. Mio-Pliocene mammals are known from the sediments of Siwalik Hills, Piram Island (Bhavnagar district, Gujarat), and alluvium of Narmada valley.

Fossils records of Miocene–Pliocene sediments from Rajasthan are very poor. No vertebrate fossil of Miocene–Pliocene period has been found in Rajasthan. Only a few angiosperms are known from the freshwater china-clay deposits of Inda ka Bala area (4 km northeast of Shrikolayatji) in Barmer district. They belong to the family Clusiaceae, Sterculiaceae, Rutaceae, Rhamnaceae, Tiliaceae, Leguminosae (Fig. 5.19), Ebenaceae, Lauraceae, and Moraceae. The floral assemblage is dominated by the families Sterculiaceae and Moraceae. The association of Rutaceae, Rhamnaceae, and Tiliaceae in the assemblage is suggestive of moist to dry tropical semi-evergreen deciduous forest conditions in the area during late Neogene times [98]. There are quite a few petrified woods recorded from the Tertiary rocks of Jaislamer and Bikaner of Rajasthan belonging to the families Lecythidaceae, Combretaceae, Lythraceae, Anacardiaceae, Sonneratiaceae, Meliaceae, and Araucariaceae [99–102]. They are of not much significance as they cannot be used with confidence for paleoclimatic interpretation or age fixation. Their importance lies in the fact that they suggest the presence of forest in the area in the past.

Pleistocene–Holocene (~1.8 Million Years Ago to Present Day)

The beginning of the Quaternary era (subdivided into Pleistocene and Holocene) was marked by the onset of extensive cooling all over the Earth. The climate fluctuated repeatedly between glacial and warm (also called "Interglacial"). The last glacial period started ~25,000 years ago, reached its peak ~18,000 years ago, and ended 10,000 years ago. The period since then is known as Holocene. At present, we are in an Interglacial period when the climate is warming up as a part of natural process added by anthropogenic activities.

The Pleistocene was the age of primates, when the hominid primates evolved into *Homo sapiens*. Fossils of hominid are found in the Siwalik rocks of Himalaya,

Fig. 5.19 An angiosperm leaf from Srikolayatji, Bikaner district

but none of them is known from Rajasthan. Early man had set his foot on the Indian soil about 1.27 ± 0.17 million years ago as is evident from the records of Lower Paleolithic Acheulian tools found in Isampur, Gulbarga district, Karnataka [103]. However, the only fossil of early man (*Homo erectus*) is known from Narmada valley Alluvium of Hoshangabad district (Madhya Pradesh) [104].

The antiquity of the Thar Desert of Rajasthan is not yet established but it goes back to at least 200,000 years before present. The pollen studies of the Sambhar, Lunkaransar, and Didwana lake sediments have helped in establishing the following climatic history of the region [105–107].

10 thousand years before present—Cold-dry
10–4.5 thousand years before present—Warm humid with frequent floods
4.5–3.5 thousand years before present—Warm humid with less extremes of climate
3.5 thousand years before present to Present—Warm-dry with frequent droughts and seasonal extremes

The pollen analysis sediments from the fresh-water Punlota Lake situated at the outskirt of Degana township in district Nagour (eastern Rajasthan) has revealed that the region had predominantly non-arboreal vegetation in the late Holocene [107]. Around 4,200 years before present, savannah type-vegetation covered the region. Increases in the frequency of ferns as well as fungal spores suggest warm and moist conditions during this period. Around 3,000 years before present, savannah vegetation witnessed reduction in tree taxa, ferns, and grasses with simultaneous increase in Cyperaceae, Chenopodiaceae, etc., depicting decrease in the warm and moist conditions. The pollen grains of the families Chenopodiaceae, Amaranthaceae and Artemisia suggest that during the Little Ice Age (AD 1550–1800), the lakes of western Rajasthan were hypersaline as a result of low precipitation due to weak summer monsoon [105].

The dune sands contain at places a significant quantity of benthic and planktonic foraminifers up to 800 km inland from the Great Rann in Gujarat [108, 109]. Eighty two species of marine foraminifers in the surface sands are derived from the Arabian Sea coast by wind action.

The gypsum deposits of Jamsar (Nagour district) are one of the most important evaporates that have been studied for microfossils [110]. They contain freshwater ostracodes [111], gastropod, Chara fruits, and wind derived marine foraminiferans. On the basis of the fauna in the Jamsar gypsum deposits it is concluded that the beds were deposited in a brackish water environment with salinity ranging from 3.0 to 20‰.

Conclusion

In Rajasthan, the evidences of life during the oldest period of Proterozoic eon, viz., Paleoproterozoic, are found in abundance in the form of the organo-sedimentary structures (known as stromatolites) that were formed by blue-green algae ~1,500 million years ago. Fossil records of Mesoproterozoic consist of only trace fossils of uncertain affinity. The next younger geological period, viz., Neoproterozoic, has yielded stromatolites, MISS, and trace fossils of trilobite origin. The last named fossils are of great interest as they suggest deposition just before the beginning of Cambrian period in Rajasthan.

The sediments of greater part of the Paleozoic era are not known from Rajasthan. The fauna of Permian Period, which saw drastic climatic and faunal changes, is represented by a number of marine forms. There is no record of Triassic period, and the geological sequence of the following Jurassic and Cretaceous periods of the Mesozoic era are well known. The ammonoids, the relatives of the squids, nautilids, and octopi of today, were the most dominant form of life in the seas of the Jurassic and Cretaceous. They are of great correlative value. The fossils of dinosaurs and other reptilian remains that make an interesting history in the adjoining states of Gujarat and Madhya Pradesh are poorly represented in western Rajasthan, in spite of a rich forest cover in dinosaur era. The fossiliferous marine sediments of Eocene period contain a variety of index fossils of foraminifers, and other invertebrates. The marine Oligocene and Miocene sediments, that are natural oil and gas bearing in Gujarat and Pakistan, are not exposed in Rajasthan. The poorly known fossils of fauna and flora of the younger Pliocene–Pleistocene–Holocene sediments are of little importance.

Rajasthan has many geological heritage sites with a rich fauna and flora. Three National Fossil Parks have been established, which include Stromatolite Fossil Parks at Jhamarkotra and Bhojunda in Udaipur and Chittourgarh districts respectively; and the Akal Fossil Wood Park in Jaisalmer district (www.portal.gsi.gov.in). The last one is attracting domestic as well as international tourists visiting Jaisalmer, and the former two needs to be further developed, and advertised to attract geo-tourism. A few more sites are worthy of consideration for preservation as geological heritage sites after more work is done, particularly near Kanak Vrindawan–Amer Fort (Jaipur district) for trace fossils in Proterozoic rocks; near Sursagar (Jodhpur district) for MISS; near Kuldhar and Ramgarh (Jaisalmer district) for Jurassic invertebrates and larger foraminifers, respectively. Please see Chaps. 1–3 for relevant details and pictures.

Acknowledgments The author is thankful to Dr. D.K. Pandey, Professor, Department of Geology, University of Rajasthan, Jaipur for fruitful discussions, Dr. Tej Bahadur for designing the illustrations for the paper, and their help in various ways. Thanks are also due to anonymous scrutinizer, who carefully edited the manuscript and improved the contents.

References

1. Barman G, Verma KK, Puri SN (1978) Biostratigraphic zonation of stromatolite bearing horizons of Udaipur district, Rajasthan. J Geol Soc Ind 19(6):264–267
2. Barman G (1973) Algal stromatolites from the Aravallis of Udaipur district, Rajasthan-aid to geological mapping. Abst Semin Recent Adv Geol Rajasthan, Gujarat Sect. 13:73
3. Chauhan DN (1989) Microbial activities and genesis of Aravalli Phosphorite, Udaipur, Rajasthan. Wadia Inst Himalayan Geol Publ 13:39–52
4. Banerjee DM (1971) Aravallian stromatolites from Udaipur, Rajasthan. J Geol Soc Ind 12(4):349–355
5. Bose U (1977) Fusiform structures in Sullavai Sandstone- biogenic or abiogenic? J Geol Soc Ind 18:60–63
6. Singh SP, Bose U (1985) The occurrence of trace fossils in the Delhi Supergroup of north-eastern Rajasthan. J Geol Soc Ind 26:422–425
7. Kaila P, Bhagwat RJ, Pandey PK, Trivedi V (1992) Probable fossils from Alwar Quartzites, Aravalli Range, North India. Curr Sci 62(5):427–430
8. Negi R, Ravindra R (1981) On the occurrence of stromatolites in Kushalgarh formation of Delhi supergroup from Barand, Alwar Dist., Rajasthan. Geol Surv Ind Misc Publ 44:90–95
9. Maithy PK (2003) Pre-phanerozoic evidences of life from central India: implication to biostratigraphy and evolution. Gondwana Geol Mag Spec Vol. 7:401–412
10. Sharma M, Mishra S, Dutta S, Banerjee S, Shukla Y (2009) On the affinity of *Chuaria–Tawuia* complex: a multidisciplinary study. Precambrian Res 173(1–4):123–136
11. Raja Rao CS, Mahajan VD (1965) Notes on the stromatolites and probable correlation of the Bhagwanpura Limestone, Chittorgarh district, Rajasthan. Curr Sci 34:82–83
12. Rastogi KK, Srivastava DK (1992) Search for evidences of primitive life in Vindhyan sediments. Rec Geol Surv Ind 124(7):106–109
13. Pareek HS (1981) Basin configuration and sedimentary stratigraphy of western Rajasthan. J Geol Soc Ind 22:517–527
14. Shanker R, Maithy PK, Singh G, Kumar G (2000) Palaeogeographic evolution and biotic changes of the Indian subcontinent from the late proterozoic to early palaeozoic. Proceedings of international seminar precambrian crust in eastern and central India. IGCP Project 368, Bhubaneswar. Geol Surv Ind Spec Publ 57:34–48

15. Barman G (1980) An analysis of the Marwar Basin in the light of stromatolite study. Proc workshop on stromatolites: characteristics and utility. Geol Surv Ind Misc Publ 44:292–297
16. Barman G (1987) Stratigraphical position of the Marwar supergroup in the light of stromatolite study. Geol Surv Ind Spec Publ 11(1):72–80
17. Verma KK, Barman G (1980) Lower phanerozoic stromatolites of Rajasthan, western India. In: Monty CLV (ed) Stromatolites. Springer, Germany, pp 19–24
18. Khan EA (1973) Discovery of fossil brachiopod from Trans-Aravalli sequence near Jodhpur, Rajasthan. Semin Recent Adv Geol Rajasthan and Gujarat, Jaipur, Abst Sec 13:74
19. Kumar G, Shankar R, Maithy PK, Mathur VK, Bhattacharya SK, Jani RA (1997) Terminal proterozoic cambrian sequences in India: a review with special reference to precambrian-cambrian boundary. Palaeobotanist 46(1, 2):19–31
20. Shankar R, Mathur VK (1992) Precambrian-cambrian sequence in Krol belt and additional ediacaran fossils. Geophytology 22:27–39
21. Raha PK, Moitra AK, Das Sarma DC, Ashok Kumar P, Rama Rao M (1991) Search for microfossils in the Bhima Kaladgi-Badami sequence of south India. Rec Geol Surv Ind 124(2):10
22. De C (2006) Ediacara fossil assemblage in the upper Vindhyans of central India and its significance. J Asian Earth Sci 27(5):660–683
23. Raghav KS, De C, Jain RL (2005) The first record of Vendian medusoids and trace fossil-bearing algal metagrounds from the basal part of the Marwar supergroup of Rajasthan, India. Indian Miner 59(1–2):23–30
24. Azmi RJ (1998) Discovery of lower cambrian small shelly fossils and brachiopods from the lower vindhyan of Son Valley central India. J Geol Soc Ind 52:381–389
25. Bhatt DK, Mathur AK (1990) Small shelly fossils of precambrian boundary beds from the Krol-Tal succession in the Nainital syncline, lesser Himalaya. Curr Sci 59(4):218–221
26. Bhatt DK (2003) On the report of the small shelly fossils and brachiopoda from the Vindhyan strata. J Pal Soc Ind 48:225–229
27. Bendtson S, Belivanova V, Rasmussen B, Whitehouse M (2009) The controversial "cambrian" fossils of the Vindhyan are real but more than a billion years older. Proc Natl Acad Sci 106(19):7729–7734
28. Azmi RJ, Paul SK (2004) Discovery of precambrian–cambrian boundary protoconodonts from the Gangolihat Dolomite of inner Kumaun lesser Himalaya: implication on age and correlation. Curr Sci 86(12):1653–1660
29. Meera T (1989) Discovery of pre-trilobite small shelly fossils and the position of precambrian-cambrian boundary in Tethyan sequence of northwestern Kashmir. Curr Sci 58(15):839–843
30. Noffke N, Beukes N, Gutzmer J, Hazen R (2006) Spatial and temporal distribution of microbially induced sedimentary structures: a case study from siliciclastic storm deposits of 2.9 Ga Witwatersrand supergroup, south Africa. Precambrian Res 146:35–44
31. Sarkar S, Bose PK, Samanta P, Sengupta P, Eriksson PG (2008) Microbial mat mediated structures in the ediacaran Sonia sandstone, Rajasthan, India, and their implications for proterozoic sedimentation. Precambrian Res 162(1–2):248–263
32. Rasmussen B, Bose PK, Sarkar S, Banerjee S, Fletcher IR, McNaughton NJ (2002) 1.6 Ga age for the Chorhat sandstone: implications for the early evolution of animals. Geology 30:103–106
33. Knoll AH, Walter MR, Narbonne G, Christie-Blick N (2004) A new period for the geologic time-scale. Science 305:621–622
34. Rathore SS, Venkatesan TR, Srivastava RK (1996) Rb–Sr and Ar– Ar systematics of Malani volcanic rocks of southwest Rajasthan: Evidence for a younger post-crystallization thermal event. Proc Ind Acad Sci (Earth Planet Sci) 105:131–141
35. Rathore SS, Venkatesan TR, Srivastava RK (1998) Rb-Sr isotope dating of neoproterozoic (Malani group) magmatism from southwest Rajasthan, India: evidence of younger Pan-African thermal event by 40Ar39- Studies. Gond Res 2(2):271–281
36. Paliwal BS (1998) Felsic volcanics interlayered with sediments of the Marwar super group at Chhoti Khatu, District Nagour, Rajasthan. J Geol Soc Ind 52:81–86
37. Kumar S, Misra PK, Pandey SK (2009) Ediacaran megaplant fossils with Vaucheriacean affinity from the Jodhpur sandstone, Marwar supergroup, western Rajasthan. Curr Sci 97(5):701–705

38. Kumar S, Pandey SK (2008) Discovery of trilobite trace fossils from the Nagaur sandstone, the Marwar supergroup, Dulmera area, Bikaner District, Rajasthan. Curr Sci 94(8):1081–1085
39. Kumar S, Pandey SK (2009) Note on the occurrence of *Arumberia banksi* and associated fossils from the Jodhpur sandstone, Marwar supergroup, western Rajasthan. J Pal Soc Ind 54:1081–1085
40. Pandey DK, Bahadur T (2009) A review of the stratigraphy of Marwar supergroup of west-central India. J Geol Soc Ind 73(6):747–758
41. Prasad B, Asher R, Borgohai B (2010) Late neoproterozoic (ediacaran)–early palaeozoic (cambrian) acritarchs from the Marwar supergroup, Bikaner—Nagaur Basin, Rajasthan. J Geol Soc Ind 75:415–431
42. Misra JS, Shrivastava BP, Jain SK (1961) Discovery of marine, permo-carboniferous in the western Rajasthan. Curr Sci 30:262–263
43. Sastry MVA, Shah SC (1964) Permian marine transgression in peninsular India. 22nd Int Geol Congress, New Delhi 19:141–142
44. Dickens JM, Shah SC (1965) The pelecypods *Undulomya, Cosmomia* and *Palaeocosmomya* in the permian of India and western Australia. J Geol Soc Aust 12(2):253–260
45. Dickens JM, Shah SC (1979) Correlation of the Permian marine sequences of India and Western Australia. In: Laskar B, Rao CS (eds). Fourth International Gondwana Symposium, Calcutta 1977, II. Hindustan Publishing Corporation, pp. 387–408
46. Ranga Rao A, Dhar CL, Obergfell FA (1979) Badhaura Formation of Rajasthan—its stratigraphy and age. In: Laskar B, Rao CS (Eds). Fourth International Gondwana Symposium, Calcutta 1977, II. Hindustan Publishing Corporation, pp. 481–490.
47. Waterhouse JB, Ranga Rao A (1989) Early permian brachiopod and molluscan species from the Bap formation of peninsula India. Palaontologische Zeitschrift 63(1–2):25–39
48. Archibold NW, Shah SC, Dickins JM (1996) Early permian brachiopod faunas from peninsular India: their gondwanan relationships. Hist Biol 11:125–135
49. Kulkarni KG, Borkar VD (1999) Record of a nesting burrow from the Badhaura formation (Permian), Rajasthan. Curr Sci 77(5):7–8
50. Jain RL, Kumar Ravindra (2009) Paleontological study of Permian fossils from the Bap and Badhaura Formations exposed in the Bikaner and Jodhpur district of Rajasthan. In: Shrivastava P. Status of Permo-Carboniferous sediments of Rajasthan—A case study from the Bap-Badhaura- Nokhra area, Jodhpur and Bikaner districts, Rajasthan. http://www.portal.gsi.gov.in/gsiDoc/pub/cs_bap_badhaura.pdf; 2010. pp. 1–4.
51. Srivastava SK (1966) Jurassic microflora from Rajasthan. Micropaleontology 12(1):87–107
52. Moser M, Mathur UB, Fursich FT, Pandey DK, Mathur N (2006) Oldest camarasauromorph sauropod (Dinosauria) discovered in the middle jurassic (Bajocian) of the Khadir Island, Kachchh, western India. Paläontologische Zeitschrift, Stuttgart 80(1):34–51
53. Mathur UB, Pant SC, Mehra S, Mathur AK (1985) Discovery of dinosaurian remains in middle jurassic of Jaisalmer, Rajasthan, western India. Bull Indian Geol Assoc 18(2):59–65
54. Das Gupta SK (1975) A revision of the mesozoic and tertiary stratigraphy of the Jaisalmer basin. Indian J Earth Sci 2(1):77–94
55. Kachchara RP, Jodhawat RL (1981) On the age of Jaisalmer formation, Rajasthan, India. Proc 9th Indian Coll Micropal Strat., Agrawal Printers, Uaipur:235–247
56. Dave A, Chatterjee TK (1996) Integrated foraminiferal and ammonoid biostratigraphy of jurassic sediments in Jaisalmer basin, Rajasthan. J Geol Soc India 47:477–490
57. Pandey B, Krishna J (1996) New ammonoid data: remarks on the age of Bhadasar formation of Jaisalmer basin, Rajasthan. Bull Pure Appl Sci 15F(1):31–36
58. Pandey B, Krishna J (2002) Ammonoid biostratigraphy in the tithonian (late jurassic) of Jaisalmer, western India. Geophytology 30(1, 2):17–25
59. Surendra P (2006) Ammonite biostratigraphy of the middle-late jurassic rocks of Jaisalmer district, Rajasthan, India. Geol Surv Ind Palaentol Indica 52:1–160
60. Lubinova PS, Guha DK, Mohan M (1960) Ostracodes of jurassic and tertiary deposits from Kutch and Rajasthan (Jaisalmer), India. Bull Geol Min Met Soc Ind 22:1–61
61. Subbotina NN, Datta AK, Srivastava BN (1960) Foraminifera from upper jurassic deposits of Rajasthan (Jaisalmer) and Kutch, India. Bull Min Met Soc Ind 123:1–48

62. Kalia P, Chowdhary S (1983) Foraminiferal biostratigraphy. Biogeography and environment of the callovian sequence, Rajasthan, northwestern India. Micropalaeontology 29(3):223–254
63. Rai J, Garg R (2007) Early callovian nannofossils from the Kuldhar section, Jaisalmer, Rajasthan. Curr Sci 92(6):816–820
64. Mukherjee D (2010) New record of plectoidothyris from the middle jurassic sequence of Jaisalmer basin, western India: implications on the easterly brachiopod migration. J Geol Soc Ind 76(3):267–274
65. Qiang J, Zhexi L, Chongxi Y, Tabrum AR (2006) A swimming mammaliaform from the middle jurassic and ecomorphological diversification of early mammals. Science 311:1123–1127
66. Kishna J (1987) An overview of the mesozoic stratigraphy of Kachchh and Jaisalmer basin. J Pal Soc Ind 32:136–149
67. Pandey DK, Bhadu B (2008) Sequence Stratigraphy a Tool for Inter-Basinal Correlation: A Case Study of Paleogene Sediments of Jaisalmer and Barmer Basins. Geo India. 2008 Conference & Exhibition, Greater Noida, New Delhi, India
68. Das Gupta SK (1975) A note on the occurrence of plant fossils in the Pariwar formation, Jaisalmer district, Rajasthan. Sci Cult 61(5):234–237
69. Bhatia SB, Srinivasan S, Bajpai S, Jolly A (1990) Microfossils from Deccan Intertrappean beds at Mamoni, District Kota, Rajasthan. In: Sahni A Lolly (eds). Cretaceous Event Stratigraphy and Correlation of the Indian Non-marine Strata, Contributions from the Seminar cum Workshop, I.G.C.P. 216 & 245, Chandigarh, India, pp. 118–119.
70. Mathur AK, Verma KK (1988) Freshwater ostracodes from the intertrappean of southeastern Rajasthan. Geol Surv Ind Spec Publ 11(2):112–113
71. Barooah SK (1950) Fossil fish and crabs in the Fuller's earth bed in Kapurdi, Jodhpur, Rajasthan. Curr Sci 19(5):165
72. Glaessner MF, Rao VR (1960) A new species of crab from early tertiary Fuller's earth deposits of Kapurdi, Rajasthan, Western India. Rec Geol Surv Ind 86:675–683
73. Tiwari KK (1962) Occurrence of a penaeid prawn from the Fuller's earth deposits of Kapurdi (Barmer district), Rajasthan. Sci Cult 28(5):244–245
74. Tiwari KK (1963) Lower tertiary penaeid shrimp from Kapurdi (Barmer district), Rajasthan. Crustaceana 6(3):205–212
75. Prasad KN (1966) Decapod crustacea from the Fuller earth deposits of Kapurdi, Rajasthan. Rec Geol Surv Ind 94(2):313–316
76. Mathur UB (1988) *Goniocypoda sastri*—a new hexapodinae (crab) from eocene of Rajasthan, India. Geol Surv Ind Spec Publ 11(2):255–259
77. Srivastava DK, Mathur UB (1996) New fibulariid echinoid from the middle eocene rocks of Rajasthan, India. J Geol Soc Ind 41:53–56
78. Singh SN (1951) Kirthar foraminifera from Rajasthan. Curr Sci 20(9):230
79. Singh SN (1960) Two aberrant types of *Nummulites* from the eocene of Rajasthan, India. J Pal Soc Ind 2:209–212
80. Singh SN, Misra PC (1968) A new genus and species of ostracodes from Fullers earth, Kolayatji, Bikaner, Rajasthan. J Pal Soc Ind 11:26–37
81. Khosla SC (1967) A note on the stratigraphy and microfaua of the Kirthar beds of Jaisalmer area. Curr Sci 36(24):670–671
82. Khosla SC (1972) Ostracodes from the eocene bed of Rajasthan, India. Micropalaeontology 14(4):467–507
83. Bhatia SN, Khosla SC (1978) Some lower eocene mollusca from Rajasthan. Recent researches in geology. A collection of papers in honour of Prof. G.W. Chiplonkar 4:225–247
84. Kalia P (1978) Benthonic foraminiferan assemblage from the Kirthar beds of Srikolayatji, Bikaner. Proc VII Indian Colloquium on Micropal and Strat., Manorama Press, Madras:241–256
85. Mathur UB (1983) Report on the study of nannofossils from Tertiary of Bikaner area, Rajasthan. Geol Surv Ind progress report for the field season: 1980–81 (unpublished)
86. Sahni A, Chaudhary NK (1972) Lower eocene fishes from Barmer, southwestern Rajasthan, India. Proc Ind Natl Sci Acad 38(Pt. A.3–4):97–102

87. Rao SRN, Misra SS (1949) An oil bearing alga from palana lignites (?eocene), Bikaner of Rajasthan. Curr Sci 18(10):380–381
88. Rao SRN, Vimal KP (1952) Tertiary pollen from lignites of palana (eocene), Bikaner. Proc Nat Inst Sci Ind 18(6):595–601
89. Vimal KP (1953) Tertiary pollen from lignite of palana (Bikaner). Proc Ind Sci Cong 40(8):27
90. Rao AR (1957) Algal remains from tertiary lignites of palana (Eocene), Bikaner district, Rajasthan. Curr Sci 26:177–178
91. Sah SDC, Kar RK, Singh RY (1971) Stratigraphic range of dandotiospora gen. nov. in the lower eocene sediments in India. Geophytology 1(1):54–63
92. Jain KP, Kar RK, Sah SCD (1973) A paleological assemblage from Eocene, Rajasthan. Geophytology 3(2):150–165
93. Tripathi SKM, Singh UK, Sisodia MS (2003) Paleological investigation and environmental interpretation on Akli formation (Late Palaeocene) from Barmer basin, western Rajasthan, India. Palaeobotanist 52:87–95
94. Kaul KN (1951) A palm fruit from Kapurdi (Jodhpur, Rajasthan desert) *Cocos sahnii* sp nov. Curr Sci 20(5):138
95. Lakhanpal RN (1964) Specific identification of guttifereae leaves from the Tertiary of Rajasthan. Palaeobotanist 12(3):265–266
96. Kar RK, Sharma P (2001) Paleostratigraphy of late palaeocene and early eocene sediments of Rajasthan, India. Palaeontographica B256:123–157, Abt
97. Rana RS, Kumar K, Singh H, Rose K (2005) Lower vertebrates from the late palaeocene–earliest eocene Akli formation, giral lignite mine, Barmer District, western India. Curr Sci 89(9):1606–1613
98. Mathur UB, Mathur AK (1996) A neogene flora from Bikaner, Rajasthan. Geosci J XIX(2):129–144
99. Guleria LS (1984) Occurrence of anacardiacous wood in the tertiary of western India. Palaeobotanist 32:35–43
100. Guleria LS (1990) Fossils dicotyledonous wood from Bikaner, Rajasthan, India. Geophytology 19:182–188
101. Harsh R, Sharma BD (1988) *Auriocarioxylon bikanerensis* sp.nov. from the tertiary of Bikaner, Rajasthan. India. Phytomorphology 38:111–115
102. Harsh R, Sharma BD, Suthar OP (1992) Anatomy of petrified woods of lecythidaceae and combretaceae from Bikaner (Rajasthan) India. Phytomorphology 42(1&2):87–102
103. Paddayya K, Blackwell BAB et al (2002) Recent findings on the acheulian of the hunsgi and baichbal valleys, Karnataka, with special reference to the isampur excavation and its dating. Curr Sci 83(5):641–647
104. Sonakia A, de Lumley H (2006) Narmada *Homo erectus*—a possible ancestor of the modern Indian. CR Palevol 5(1–2):353–357
105. Singh G, Joshi RD, Chopra SK, Singh AB (1974) Late quaternary history of vegetation and climate of the Rajasthan desert, India. Phil Tran R Soc London B267:467–501
106. Singh G, Wasson RJ, Agrawal DP (1990) Vegetational and seasonal climatic changes since the last full glacial in the Thar desert, northwestern India. Rev Palaeobot Palynol 64:351–358
107. Sharma C, Srivastava C, Yadav DN (2003) Holocene history of vegetation and climate of fresh water punlota (degana) lake in eastern Rajasthan, India. Palaeobotanist 52(1–3):127–135
108. Goudie AS, Sperling CHB (1977) Long distance trans- port of foraminiferal tests by winds in the Thar desert, NW India. J Sedement Petrol 47:630–633
109. Kameswara Rao K, Wasson RJ, Krishnan K (1989) Foraminifera from late quaternary dune sands of Thar Desert. Palaios 4:168–180
110. Jacob K, Sastry MVA, Sastri VV (1952) A note on the microfauna of the impure gypsum from the jamsar mine, Bikaner, and the possible origin of gypsum. Bull Natl Inst Sci Ind 1:68–69
111. Bhatia SN, Khosla SC (1970) *Cyprideis westi* sp. nov. from the jamsar deposits, Rajasthan. West commemoration volume. Today & Tomorrow Printers and Publishers, Faridabad, India, pp 359–363

Part II
Faunal Ecology: An Insight – Piscifauna and Herpetofauna

Chapter 6
Ichthyofauna of Rajasthan

Devendra Mohan and Ramkishor

Abstract This chapter presents the fish diversity of Rajasthan recorded from a large number of seasonal and perennial ponds, freshwater and saline lakes, rivers, and canals, the most prominent being the Indira Gandhi Nahar Pariyojana (IGNP). A large number of freshwater fish inhabit seasonal and perennial rivers and canals. All in all, 160 fish species belonging to 9 orders, 30 families, and 75 genera have been reported from the state. The order cypriniformis with 95 species and family Cyprinidae belonging to this order with 81 species are the most dominating ones. Ten exotic fish species have also been reported from various water bodies. According to IUCN (2012) status, four species are Endangered, eight Near-threatened, four Vulnerable, 87 Least Concern, three Data Deficient and 55 species are in the category of Not Evaluated. An urgent need to fill gaps in research has also been emphasized in the text. The authors point out that the fish species inhabiting rivers and reservoirs of Rajasthan are facing problems related to water pollution, habitat alteration, introduction of exotic species, and destruction of brood stock along with other usual anthropogenic disturbances.

Introduction

Despite being an arid state, Rajasthan has many water bodies and a few rivers too. Based on the drainage and water regime, the state can be divided into six distinct geographic regions while the Western desert region is devoid of any natural reservoir. However, Indira Gandhi Nahar Pariyojna (IGNP) carrying water from the Bhakra Dam passes through Sri Ganganagar, Bikaner, and Jaisalmer districts. The semiarid region between the Aravalli Ranges and the western desert runs across the state from Jalore to Jhunjhunu districts. Southern part of this region is drained

D. Mohan (✉) • Ramkishor
Department of Zoology, Jai Naraian Vyas University, Jodhpur, India
e-mail: mohandevendra@hotmail.com; ram_k1@yahoo.com

B.K. Sharma et al. (eds.), *Faunal Heritage of Rajasthan, India: General Background and Ecology of Vertebrates*, DOI 10.1007/978-1-4614-0800-0_6,

by the Luni River while the northern part remains dry. A large number of small impoundments have been created in the undulating Aravalli Hill area in its uneven terrain especially in the districts of Pali, Udaipur, and Sirohi. The eastern region is extensively drained by the Banas River and its tributaries and comprises a large number of reservoirs in Sawai Madhopur, Bundi, Alwar, and Bharatpur districts. The Southern region consists of stony uplands setting ideal sites for water resource developments with the maximum numbers of man-made lakes. The Chambal ravine region lies along the River Chambal forming boundary between Rajasthan and the adjoining Madhya Pradesh state. From a hydrological point of view, the state is divided in to four major river basins:

1. Chambal River basin covers the eastern and northeastern part of the state with tributaries like Bedach, Banas, and Kalisindh. The river originates from Madhya Pradesh, flows through Rajasthan, and finally merges with Yamuna River in the State of Uttar Pradesh. The water of this basin debouches in Bay of Bengal through River Ganga.
2. Mahi River basin and its main tributaries Jakham, Som, and Anas cover the southern part of the state. The river originates from Madhya Pradesh and finally debouches in Arabian Sea after flowing through the states of Rajasthan and Gujarat.
3. Luni River basin covers the central and southern desert part of the state. This is the major river of the Thar which originates from Ajmer district and reaches up to Rann of Kutch in Gujarat.
4. Ghaggar River basin covers the Western Desert. The excess water of Satluj River is flown into Ghaggar through Indira Gandhi Canal. The river finally debouches in Arabian Sea after passing through Pakistan.

A total of 1,80,280 ha area is covered by lentic water bodies present in Rajasthan in the form of 3,244 impoundments. Out of them, 2,373 water bodies thrive in the form of small seasonal ponds with an area of less than 10 ha making a total of 8,137 ha while, about 816 water bodies are long seasonal and perennial type and covers an area between 10 and 500 ha making a total of 58,154 ha. In addition, the remaining 55 large perennial water bodies located in the eastern and southern part of the state with an area of more than 500 ha (making a total of 113,989 ha) are in the form of reservoirs, dams and lakes. Rawatbhata (19,600 ha) in Chittourgarh, Bajaj Sagar (13,500 ha) in Banswara, Jaisamand (7,286 ha) in Udaipur and Kadana (9,000 ha) in Banswara and Dungarpur districts are the largest water bodies covering about 36% of the total lentic water area of the state. Due to variable climatic conditions and catchment connections, there is a great variation in the fish fauna of the state. A rich faunal diversity was reported in the arid and semiarid region with an expansion of IGNP and its escape reservoirs like RD 507, RD 750, RD 1240, RD 1365, Digha Lake, depression of Ghaggar River and the Gang canal. Apart from the above, the saline lakes, namely, Sambhar Salt Lake, Pachpadra Lake, and Lunkaransar Lake covering about 25,000 ha area with a salinity of up to 250 PSU have no fish fauna at all but may harbor a few freshwater fishes during rainy season or occasional floods.

Fish Faunal Diversity in Rajasthan: A Review

Investigation on the fish fauna of Rajasthan was actually initiated in an organized manner in the early 1950s [1–13]. The systematic position and the IUCN status of the fish fauna of Rajasthan are summarized in Table 6.1. Accordingly, the order Cypriniformes constitutes 60% (95 species) of the fishes followed by Siluriformes 18.5%, Persiformes 10%, Cyprinodontiformes and Synbrachiformes 3% each, Mugiliformes 2.5%, Osteoglossiformes and Beloniformes 1.2% each, and Clupeiformes 0.6%. Cyprinidae with 81 species has been reported as the most dominating family (Fig. 6.1).

According to the IUCN (2012) status, four fishes are Endangered (EN), eight Near-Threatened (NT), four Vulnerable (VU), 87 Least Concern (LC), three Data-Deficient (DD) and 55 species Not Evaluated (NE). The ten exotic species reported from the state include *Ctenopharyngodon idellus, Hypophthalmichthys molitrix, Hypophthalmichthys nobilis, Cyprinus carpio, Carassius carassius, Tor khudree, Tor putitora*, and *Tor tor* which were introduced for experimental fish culture while *Gambusia affinis* was mainly introduced to control the spread of mosquito larva. *Tilapia mossambica* entered along with contaminated fish seeds and spread to a large number of water bodies in Rajasthan and started competing with cultured carps. *Pterygoplichthys disjunctivus,* well established in the wetlands of East Kolkata, was reported for the first time in Rajasthan from the Lake Kailana which must have entered along with the contaminated fish seed. Datta and Majumdar [14] reported 63 species from 14 districts of Rajasthan belonging to 36 genera and 16 families. Interestingly, eleven species, namely, *Gudusia chapra, Barilius barila, Danio rerio, Amblypharyngodon mola, Cabdio morar, Chagunius chagunio, Labeo dyocheilus, Botia lohachata, Puntius vittatus, Mystus vittatus,* and *Mastacembelus pancalus* have been reported for the first time from Rajasthan.

Johal *et al.* [11] reported 95 species of fishes belonging to 52 genera, seven orders, five super orders and two cohorts from the rivers Chambal, Banas, and Luni apart from Ranapratap Sagar Dam, Jaisamand Lake, and the Indira Gandhi Canal. Among these, 15 species were reported for the first time. Eighteen species have been reported to have disappeared due probably to the altered ecological conditions. In fact, the fish fauna of the state is a blend of western Himalayas, Aravallis, and Peninsular elements. Johal [15] prepared a field key of the fishes of Ganganagar district while Johal and Chahal [16] described the importance of further studies for fisheries development. In addition, Sharma and Johal [17] reported 41 species of fishes including eight new records from the Jaisamand Lake, Udaipur. *L. dussumier* and *L. potail,* inhabitants of Haryana and Punjab state are also found in Rajasthan on account of the close proximity of the region or introduction by man [18, 19]. *Puntius parrah, Liza parsia, Acentrogobius viridipunctatus* and *Salmostoma phulo orissaensis* also occur in Rajasthan as reported earlier [20] from the lower reaches of Mahanadi River in Orissa [20]. Occurrence of these species can be correlated with the flow of water from east to west during the Pleistocene period (pre-tilt period) [21]. Johal and Sharma [22] reported 67 species representing eight orders, 16 families

Table 6.1 Fish fauna of Rajasthan and its IUCN status

S. No.	Species name	Fauna of Rajasthan, India Datta & Majumdar [14]	Ichthyofauna of Rajasthan State (India) Johal *et al.* [11]	Fish diversity in the Thar Desert Yazdani [25]	Fish fauna diversity of Thar Desert of Rajasthan. Mohan and Singh [12]	Conservation and Management of Fish diversity of Rajasthan. Sharma & Choudhary [13]	IUCN (2012) status
I	Order—Osteoglossiformes Family—Notopteridae						
1	*Notopterus notopterus* (Pallas)	+	+	+	+	+	LC
2	*Chitala ornata* (Gray, 1831)		+	+	+	+	LC
II	Order—Clupeiformes Family—Clupeidae						
3	*Gudusia chapra* (Ham-Buch)	+	+	+	+	+	LC
III	Order—Cypriniformes Family—Cyprinidae Subfamily—Cyprininae						
4	*Carassius carassius* (Lin)	+		+		+	LC
5	*Catla catla* (Ham-Buch)	+	+	+	+	+	–
6	*Chagunius chagunio* (Ham-Buch)	+		+	+	+	LC
7	*Cirrhinus mrigala* (Ham-Buch)	+	+	+	+		LC
8	*Cirrhinus cirrhosus (Bloch)*					+	LC
9	*Cirrhinus reba (Ham-BUch)*	+	+	+	+	+	VU
10	*Cyprinus carpio* (Lin)		+	+	+	+	LC
11	*Ctenopharyngodon idellus* (Val)		+	+	+	+	VU
12	*Labeo angra* (Ham-Buch)	+	+	+	+	+	–
13	*Labeo bata* (Ham-Buch)	+	+	+	+	+	LC
14	*Labeo boga* (Bloch)	+	+	+	+	+	LC
15	*Labeo boggut* (Sykes)	+	+	+	+	+	LC
16	*Labeo calbasu* (Ham-Buch)	+	+	+	+	+	LC

17	*Labeo dero* (Ham-Buch)	+		+	+	+	LC
18	*Labeo dussumieri* (Val)		+	+	+	+	LC
19	*Labeo dyocheilus* (Mc Clelland)	+		+	+	+	LC
20	*Labeo fimbriatus* (Bloch)	+	+	+	+	+	LC
21	*Labeo gonius* (Ham-Buch)	+	+	+	+	+	LC
22	*Labeo microphthalmus* (Day)					+	LC
23	*Labeo nigripinnis* (Day)	+					–
24	*Labeo pangusia* (Ham-Buch)		+	+	+	+	NT
25	*Labeo potail* (Ham-Buch)		+	+		+	EN
26	*Labeo rajasthanicus* (Datta & Majumdar)	+					–
27	*Labeo rohita* (Ham-Buch)	+	+	+	+	+	LC
28	*Labeo sindensis* (Day)			+	+	+	–
29	*Osteobrama cotio cotio* (Ham-Buch)	+	+	+		+	LC
30	*Puntius arulius* (Jerdon)					+	EN
31	*Puntius amphibius* (Val)	+	+	+	+	+	LC
32	*Puntius chola* (Ham)	+	+	+	+	+	LC
33	*Puntius conchonius* (Ham-Buch)	+	+	+	+	+	LC
34	*Puntius dorsalis* (Jerdon)	+		+	+	+	LC
35	*Puntius gugaio* (Ham-Buch)					+	–
36	*Puntius punjabensis* (Day)					+	–
37	*Puntius parrah* (Day)		+				LC
38	*Puntius sarana sarana* (Ham-Buch)	+	+	+	+	+	LC
39	*Puntius sophore* (Ham-Buch)/*P. chrysopterus* (McClelland)	+	+	+	+	+	LC
40	*Puntius terio* (Ham-Buch)		+	+	+	+	LC
41	*Puntius ticto* (Ham-Buch)	+	+	+	+	+	LC
42	*Puntius vittatus* (Day)	+		+	+	+	LC
43	*Puntius waggeni* (Day)					+	–
44	*Oxygaster clupeoides* (Bloch)	+					–

(continued)

Table 6.1 (continued)

S. No.	Species name	Fauna of Rajasthan, India Datta & Majumdar [14]	Ichthyofauna of Rajasthan State (India) Johal *et al.* [11]	Fish diversity in the Thar Desert Yazdani [25]	Fish fauna diversity of Thar Desert of Rajasthan. Mohan and Singh [12]	Conservation and Management of Fish diversity of Rajasthan. Sharma & Choudhary [13]	IUCN (2012) status
45	*Oxygaster bacaila* (Ham-Buch)	+			+		LC
46	*Oxygaster gora* (Ham)	+			+		LC
47	*Oxygaster phulo* (Ham-Buch)	+					–
48	*Tor khudree* (Sykes)	+		+	+	+	EN
49	*Tor putitora* (Ham-Buch)		+	+	+	+	EN
50	*Tort tor* (Ham-Buch)		+	+	+	+	–
	Subfamily—Cultrinae						
51	*Chela cachius* (Ham-Buch)		+	+		+	LC
52	*Chela laubuca* (Ham-Buch)					+	–
53	*Salmostoma acinaces* (Val)					+	LC
54	*Salmostoma bacaila* (Ham-Buch)		+	+		+	LC
55	*Salmostoma clupeoides* (Bloch)			+		+	LC
56	*Salmostoma phulo* (Ham-Buch)/*orissaensis*		+	+		+	–
57	*Salmostoma phulo punjabensis* (Day)		+	+		+	–
58	*Securicula gora* (Ham-Buch)		+	+		+	LC
	Subfamily—Hypophthalmichthyinae						
59	*Hypophthalmichthys molitrix* (Val)			+	+	+	NT
60	*Hypophthalmichthys nobilis* (Val)					+	DD
	Subfamily—Rasborinae						
61	*Amblypharyngodon microlepis* (Bleeker)		+	+		+	LC
62	*Amblypharyngodon mola* (Ham-Buch)	+	+	+	+	+	LC
63	*Aspidoparia morar* (Ham-Buch)	+	+	+		+	LC
64	*Barilius barna* (Ham-Buch)	+		+		+	LC

65	*Barilius barila* (Ham-Buch)	+		+		+	LC
66	*Barilius bendelisis bendelisis* (Ham-Buch)	+	+	+		+	LC
67	*Barilius bola* (Hamilton)	+	+				–
68	*Barilius modestus* (Day)					+	–
69	*Barilius vagra vagra* (Ham-Buch)	+	+	+		+	–
70	*Brachydanio rerio* (Ham-Buch)/*Danio rerio*	+		+		+	LC
71	*Bengala elanga* (Ham-Buch)/*Rasbora elanga*	+		+		+	–
72	*Danio acquipinnatus* (Mc Clelland)	+		+	+	+	–
73	*Danio devario* (Ham-Buch)	+	+	+	+	+	–
74	*Esomus danricus* (Ham-Buch)	+	+	+		+	LC
75	*Raiamas bola* (Ham-Buch)			+		+	LC
76	*Rasbora daniconius* (Hamilton)	+					LC
	Subfamily—Schizothoracinae						
77	*Schizothorax richardsonii* (Gray)					+	VU
78	*Schizopygopsis stoliczkae* (Steindachuer)					+	–
	Subfamily—Garrinae						
79	*Crossocheilus latius diplocheilus* (Heckel)			+		+	–
80	*Crossocheilus latius punjabensis (Mukherji)*						–
81	*Crossocheilus latius latius* (Ham-Buch)		+			+	DD
82	*Garra gotyla gotyla* (Gary)	+	+	+	+	+	LC
83	*Garra lamta* (Ham-Buch)		+	+	+	+	LC
84	*Garra mullya* (Sykes)	+		+	+	+	LC
	Family—Psilorhynchidae						
85	*Psilorhynchus balitora* (Ham-Buch)		+	+		+	LC
	Family—Balitoridae						
	Subfamily–Nemacheilinae						

(continued)

Table 6.1 (continued)

S. No.	Species name	Fauna of Rajasthan, India Datta & Majumdar [14]	Ichthyofauna of Rajasthan State (India) Johal *et al.* [11]	Fish diversity in the Thar Desert Yazdani [25]	Fish fauna diversity of Thar Desert of Rajasthan. Mohan and Singh [12]	Conservation and Management of Fish diversity of Rajasthan. Sharma & Choudhary [13]	IUCN (2012) status
86	*Nemacheilus baluchiorum* (Zugmayer)			+		+	–
87	*Nemacheilus botia botia* (Ham-Buch)	+	+	+	+	+	LC
88	*Nemacheilus carletoni* (Fowler)					+	–
89	*Nemacheilus corica* (Ham-Buch)		+	+		+	LC
90	*Nemacheilus denisoni denisoni* (Van Hasselt)	+		+	+	+	LC
91	*Nemacheilus horai* (Menon)					+	–
92	*Nemacheilus montanus* (Mc Clelland)					+	–
93	*Nemacheilus rupicola* (Mc Clelland)					+	–
94	*Triplophysa gracilis* (Day)					+	–
	Family—Cobitidae						
	Sub–family—Cobitinae						
95	*Lepidocephalus guntea* (Ham-Buch)	+	+	+		+	LC
	Subfamily—Botiinae						
96	*Botia birdi* (Chaudhuri)		+	+		+	–
97	*Botia dayi* (Hora)					+	–
98	*Botia lohachata* (Chaudhuri)	+		+		+	–
IV	Order—Siluriformes						
	Family—Bagridae						
99	*Aorichthys aor* (Ham-Buch)/*Mystus aor*		+	+	+	+	LC
100	*Aorichthys seenghala* (Sykes)/M. *seenghala*	+	+	+	+	+	–
101	*Mystus bleekeri* (Day)	+	+	+	+	+	LC
102	*Mystus cavasius* (Ham-Buch)	+	+	+	+	+	LC

103	*Mystus horai* (Jayaram)					+	–
104	*Mystus tengara* (Ham-Buch)				+		LC
105	*Mystus vittatus* (Bloch)	+	+	+	+	+	LC
106	*Rita rita* (Ham-Buch)		+	+	+	+	LC
	Family—Siluridae						
107	*Ompok bimaculatus* (Bloch)	+	+	+	+	+	NT
108	*Ompok pabda* (Lace)				+		NT
109	*Wallago attu* (Schn)	+	+	+	+	+	NT
	Family—Scheibeidae						
	Subfamily—Aillinae						
110	*Ailia coila* (Ham-Buch)		+	+		+	NT
	Subfamily—Schilbeinae						
111	*Clupisoma gaura* (Ham-Buch)		+	+		+	–
112	*Clupisoma montana* (Hora)					+	LC
113	*Eutropiichthys vacha* (Ham-Buch)		+	+	+	+	LC
114	*Silonia silondia* (Ham-Buch)	+		+		+	LC
	Family—Amblycipitidae						
115	*Amblyceps mangois* (Ham-Buch)		+	+		+	LC
	Family—Sisoridae						
	Subfamily—Sisorinae						
116	*Bagarius bagarius* (Ham-Buch)	+	+	+		+	NT
117	*Bagarius yarrelli* (Sykes)					+	NT
118	*Nangra nangra* (Ham)		+	+	+	+	LC
119	*Nangra viridescens* (Ham-Buch)				+	+	–
120	*Gagata cenia* (Ham-Buch)					+	LC
121	*Sisor rhabdophorus* (Ham-Buch)				+		–
	Subfamily—Glyptosterninae						
122	*Glyptothorax pectinopterus* (Mc Clelland)		+	+	+	+	LC
123	*Glyptothorax telchitta* (Ham-Buch)			+			LC

(continued)

Table 6.1 (continued)

S. No.	Species name	Fauna of Rajasthan, India Datta & Majumdar [14]	Ichthyofauna of Rajasthan State (India) Johal *et al.* [11]	Fish diversity in the Thar Desert Yazdani [25]	Fish fauna diversity of Thar Desert of Rajasthan. Mohan and Singh [12]	Conservation and Management of Fish diversity of Rajasthan. Sharma & Choudhary [13]	IUCN (2012) status
	Family—Erethistidae						
124	*Erethistes pussilus* (Muller&Troschel)		+			+	–
	Family—Clariidae						
125	*Clarias batrachus* (Lin)		+	+	+	+	LC
126	*Clarias gariepinus* (Lin)					+	–
	Family—Heteropneustidae						
127	*Heteropneustes fossilis* (Bloch)/*Sacchobranchus fossilis*	+	+	+	+	+	LC
	Family—Loricariidae						
	Subfamily—Hypostominae						
128	*Pterygoplichthys disjunctivus* (Weber)						–
V	Order—Beloniformes						
	Family—Belonidae						
129	*Xenentodon cancila* (Ham-Buch)	+	+	+	+	+	LC
	Family—Adrianichthyidae						
	Subfamily—Oryziinae						
130	*Oryzias melastigma* (Mc Clelland)			+		+	LC
VI	Order—Cyprinodontiformes						
	Family—Aplocheilidae						
131	*Aplocheilus blocki* (Arnold)			+		+	–
132	*Aplocheilus lineatus* (Val)	+		+		+	LC
133	*Aplocheilus panchax* (Ham-Buch)					+	LC
	Family—Poeciliidae						
134	*Gambusia affinis* (Baird & Girard)	+	+	+	+	+	–

	Family—Cyprinodontidae						
135	*Aphanius dispar* (Ruppell)			+		+	–
VII	Order-Synbranchiformes						
	Family—Synbranchidae						
136	*Monopterus cuchia* (Ham-Buch)					+	LC
	Family—Mastacembelidae						
137	*Macrognathus aral* (Bloch & Schn)			+	+	+	LC
138	*Mastacembelus armatus armatus* (Lacepede)	+	+	+	+	+	LC
139	*Mastacembelus pancalus* (Ham)	+	+	+	+	+	–
VIII	Order—Perciformes						
	Family—Cichlidae						
140	*Tilapia mossambica* (Peters)				+	+	–
	Family—Ambassidae						
141	*Chanda baculis* (Ham-Buch)/A*mbassis baculis*	+	+	+	+	+	–
142	*Chanda nama* (Ham-Buch)/*Ambassis nama*	+	+	+	+	+	LC
143	*Chanda ranga* (Ham-Buch)/*Ambassis ranga*	+	+	+	+	+	LC
	Family—Nandidae						
	Subfamily—Nandinae						
144	*Nandus nandus* (Ham-Buch)		+	+	+	+	LC
	Family—Anabantidae						
145	*Anabas testudineus* (Bloch)					+	DD
	Family—Gobiidae						
	Subfamily—Gobiinae						
146	*Acentrogobius viridipunctatus* (Val)		+	+		+	–
147	*Glossogobius giuris giuris* (Ham-Buch)	+	+	+	+	+	–
	Family—Osphronemidae						
148	*Colisa fasciatus* (Schn)	+	+	+	+	+	–

(continued)

Table 6.1 (continued)

S. No.	Species name	Fauna of Rajasthan, India Datta & Majumdar [14]	Ichthyofauna of Rajasthan State (India) Johal *et al.* [11]	Fish diversity in the Thar Desert Yazdani [25]	Fish fauna diversity of Thar Desert of Rajasthan. Mohan and Singh [12]	Conservation and Management of Fish diversity of Rajasthan. Sharma & Choudhary [13]	IUCN (2012) status
149	*Colisa labiosus* (Day)					+	–
150	*Colisa lalia* (Ham-Buch)		+	+	+	+	–
151	*Osphronemus goramy* (Lace)	+		+	+	+	LC
	Family—Channidae						
152	*Channa gachua* (Ham)	+	+	+	+		LC
153	*Channa marulius* (Ham-Buch)	+	+	+	+	+	LC
154	*Channa orientalis* (Bloch&Schn)			+	+	+	–
155	*Channa punctatus* (Bloch)	+	+	+	+	+	–
156	*Channa striatus* (Bloch)	+	+	+	+	+	–
IX	Order—Mugiliformes						
	Family—Mugilidae						
157	*Liza parsia* (Ham-Buch)	+	+	+	+	+	–
158	*Mugil cephalus* (Lin)		+	+	+	+	LC
159	*Rhinomugil corsula* (Ham-Buch)		+	+		+	LC
160	*Sicamugil cascasia* (Ham-Buch)					+	LC

Fig. 6.1 Photographic plate showing major fish species found in Rajasthan (Scientific names mentioned at S. No. 1-36 in Table 6.1)

and 40 genera with new records of five species from Sawai Madhopur district. The Thar Desert receives Himalayan water through Gang Canal, Indira Gandhi Canal and River Ghaggar. Earlier due to prolonged droughts, fish species diversity was scanty but floods in the recent past have increased the diversity by adding 21 more fish species to the region. Johal and Dhillon [23] reported 57 species of the fish from

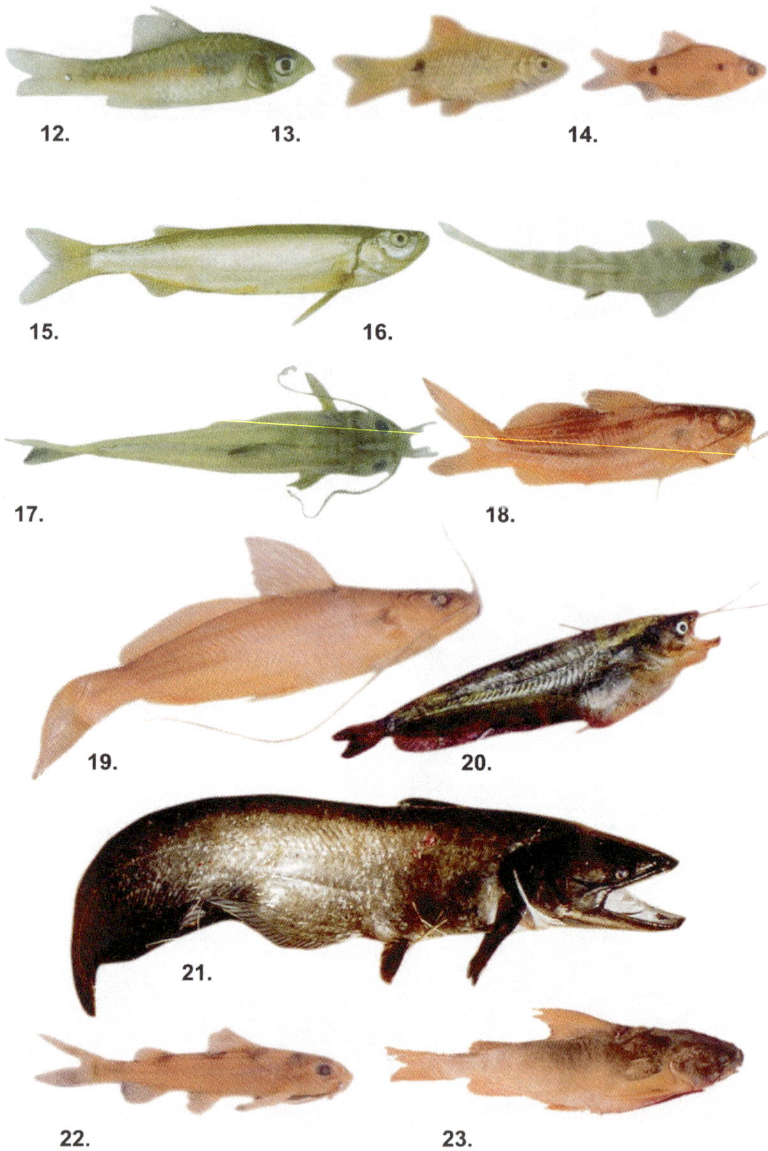

Fig. 6.1 (continued)

Ganganagar district belonging to 30 genera and also reported 19 fish species as new records. Durve [24] described the fish fauna of Jaisamand Lake of Udaipur and its potential to develop fisheries.

Yazdani studied the fish fauna of the Thar Desert of Rajasthan, Gujarat, Punjab, and Haryana and reported a total of 142 species out of which 112 species belonging to 64 genera 26 families and six orders were alone reported from the Thar region of

Fig. 6.1 (continued)

Rajasthan [25]. Family Cyprindae was the most dominating family with 58 species followed by 12 species of cat fishes namely, Bagridae, Schilbeidae, and Sisoridae. Mohan and Singh [12] reported eighty species belonging to six orders, 20 families and 37 genera from thirteen districts of the Thar Desert. The districtwise distribution shows a maximum of 43 species from Jaisalmer followed by Pali with 39 species including eight exotic and nine larvivorous fishes. Two species, namely, *Mystus tengara* and *Nandus viridesceus*, were reported for the first time from the Thar Desert. The maximum number of fish species and their abundance was recorded from the Jaisalmer District due mainly to the extension of IGNP and the presence of four escape reservoirs, namely, Digha, Sultana, RD 1356, and RD 1240. These reservoirs with a maximum depth of seven meters and abundance of macro vegetation provide suitable habitat for fish growth and breeding. The species diversity of fishes was higher in Pali (39) and Sirohi districts (20) on account of the three perennial water bodies, namely, Jawai Dam, Raipur dam (Pali district), and Banas (Sirohi). Mohan and Singh [12] reported eight exotic species against nine as reported by Yazdani [25]. *Tilapia mossambica* was not reported by Yazdani, while *Tor putitora* and *Carassius carassius* were not reported by Mohan and Singh. *Tor tor* and *T. khudree* are exotic cold water edible fishes, the latter entered in Thar Desert through IGNP while *T. tor* might have entered in to the Jawai dam along with the seeds of major carps. Johal *et al.* [11] reported the disappearance of 18 already known fish species out of which *Labeo dyocheilus, Chagunius chagunio*, and *Garra mullya* reportedly reappeared [12].

Dhawan [26] reported 35 species of fishes from the lakes of Udaipur belonging to 9 families dominated by Cyprinidae, followed by Siluridae and Cobitidae. Ajith Kumar and Vijayan [9] recorded 40 species from the Keoladeo National Park, Bharatpur and included all 40 species of the park in their distribution records with an addition of 13 new species to the fish fauna of Rajasthan thereby, increasing the number to 88 species at that time. Datta and Majumadar [14] reported *Labeo rajasthanicus* as a new record; however, this species which closely resembles *L. boggut* was not reported by subsequent workers. The exotic fish *Hypophthalmichthys molitrix* (Silver Carp) was introduced in Ummaid Sagar of Jodhpur in the year 2000 for experimental culture [27]. Another exotic carp, *Cyprinus carpio* was introduced in a few ponds of Pali and Jodhpur for composite fish culture [28, 29]. The occurrence of *T. khudree* in Rajasthan is interesting as this species commonly occurs in Tamil Nadu and Kerala [30], Orissa, Karnataka, and Maharashtra. [31]. *T. khudree* was collected from Bikaner while *T. tor* was found in Jodhpur, Sri Ganganagar, and Jawai dam of Pali [12]. Dhanze and Dhanze had earlier reported *T. tor* from different hill streams of Himachal Pradesh [32].

Fish faunal diversity of the Thar Desert contains a mixture of four elements, namely Western Himalayas, Aravallis, Peninsular India, and the Middle East [25]. The fishes of western Himalayas, particularly those of the Indus' drainage are found in the Thar Desert owing to the fact that Sutlej and Ravi flow through the north-western part of this desert [33]. The fish fauna of the Aravallis reached other areas of the desert through small seasonal rivers flowing westwards from the Aravallis and ending either in the desert sand or the Rann of Kutch. The peninsular species could have reached the Thar Desert similarly, since, the Aravallis, during the recent geological epoch, tilted to the north and, thus, carried the peninsular fauna more particularly of the Satpura northwards [34]. Due to the expansion of lift canal and its various distributaries, the fish fauna of Aravallis has intermingled with the fauna of Western Himalayas. This event is clearly observed in Lake Kailana of Jodhpur which was connected with Jawai dam (receiving water of Aravalli catchments) until

October 1993 when it was connected to the IGNP lift canal. Datta and Majumadar [35] and Mohan and Singh [36] reported 10 and 16 species of fishes respectively from this lake, but after the connection with IGNP, 30 species were reported [10]. Twelve new species entered through the canal in this lake which include some carnivorous and omnivorous species like *Xenentodon cancila, Mystus vittatus, Mystus cavasius*, and *Tilapia mossambica*.

The occurrence of *Aphanius dispar* in the Luni River of Jodhpur gives a strong support to the view that the fishes occurring in the coastal zones of the Middle East have immigrated along the coasts to the Indian region [25]. The occurrence of *Nemacheilus baluchiorum* in Baluchistan and Jodhpur can also be explained by the fact that, the Sindh hills were once connected to Aravallis through the Sangla Hills in Punjab [34]. The occurrence of peninsular species, namely, *Salmophasia balookee* and *Garra mullaya*, in the Thar Desert area support the view that fauna of Aravalli Hills is derived from peninsular India [25].

It may be concluded that due to the intermingling of fish fauna of Western Himalaya and Aravalli, the fish diversity of Rajasthan might have increased. Order cypriniformes showed the highest abundance not only in the Thar region but also in the Aravalli foot hill region due to escape reservoirs and dams respectively. In Barmer, Sikar, Jhunjhunu, and Nagour districts, the unavailability of perennial water bodies may be a main cause of lower fish diversity. A few species, like *Rita pavimentcita*, *Silonia silondia*, *Haplochellus lineatus* [4], and *Aplocheilus blochii* [37] reported earlier, do not appear in the list of freshwater fishes of India [38] and may not be considered as established species. Some species reported earlier have been synonymized, for example *Puntius stoliczkanus* with *P. ticto*, *P. tetrdrupagus* with *P. chola*, *P. stigma* with *P.* sophore, *Labeo microphthalmus* with *L. dero*, *Rasbora daniconius* with *Esomus daniconius, Rasbora elanga* with *Bengala elanga, Puntius parrah with Puntius sophore, Mystus seenghala* with *Aorichthys seenghala, and Sperata aor* with *Aorichthys aor* [38, 39].

Recommendations

As compared to the neighboring states of Gujarat, Punjab, and Haryana, the fish faunal diversity of Rajasthan is more diverse because of the expansion of IGNP canal system and entry of some of the exotic fish species along with the seed of culturable carps. Ten exotic fishes have also made their appearance, some of which became an important component of composite fish culture while some others are competing with the indigenous fish species. Some of the suggested measures to protect the threatened species are as follows: selected perennial water bodies should be developed as fish sanctuaries; care should be taken to protect the breeding ground of fish; regular monitoring of fish diversity; and sample collection and restoration of threatened species by stocking of yearlings along with *in situ* conservation. In addition, suitable infrastructure should be developed in order to regulate the restoration program and to record the current fish faunal diversity in various regions of the state.

References

1. Mathur BBL (1952) Notes on the fishes from Rajasthan. Indian Rec Mus Delhi 50(1):105–110
2. Hora SL, Mathur BBL (1952) On certain Palaeagraphical features of Rajasthan as evidenced by the distribution of fishes. Bull Nation Inst Set India, Delhi. (I):32–36
3. Moona JC (1963) Notes on the fishes of Bharatpur district Rajasthan. Rec Indian Mus 58:59–66
4. Datta Gupta AK, Menon PKB, Nair CKG, Das CR et al (1961) An annotated list of fishes of Rajasthan. Proc Rajasthan Acad Set Pilani 8(1&2):120–131
5. Dubey GP, Mehra RK (1962) Fish and fisheries of Chambal River. Proc All India Cong Zool Jabalpur 11(Pt 1):647–665
6. Yazdani GM, Bhargava RN (1969) On a new record of Minnow *Aphanius dispar* from Rajasthan. Lab Dev J Sci Tech India Kanpur 7-B(4):332–333
7. Krishna D, Menon CB (1958) A note on the fishes of Jodhpur (Rajasthan). Vijnan Parishad Anushandhan Patrika, Allahabad 1(4):207–209
8. Gupta SN, Kulshreshta SD (1985) Fish and fisheries of the district Jhalawar, Rajasthan. Int J Acad Ichthyol 6:91–96
9. Ajith Kumar CR, Vijayan VS (1988) On the fish fauna of Keoladeo National Park, Bharatpur (Rajasthan). J Bomb Nat Hist Soc 85(1):44–49
10. Mohan D, Singh S (2004) Impact of IGNP lift canal on fish fauna of Kailana lake. Fish diversity in protected habitats. Nat Conservat Publ 8:255–259
11. Johal MS, Chahal IS, Tandon KK et al (1993) Ichthyofauna of Rajasthan State (India). J Bomb Nat Hist Soc 90(3):404–411
12. Mohan D, Singh S (2006) Fish faunal diversity of Thar Desert of Rajasthan. J Nat Cong 18(2):261–270
13. Sharma LL, Choudhary CS (2007) Conservation and Management of fish diversity in Rajasthan. In: Lakra WS, Sarkar UK (eds) Fresh water fish diversity of Central India. NBFGR (ICAR) Publication, Lucknow, pp 132–141
14. Datta AK, Majumdar N (1970) Fish fauna of Rajasthan India. Part 7. Fishes Rec Zoo Surv India 62(1–2):63–100
15. Johal MS (1982) Field key to the fishes of the Ganganagar district (Rajasthan). Res Bull Punjab Univ 33((iii-iv)):43–50
16. Johal MS, Chahal JS (1988) Importance of faunal studies of lakes in fisheries development. In: Kulshrestha SK, Adholia UN, Jain OP, Bhatnagar Anita (eds) Proc Nat Symp. Past, Present and Future of Bhopal Lakes, pp 115–121
17. Sharma KP, Johal MS (1982) On the fish and fisheries of Jaisamand lake Rajasthan. Vest Cs Spolec Zool 46:56–69
18. Johal MS, Tandon KK (1979) Monograph on the fishes of re-organized Punjab. Part I. Pb Fish Bull 3(2):1–44
19. Johat MS, Tandon KK (1980) Monograph on the fishes of re-organized Punjab. Part II. Pb Fish Bull 4(1):39–70
20. Banarescu P (1968) Revision of the Indo-Burmese genus *Salmostoma* Swainson (Pisces, Cyprinidae) with description of a new subspecies. Biologie Zoologie 13(1):3–14
21. Menon AGK (1951) The role of Eastern Ghats in the distribution of the Malayan fauna and flora to Peninsular India. Proc Nat Inst Sci India 17(6):475–497
22. Johal MS, Sharma KP (1986) Fish fauna of Sawai Madhopur district–Rajasthan State, India. Vest Cs Spolec Zool 50:112–119
23. Johal MS, Dhillon KS (1981) Ichthyofauna of Ganganagar district (Rajasthan) India. Res Bull Punjab Univ 32:105–110
24. Durve VS (1976) The fisheries of the Lake Jaisamand, Rajasthan. J Inland Fish Soc India 8:19–23

25. Yazdani GM (1996) Fish diversity in Thar Desert. In: Ghosh AK, Baqri QH, Prakash I (eds) Faunal diversity in the Thar-Desert. Gaps in Research. Scientific Publication Jodhpur, India, pp 285–295
26. Dhawan P (1969) Fish Fauna of Udaipur lakes. J Bomb Nat Hist Soc 66(1):190–194
27. Singh P, Mohan D (2003) Length-weight relationship of *Hypophthalmichehys molitrix* from Ummaid Sagar pond. Aquacult 4(2):221–224
28. Mohan D, Saraswat P (2000) Length-weight relationship of *Cyprinus carpio* in Rawati Pond, Jodhpur. J Nat Con 13(2):123–130
29. Mohan D, Jhanjhria A (2001) Comparative growth studies on the fries of *Labeo rohita*, *Cirrhinus mrigala* and *Cyprinus carpio* for different food variables. J Appl Zool Res 13(1):80–89
30. Jhingran VG, Sehgal KL (1978) Cold water fisheries of India. Indian Fisheries society of India, Barrackpore, Publication, p 239
31. Jhingran VG (1991) Fish and fisheries of India. Hindustan Pub. Co., New Delhi, p 727
32. Dhanze R, Dhanze JR (2004) Fish diversity of Himachal Pradesh. In: Ayyappan S, Malik DS, Dhanze R, Chauhan RS (eds) Fish diversity in protected habitats. NATCON, vol 8. p 39–60
33. Menon AGK (1963) A distributional list of fishes of the Himalayas. J Zool Soc India Calcutta 14(1):23–32
34. Krishnan MS (1952) Evolution of the Desert: geological history of Rajasthan and its relation to present day condition. Bull Nation Inst Sci India Delhi 1:19–31
35. Datta AK, Majumdar (1964) Fauna of Rajasthan, India. Part 7. Fishes Rec Indian Mus Delhi. 62(1&2):63–100.
36. Mohan D, Singh G (1985) Fisheries potential of Kailana lake (Prarap Sagar & Takhat Sagar). Zoological Orientalis 2(182):19–23
37. Mathur DS, Yazdani GM (1973) Additional record of the fish from Jodhpur with a list of species occurring in the district. Sci Cult Calcutta 39:87–89
38. Jayaram KC (1981) Freshwater fishes of India, Pakistan, Bangladesh, Burma and Sri Lanka, A Handbook. Zoological Survey of India, Calcutta, p 475
39. Menon AGK (1974) A check-list of fishes of the Himalayan and Indo-Gangetic plains. Special Publication No. 1, Inland Fisheries Society of India, Barrackpore, India Publication, p 136

Chapter 7
Chelonian Status and Conservation in Rajasthan

Subramaniam Bhupathy and Reena Mathur

Abstract This chapter describes Rajasthan as the abode of 11 species of chelonians, out of them, 10 are freshwater species and one is a land form. The Indo-Gangetic plains bordering the eastern and southern parts of the state, especially the Chambal River, favours their survival. The Indian Flapshell Turtle *Lissemys punctata* is distributed all over the state and many species, especially the hard-shelled turtles are restricted to the Chambal River. The distribution, behaviour, biology, life cycle and conservation status of each species has been concisely described. Like other parts of the country, data on the ecology of turtles is scanty in Rajasthan too. Information on the biology of Indian flapshell and a few species of hard-shelled turtles is largely known from studies conducted at the Keoladeo National Park, Bharatpur, and the Chambal River. Unlike other animals, turtles are not much exploited in Rajasthan. However, nomads and some underprivileged communities do consume turtles for subsistence. The Indian Star Tortoises are reportedly collected illegally for trade from the arid zones of the state; therefore, vigilance by enforcing authorities is suggested. All soft-shelled and a few hard-shelled species are protected by the Indian Wildlife (Protection) Amendment Act, 2006. Conservation status, habitat protection and further research on turtles in the state are also briefly discussed.

S. Bhupathy (✉)
Sálim Ali Centre for Ornithology and Natural History, P.O. Anaikatti,
Coimbatore 641108, India
e-mail: bhupathy.s@gmail.com

R. Mathur
Formerly, Department of Zoology, University of Rajasthan, Jaipur 302004, India
e-mail: rmzooprof@gmail.com

B.K. Sharma et al. (eds.), *Faunal Heritage of Rajasthan, India: General Background and Ecology of Vertebrates*, DOI 10.1007/978-1-4614-0800-0_7,

Introduction

Chelonians are considered the oldest terrestrial vertebrate due to their close affinity to cotylosaurians—the stem reptiles and precursor of all other vertebrates including birds and mammals. It is reported that turtles evolved about 250 million years ago and have coexisted with the dinosaurs. However, chelonians are still extant, while all dinosaurs became extinct due to various reasons. The major reason attributed to the successful survival of turtles till date is the presence of boxlike shell, carapace, the (dorsal) top one, and plastron, the ventral or bottom one. Relationship between turtles and humans in India dates back to time immemorial. Followers of all major religions worship turtles. For instance, in Hindu mythology, turtles are considered as one of the ten incarnations of Lord Vishnu. Chelonians are variously termed as "turtle" (highly aquatic forms), "terrapin" (semi-aquatic forms) and "tortoise" (terrestrial (land) forms). Most of the turtles have web between digits, but, it is modified as paddles in the highly aquatic sea turtles to facilitate easy movement. The land forms lack the web and their open soft body parts have tubercles and small cornified structures.

India has one of the most diverse chelonian fauna of the world with 33 species including five marine and 28 non-marine species. Among the 28 non-marine species, four are terrestrial and the rest are freshwater species. This high diversity of chelonian fauna of India is largely due to the location of the country at the crossroads of Ethiopian, Palaearctic and Oriental Realms. In the Oriental Realm, the country form parts of the Indian and Indo-Malayan faunal subregions. Several Indo-Malayan turtle fauna have extended their ranges into Indian region through northeast India, especially the lowland hill forests. Availability of various habitat types (lowland forests, marshy grasslands, perennial rivers, oxbow lakes) and climate are among the important factors that probably resulted in the high diversity. Relatively little is known about the turtles of Rajasthan. In this chapter, we provide information on the distribution pattern of turtles and major aspects of conservation issues pertaining to this state of India.

Turtle Fauna

A total of 11 chelonian species including one land and 10 freshwater forms have been reported so far from Rajasthan (Table 7.1). The freshwater turtles may broadly be divided into two: Soft-shelled turtles and Hard-shelled turtles. Soft-shelled turtles have relatively thin shells which lack scutes, whereas the hard-shelled turtles have thick shells with well-marked scutes. The soft-shelled turtles are represented by four species and the hard-shelled turtles by six. The number and distribution pattern of turtle species increases from west to east or southeast in the state of Rajasthan due probably to the presence of perennial rivers such as Yamuna and Chambal in these areas. Many turtle species are solely restricted to the Chambal—a perennial river of the state.

Table 7.1 Chelonian fauna of Rajasthan, India

S. No.	Common name	Scientific name
1	Indian Star Tortoise	*Geochelone elegans*
2	Indian Soft-shelled Turtle	*Nilssonia (Aspideretes) gangeticus*
3	Indian Peacock Soft-shelled Turtle	*Nilssonia (Aspideretes) hurum*
4	Narrow-headed Soft-shelled Turtle	*Chitra indica*
5	Indian Flapshell Turtle	*Lissemys punctata*
6	Spotted Pond Turtle	*Geoclemys hamiltonii*
7	Crowned Pond Turtle	*Hardella thurjii*
8	Red-crowned Roofed Turtle	*Batagur (=Kachuga) kachuga*
9	Three-striped Roofed Turtle	*Batagur (=Kachuga) dhongoka*
10	Indian Roofed Turtle	*Pangshura (=Kachuga) tecta*
11	Indian Tent Turtle	*Pangshura (=Kachuga) tentoria*

Indian Star Tortoise *Geochelone elegans*

This is the only tortoise species found in Rajasthan which may be identified from other species based on limbs without web, soft parts with tubercles and shell with black and white starlike radiating marks on each scute. It grows up to 40 cm in terms of shell length. The Indian Star Tortoise has been reported from habitats with thorny scrubs in the arid and semiarid zones. It is reported that the star tortoises listed in Schedule IV of the Indian Wildlife (Protection) Amendment Act, 2006 are heavily traded.

Soft-shelled Turtles

Four species of soft-shelled turtle are distributed in Rajasthan, namely, Indian Soft-shelled Turtle, Indian Peacock Soft-shelled Turtle, Narrow-headed Soft-shelled Turtle and Indian Flapshell Turtle. Soft-shelled Turtle may be differentiated from Hard-shelled Turtle based on relatively thinner and flatter shells lacking scutes and presence of three claws in each limb.

Indian Soft-shelled Turtle *Nilssonia gangeticus*

The Indian Soft-shelled Turtle may be identified by a relatively large head with three oblique black streaks along the tympanum. The head and dorsal shell are olive green. The species grows up to 100 cm in shell length and inhabits rivers, reservoirs and temple ponds of the south, central and eastern parts of the state. Several temple ponds in the southern and eastern Rajasthan hold good number of them, which are fed by people. The Indira Gandhi Canal that cuts across the arid zone of the state might extend this species range into desert region. It is omnivorous in food habits, thriving on both live and dead plant and animal matters. The Indian Soft-shelled Turtle breed during monsoon and may lay up to 50 eggs. Low-income-group people consume this species occasionally. This species is listed in Schedule I of the Indian Wildlife (Protection) Amendment Act, 2006.

Indian Peacock Soft-shelled Turtle *Nilssonia hurum*

The Indian Peacock Soft-shelled Turtle gets its name as the juveniles have four to six ocelli on the carapace, which resemble the ocelli found in the peacock feather. These markings disappear as the turtle grows. The shell and soft parts are largely dark green or black. The head of this species has yellow/orange spots instead of three oblique black streaks along the tympanum as seen in the Indian Soft-shelled Turtle. It grows up to 60 cm in shell length. This species has reliably been reported only from the wetlands of Keoladeo National Park, Bharatpur from Rajasthan, and in the Chambal River as well. Biology of this species is poorly known and it may breed during monsoon similar to that of the other soft-shelled turtles. The Indian Peacock Soft-shelled Turtle is also listed in Schedule I of the Indian Wildlife (Protection) Amendment Act, 2006.

Narrow-headed Soft-shelled Turtle *Chitra indica*

Small narrow head and long neck, placement of eyes closer to the nostril, olive-green shell and soft parts with prominent markings differentiate this turtle from other soft-shelled turtles. Beautiful markings on the shell appear like art/drawing, which brings the genus name *Chitra* (*chitram*=drawing). These markings may provide camouflage to the burring turtles at the river bottom. It is a large species, may grow up to 1.8 m in shell length and inhabits flowing rivers with sandy bottom. It is rare and reported only from the Chambal River with respect to Rajasthan. The Narrow-headed Soft-shelled Turtles reportedly feed on small fishes. It breeds during August–September, and a maximum of 180 eggs may be laid at a time in sandy beaches of the river. Sand mining appears to be one of the major threats for the decline of this species. It is listed in Schedule I of the Indian Wildlife (Protection) Amendment Act, 2006.

Indian Flapshell Turtle *Lissemys punctata*

It is the smallest of all soft-shelled turtles found in India, which grows up to 37 cm in shell length. Presence of marginal bones in the posterior side of the carapace and seven well-developed callosities in the plastron differentiate this species from others. The shell and soft parts are largely olive green. Two subspecies (*L. punctata punctata*, shell and head with black spots or blotches, and *L. p. andersoni*, shell and head with yellow spots and blotches) have been reported. Most of the flapshells of Rajasthan belong to *L. p. andersoni*. Intergrades of these subspecies have been reported from Shri Mahavir Ji and Chambal. This species is widely distributed in both arid and semiarid parts of the wetland state, which affects rivers, reservoirs and ponds. This species may undertake overland movements in response to drought conditions or increased water levels in waterbodies during the rainy season. Turtles burrow and aestivate when shallow ponds and lakes dry up in the summer (May). The maximal duration of aestivation recorded for the wild turtles is 160 days. The species is an opportunistic omnivore which feeds on live and dead matters of both plants and animals. Nesting generally occurs in late summer, extending to the monsoon (July–November). Clutch size varies from 4 to 15 eggs and has prolonged incubation period (over 300 days). Due to the continued and widespread exploitation on this species in most parts of the country, it is protected in Schedule I of the Indian Wildlife (Protection) Amendment Act, 2006.

Hard-shelled Turtles

Six species of hard-shelled turtle have been reported from Rajasthan, namely, Spotted Pond Turtle, Crowned River Turtle, Red-crowned Roofed Turtle, Three-striped Roofed Turtle, Indian Roofed Turtle and Indian Tent Turtle. Thicker and largely raised shell with demarcated scutes and limbs with four or five claws differentiate hard-shelled turtle from soft-shelled turtle.

Spotted Pond Turtle *Geoclemys hamiltonii*

This hard-shelled turtle may be identified from other species by the presence of three keels on the upper shell (carapace), down-curved upper jaw, shell with yellow radiating streaks in each scute and head and soft parts with several yellow or orange spots. Males have concave plastron and longer tail that extends well beyond the shell. The spotted pond turtle grows up to 50 cm in shell length. This species is known from a few localities in Rajasthan. It was relatively common in the stagnant waterbodies with rich aquatic plants in Keoladeo National Park, Bharatpur, but the current status of the species has been reported as rare owing to poor water input. It is a specialist feeder on molluscs, and the jaws are stronger and adapted to crush and break in order to open the shells. Breeding season of this species in Rajasthan and other western parts of its distribution is poorly known. The Spotted Pond Turtle is listed in Schedule I of the Indian Wildlife (Protection) Amendment Act, 2006.

Crowned River Turtle *Hardella thurjii*

The shell is thick and heavy in adults, moderately depressed with a weak vertebral keel (in the turtles) found in Rajasthan (the eastern subspecies). The carapace is dark brown, with grey-black keel and costo-marginal juncture is marked with an orange-yellow band. The plastron is yellow and each scute has a large black blotch. The head has four orange-yellow stripes on each side and sometimes a short crossbar of the same colour on the forehead. Sexual size dimorphism is extreme; males grow up to 21 cm in shell length, whereas females grow up to 60 cm. The tail of the male is longer and thicker than that of the female, but the adult males show no plastral concavity. In Rajasthan, this species is common in Keoladeo National Park, Bharatpur, and is reported to be a shy species feeding largely on aquatic vegetation. Information on the ecology of this species is scanty, and further studies on the biology of the species are required. This species is not protected by the Indian Wildlife (Protection) Amendment Act, 2006.

Red-Crowned Roofed Turtle *Batagur kachuga*

This species has a thick and moderately elevated shell with a weak vertebral keel, dark-brown carapace and uniform cream or yellow plastron. Adult males have blue, cream and red patches on head and neck, whereas in females, it is more or less uniform. It grows up to 60 cm in shell length. In Rajasthan, this species is reported only from Chambal River and is restricted to the river throughout its distribution range. The Red-crowned Roofed Turtle nests are found in the sandy beaches of the rivers during February–April. Barring some information from Chambal River, biology of this species is poorly known. It is listed in Schedule I of the Indian Wildlife (Protection) Amendment Act, 2006.

Three-Striped Roofed Turtle *Batagur dhongoka*

This species has thick and moderately elevated shell with three weak dorsal keels, pale-brown carapace with dark-brown keels and yellow plastron without spots or blotches. It grows up to 50 cm in shell length. This species is also reported only from Chambal River in Rajasthan. It also nests during March–April along the sandy river banks. This species is listed in Schedule I of the Indian Wildlife (Protection) Amendment Act, 2006. The biology of Three-striped Roofed Turtle is poorly known.

Indian Roofed Turtle *Pangshura tecta*

The shell is elevated like that of the tiled rooftop with elevated vertebral keel spiked on the third vertebral scute. The brown carapace with orange vertebral stripe, marginal border with yellow, plastron having 2 or 3 blotches in each scute and reddish crescent-shaped marking between eye and tympanum differentiate this species from other species, especially from that of the Indian tent turtle. This species may grow up to 25 cm in shell length. It is common in many stagnant waterbodies, such as ponds and reservoirs with abundant aquatic vegetation but seldom found in rivers. The ecology of Indian roofed turtle is poorly known. In Keoladeo National Park, it nests during February–March. The Indian Roofed Turtle is protected by the Indian Wildlife (Protection) Amendment Act, 2006.

Indian Tent Turtle *Pangshura tentoria*

The common name of this species, Tent Turtle, is due to its tentlike shell. It has a vertebral keel, which is spiked on the third vertebral scute. Three subspecies have been recognised out of them; the Pink-ringed Tent Turtle (*Pangshura*=*Kachuga tentoria circumdata*) is distributed in Rajasthan. The brown carapace with pink stripe between marginal and pleural scutes, yellow plastron with single large blotch, pink spot and bar behind the eye differentiate this species from other species, especially from its closest relative, the Indian Roofed Turtle. This species may grow up to 27 cm in shell length. The Indian Tent Turtle largely inhabit rivers and is commonly found in the Chambal River. A record of this species has also been obtained from the wetlands of Keoladeo National Park, Bharatpur. Both animal and plant matters are accepted as food and it nests during winter.

Research

India has one of the most diverse chelonian fauna of the world [1], but it is little studied and the state of Rajasthan is no exception in this regard. Most of the reports on the turtles are either anecdotal or based on short studies [2–8]. Ecological information on the turtles of Rajasthan is largely based on studies at Keoladeo National Park, Bharatpur, and the Tri-state (Madhya Pradesh, Rajasthan and Uttar Pradesh) Chambal Sanctuary. The Wildlife Institute of India, Dehra Dun has conducted a study on the turtles inhabiting the Chambal River during 1985–1990 [9] which revealed the habitat use, breeding season, habitat and clutch size of various hard-shelled turtles, especially those belonging to the genera *Batagur* and *Pangshura* (=*Kachuga*) [9, 10]. The Bombay Natural History Society conducted short studies on the ecology of turtles, especially of soft-shelled turtles at Keoladeo National Park, Bharatpur. Information on the aestivation, predation and morphometry of Flap-shell Turtle (*Lissemys punctata*) is fairly well known [11–13]. Studies on the food habits of Indian Flapshell and Indian Soft-shelled Turtle revealed their omnivory [14, 15]. Unfortunately, many aspects of ecology of turtles both in the wild and captivity are poorly known, which warrant further in-depth studies.

Conservation Status

Exploitation

Many of the turtle species are listed as endangered in the Indian Wildlife (Protection) Amendment Act, 2006. Of the 11 species of turtles found in Rajasthan, eight species are listed in Schedule I and one in Schedule IV. Complete ban on collection and exploitation is enforced on species listed in Schedule I (endangered species). Being a state of conservative communities, turtles are not much exploited in Rajasthan. However, nomads and some underprivileged community people do consume turtles for subsistence. All soft-shelled turtles and large hard-shelled turtles are consumed on availability. It is suspected that during fishing season (summer), turtles are being collected from various reservoirs of south and south-eastern parts of the state and sent to markets in eastern India, especially to Kolkata. Similarly, there are unconfirmed reports available that star tortoises are collected illegally from the foot hills of Aravalli and other parts of the arid zone and exported. Vigilance on the exploitation is thus required by the State Forest Department and other implementing agencies.

Habitat Conservation

It is reported that sand mining affects the nesting habitats of turtles inhabiting the Chambal River [9, 10]. Efforts should be made to control sand mining for conserving river turtles in the state and elsewhere. Most parts of Rajasthan depend on monsoon and support a large number of temporary wetlands. At least two species of soft-shelled turtle (Indian Soft-shelled Turtle and Flapshell Turtle) and one species of hard-shelled turtles (Indian Roofed Turtle) are common in these wetlands which go dry during summer and many of them are de-silted periodically. Turtles inhabiting the temporary wetlands may get struck or aestivate in drying mud; hence, entangled turtles must be removed prior to de-silting programmes. For instance, in Keoladeo National Park, the Forest Department relocated 618 turtles from Manasarovar—the deepest waterbody of the park, prior to de-silting of the same during 1987 [16].

Further Research

Conservation planning without understanding the ecology of the species would be futile [17]. Hence, research becomes an important component in planning species conservation. This includes research, both in wild and in captivity. Information on the distribution of turtles in Rajasthan is poorly known, except the eastern and south-eastern parts [7]. Urgent surveys are required in other parts for preparing species distribution maps. Ecology of Indian chelonians is poorly known, and this is true for Rajasthan as well, though aspects of the ecology of Indian Flapshell [7, 13, 14] and river turtles [9, 10] distributed in Rajasthan are available. Further data on habitat use, food habits and breeding biology of turtles is required for preparing effective species conservation plans. With respect to laboratory and captive studies, a lot needs to be done. Physiological adaptations (energetic) of Indian turtles are not known, especially those living in temporary waterbodies. It is to be noted that we do not know the chromosome number of many species of turtles. Hence, many laboratory studies are to be initiated to know this and other basic details of this oldest lineage of land vertebrates.

Acknowledgements We are grateful to the Bombay Natural History Society, Mumbai (Dr V. S. Vijayan and Mr J. C. Daniel), Wildlife Institute of India, Dehra Dun (Mr B. C. Choudhury), and U.S. Fish and Wildlife Service, USA (Mr David Ferguson), for support at various levels. We thank the Rajasthan Forest Department and University of Rajasthan for support and encouragements.

Please see Chap. 2 for pictures.

References

1. Moll EO (1986) Survey of the freshwater turtles of India part I: the genus *Kachuga*. J Bomb Nat Hist Soc 83:538–552
2. Smith MA (1931) The Fauna of British India, including Ceylon and Burma. Reptilia and Amphibia, vol I. Testudines. Taylor and Francis, London, p 185
3. Tikadar BK, Sharma RC (1985) Handbook Indian Testudines. Zoological Survey of India, Calcutta, p 156
4. Das I (1995) Turtles and Tortoises of India. Oxford Univ. Press, Bombay, India, p 179
5. Daniel JC (2002) The book of Indian reptiles and amphibians. Oxford University Press, Oxford House, Mumbai, p 238
6. Choudhury BC, Bhupathy S (1993) Turtle trade in India: a study of tortoises and freshwater turtles. World Wildlife Fund-India (prepared by TRAFFIC-India), New Delhi, p 50
7. Bhupathy S, Vijayan VS (1991) The freshwater turtle fauna of Eastern Rajasthan. J Bomb Nat Hist Soc 88(1):118–122
8. Sharma RC, Vazirani TG (1977) Food and feeding habits of some reptiles of Rajasthan. Rec Zool Surv Ind 73:77–93
9. Rao RJ (1990) Ecological relationships among freshwater turtles in the National Chambal Sanctuary. Final Report. Wildlife Institute of India, Dehra Dun
10. Rao RJ, Singh LAK (1987) Notes on comparative body size, reproductive effort and areas of management priority for three species of Kachuga (Reptilia, Chelonia) in the National Chambal Sanctuary. J Bomb Nat Hist Soc 84:55–65
11. Auffenberg W (1981) Behaviour of *Lissemys punctata* (Reptilia, Testudinata, Trionychidae) in a drying lake in Rajasthan, India. J Bomb Nat Hist Soc 78(3):487–494
12. Bhupathy S (1989) Morphometry of the Indian flapshell turtle (*Lissemys punctata andersoni*). J Bomb Nat Hist Soc 86(2):252
13. Bhupathy S, Vijayan VS (1993) Aspects of the feeding ecology of *Lissemys punctata* (Reptilia: Trionychidae) in Keoladeo National Park, Bharatpur, India. Hamadryad 18:13–16

14. Bhupathy S, Vijayan VS (1994) Aestivation of turtles in Keoladeo National Park, Bharatpur with special reference to *Lissemys punctata* (Reptilia: Trionychidae). J Bomb Nat Hist Soc 91(3):398–402
15. Bhupathy S (1990) Observations on the food of the Ganges soft-shell turtle *Trionyx gangeticus* in Keoladeo National Park, Bharatpur. J Bomb Nat Hist Soc 87(3):460–461
16. Vijayan VS (1987) Keoladeo National Park Ecology Study. Bombay Natural History Society, Bombay, p 131
17. Frazier J (1992) Management of the tropical Chelonian: Dream or Nightmare? In: Singh JP (ed) Tropical ecosystem: ecology and management. Willy Eastern Ltd., New Delhi, pp 125–133

Chapter 8
Important species of Lizards in the Thar Desert of Rajasthan

Shalini Gaur, Sanjeev Kumar, and N.S. Rathore

Abstract In this chapter, the authors have provided an interesting account of lizards found in the Thar Desert with an exploration of the typical faunal representation of the area. Adaptability of these reptiles to the harsh climatic conditions and their activities has been highlighted. In general, these species are scattered in the Thar Desert, but in some districts, their population is dense. A total of 20 species have been reported from different areas of the state. The habitat and distribution of these species in the Thar Desert of Rajasthan has been systematically presented in this chapter. Though, people in the remote areas of Rajasthan are religious and consider animals as part of their families, at times, they kill these varanids thinking them to be poisonous and dangerous and being aware of the related myths. Monitor lizards are consumed by some of the tribal communities. In the present study, efforts have been made to create awareness about these lizards along with conservation strategies.

Introduction

A desert species is naturally modified to cope with a multitude of xeric conditions. The saurian fauna of the Thar Desert of Rajasthan has developed suitable modifications pertaining to sense of hearing, photoperiodism, sight, water balance, thermoregulation, predation, mimicry and defence. Many workers have contributed

S. Gaur (✉) • S. Kumar • N.S. Rathore
Desert Regional Station, Zoological Survey of India, Jhalmand, Pali Road, Jodhpur, Rajasthan, India
e-mail: drshalini2007@yahoo.com

B.K. Sharma et al. (eds.), *Faunal Heritage of Rajasthan, India: General Background and Ecology of Vertebrates*, DOI 10.1007/978-1-4614-0800-0_8,

on the studies of reptiles [1–14]. This chapter is based on prospective field studies and exhaustive surveys in the deserts of Rajasthan over a period of 10 years (1994–2004).

Methods

Time-constrained searches and pitfall trap methods were employed for the study of lizard species in the area. Time-constrained searches include search efforts by 2–3 persons while moving around the area randomly examining all the microhabitats encountered, raking through litter, turning rocks and logs, tearing open decomposed logs, probing in vegetation and so on. Pitfall traps consist of trap grids with 36 traps spaced at 15-m intervals in a 6×6 arrangement. Traps were made of a couple of number ten tins taped together and buried with the lip at the ground line and concealed by the cover of the bark propped above the ground.

Results and Discussion

Gekkonids

A total of five species of gekkonids have been observed in the Thar Desert of Rajasthan which include *Crossobamon orientalis*, *Cyrtodactylus scaber*, *Hemidactylus brookii*, *Hemidactylus leschenaultii* and *Hemidactylus flaviviridis*. *Hemidactylus brookii* prefers to live under the bark of tree, under the stone and lonely areas away from human-dwellings. *Hemidactylus leschenaultii* was mostly observed inhabiting rocks scattered along with angiosperm trees.

Agamids

Commonly observed agamids in the Thar Desert of Rajasthan are *Calotes versicolor*, *Agama agilis* and *Phrynocephalus laungwalensis. Trapelus agilis* stays in sandy areas with some xerophytic vegetation, climbing on small bushes. It has also been observed resting on the ground under the shade of small bushes. *Phrynocephalus laungwalensis* prefers to stay in the extreme desert part of Rajasthan, particularly in the district of Jaisalmer where there are barren shifting sand-dunes of 5–20 m height. The lizard does not make burrow and stays hidden under the sand. The dunal areas with tourist activities have very few of these lizards. A sizeable part of the desert in Rajasthan is occupied by dry undulating land with hardened sand, while the rest consists of rolling shifting sand-dunes. Plantation may be essential for making the dunes stabilised, but some species like *Phrynocephalus laungwalensis* which preferably live within the sand-dunes face the direct threat of habitat destruction. *Uromastyx hardwickii* lives in the hard gravel soil with sparse xerophytic vegetations. Oil is extracted from *Uromastyx hardwickii* by quacks for treatments of various diseases; hence, it is hunted extensively despite a total ban. This species is still fairly common in the land with hard soil surrounded by grassy patches especially in Barmer, Jaisalmer, Bikaner, Sri Ganganagar and some parts of Pali District.

Chamaeleon

Chamaeleo zeylanicus is the only genus of the family Chamaeleonidae observed in the Thar Desert, particularly in the villages of Ratkuria and Desuri located in the districts of Jodhpur and Pali, respectively. The species was rarely noticed elsewhere. The chameleon is insectivorous and arboreal and is facing the threat of habitat destruction due to urbanisation and killing by people on account of lack of awareness.

Skinks

Commonly observed skinks are *Eutropis carinata*, *Eutropis macularia*, *Riopa punctata* and *Ophiomorus raithmai*. In sand dune areas, the sand-dwelling species *Ophiomorus raithmai* is commonly found.

Lacertids

Ophisops jerdonii, *O. microlepis* and *Acanthodactylus cantoris* are commonly observed lacertids in the Thar Desert of Rajasthan.

Varanids

Varanus bengalensis and *Varanus griseus* are the monitor lizards commonly found in the Thar Desert of Rajasthan. Among the observed saurian fauna, lizards like *Crossobamon orientalis* of family Agamidae; *Phrynocephalus laungwalensis* and *Uromastyx hardwickii*, family Agamidae; *Riopa punctata*, family Scincidae; and *Ophiomorus raithmai* and *Ophisops microlepis*, family Lacertidae are rare, whereas

Table 8.1 Species of lizards observed in the Thar Desert of Rajasthan

S. No.	Family	Scientific name	Sighting
1	Gekkonidae	*Crossobamon orientalis* Blanford	Rare
2		*Cyrtodactylus scaber* (Heyden)	Common
3		*Hemidactylus brookii* Gray	Very common
4		*Hemidactylus leschenaultii* Dum. & Bib.	Common
5		*Hemidactylus flaviviridis* Ruppell	Common
6	Agamidae	*Calotes versicolor* (Daudin)	Rare in winter
7		*Trapelus agilis* Olivier	Rare in summer
8		*Phrynocephalus laungwalensis* Sharma	Rare
9		*Uromastyx hardwickii* Gray	Common
10	Chamaeleonidae	*Chamaeleo zeylanicus* Laurenti	Rainy season
11	Scincidae	*Eutropis carinata* (Schneider)	Common in rainy season
12		*Eutropis macularia* (Blyth)	Common in rainy season
13		*Riopa punctata* (Linnaeus)	Rare
14	Lacertidae	*Acanthodactylus cantoris* Gunther	Common
15		*Ophiomorus raithmai* Anderson and Leviton	Rare
16		*Ophisops jerdonii* Blyth	Common
17		*Ophisops microlepis* Blanford	Rare
18	Varanidae	*Varanus griseus Koniecznyi* Mertens	Rare
19		*Varanus bengalensis* (Linnaeus)	Common

Hemidactylus brookii, family Gekkonidae, was quite common. Similarly, *Cyrtodactylus scaber*, *Hemidactylus leschenaultii* and *Hemidactylus flaviviridis*, family Gekkonidae; *Acanthodactylus cantoris* and *Ophisops jerdonii*, family Lacertidae; and *Varanus bengalensis*, family Varanidae, were observed commonly throughout all the surveys. Some lizards could only be observed in a particular season such as *Chamaeleo zeylanicus*, family Chamaeleonidae, which was observed only in rainy season. On the other hand, *Calotes versicolor* and *Trapelus agilis* were observed rarely during the winter and summer seasons (Table 8.1; Table 8.2). Please see Chap. 2 for pictures.

Table 8.2 Species of lizards and their localities in the Thar Desert of Rajasthan

S. No.	Family	Species	Time of activity	Locality	Remarks
1	Gekkonidae	*Crossobamon orientalis* Blanford	Night	Jaisalmer, Jodhpur, Churu, Sri Ganganagar, Barmer,	Barren sand dune areas
2		*Cyrtodactylus scaber* (Heyden)	Evening	Jodhpur and Jalore	Dry grass lands
3		*Hemidactylus brookii* Gray	Afternoon	Bikaner, Jodhpur, Jaisalmer and Hanumangarh	Under stone
4		*Hemidactylus leschenaultii* Dum. & Bib.	Afternoon	Jodhpur, Jaisalmer, Nagour, Pali, Jalore, Sirohi	Buildings, crevices
5		*Hemidactylus flaviviridis* Ruppell	Night	Barmer, Nagour, Jaisalmer and Nagour, Jalore, Churu, Sikar, Bikaner	Wall of human habitations
6	Agamidae	*Calotes versicolor* (Daudin)	Afternoon	Barmer, Bikaner, Churu, Jaisalmer, Jodhpur, Pali, Sikar, Churu	Garden, agriculture field
7		*Trapelus agilis* Olivier	Afternoon	Barmer, Jalore Nagour, Bikaner, Jaisalmer, Pali, Sirohi, Jodhpur	Rocks near agriculture fields
8		*Phrynocephalus laungwalensis* Sharma	Evening	Jaisalmer	Sandy barren desert parts, dunes

(continued)

Table 8.2 (continued)

S. No.	Family	Species	Time of activity	Locality	Remarks
9		*Uromastyx hardwickii* Gray	Morning	Jaisalmer, Churu, Hanumangarh, Jaisalmer, Jalore	Burrow in hardened sand surrounded by xerophytic vegetation
10	Chamaeleonidae	*Chamaeleo zeylanicus* Laurenti	Afternoon	Pali, Jodhpur	Arboreal
11	Scincidae	*Eutropis carinata* (Schneider)	Afternoon	Pali, Barmer, Bikaner, Sri Ganganagar	Terrestrial grassy vegetated areas
12		*Eutropis macularia* (Blyth)	Afternoon	Barmer, Pali and Jodhpur	Terrestrial grassy vegetated areas
13		*Ophiomorus raithmai* Anderson and Leviton	Evening	Jaisalmer	Sand-dunes
14	Lacertidae	*Acanthodactylus cantoris* Gunther	Afternoon	Bikaner, Churu, Hanumangarh, Jaisalmer and Barmer	Sandy areas in dense xerophytic bushes
15		*Ophisops jerdonii* Blyth	Evening	Jaisalmer, Jodhpur	Soft soil areas in dense clusters of vegetation
16		*Ophisops microlepis* Blanford	Evening	Sirohi	Soft soil areas in dense clusters of vegetation
17	Varanidae	*Varanus griseus Koniecznyi* Mertens	Afternoon	Hanumangarh, Jaisalmer and Pali	In burrows, nullahs, under stones, rock crevices, dense vegetations around marshes, ponds, canals and hollow trees
18		*Varanus bengalensis* (Linnaeus)	Afternoon	Bikaner Jodhpur, Pali, Nagour, Bikaner, Sirohi and Jaisalmer	In burrows, nullahs, under stones, rock crevices, in dense vegetations around marshes

Acknowledgements Authors are thankful to the officers in-charge of Desert Regional Station, Zoological Survey of India, Jodhpur during the study period for providing necessary facilities and cooperation.

References

1. Biswas S, Sanyal DP (1977) Fauna of Rajasthan, reptilia. Rec Zool Surv India 77(1–4):247–269
2. Blanford WT (1879) Notes on a collection of reptiles made by Major O.B. St John, R.E. at Ajmer in Rajputana. J Asiat Soc Bengal 48:119–127
3. Boulenger GA (1890) The fauna of British India, including Ceylon and Burma. Reptilia and batrachia. Taylor & Francis, London, p 541
4. Dave KC. Contribution to the systematic, distribution and ecology of the reptiles of the desert of Rajasthan, with special reference to the ecology of certain lizards. Proc. 47th Indian Sci Congr; Calcutta. 1960. Pt. 3 (Abstract), p 482–483
5. Minton SA (1966) A contribution to the herpetology of west Pakistan. Bull AMNH (New York) 134(2):27–184
6. Prakash I (1972) Notes on little known lizards from Rajasthan desert. J Bomb Nat Hist Soc 69(2):424–428
7. Prakash I (1974) The ecology of vertebrates of the Indian desert. In: Mani MS (ed) Ecology and biogeography of India. Hague, Junk Publishers, pp 369–420
8. Rathore NS (1970) Movements, homing behavior and territory of the Indian sand lizard, *Ophiomorus tridactylus* (Blyth) Boulenger. Jap J Ecol 20(5):208–210
9. Roonwal ML (1982) Fauna of the great Indian desert. Desert, resources and technology, 1. Scientific Publication, Jodhpur, pp 1–86
10. Sharma RC (1996) Herpetology of the Thar desert. In: Ghosh AK, Baqri QH, Prakash I (eds) Faunal diversity in Thar desert: gaps in research. Scientific publisher, Jodhpur, pp 207–306
11. Sharma RC (2002) The fauna of India and the adjacent countries—reptilia (Sauria), vol II. Director, Zoological Survey of India, Kolkata, p 430
12. Sharma RC, Vazirani TG (1977) Food and feeding habits of some reptiles of Rajasthan, India. Rec Zool Surv India 73:77–93
13. Smith MA (1935) The fauna of British India including Ceylon and Burma: reptilia and amphibia-sauria, vol 2. Taylor & Francis, London, p 440
14. Tikader BK, Sharma RC (1992) Handbook: Indian lizards. Zoological survey of India, Calcutta, p 250

Chapter 9
Natural History Observations on the Indian Spiny-tailed Lizard *Uromastyx hardwickii* in the Thar Desert

Madhuri Ramesh and Ravi Sankaran†

Abstract Members of the genus *Uromastyx* are a fascinating group of lizards which have successfully survived in a variety of arid microhabitats despite the challenges posed by extreme temperatures and sparse resources. This group is represented by a single species *Uromastyx hardwickii* on the Indian subcontinent. The largest population of Indian spiny-tailed lizards is found in the Thar Desert of Rajasthan principally within the western districts while smaller populations have been reported from western Gujarat. Locally known as the *sanda*, it was earlier hunted in large numbers for its meat and oil obtained from the fat (*sanda ka tel*). This species shelters in burrows that each lizard excavates for itself, and the burrows are usually found clustered together. It is uniquely herbivorous but little is known about this lizard. Here, we present our observations on the natural history of this species and summarize the available information. In addition, importance of the proposed Desert National Park in the conservation of desert fauna, particularly reptiles, has been discussed.

M. Ramesh (✉)
Group for Nature Preservation and Education, New No. 30, Block II, Gandhi Mandapam Road, Kotturpuram, Chennai 600085, India

Ashoka Trust for Research in Ecology and the Environment, Royal Enclave, Sriramapura, Jakkur Post, Bangalore, Karnataka 560064, India
e-mail: madhurir@hotmail.com

R. Sankaran (Late)
Formerly, Salim Ali Centre for Ornithology and Natural History (SACON), P.O. Anaikatty, Coimbatore 641108, India

B.K. Sharma et al. (eds.), *Faunal Heritage of Rajasthan, India: General Background and Ecology of Vertebrates*, DOI 10.1007/978-1-4614-0800-0_9,

Introduction

The genus *Uromastyx* comprises of spiny-tailed agamids which are restricted to the hot deserts of the Old World [1, 2]. These lizards are diurnal, ground-dwelling, primarily herbivorous and live in burrows. They have a Saharo-Sindic distribution, ranging from the Sahara Desert in Africa to the Thar Desert in Asia, across the Arabian Peninsula, where typical *Uromastyx* habitats have high temperatures, scanty rainfall, sparse vegetation and firm substrates. Well adapted for surviving in arid regions, these lizards have occupied different microhabitats from rock caves (e.g., *U. ocellata*) to compact sandy plains (e.g., *U. aegyptia*) but are not found on sand-dunes probably due to the difficulty of burrowing in these areas [2].

They have a compact, dorsoventrally flattened body with short, strong limbs and a distinctive tail covered by whorls of spiny scales. The tympanum is externally visible. Dentition is acrodont, but teeth are not continuously replaced; therefore, in older lizards, when the teeth get worn-out completely, the anterior jaw margins act as cutters [2, 3]. To cope with their unusual herbivorous diet, *Uromastyx* have nasal salt glands [2, 4], enlarged colons with colic valves [1] and microbial fermentative processes to digest cellulose [5].

In total, 17 species of *Uromastyx* are recognized [2, 3] and the conservation status of all of which is of concern because they are heavily exploited for the pet trade or for medicine. The entire genus is listed in Appendix II of CITES [6]. On the whole, knowledge of *Uromastyx* is based mostly on anecdotes of pet-breeders, though, a few studies have been carried out in the Middle East, for example on *U. aegyptius* [7, 8]. In fact, the Indian Spiny-tailed Lizard *Uromastyx hardwickii* (Fig. 9.1) is one of the relatively better-known species with descriptive observations dating back to almost a century ago [9]. Information gathered during a preliminary ecological study conducted in 2008 in Jaisalmer district of western Rajasthan has been presented in the chapter.

Fig. 9.1 The Indian Spiny-tailed Lizard *Uromastyx hardwickii* from Jaisalmer district, Rajasthan

Distribution

The Indian Spiny-tailed Lizard *Uromastyx hardwickii* is the oldest member of the genus [10, 11] and the only herbivorous lizard of the Indian subcontinent. Earlier reported to be widespread but patchily distributed in the arid zone of Indo-Pakistan extending from western Uttar Pradesh to eastern Sind, including parts of Gujarat and Punjab [9, 12–17], it is now considered "Vulnerable" [18] in India and "Endangered" in Pakistan [19]. A desert-adapted species, the last large populations of this unique lizard are probably confined to the remnants of the truly arid tracts of India (Fig. 9.2) namely the Kutch district in Gujarat [20] and Jaisalmer district in Rajasthan [21].

Habits and Habitat

In western Rajasthan, *U. hardwickii* usually occurs in the gravel plains or "thalar" areas of the desert (Fig. 9.3). These plains have characteristically short vegetation cover (<2 ft in height) consisting of herbs and grasses (e.g., "dhamasa" *Fagonia* spp., "chapri" *Neurada procumbens*, "ghantiya" *Dactyloctenium* spp.), and sometimes shrubs of *Capparis decidua* (kair) may also be present.

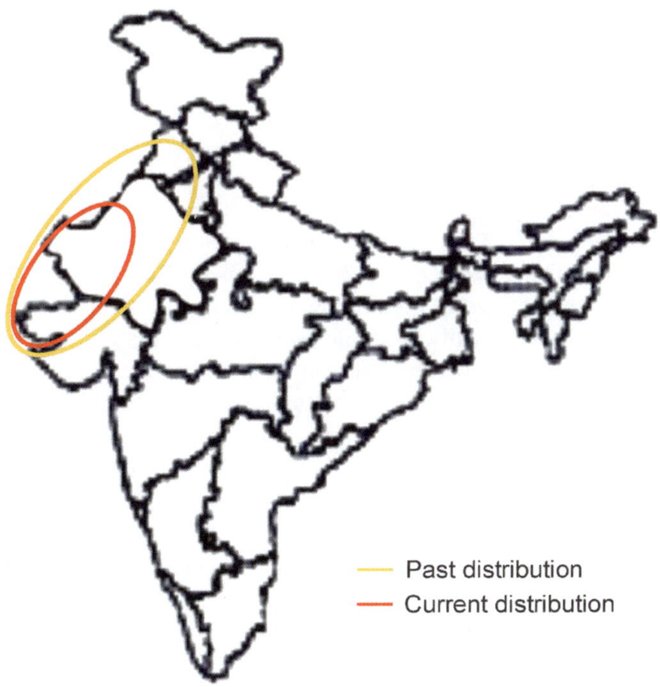

Fig. 9.2 Distribution of the Indian Spiny-tailed Lizard *Uromastyx hardwickii*, in India (Smith, 1935)

Fig. 9.3 Typical *Uromastyx* habitat—*thalar* covered by herbs and short grasses

U. hardwickii excavate long, curved burrows in *thalar* for shelter. The burrows have an elliptical mouth, flush with the ground, which leads into a tunnel that gradually slopes downward. After a few feet, the tunnel may curve before ending in a small chamber. Some burrows are zigzag and extend to several feet in length. For instance, Purves [9] excavated a burrow that sloped more than 1 m down with a tunnel almost 3 m long. The burrow entrance is marked by a small and triangular patch of bare ground where the lizard usually sits and basks in the morning. While basking, most *Uromastyx* sit at the burrow entrance facing the direction in which the tunnel slopes. This probably facilitates a quick retreat into the burrow. Burrows are found irregularly clustered together and inter-burrow distance may be less than 1 m in densely populated clusters (e.g., Khuri, Desert National Park). Each lizard excavates a burrow for itself and shows a strong homing ability. The active lizards can run fast either in a single stretch or in short spurts, moving more than 40 m away from their burrow during the breeding season. If threatened, when far from the burrow, they usually dash directly back to their own burrow, rather than simply entering the nearest one. The mouth of the burrow is plugged completely with soil when the lizard has retired for the night, or is inactive during unfavorable weather conditions such as a cold spell or sandstorm (Fig. 9.4). The plug may also assist in protection from predators such as snakes.

U. hardwickii are most active when the air temperature ranges from 30 to 45 °C, though a few individuals may be active for a short span even at 50 °C (Fig. 9.5).

Fig. 9.4 Burrow of an adult *U. hardwickii* plugged with soil

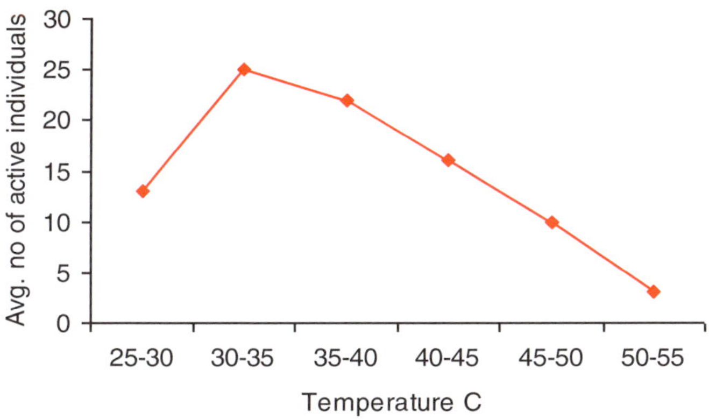

Fig. 9.5 Temperature range of activity of *Uromastyx hardwickii* in Jaisalmer district (March–September, 2008) determined by scan sampling ($n = 101$ h)

They usually bask first with the torso or just the head visible outside the burrow, before completely emerging from it (Fig. 9.6). On emerging from the burrow, these lizards are dark grayish to olive-brown in color and sluggish in their movements. During the cooler months, preceding hibernation (such as October), the smaller individuals bask by completely emerging from the burrow and tilting the dorsum towards the direction of sunlight. Apart from hibernating in winter [9, 16], our observations suggest that some individuals may also go into torpor during peak summer.

In several species of *Uromastyx* the body color changes with increase in temperature [2], though the underlying mechanism by which this occurs is yet to be investigated. In the case of *U. hardwickii*, as the lizard gets warmer, the body color

Fig. 9.6 *U. hardwickii* emerging from its burrow

becomes light brown or beige with minute orange flecks on the dorsum and the tail appears sky blue. On very hot days, the body looks almost white. However, molting individuals look considerably duller than others, especially at the tail.

Diet

U. hardwickii are known to be primarily herbivorous, though juveniles have been reported to consume locusts in captivity [13]. We found adults to be folivorous for the most part, feeding on the small leaves of herbs such as *Fagonia* spp., *Neurada procumbens*, *Barleria acanthoides* and fine stalks and new shoots of *Dactyloctenium* spp. They have also been reported to feed on flowers and berries of *Capparis decidua* and *Salvadora persica* [9]. In summer, *U. hardwickii* were occasionally observed climbing 2.5 m up *kair* shrubs to feed on the flowers.

Scat

The scat of *U. hardwickii* is spindle-shaped, dark green when fresh and straw-colored or dirty white when bleached by the sun for several days (Fig. 9.7). These lizards do not have a particular defecation spot. Coprophagy was noted on three instances when lizards consumed fresh cow dung. Examination of the cow dung

Fig. 9.7 Fresh scat of *U. hardwickii*

revealed the presence of undigested seeds (*guar*), perhaps it was these that the lizards were feeding on.

Courtship and Combat

The breeding season commences as soon as the lizards emerge from hibernation [9]. We were unable to identify the sexes by sight alone, due to the absence of distinguishing field characters, contrary to earlier reports [12, 20] but consistent with observations of pet breeders [2] . From the preliminary observations by the authors it seems likely that some form of polygamy exists and for the convenience of description here, we have chosen to assume territorial individuals as "males" and these individuals defended access to several other burrows, the occupants of which will henceforth be referred to as "females."

Courtship and mating period lasts for about 3 months, from February to April. During this time, the "males" patrol certain patches and chase off other "males" if they approach burrows within the patch. While chasing the intruder, the "male" arches its tail above its back (Fig. 9.8).

Courtship, consists of "males" peeping into the "female" burrows and occasionally, if the "female" has partially emerged from the burrow (i.e., the head and half the torso outside), they nudge or try to extract the "female" completely by pulling

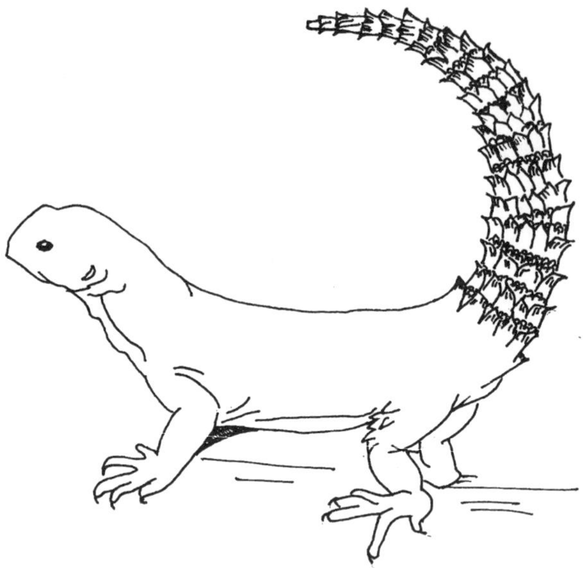

Fig. 9.8 Diagrammatic representation of arched tail posture of a patrolling *U. hardwickii*

the skin at the nape of the neck. The "males" also perform a "belly drag walk" around the "female" burrows that they defend; this consists of circling the mouth of the burrow once or several times, with the pelvic region rubbing against the ground. This may result in some sort of scent-marking, since the males have well-developed femoral scent glands [16]. Scent-marking has been known to assist in territory maintenance as well as for sex and individual recognition in some lizards, such as the Desert Iguana *Dipsosaurus dorsalis* and the Green Iguana *Iguana iguana* and could serve a similar purpose in *U. hardwickii* [22].

During mating, the "male" approaches the "female," attempts to position itself on the "female" and immobilizes it by biting or pressing down on the back of the neck and simultaneously entwining its tail around that of the "female." Often during the second stage, the "female" flips the "male" over its back, while the "male" tries to regain its position. This results in the lizards rolling one over the other. They often "wrestle" with such force that the toppled lizard lands with a thud that can easily be heard 20 m away. Mating attempts last up to 5 min. Sometimes, for a few minutes after separation of the adults, the male's tail remains arched in the middle and the hemipenis is visible, presumably indicating successful copulation.

Combat was also observed and consists of three stages. The first consists of inflation and sidling in which the opponents inflate their bodies, tilt the farther side of the body up (probably in order to look taller) and walk with a stiff-legged gait, often sideways. This is followed by the lunge in which one or both lizards charge towards the other, usually with open gape but rarely bite. They sometimes use their head to shove the opponent aside. A combat session ends when either or both lizards move apart and deflate their bodies.

Juveniles

Hatchlings are first seen in June, when up to 16 young ones emerge from a single adult burrow (Fig 9.9). They are overall dark in appearance with black vermiculations on a brown background. This species probably provides some form of parental care because "females" are believed to lay eggs in a branch of the main burrow [15] and can be seen basking at the entrance along with several of their young ones. They all dive in one after the other at the appearance of a raptor. Neither feeding nor defecation was observed in the hatchlings at this stage. They rarely moved more than 2 m beyond the burrow, and seemed to show strong burrow fidelity even when adjacent burrows were less than 1 m apart.

The young begin to excavate their own burrows only in July, probably because by then the compact *thalar* soil becomes soft after the first showers. They take up to a week to excavate the burrow and shift permanently from the shared adult burrow to their individual one. Young burrow mouths are small and round, the size of a one rupee coin, and the triangular "doorstep" is indistinct (Fig. 9.10). The young too plug the mouth of their own burrow with soil. The monsoon results in the availability of fresh forage and hatchlings, first seen feeding at this time. They initially consumed shoots of *ghantiya* and a few weeks later, fresh leaves of *chapri*. Even this tender forage required the use of a considerable amount of force and it was not unusual to see hatchlings tug at a shoot with such force that they staggered backwards.

After they begin to live independently, the young undergo what is probably their first molt. This results in the body color lightening to a medium brown but the dark vermiculations on the head are still present. The tail spines are lighter towards the sides, but the sky blue coloration develops only later, probably after a year or more.

Fig. 9.9 A juvenile *U. hardwickii*

Fig. 9.10 Burrow of a juvenile *U. hardwickii*

Hatchlings and juveniles show combat behavior with opponents of a similar size but run away if chased by a bigger individual.

Predators

Predators of *U. hardwickii* include raptors such as the Laggar Falcon (*Falco jugger*), Tawny Eagle (*Aquila rapax*) and Steppe Eagle (*Aquila nipalensis*). The eagles were observed feeding on the head and body parts, but not the tail (Fig. 9.11). Crows (*Corvus splendens*) also occasionally prey on hatchlings of *Uromastyx* if the cluster is located near habitation. Desert Foxes (*Vulpes vulpes pusilla*) and feral dogs were the only mammalian predators (the latter solely near habitation, especially after the rains). Among the reptiles, Desert Monitors (*Varanus griseus*) were observed capturing and consuming *U. hardwickii*. Earth Boas (*Eryx johnii*) may also feed on these lizards because on two occasions, boas were seen emerging from open *U. hardwickii* burrows and when probed with a flexible twig, the burrows appeared to be empty.

Trade and Exploitation

Through most of their range, *Uromastyx* lizards are hunted for their meat and lizard eating appears to be a hoary practice, there are references to it in some versions of the ancient epic poem of Persia, the Shahnameh (Fig. 9.12). These most probably refer to *Uromastyx* lizards and not monitors, since people of many cultures believe that consuming the meat of the former confers strength and virility.

Fig. 9.11 Eagles prey on *U. hardwickii* but do not consume the spiny tail

Locally known as the *sanda*, *U. hardwickii*, too, is hunted for the meat and the oil obtained from the fat, the latter is known as *sanda ka tel* and is believed to have medicinal properties. It is used to treat joint pains and is also considered an aphrodisiac [9]. Earlier, it was hunted in large numbers and the extensive trade and exploitation posed a serious threat [17, 18]. However, from a survey conducted in 2007 in western Rajasthan [21], it appears that the volume of trade in *U. hardwickii* has declined considerably in this region primarily due to the strict enforcement of Wildlife Protection laws by the Rajasthan Forest Department. A few communities do continue to hunt this species, but largely for immediate household consumption which is mainly for their allegedly medicinal properties, than as a major source of protein. Normally, only two or three lizards are caught at a time so this level of exploitation does not pose a significant threat. Information on hunting techniques and modes of utilization was also collected during this survey and some of the details are described below:

Hunting Techniques

The techniques that are used to hunt the *sanda* in western Rajasthan, mainly rely on the fact that all burrows have only one opening and consist of a single long tunnel. The post-monsoon season (around September) is considered as the best time to capture these lizards because forage is abundant and they start building up body fat in preparation for hibernation in the following winter months. Both men and women hunt this lizard; sometimes children may do so all by themselves since it is a harmless species which at the most will thrash its spiny tail when handled. Except one, all hunting techniques involve destruction of the burrow.

Fig 9.12 Verses (in red) from the Shahnameh that refer to "Arabs eating lizards" (most probably *Uromastyx*)

به من بازگوی آنک شاه تو کیست
چه مردی و آیین و راه تو چیست

به نزد که جویی همی دستگاه
بر هنه سپهبد بر هنه سپاه

بنانی تو سیری و هم گرسنه
نه پیل و نه تخت و نه بارو بنه

به ایران تو را زندگانی بس است
که تاج و نگین بهر دیگر کس است

که با پیل و گنجست و با فروجاه
پدر بر پدر نامبردار شاه

به دیدار او بر فلک ماه نیست
به بالای او بر زمین شاه نیست

هر آن گه که در بزم خندان شود
گشاده لب و سیم دندان شود

به بخشد بهای سر تازیان
که بر گنج او زان نیاید زیان

سگ و یوز و بازش ده و دو هزار
که با زنگ و زرند و با گوشوار

به سالی هم دشت نیزه وران
نیابند خورد از کران تا کران

که او را به باید به یوز و به سگ
که در دشت نخچیر گیرد به تگ

سگ و یوز او بیشتر زان خورد
که شاه آن به چیزی همی‌نشمرد

ز شیر شتر خوردن و سوسمار
عرب را به جایی رسیدست کار

که تاج کیان می کند آرزو
تفو بر تو ای چرخ گردون تفو

شما را به دیده درون شرم نیست
ز راه خرد مهر و آزرم نیست

بدان چهره و زاد و آن مهر و خوی
چنین تاج و تخت آمدت آرزوی

جهان گر بر اندازه جویی همی
سخن بر گزافه نگویی همی

1. Excavation: First a flexible twig is inserted into the burrow to check if the burrow is occupied and to gauge the depth at which the lizard is resting. Then the lizard is simply dug out using a crowbar, a sharp wooden stake or a spade. This appears to be the most common and fastest technique, particularly after the rains when the ground is soft and easy to dig. A modification of this technique was reported by the *Baori* people—late in the evening, the anterior end of the burrow is filled with sand. Then, the next morning, as the *sanda* emerges, the sand gets pushed behind, thereby, reducing the effective length of the tunnel. Hence, the lizard cannot retreat very far underground when disturbed and can be dug out easily by the hunter.
2. Smashing: Using a flexible twig, the position of the *sanda* in the burrow is determined. Then a large rock is thrown with great force on the ground at a point just

behind the position of the lizard. This causes the burrow to collapse behind the lizard and prevents it from retreating further into the tunnel so that it can be dug out easily.

3. Noosing: One end of a short length of rope (less than 0.5 m) is made into a noose and set at the mouth of the burrow, while the other end is weighed down by a heavy rock. The noose is usually set early in the morning and as the lizard emerges to bask or forage, it gets caught in the noose. Since, the rope is short and weighed down, the lizard can retreat only a few inches into the burrow. According to the hunters, the noosed lizard cannot simply be pulled out because it will inflate its body and wedge itself tightly into the burrow. Instead, it has to be dug out carefully.
4. Flooding: The lizard is flushed out by pouring water into the burrow and is then captured by hand. Since, the burrows are fairly long and the substrate dry, large volume of water is required to flood each burrow (up to 5 L). Hence, this method is often impractical or at the best tedious. Several hunters opined that wetting the substrate actually helps the lizard to dig further and extend the burrow, thereby evading capture.
5. Snake Mimicry: This unusual method is now used only by some hunters of the *Nayak* community. Here, the hunter approaches the burrow in the evening as soon as the lizard has retired for the day, and brushes the ground in front of the mouth of the burrow with a bundle of dry grass or peacock feathers. This makes a rustling sound, similar to that of a snake. In order to defend itself, the lizard supposedly thrashes its spiny tail close to the mouth of the burrow and can be grabbed by the hunter while doing so.
6. Chasing: At least two people are required to implement this technique. When a *sanda* is foraging some distance away from its burrow, one person stealthily approaches the burrow and blocks the opening with his/her foot while the other person attempts to catch the lizard by hand. In most cases, the lizard will run straight back to its own burrow and gets caught by the person waiting there. This method is quite strenuous, since, *Uromastyx* can run very fast and in most cases, will dive into their burrows well before the hunter can block off the entrance. It is usually used only by children, more as a game than as a serious attempt to hunt.

Consumption

In winters, the *Bhils* make a *sabzi* (curry) of the meat and eat it with *bajra roti* (millet chapatti) because they believe that it helps to ward off the cold. The fat bodies (one on either side of the pelvis) melt into oil when heated, are used as *tadka*, or kept aside to be used as medicine.

The *Baori* on the other hand, prefer to eat the meat after roasting it over the fire. Sometimes, this can make the skull explode so it needs to be done carefully. After roasting, the belly skin is completely charred and can be scraped off easily.

The entrails, fat bodies and nails are removed and the animal is roasted again till it is fully cooked. Finally, the tail spines are removed and the lizard is eaten along with the thin dorsal skin. The fat bodies are dissolved into oil to be eaten with meat or are used later as medicine. Sometimes, the oil is used to make collyrium (*kajal*): the oil is poured into a small lamp which is lit and then covered with a vessel (usually a *paranth*). The soot deposited on the vessel is scraped off and brought to the consistency of an ointment by the addition of more oil. This is later used as collyrium and is believed to improve vision and makes the eyes lustrous.

Trade

As mentioned earlier, *sanda ka tel* is used as a remedy for joint pains and as an aphrodisiac; it is claimed that a lizard weighing 100 g can yield about 10 g of fat. This is considered a "lowly" animal so while only members of the hunting communities will capture and eat the meat of this species, many others including educated city-dwellers purchase the oil for its medicinal property. Therefore, unlike the meat, the oil is sold surreptitiously in many places and may be available hundreds of kilometers away. In the smaller villages, it is usually obtained by word of mouth.

Sanda ka tel currently fetches upwards of Rs 100 per *tola* (1 *tola* = 11.6 g) but the price varies greatly depending upon the effort that went into extraction and smuggling of the oil—the latter being proportional to distance from the nearest colony of *Uromastyx*. The pure oil is supposedly yellowish in color, odorless and causes a warm sensation when rubbed into the skin. It does not congeal even in the desert winter. Because of its similarity, mustard oil is the preferred adulterant in *sanda ka tel*. In fact, most hunters opined that traded oil usually had negligible amounts of the original ingredient. Some respondents mentioned that the oil is also bought by middlemen who later sell it in states as far away as Maharashtra, Bihar (where it is reddish in color), Punjab, and Uttar Pradesh (especially in New Delhi at the market in front of the Jama Masjid). In Bikaner and Sriganganagar, the middlemen are often truck drivers. One person mentioned that the oil is now also supplied to Bengalooru (Karnataka).

Conservation

While trade and exploitation has clearly declined at least in western Rajasthan where some of the largest populations of *U. hardwickii* are found, habitat loss continues to be a serious threat. Since *thalar* forms a stable substrate, large expanses are continuously being lost due to agricultural and developmental activities including afforestation and irrigation. In addition, the thriving tourism industry has resulted in previously undisturbed *thalar* areas now being criss-crossed by resort-builders, campers, and off-road safari goers.

Given this scenario of rapid change, Protected Areas play a crucial role, not only in the long-term conservation of this iconic reptile of Rajasthan, but also in the maintenance of arid zone biodiversity [23]. As of now, there are only two Protected Areas in the Thar Desert—the Desert National Park (3,162 sq.km) and Churu Wildlife Sanctuary (7 sq.km). Many fauna typical of the arid region are found here including the Great Indian Bustard (*Ardeotis nigriceps*), Desert Fox (*Vulpes vulpes pusilla*), Afro-Asian Sandsnake (*Psammophis schokari*), and the Laungwala Toad-headed Lizard (*Bufoniceps laungwalaensis*).

Moreover, the desert ecosystem is probably one of the least studied and most threatened ecosystems in the world. Though climatologically the desert region is vast and covers much of Western India, in ecological terms, the real desert is now probably confined to pockets of Jaisalmer district (Rajasthan) and Kutch district (Gujarat). Therefore, to conserve the Indian Spiny-tailed Lizard and other desert fauna, it is imperative to ensure that sufficiently large areas of the desert remain protected and that these areas span the entire range of habitats found in the Thar including barren sand-dunes, *sewan* grasslands, *muggra* (rocky) plains, and *doongar* (rocky hillocks). In addition, given that the desert is probably the largest grazing ground in India, the maintenance and sustainable use of its grasslands in particular becomes very important. To this end, the implementation of the recommendations made by the XI Planning Commission's Task Force on Grasslands and Deserts (2006) is crucial.

Box 9.1 Status of the Indian Spiny-Tailed Lizard *Uromastyx hardwickii*

National Status: "Vulnerable" in India (large populations mainly in Jaisalmer district)

Regional Status: Restricted to the arid zone of northwestern India and eastern Pakistan

(Great Indian Desert)

International Status: Data deficient

Acknowledgments We thank the Rajasthan Forest Department for permission and support to work in the western Rajasthan, particularly PCCF and CWW Shri R. N. Mehrotra for his encouragement. MR would also like to thank the staff of the Desert National Park especially Shri Hazara Ram, Shri Chain Singh Khichi and family, Dr and Mrs Verma, Mr and Mrs Devendra Singh Rathore for their help and hospitality, and Musé Khan for braving two summers of fieldwork. Drs Elaheh Kheirandish and Arash Afraz kindly helped to find the relevant verses of the Shahnameh. MR's fieldwork was supported by the Rufford Small Grants Foundation and IdeaWild.

References

1. Iverson JB (1982) Adaptations to herbivory in Iguanine lizards. In: Burghardt GM, Rand AS (eds) Iguanas of the World. Noyes, Park Ridge, NJ, pp 60–76
2. Wilms T. *Uromastyx*. Natural History, Captive Care, Breeding. Herpeton, Germany. 2005; pp 143

3. Wilms T, Schmitz A (2007) A new polytypic species of the genus Uromastyx Merrem 1820 (Reptilia: Squamata: Agamidae: Leiolepidinae) from southwestern Arabia. Zootaxa 1394:1–23
4. Hazard LC (2004) Sodium and Potassium secretion by Iguana salt glands: Acclimation or adaptation? In: Alberts AC, Carter RL, Hayes WK, Martins EP (eds) Iguanas: Biology and Conservation. University of California, Oakland, CA
5. Foley WJ, Bouskila A, Shkolnik A, Choshniak I (1992) Microbial digestion in the herbivorous lizard *Uromastyx aegyptius*. J Zool 226:387–398
6. UNEP-WCMC. (2004) Review of significant trade: analysis of trade trends with notes on the conservation status of selected species. Annex C: Reptiles and Amphibians; 161–183
7. Robinson MD (1995) Food plants and energetic of the herbivorous lizard, Uromastyx aegyptius microlepis, in Kuwait. J Univ Kuwait (Sci) 22:255–261
8. Cunningham PL (2000) Daily activity pattern and diet of a population of the Spiny-tailed Lizard, *Uromastyx aegyptius microlepis*, during summer in the United Arab Emirates. Zool in Mid East 21:37–46
9. Purves EH (1915) The thorny-tailed lizard. J Bomb Nat Hist Soc 23:780–784
10. Joger U (1991) A molecular phylogeny of Agamid Lizards. Copeia 3:616–622
11. Amer SAM, Kumazawa Y (2005) Mitochondrial DNA sequences of the Afro-Arabian spiny-tailed lizards, (genus *Uromastyx*; family Agamidae): phylogenetic analyses and evolution of gene arrangements. Biol J Linnean Soc 85:247–27
12. Abdulali H (1960) Notes on the Spiny-tailed lizard, *Uromastix hardwicki* [sic] Gray. J Bomb Nat Hist Soc 52:421–423
13. Bhatnagar RR, Bhanotar RK, Srivastava YN, Mahto Y (1973) Observations on colony pattern in *Uromastix* [sic] *hardwickii* Gray. Entomologists' Newsletter 3(4):27
14. Gray JE (1827) *Uromastix* [sic] *hardwickii*. Zool J 3:219
15. Minton SA (1966) Contributions to the herpetology of Pakistan. Bull Amer Nat Hist Mus 134:27–184
16. Smith MA (1935) The fauna of British India. Vol.II: Sauria. Francis & Taylor, London, pp 242–247
17. Vyas R (1991) Notes on capture of the Spiny-tailed Lizard (*Uromastyx hardwickii*) in Gujarat. Hamadryad 15:28
18. Molur S, Walker S (1998) Report of the workshop 'Conservation Assessment and Management Plan for reptiles of India'. BCPP—Endangered species project, Zoo Outreach Organisation, CBSG, India. Coimbatore, India, pp 175
19. Khan MZ (2004) N Mahmood. Study of Population status and natural history of Agamid lizards of Karachi. Pakist. J Biol Sci 7:1942–1945
20. Datta S, Jhala Y (2007) Ecological aspects of the Indian spiny-tailed Lizard *Uromastyx hardwickii* in Kutch. J Bomb Nat Hist Soc 104:255–265
21. Ramesh M, Ishwar NM (2008) Status and distribution of the Indian Spiny-tailed Lizard *Uromastyx hardwickii* in the Thar Desert, western Rajasthan. Technical Report T 01. Group for Nature Preservation and Education, pp 50
22. Alberts A (1991) Phylogenetic and adaptive variation in lizard femoral gland secretions. Copeia 1:69–79
23. Sankaran R, Rahmani AR (1998) The role of grazing enclosures of the Desert National Park in the conservation of wildlife in the Thar Desert, India. Research and management options for Protected Areas: Proceeding of the First International Symposium and Workshop on Arid Zone Environments, pp 145–161

Further Reading

Pianka E, Vitt LJ (2003) Lizards: Windows to the Evolution of Diversity. University of California Press, Oakland, CA, p 310

Chapter 10
Ophidians of Rajasthan

Chhaya Bhatnagar, Satish Kumar Sharma, Meenakshi Mathur, and B.K. Sharma

Abstract The chapter briefly presents diversity and distribution of the Ophidiofauna of Rajasthan in various ecological regions. Out of a total of 275 snake species occurring in India, 34 species belonging to six families, namely, Typhlopidae, Pythonidae, Boidae, Colubridae, Elapidae, and Viperidae have been reported from Rajasthan. Of them, only six species are venomous, seven are mildly venomous, and 21 species are of the nonvenomous variety. IUCN categorizes eight snakes as Low Risk Least Concern (LR-LC), another 22 as Low Risk but considered Near Threatened (LR-NT), and one species as Vulnerable (VU). Common serpent taxa like Brahminy Worm Snake, Common Sand Boa, Red Sand Boa, Indian Rat Snake, Common Wolf Snake and three venomous species, namely, Common Krait, Spectacled Cobra, and Saw-scaled Viper are widespread in the state of Rajasthan. Keoladeo National Park (KNP), Bharatpur is an excellent site to observe the Indian Rock-Python. Conservation threats such as increase in agricultural and urban areas and killing of snakes out of fear and recommendations for their protection have also been discussed along with the myths about snakes. No distribution mapping and proper taxonomical studies of the snake species found in the state have been done so far in the recent past.

C. Bhatnagar (✉) • M. Mathur
Department of Zoology, Mohanlal Sukhadia University, Udaipur, Rajasthan, India
e-mail: bhatnagarchhaya@yahoo.co.in

S.K. Sharma
Sajjangarh Wildlife Sanctuary, Department of Forests and Wildlife, Government of Rajasthan, Udaipur, Rajasthan, India
e-mail: sksharma56@gmail.com

B.K. Sharma
Department of Zoology, R.L. Saharia Government P.G. College, Kaladera (Jaipur), Rajasthan, India
e-mail: drbksharma@hotmail.com

B.K. Sharma et al. (eds.), *Faunal Heritage of Rajasthan, India: General Background and Ecology of Vertebrates*, DOI 10.1007/978-1-4614-0800-0_10,

Introduction

Reptiles are believed to have originated during the Upper Carboniferous period. Approximately 530 species of reptiles are believed to occur in India. Of these a total of 197 are endemic to India [1] with 98 being endemic to the Western Ghats alone [2]. Being an arid and semiarid state, reptiles are well represented in terrestrial, aquatic, and arboreal life forms in Rajasthan. Snakes belonging to the suborder Ophidia are colorful, sinuous, and graceful and form a vital component of our ecosystem. This taxon of silent creepers is also well distributed in all types of habitats. Many workers [3–41] have contributed about various aspects of snakes of Rajasthan. It was concluded in the present study that out of the 34 species of snakes distributed in the varied habitats of the state, 21 are nonvenomous, seven are mildly venomous, and six are venomous. The identification features [1] and general habitats of the ophidians of Rajasthan are described below:

1. **Brahminy Worm Snake *Ramphotyphlops braminus*** IUCN Status (2012)-Not Evaluated
   Identification—A division line (suture) of nasal shield touching the pre-ocular shield instead of the second labial shield.
   Maximum Length—230 mm
   Habitat—Spends most of its life lying beneath stones, flower pots or digging in to slightly damp earth.
2. **Beaked Worm Snake *Rhinotyphlops acutus*** (IUCN Status (2012)-Not Evaluated
   Identification—The largest among Indian worm snakes with a beaked snout (Fig. 10.1)
   Maximum Length—600 mm
   Habitat—Spends most of its life underground.

Fig. 10.1 Beaked Worm Snake *Rhinotyphlops acutus*

Fig 10.2 Indian Rock Python *Python molurus. Courtesy: Devendra Bhardwaj*

3. **Asiatic Rock Python *Python molurus molurus*** IUCN Status: Lower Risk/ Near Threatened
   Identification—Sensory pits on the rostral (snout shield) and first two labials (lip shields) distinguish it from all other Indian snakes (Fig. 10.2).
   Maximum Length—7,620 mm
   Habitat—Inhabits estuarine mangrove forests, arid scrub jungle, rain forest, and grasslands.

4. **Common Sand Boa *Gongylophis conicus*** IUCN Status (2012)-Not Evaluated
   Identification—Absence of a "groove" beneath the chin and angular transverse ridge on rostral shield. A markedly stout snake with a short yet heavy tail which tapers to form a conical shape.
   Maximum Length—1,000 mm
   Habitat—Lives in burrows.

5. **Red Sand Boa *Eryx johnii*** IUCN Status (2012)-Not Evaluated
   Identification—A stout, heavy, and muscular snake, more or less uniform in girth from head to tail with little constriction at the neck. Presence of a mental (chin) groove, the pronounced angular ridge on muzzle, blunt tail, and larger number of costal scales are the chief identifying characters (Fig. 10.3).
   Maximum Length—1,000 mm
   Habitat—Lives in rodent burrow.

6. **Common Trinket Snake *Coelognathus helena helena*** IUCN Status (2012)-Not Evaluated
   Identification—Fore-body is of variable length, beautifully ornamented with oscillated crossbars of a pattern peculiar to this snake.

Fig. 10.3 Red Sand Boa *Eryx johnii* with baby

Maximum Length—1,680 mm

Habitat—Found in termite mounds, rock piles, and crevices

7. **Indian Rat Snake *Ptyas mucosa*** IUCN Status (2012)-Not Evaluated

   Identification—Head is rather elongated, eyes large and lustrous, neck distinctly constricted. Body is robust, compressed, and tapering towards both ends. Skin is blackish dorsally mottled with fawn or whitish transverse streaks and hidden by scales (Fig. 10.4).

   Maximum Length—3,500 mm

   Habitat—Inhabits a wide range of habitats from coastal, arid, wet, mountains, open fields to forests, rat holes, and termite mounds.

8. **Glossy-bellied Racer *Platyceps ventromaculatus*** IUCN Status (2012)-Not Evaluated

   Identification—Pale grayish-white, sand colored, or pale reddish brown black snake with zigzag, short, darker grayish-brown cross bands of variable width and edges are black.

   Maximum Length—1,260 mm

   Habitat—Found in sandy desert areas and adjacent patches of scrub, even *Euphorbia* clumps.

9. **Slender Racer *Coluber gracilis*** IUCN Status (2012)-Not Evaluated

   Identification—Slender bodied snake that has two pale brown, black edged, forward pointing V-shaped marks on top of the head continuing throughout the body in a similar pattern. Towards the hind end of the body, the bands are

Fig. 10.4 Indian Rat snake *Ptyas mucosa*

Fig. 10.5 Slender Racer *Coluber gracilis*

replaced by narrow, sometimes broken blackish cross lines. A black stripe is present below each eye at meeting line of sixth and seventh supralabials (Fig. 10.5).

Maximum Length—930 mm

Habitat—It is usually seen in hilly and highly degraded deciduous forest.

Fig. 10.6 Common Kukri Snake *Oligodon arnensis*

10. **Banded Racer** ***Argyrogena faisciolata*** IUCN Status (2012)-Not Evaluated
    Identification—Brown or olive brown snake. Body is elongated, fairly robust and cylindrical, tapering slightly at the neck.
    Maximum Length—1,347 mm
    Habitat—Found in rodent burrows and heavy bushes.

11. **Russell's Kukri Snake** ***Oligodon taeniolatus*** IUCN Status (2012): Least Concern
    Identification—Only peninsular snake having 15 scale rows throughout the length of body and having anal shield divided and different type of color and markings.
    Maximum Length—590 mm
    Habitat—Found in forests as well as near human habitations.

12. **Common Kukri Snake** ***O. arnensis*** IUCN Status (2012)-Not Evaluated
    Identification—Scale rows 17:17:15 and 4 or 5 infralabials. Neck is slightly indicated. It has loreal shields and 7 supralabials. A large number of bars are found on the body (Fig. 10.6).
    Maximum Length—700 mm
    Habitat—Found in termite mounds, caves, crevices, tree holes, and old houses.

Fig. 10.7 Common Wolf Snake *Lycodon aulicus*

13. **Common Bronze-back Tree Snake *Dendrelaphis tristis*** IUCN Status (2012)-Not Evaluated
    Identification—Enlarged vertebrals and ridged ventrals are present. Neck and fore-body usually with a series of black streaks often paired and disjointed. A yellow black bordered flank stripe is found from neck to vent. A round yellow spot is present on parietal suture. An obscure black post-ocular (eye) streak is also found.
    Maximum Length—1,690 mm
    Habitat—Arboreal

14. **Barred Wolf Snake *Lycodon striatus*** IUCN Status (2012)-Not Evaluated
    Identification—Belly and upper lip is white. It is a timid snake which hides its head beneath its coils if disturbed.
    Maximum Length—488 mm
    Habitat—Usually terrestrial

15. **Common Wolf Snake *L. aulicus*** IUCN Status (2012)-Not Evaluated
    Identification—Glossy slender snake with a pear-shaped depressed head. Eye is black and the tongue is pinkish bearing a white tip. Bars may be present on whole length of the body or confined to anterior region of the body (Fig. 10.7).
    Maximum Length—800 mm
    Habitat—Found in and around caves, wells, stone piles, hollow trees and often in houses.

16. **Dumeril's Black-headed Snake *Sibynophis subpunctatus*** IUCN Status (2012)- Not Evaluated
    Identification—Slender bodied, smooth scaled snake. Body and tail brown in colors with series of small black dots along mid-back. Head grayish-brown with tiny brown or black spots and two black cross bands.
    Maximum Length—460 mm
    Habitat—Usually terrestrial

Fig. 10.8 Green Keelback *Macropisthodon plumbicolor*

17. **Checkered Keelback Snake *Xenochrophis piscator*** IUCN Status (2012)-Not Evaluated
    Identification—It has strongly keeled scales and five rows of black spots on a yellowish or olivaceous back ground, spots varying in size sometimes occupying most of the back.
    Maximum Length—1,750 mm
    Habitat—Found in paddy fields, ponds, thick grass, bushes, and gardens.
18. **Buff-striped Keelback Snake *Amphiesma stolatum*** IUCN Status (2012)-Not Evaluated
    Identification—A pair of conspicuous buff stripes covering one whole or two half row of scales from neck or fore-body to tip of tail.
    Maximum Length—800 mm
    Habitat—Found in paddy fields, ponds, thick grass, bushes, and gardens
19. **Green Keelback Snake *Macropisthodon plumbicolor*** IUCN Status (2012)-Not Evaluated
    Identification—Uniform grass—green above with a few black spots which may show a tendency to form transverse bars (Fig. 10.8).
    Maximum Length—940 mm
    Habitat—Found in dense forest

20. **Common Cat Snake** ***Boiga trigonata*** IUCN (2012) Status: Least Concern
    Identification—Costals or body shields 21:21:15. Eyes are large with mustard yellow iris and vertical pupil. It is yellowish brown in color. Dorsally a series of dark Y-shaped marks are present which meet at the center and resemble arrowheads.
    Maximum Length—1,250 mm
    Habitat—Arboreal

21. **Forsten's Cat Snake** ***B. forsteni*** IUCN (2012) Status: Not Evaluated
    Identification—It is dirty whitish in color with a series of large brown crossbars which are distinct at anterior end.
    Maximum Length—2,313 mm
    Habitat—Arboreal

22. **Afro-Asian Sand Snake** ***Psammophis schokari*** IUCN (2012) Status: Not Evaluated
    Identification—Slender bodied, smooth scaled snake with outer edges of belly scales that are round in shape. It is cream, pale brown, or grayish black without broad dark brown stripes, only their dotted, black outlines remain.
    Maximum Length—1,280 mm
    Habitat—Found in dry habitats, like sandy desert

23. **Leith's Sand Snake** ***P. leithi*** IUCN (2012) Status: Not Evaluated
    Identification—Slender bodied, smooth scaled snake with outer edges of belly scales rounded. It has pale yellowish brown back with four dark brown stripes that begin on head and run along the length of body. Body stripes are dotted on line on both sides with black.
    Maximum Length—895 mm
    Habitat -Terrestrial, arboreal, found in marshes, grassland, or sandy desert.

24. **Common Vine Snake** ***Ahaetulla nasuta*** IUCN (2012) Status: Not Evaluated
    Identification—Parrot green in color with horizontal pupil and pointed snout. The iris is powdered with gold.
    Maximum Length—2,000 mm
    Habitat—Arboreal

25. **Isabelline Vine Snake** ***A.n. var. isabellinus*** IUCN (2012) Status: Not Evaluated
    Identification—Brown in color with horizontal pupil and pointed snout, iris powdered with gold. Head scales have black dots (Fig. 10.9).
    Maximum Length—2,000 mm
    Habitat—Arboreal

26. **Black-headed Royal Snake** ***Spalerosophis diadema atriceps*** IUCN (2012) Status: Not Evaluated
    Identification—Scales weakly keeled, light yellowish brown or orange with irregular blue-black or dark brown markings that look like "spattered tar" on the back (Fig. 10.10).
    Maximum Length—2,000 mm
    Habitat—Found in rodent burrows, rocky areas, and crevices

Fig. 10.9 Isaballine Whip Snake *A. nasuta* var. *isabellinus*

Fig. 10.10 Black-headed Royal Snake *Spalerosophis atriceps*

27. **Red-spotted Royal Snake *S. arnarius*** IUCN (2012) Status: Not Evaluated
    Identification—Scales are weakly keeled, ash grey, cream, or pale brown back with caramel, reddish brown markings which are arranged in 5–7 alternating series and merge to form stripes on tail.
    Maximum Length—1,280 mm
    Habitat—Dry area
28. **Sind Awl-headed Snake *Lytorhynchus paradoxus*** IUCN (2012) Status: Not Evaluated
    Identification—Grey or pale brown snake with a series of H or X-shaped dark brown or sooty-black marks connected by a white stripe running down the mid-back.
    Maximum Length—380 mm
    Habitat—Inhabits sand-dunes

29. **Sindh Krait *Bungarus sindanus sindanus*** IUCN (2012) Status: Not Evaluated
    Identification—Smooth scaled, glossy black, grayish black or brownish black with thin, unpaired white cross lines on body and tail. On fore-body, cross lines are broken into a chain of white dots.
    Maximum Length—1,518 mm
    Habitat—Found in deserts

30. **Common Krait *Bungarus caeruleus*** IUCN (2012) Status: Not Evaluated
    Identification—The enlarged hexagonal vertebral scales, entire subcaudals, uniform white belly, and narrow white crossbars on the back, more or less distinctly in pairs.
    Maximum Length—1,750 mm
    Habitat—Found resting in termite mounds, rodent burrows, piles of bricks and rubble, even in rolled up carpets, often found in farms and garden.

31. **Spectacled Cobra *Naja naja*** IUCN (2012) Status: Not Evaluated
    Identification—Small "cuneate" scale is present between the fourth and fifth infralabials. Preocular touches the internasal. It has third supralabial in contact with the eye.
    Maximum Length—2,200 mm
    Habitat—Found in fields, near streams, rock piles, trees, graperies, and grain shops.

32. **Central Asian Cobra *N. oxiana*** IUCN (2012) Status: Data Deficient
    Identification—Black colored, spectacled cobra without hood.
    Maximum Length—2,200 mm
    Habitat—Found in fields, near streams, rock piles, trees, graperies, and grain shops.

33. **Eastern Russell's Viper *Daboia siamensis*** IUCN Status (2012): Least Concerned [1]
    Identification—Ground color brown of varying shades with three series of large ovate spots, one ventral and two costal spots, brown in the center and margined successively by black and white or buff. A conspicuous white, buff or pink line from gape converges to form a "V" above snout. Maximum Length—1,800 mm
    Habitat—Found resting in termite mounds, rodent burrows, piles of bricks and rubble, even in rolled up carpets, often found in farms and gardens.

34. **Saw-scaled Viper *Echis carinatus*** IUCN Status (2012): Not Evaluated
    Identification—Absence of shields on the head, the broad ventrals cover the whole belly and undivided subcaudal shields. A cruciform or trident shaped mark on crown.
    Maximum Length—800 mm
    Habitat—Found in open dry, sandy, and rocky terrain.
    The status and the distribution of all the above mentioned snakes along with Latin names and names in local dialect have been presented in Table 10.1. The table also shows venomous, nonvenomous, and mildly venomous snakes.

Table 10.1 Distribution and occurrence of the Ophidiofauna in Rajasthan

S. No.	Family	Latin name	Common name	Local name	Venomous/ nonvenomous	Status in the state	Precise location of occurrence
1	Typhlopidae	*Ramphotyphlops braminus*	Brahminy Worm Snake	Andha Samp, Kana	N	C	Jodhpur, Bikaner and Ajmer [20]; Kumbhalgarh WLS [13], Baghdarrah [5], Mt. Abu [15], KNP [7], Phulwari WLS [30] Personal records: Nahargarh WLS, Arboretum (Jaipur), Sariska Tiger Reserve, Sajjagnarh WLS, Sitamata WLS
		Rhinotyphlops acutus	Beaked Worm Snake	Andha Samp, Kana	N	R	Baghdarrah [5], Jodhpur [20] Personal records: Sajjagnarh WLS, Neemach Mata Hills, Jaisamand WLS (All in Udaipur District)
2	Pythonidae	*Python mlurus molurus*	Indian Rock Python	Ajgar, Ijgar, Agar	N	LC	Kumbhalgarh WLS [13], Baghdarrah [5], KNP [7, 9, 36], Mt. Abu [15], Recently in 2010 a Python was seen in Jodhpur city. It was probably an escape case Personal records: Bandipul (Sariska), Bassi WLS, Bhainsroadgarh WLS, Sitamata WLS, Borawas (Kota)
3	Boidae	*Gongylophis conicus*	Common Sand Boa	Dumbi, Bahri, Dhanrai, Kamboi	N	C	Kumbhalgarh WLS [13], Baghdarrah5, KNP [7, 17], Phulwari WLS [30] Personal records: Niwai (Tonk), Sitamata WLS, Sariska Tiger Reserve, Bassi WLS, Ranthambhore Tiger Reserve

		Eryx johnii	Red Sand Boa	Dumbi, Bahri, Dhanrai, Kamboi, Chakland, Bogi	N	C	Jaisalmer and Barmer [10], DNP [8], Baghdarrah [5], KNP [7], Phulwari WLS [30], Jodhpur, Pali, Bikaner, Ajmer and Jaipur [20] Personal records: Sitamata WLS, Sariska Tiger Reserve, Bassi WLS
4	Colubridae	*Coelognathus helena helena*	Common Trinket Snake		N	C	Baghdarrah [5], Mt. Abu [15], Sirohi and Ajmer [20], Phulwari WLS [30]
		C. h. monticollaris	Montane Trinket Snake		N	R	Phulwari WLS, Mt. Abu WLS, Ranthambhore [32]
		Ptyas mucosa	Indian Rat Snake	Dhaman, Kumbhra	N	C	Baghdarrah [5], KNP, [7] Phulwari WLS [30], Jodhpur, Pali and Ajmer [20]
		Platyceps ventromaculatus	Glossy-bellied Racer	Girawa, Gurawa	M	LC	Sikar [39], Miazlar and Sam in Jaisalmar district [35], Churu [10], Jodhpur, Bikaner, and Jaisalmer [20] Personal records: I have seen this species many times in World Forestry Arboretum, Jaipur between 1988 and 1992. One snake was captured from Goverdhan Villas (Udaipur) towards southern end of Machhla Magra hills on August, 22, 2010
		Argyrogena fasciolatus	Banded Racer		N	R	Sitamata WLS and other forested area of Pratapgarh Forest Division (Dr. Dharmendra Khandal, Pers. Com. 2009)

(continued)

Table 10.1 (continued)

S. No.	Family	Latin name	Common name	Local name	Venomous/ nonvenomous	Status in the state	Precise location of occurrence
		Coluber gracilis	Slender Racer		N	R	Jhadol [34] Personal records: Sajjangarh WLS, Jaisamand WLS, Keora Ki Nal, Chirwa Ghata (All in Udaipur)
		Diadema atriceps	Black-headed Royal Snake	Ghoda Pachhad, Ragit Bansar	N	LC	Drier parts of Rajasthan [39], Nagour [20] Personal records: Jaipur, Nahargarh WLS, Sariska Tiger Reserve, Ajmer, Jhunjhunu
		S. arnarius	Red-spotted Royal Snake	Lal Phoonki	N	R	Sikar [39] Ratangarh (Churu), Fakiron ki Basti (Jaisalmer), and Alamsar (Barmer) [10]
		S. diadema	Blotched Royal Snake	Ragit Bansar, Rajat Bansi	N	R	Jodhpur, Nagour, Jaipur, and Ajmer [20]
		Lytorhynchus paradoxus	Sind Awl-headed Snake		N	R	Sikar [39], Churu [4]
		Oligodon taeniolatus	Russell's Kukri Snake		N	LC	Baghdarrah [5], Phulwari WLS [30] Personal records: Sajjangarh WLS, Jaisamand WLS, Bassi WLS, Kumbhalgarh WLS, Sitamata WLS
		O. arnensis	Common Kukri Snake		N	C	Baghdarrah, [5] KNP [7], Phulwari WLS [30] Personal records: Sajjangarh WLS, Jaisamand WLS, Bassi WLS, Kumbhalgarh WLS, Sitamata WLS, Mt. Abu WLS

Dendrelaphis tristis	Common Bronze-back Tree Snake	Udni	N	R	Baghdarrah [5], Jhadol and Ogna [23] Phulwari WLS [30] Personal records: Sajjangarh WLS, Jaisamand WLS, Bassi WLS, Kumbhalgarh WLS, Sitamata WLS, Mt. Abu WLS, Torna and Kamalnath Forest of Udaipur district
Lycodon striatus	Barred Wolf Snake		N	LC	Kamalnath, district Udaipur [25–27]. Phulwari WLS [30] Personal records: Bassi WLS, Nahargarh WLS, Jaipur
L. aulicus	Common Wolf Snake		N	C	Jodhpur and Ajmer [20], Baghdarrah [5], Mt. Abu [15], Jhadol [28], KNP [7] Phulwari WLS [30] Personal records: Bassi WLS, Nahargarh WLS, Jaipur
Sibynophis subpuncatus	Dumeril's Black-headed Snake		N	R	Kamalnath, Kumbhalgarh [29], Ranthambhore (Dr. D. Khandal, Pers. Com.) Personal records: Jaisamand WLS, Keora Ki Nal (Udaipur)
Xenochrophis piscator	Checkered Keelback	Dindolia, Dindu, Paneta	N	C	Sirohi, Ajmer, Pali and Jodhpur [20], Jalore [10], Kumbhalgarh WLS [13], Baghdarrah [5], Mt. Abu [15], KNP [7, 18], Phulwari WLS [30] Personal records: Jaisamand WLS, Bassi WLS, Todgarh Raoli WLS
Amphiesma stolatum	Striped Keelback		N	R	Mahipal Forest Block in Udaipur district [23], Kota (Rakesh Vyas, Pers. Com.), Phulwari WLS [30]

(continued)

Table 10.1 (continued)

S. No.	Family	Latin name	Common name	Local name	Venomous/ nonvenomous	Status in the state	Precise location of occurrence
		Macropisthodon plumbicolor	Green Keelback	Hara Samp, Leela Samp	N	LC	Phulwari WLS [30], Sirohi district [20]
		Boiga trigonata	Common Cat Snake		M	C	Jodhpur and Jaipur [20], Baghdarrah [5], Mt. Abu [15], KNP [7], Phulwari WLS [30] Personal records: Jaisamand WLS, Bassi WLS, Todgarh Raoli WLS
		B. forsteni	Forsten's Cat Snake		M	R	Mt. Abu [15, 33], Katawali Jher in Phulwari WLS (Dr. J.Jesua, Pers. Com.)
		Psammophis schokari	Afro-asian Sand Snake		M	R	Jodhpur and Bikaner [20], Padru in Marmer [10], Jodhpur [38, 39]
		P. leithi	Leith's Sand Snake	Silak	M	R	Whole Rajasthan [38, 39], KNP [7], Phulwari WLS [30] Personal records: Achrol—Jaipur
		Ahaetulla nasuta	Common Vine Snake	Bans Ka Samp	M	R	Phulwari WLS [30], Sirohi District [20] Personal records: Todgarh Raoli WLS, Darrah WLS
		A.n. var. *isabellinus*	Isabelline Vine Snake		M	R	Phulwari WLS [30] Personal records: Bassi WLS, Mt. Abu WLS, Deogarh (Pratapgarh)
5	Elapidae	*Bungarus caeruleus*	Common Krait		V	C	Jodhpur, Bikaner and Jaisalmer [20]. Baghdarrah [5], Mt. Abu [15], KNP [7], Jalore, Barmer, Churu [10], Phulwari WLS [30] Personal records: Bhilwara, Mandal, Menal, Todgarh Raoli WLS

		B. sindanus sindanus	Sind Karit	Peewna, Paina	V	R	Rajasthan [39]. This snake is confined to western Rajasthan
		Naja naja	Spectacled Cobra	Hapni, Nag, Kalia, Kalinder	V	C	Jodhpur Ajmer, Jaisalmer, and Jaipur [20], Barmer [10], Kumbhalgarh WLS [13], Baghdarrah [5], Mt. Abu [15], KNP [7], Phulwari WLS [30]
		N. oxiana	Central Asian Cobra	Nag	V	R(?)	Aravalli foot hills [12]; Jodhpur, Jaipur, and Ajmer [20]
6	Viperidae	*Daboia russelii*[a]	Russell's Viper	Chitti, Chittod	V	LC	Baghdarrah [5], Mt. Abu [15], KNP [7], Jodhpur [20], Ranthambhore (Dr. D. Khandal, Pers. Com.), Phulwari WLS [30] Personal records: Todgarh Raoli WLS, Sariska Tiger Reserve, Bassi WLS
		Echis carintus[b]	Saw – scaled Viper	Chitti, Pad, Bandi, Kodiyala, Annapeda, Kankriwala	V	A	Jalore, Jodhpur, and Barmer [8, 10, 20], DNP [8], Kumbhalgarh WLS [12], Baghdarrah [5], Mt. Abu [15], KNP [7], Jaipur [10], Pali Bikaner, Nagour, Sirohi, and Ajmer [20], Phulwari WLS [30] Personal records: Sajjangarh WLS, Jaisamand WLS, Todgarh Raoli WLS, Sariska Tiger Reserve, Bassi WLS, Bardod (Alwar)

V Venomous, *N* Nonvenomous, *M* Mildly Venomous, *R* Rare, *LC* Less Common, *C* Common, *A* Abundant, *WLS* Wildlife Sanctuary, *KNP* Keoladeo National Park, Bharatpur, *DNP* Desert National Park

[a]According to Wuster [41], Indian species is *D. russelii russelii*

[b]According to Whitaker and Captain [39], Scochurek's Saw-scaled Viper (*Echis carintus sochurek/Echis sochurek*) from Rajasthan either merits subspecies status or may be a separate species

Few Interesting Reports from the Border Areas and Adjoining States of Rajasthan

The King Cobra (*Ophiophagus hannah*) once reported from Deesa area of Banaskantha district of North Gujarat [16] is not seen in Rajasthan now. Deesa is very close to the Phulwari Ki Nal and Mt. Abu wildlife Sanctuaries of Rajasthan. The Central Asian Cobra or Black Cobra (*Naja oxiana*) with no hood marks is usually found in the drier areas of India [41]. This species has been reported from Jammu and Kashmir, Himachal Pradesh, Punjab, and Rajasthan. According to Whitaker and Captain [39], records of this species from Rajasthan need confirmation because what is usually identified as *N. oxiana* is possibly a pattern-less black form of the Spectacled Cobra *Naja naja*. The Monocled Cobra (*Naja kaouthia*) is reported from Sonipat area of Haryana but there is no record of this species from Rajasthan. Few years back, there was free movement of snake–charmers from Haryana and Punjab to *Gogamedhi mela* (fair) of northern Rajasthan in which snake exchange is a common practice among the *kalbeliya* (snake charmers) community of Rajasthan, Haryana, Punjab, Gujarat, and other States. Snakes were generally released during rains irrespective of their nativity. Escape of snake from the possession of charmers is also possible. Due to such events, occurrence of snakes out of their distribution range is also possible. Indian Python, Checkered Keel Back, Red Sand Boa, Royal Snakes, Common Vine Snake, and Cobras are generally liked by the *kalbeliya* for display and begging. Condanarus Sand Snake (*Psammophis condanarus*) is reported from the Indian desert. This species has been recorded from Uttaranchal, Punjab, Gujarat, Maharashtra, Uttar Pradesh, Bihar, Orissa, and West Bengal; however, confirmed reports from Rajasthan, Haryana and Madhya Pradesh are still awaited [39].

Recommendations for Conservation

1. Public awareness programs and "Vehicle driver awareness" campaigns are necessary in every corner of the state, especially where roads are passing through the Protected Areas and dense vegetation.
2. Many human and cattle lives are lost every year during rains due to snake bite both in urban and rural areas. Tribals mainly depend on *bhopas* and *ojhas* (quacks) for treatment. Proper treatment facilities should be available in government hospitals to save precious human and animal lives.
3. Many snakes are totemic to many clans. Such traditions and values should be restored and encouraged, and wherever lost should be revived.
4. Excessive fishing should be banned to protect water snakes.
5. Southern Rajasthan is a mosquito infested area and DDT is usually sprayed in houses to control malaria. Utensils used in such operations are cleaned in streams

as a result of which water becomes polluted. Snake, like *Xenochrophis piscator* are very sensitive to such polluted waters [4]. Recently, many snakes have died in a stream due to cleaning of DDT utensils. Such events should be checked.

6. By launching effective rescuing service in the tribal, rural, and urban areas, we can save many snakes every year.
7. Habitat loss is the most widespread cause of species endangerment. Habitat restoration can be practiced by avoiding construction activities and clearing of area. Habitat restoration can also be carried out by leaving fallen logs, bark, and snags as such; adding vegetated borders and hedges; building dry-stack rock walls; constructing brush and rock piles or compost piles.
8. The snake charmers of India for generations have used snakes to earn a living. The charmers are cruel to these creatures and often injure them during defanging or blocking the venom ducts, which causes immense pain to the snake and sometimes even death. Raids, seizures of snakes, and arrests for illegal possession of wild animals have been helpful in rescuing the snakes to some extent. Please see Chap. 2 for more pictures.

References

1. Daniel JC (2002) The book of Indian Reptiles and Amphibians. Bombay Natural History Society and Oxford University Press, Mumbai. pp 238
2. Zoo Outreach Organization. Biodiversity Conservation Prioritization Project (BCPP) India. Report Conservation Assessment and Management; 1998
3. Bhide KA (2004) Churu experience. Sanctuary 24(6):33–37
4. Bhide KA Captain, Khandal D (2004) First record of *Lytorhynchus paradoxus* (Gunther, 1875) from the Republic of India, with notes on its distribution. Hamadryad 28(1 & 2):123–127
5. Bhatnagar C, Mathur M (2008) Reptilian fauna of Baghdarrah Nature Park, Udaipur Wildlife Division, Udaipur, Rajasthan, India. Cobra 2(2):17–20
6. Bhupty S, Haque MN (1986) Association of Rock Python (*Python molurus*) with Porcupine (*Hystrix indica*). J Bom Nat Hist Soc 83(2):449–450
7. Bhupthy S (1996) Reptiles of Keoladeo National Park, Bhartpur, Rajasthan. J Bom Nat Hist Soc 96(3):475–477
8. Das SK, Rathore NS (2004) Herpetofauna of Desert National Park, Rajasthan, India. Zoos' Print J 19(9):1626–1627
9. Duby OP (1984) A Coucal-Python incident. J Bom Nat Hist Soc 81(3):711
10. Gala M, Khandal D (2010) A rapid survey of squamate reptile in the Thar desert of Rajasthan, India. Cobra IV(I):28–39
11. Gaur S (2002) Studies on the snakes of the Kumbhalgarh Wildlife Sanctuary, Rajasthan, India. Tiger Paper 31(3):1–3
12. Gaur S (2009) Species richness and dominance of reptiles in the Aravalli foothills of Rajasthan. In: Sivaperuman S, Baqri QH, Naseema M (eds) Faunal ecology and conservation of the Great Indian Desert. Spinger- Verlag, German. pp 54–65
13. Gaur S, Pandey VK (2007) Ecological notes on the herpetological fauna of Kumbhalgarh Wildlife Sanctuary, Rajasthan, India. Cobra 1(3):4–7
14. Kasliwal S (2008) Music of the snake charmers of Rajasthan. Cobra 2(4):7–9
15. Mc Cann C (1946) The rain comes to the Abu Hills. J Bom Nat Hist Soc 43:641–647
16. Pandey CN (2004) Gujarat's wild destinations. GEER Foundation, Gandhinagar. pp 1–81

17. Prakash V (1988) Russell's Earth Boa, *Eryx conicus*, preying on a Little Brown Dove, *Streptopelia senegalensis*. J Bom Nat Hist Soc 85(2):438
18. Prakash V, Nanjappa C (1988) An instance of active predation by Scavenger Vulture (*Neophron percnopterus gingianus*) on Checkered Keelback Water Snake (*Xenocharopis piscator*) in Keoladeo National Park, Bharatpur, Rajasthan. J Bom Nat Hist Soc 85(2):149
19. Sankaran R (1992) Inside the Thar Desert. Sanctuary 12(2):16–27
20. Sharma RC (1996) Herpetology of the Thar desert. In: Gosh AK, Baqri QH, Prakesh I (eds.), Faunal diversity in the Thar desert: Gaps in Research. Scientific Publishers. pp 297–306
21. Sharma SK (1996) Swallowing of prey "leg first" by the Cobra *Naja naja*. J Bom Nat Hist Soc 89(1):133
22. Sharma SK (1994) Snakes of Rajasthan. Vijnana Parishad Anusandhan Patrik. J Bom Nat Hist Soc 37(4):209–220
23. Sharma SK (1994) Reptiles of southern Rajasthan with comments on their etho-zoological aspects. Proceedings of seminar on eco-development, habitat and wildlife conservation in Rajasthan and adjoining areas. 21–23th Oct., 1994; Hadoti Naturalists Society, Kota. pp 11–17
24. Sharma SK (1994) Presence of the Common Indian Bronze-back Snake (*Dendrelaphis tristis*) in Rajasthan. J Bom Nat Hist Soc 91(3):462
25. Sharma SK (1995) An overview of the amphibians and reptilian fauna of Rajasthan. Fauna & Flora 1(1):47–48
26. Sharma SK (1995) Presence of Common Green Whip Snake Ahaetulla nasutus at Phulwari Ki Nal Wildlife Sanctuary in Rajasthan. J Bom Nat Hist Soc 92(1):127
27. Sharma SK (1995) Clutch size in Shaw's Wolf Snake *Lycodon striatus*. J Bom Nat Hist Soc 92(2):271
28. Sharma SK (1999) *Calotes versicolor* feeding on *Lycodon aulicus*. J Bom Nat Hist Soc 96(1):146–147
29. Sharma SK (2003) Presence of Dumeril's Black-headed Snake (*Sibynophis subpunctatus*) in Kumbhagarh Wildlife Sanctuary, Rajasthan, India. Cobra 53:17–18
30. Sharma SK (2007) Study of biodiversity and ethnobiology of Phulwari Wildlife Sanctuary, Udaipur (Rajasthan). Ph.D. Thesis. MLS University, Udaipur, India
31. Sharma SK (2008) Notes on the death of an Indian Rat Snake (*Ptyas mucosus*) in unusual circumstances. Cobra 2(2):21
32. Sharma SK (2009) Presence of Montana Trinket Snake *Coelognathus helena monticollaris* in Phulwari Ki Nal Wildlife Sanctuary, Udaipur District, Rajasthan, India. Cobra-III 4:19–20
33. Sharma SK, Chawda K, Patel S (2002) Sighting of Forsten's Cat Snake (*Boiga forsteni*) at Mt. Abu Wildlife Sanctuary, Sirohi District, Rajasthan. Cobra 48:7–8
34. Sharma SK, Nagar S (2007) First record of slender racer *Coluber gracilis* (Gunther, 1862) (Serpents: Colubridae from Rajasthan. J Bom Nat Hist Soc 104(3):355
35. Singh V, Singh M (2006) Biodiversity of the Desert National Park. Rajasthan, Botanical Survey of India
36. Sridharan U, Manohar BR (1984) A note on the Asiatic Rock Python (*Python molurus*) feeding on the Spotbill Duck (*Anas poecilorhynchus*). J Bom Nat Hist Soc 81(3):710–711
37. Tiwari SK (1985) Zoogeography of Indian and South-East Asia. CBS Publishers and distributors
38. Waltner RC (1974) Geographical and altitudinal distribution of Amphibians and reptiles in the Himalayas (Part IV). Cheetal 16(4):12–15
39. Whitaker R, Captain A (2004) Snakes of India – the field guide. Draco books, Chennai, India
40. Wuster W (1998) The genus *Daboio* (Serpents: Viperidae): Russell's Viper. Hamadryad 23(1):34–49
41. Wuster W (1998) The cobras of the genus *Naja* in India. Hamadryad 23(1):15–32

Part III
Faunal Ecology: An Insight – Avifauna

Chapter 11
Conservation of Bustards with special reference to Great Indian Bustard *Ardeotis nigriceps*: The State Bird of Rajasthan

M.K. Vijayvargiya, Rajan Mathur, and B.K. Sharma

Abstract This chapter describes status of the Critically Endangered "State bird"—the Great Indian Bustard (GIB)—and analyses problems regarding its conservation, measures and strategies to be adopted and need for further research. Bustards are the birds of grasslands. Out of the 10 genera of bustards, four are found in India, of which three are resident and one is a migratory species. Ecology, distribution and population of each species are discussed separately. The reasons for decline in the population of bustards are excessive biotic pressure, loss of breeding sites and fragmentation of their habitats which have been thoroughly discussed. Bustards in Rajasthan are almost on the brink of extinction, and the exact number of these birds is debatable. Government's inability to check poaching and habitat destruction has further slowed down the pace of conservation efforts. However, according to the Forest Department, Government of Rajasthan, a marginal increase in the bustard population has been observed very recently. Local communities in and around bustard habitats are yet to be involved in "Project Bustard", an effort for in situ conservation of GIB in the proposed Desert National Park at Jaisalmer. Likewise, conservation measures are required to be replicated in other GIB habitats, too.

Introduction

The bustards are birds of grasslands surviving for nearly 50 million years on the earth. Out of the ten genera found in the world, four are found in India, namely, Great Indian Bustard *Ardeotis nigriceps*, Lesser Florican *Sypheotides indica*, Bengal

M.K. Vijayvargiya (✉) • R. Mathur
Department of Forests and Wildlife, Government of Rajasthan, Jaipur, Rajasthan, India
e-mail: mvijayavargia@gmail.com

B.K. Sharma
Department of Zoology, R.L. Saharia Government P.G. College,
Kaladera (Jaipur), Rajasthan, India
e-mail: drbksharma@hotmail.com

B.K. Sharma et al. (eds.), *Faunal Heritage of Rajasthan, India: General Background and Ecology of Vertebrates*, DOI 10.1007/978-1-4614-0800-0_11,

Florican *Houbaropsis bengalensis* and Houbara Bustard *Chlamydotis undulata* or *C. macqueenii*. All these are resident birds except the Houbara Bustard which is a migratory species. All the Indian bustard species are solely dependent on grasslands [1]. The Great Indian Bustard is gregarious while others may be solitary. All bustards are omnivorous, consuming a large variety of seeds, fruits, insects and other small invertebrates and reptiles. The courtship display of certain species is well known and spectacular.

Great Indian Bustard *Ardeotis nigriceps*

The Great Indian Bustard is one of the indicator species of India's grasslands. It is a large and a beautiful bird inhabiting short grassland with a height below 1 m.

IUCN Red List History

The Great Indian Bustard is categorised as Critically Endangered A2a+4acd;C1 in the Red Data List of IUCN (2012).

The GIB walks slowly and prefers to freeze behind bushes in order to avoid humans. It can seldom be approached within 300 m. Formerly, it was widely distributed from Punjab in the north to Tamil Nadu in the south, and Sind (in Pakistan) in the west to Orissa and West Bengal in the east (Fig. 11.1). It was always found in the grassy plains, sometimes in highly overgrazed patches consumed by livestock or wild herbivores and strictly avoided hilly and forest regions [1, 2]. It shares its short grass plains with the Blackbuck (*Antilope cervicapra*), Chinkara (*Gazella bennettii*), Nilgai (*Boselaphus tragocamelus*), Grey Wolf (*Canis lupus*), Bengal Fox (*Vulpes bengalensis*), Red Fox (*Vulpes vulpes*), Golden Jackal (*Canis aureus*), Jungle Cat (*Felis chaus*) and a large number of bird and reptile species [4]. The main stronghold of the GIB was the Thar Desert in the north-west and the Deccan tableland of the indian Peninsula. Presently it is confined to a handful of pockets in the six states of India, namely, Rajasthan, Madhya Pradesh, Gujarat, Maharashtra, Andhra Pradesh and Karnataka. In the mid-1980s, the total population was estimated to be between 1,500 and 2,000, with Rajasthan possibly having half of the estimated numbers [2]. The males of Great Indian Bustard shows courtship behaviour by extending its gular pouch, which is developed only during the breeding season [2]. In a few species of bustards, the male remains with the female during the breeding season, and in some species, the male is polygamous. After mating, a depression in the ground is selected as the nest site, which may be under a bush or amid tall grass. The Great Indian Bustard lays only one egg, extremely rarely two, the colour is olive brown. The egg is laid on the bare ground and no attempt is made for nest building. The incubation period is over three weeks. The downy young

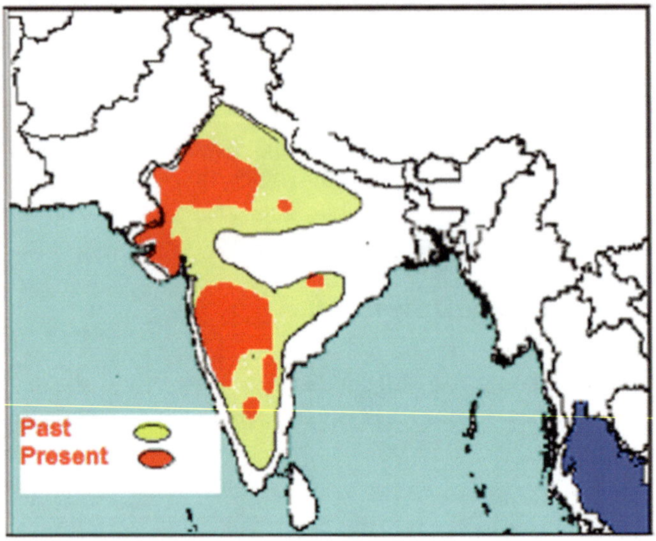

Fig. 11.1 The distribution of Great Indian Bustard (GIB) *Ardeotis nigriceps* in India: past and present

chicks are sand coloured with dark spots, and they leave the nest shortly after hatching. The young ones are reared by the female alone. Their population has greatly declined mainly due to habitat destruction and poaching, and now perhaps less than 300 are left.

Lesser Florican (*Sypheotides indicus*)

The Lesser Florican is virtually endemic to India. Earlier, it was one of the common "game birds" of the Indian grasslands, especially in Gujarat, Rajasthan, Madhya Pradesh, Maharashtra and Karnataka. Sankaran *et al.* [5] and Rahmani *et al.* [6] estimated the population of the Lesser Florican to be not more than 2,000 [6], and now perhaps less than 2,000 Lesser Floricans survive on the globe [7], making it one of the most endangered birds of India. Like the Great Indian Bustard, its habitat has also shrunk by more than 90 %. The Lesser Florican is a monsoon breeder, and with the commencement of rains, it is seen in the grasslands [15]. Sometimes, it is seen in crop fields of millet, sorghum, maize, etc., which can be considered as pseudo-grasslands. Owing to its somewhat unpredictable movement, seemingly influenced more by the monsoon than anything else, it is not an easy task to conserve the Lesser Florican, since it may/may not visit the usual habitat or visit in small numbers if the rainfall is not favourable [8, 9]. Besides, not many floricans can be saved in such habitat. Sankaran [3, 10–13] had done considerable studies on the territorial, aerial display, breeding and other aspects of the florican. The conservation of this species and natural fodder production are compatible because by the time the grass is ready for harvesting, the main breeding period of the florican is over. By delaying grass cutting by a week or leaving a small patch of the grassland uncut, pre-fledge chicks of the florican and other species, such as partridge and quail, can also be saved. However, before specific prescription can be given for their protection, long-term studies and planning have to be undertaken.

Presently, there are only two existing Lesser Florican sanctuaries in India, namely, Sailana Wildlife Sanctuary and Sardarpur Wildlife Sanctuary, both in Madhya Pradesh. With increasing human population and pressure on land, there are remote chances to develop additional grassland sanctuaries, especially the large ones. However, Florican and grassland development could be integrated, especially in the states like Gujarat and Rajasthan, where livestock husbandry is a major occupation of the rural population, and there is a tremendous demand for natural fodder. Many of the grass *vidis* of Saurashtra, which are at present in bad shape due to overgrazing and general neglect, could be saved under the rural development schemes. By protecting these grass *vidis*, the breeding grounds for the Lesser Florican can also be protected.

Houbara or MacQueen's Bustard (*Chlamydotis undulata*)

The Houbara or MacQueen's Bustard (*Chlamydotis undulata*) is a winter migrant to Rajasthan and Gujarat. They are killed in large numbers while migrating, especially in Pakistan. During winters, there is a continuous threat by poachers as they are considered good to eat. The Indira Gandhi Nahar Project (IGNP) is bringing major ecological and demographic changes to the main habitat of the Houbara Bustard.

Conservation of Indian Bustards

The GIB occurs in the Indian Subcontinent, its former strongholds being the Thar Desert in the north-west, Deccan tableland of Peninsular India and Sind, Pakistan. It is identified as an indicator of the health of grassland ecosystem of the Indian plains, and their absence is the first warning signal of the deteriorating ecology. Grasslands of western Rajasthan are the lifeline of the state, so it is imperative to save and protect them. The protection of the Great Indian Bustard and its habitat

Table 11.1 Important bustard areas (Modified by Dr. Asad R. Rahmani)

S. No.	Country/state	Name of the area
1	India/Gujarat	Banni grassland (Stray record)
2	India/Rajasthan	Desert National Park, Jaisalmer; Barmer
3	India/Rajasthan	Diyatra—Raneri area
4	India/Rajasthan	Ramdevra and adjoining areas
5	India/Rajasthan	Rasala—Devikot and adjoining areas
6	India/Rajasthan	Phalodi, Bap, and adjoining areas
7	India/Rajasthan	Didhu—Satyaya and adjoining areas
8	India/Rajasthan	Sonkhalia (<10 left)
9	India/Maharashtra	Bustard Sanctuary (Nannaj)
10	India/Madhya Pradesh	Ghatigaon Bustard Sanctuary (Stray record now)
11	India/Gujarat	Nalia, Lala Bustard Sanctuary
12	India/Gujarat	Velavadar National Park (Stray record)
13	India/Karnataka	Ranebennur Blackbuck Sanctuary (Extinct)
14	India/Andhra Pradesh	Rollapadu Wildlife Sanctuary (<10 left)
15	Pakistan/Sindh	Bijnote Bustard Game Reserve

will *suo moto* protect the Houbara Bustard. The other two floricans are not commonly reported from the desert region; hence, an emphasis is laid on the protection of the GIB. The main habitat of the GIB in Rajasthan is the Desert National Park and adjoining areas (Table 11.1).

Ecology

A high avian diversity of 250–300 species has been reported from the Thar Desert. The Thar is also known for Sewan grasslands surviving for many centuries. Currently, drastic ecological changes are believed to have been taking place in Rajasthan because of IGNP. Some other coexisting avian species of the desert are the Cream-coloured Courser (*Cursorius cursor*), Greater Hoopoe Lark (*Alaemon alaudipes*), various species of sandgrouse, raptors, larks, pipits and other birds. Lesser Florican in Rajasthan has lost to a large extent during the last 25–30 years. Now, it is reported only from a few isolated and scattered pockets. The grasslands of Indian Thar Desert are also the breeding grounds of a large number of ground-nesting bird species. Because of the free-ranging livestock, the nests of ground-living birds are trampled upon, and grasses are heavily grazed on, resulting into a loss of seeds spread. Therefore, it is imperative to protect and develop these grasslands on a sustainable basis for the rural economy and livestock and also for the protection of rare wildlife species and biodiversity conservation, in general.

Table 11.2 Population estimates of GIB in India

Protected Area	1985	2007
Karera Bustard Sanctuary, M.P.	25–30	Extinct
Ghatigaon Bustard Sanctuary	15–18	Extinct
Rannibennur Sanctuary, Karnataka	5–10	Extinct
Rollapadu, Andhra Pradesh	60	20–25
Desert National Park, Rajasthan	200	Declining
Sorsan, Rajasthan	10–15	Extinct
Bhatia, Jamnagar, Gujarat	8–10	Extinct
Sonkhalia, Ajmer, Rajasthan	80	20–25

Population of GIB in Rajasthan and elsewhere

This bird occurs in an area of 570,000 km^2 in the world and about 4000 sq km at ten locations in Rajasthan. Desert National Park in Rajasthan and Kutch and Sindh of the great Thar Desert are probably the best and the last refuge of the GIB. The current population of GIB is estimated to be less than 300 (Table 11.2). The bird is also found in the adjoining Madhya Pradesh state. Ghatigaon and Karera sanctuaries in Madhya Pradesh had sizeable populations earlier, but now there is no Great Indian Bustard seen at Karera of Shivpuri district. The GIB was identified in 1978 at Nannaj, 18 km from Solapur in Maharashtra, by Mr B.S. Kulkarni. Nearly <10 bustards are now seen in the Nannaj sanctuary. In July 2008, Santosh Martin, President of Sloth Bear Foundation, and Abdul Samad, President of the Society for Wildlife and Nature (SWAN), Hospet, reported a female GIB with an eight months old juvenile in a barren field near Siruguppa in Bellary district of Karnataka. The population of GIB is dwindling fast, and the bird has been extirpated in 90 % of its former range, as indicated by the population record in some of the GIB protected areas [4]. However, surveys from the 1990s onwards have revealed that numbers have gone down further, especially in Madhya Pradesh, Andhra Pradesh, Karnataka and Rajasthan. All the six states had taken various conservation measures, and nine existing and a couple of recently created sanctuaries have the GIB. The population has declined by about 29 % in the last ten years. The numbers of GIB observed over the years by the visitors in the desert at Sudasari area of Jaisalmer district in Rajasthan is given in Table 11.3.

Causes for the Decline in Bustard Population

According to Dharmkumarsinhji [9], although the GIB has been afforded blanket protection since 1952, shooting and poaching by trapping and falconry continued

Table 11.3 Important record of sighting of GIB at Sudasari closure, DNP, Jaisalmer

Year	Date	Sighted by	Male	Female	Chicks	Total
1981	15.11.1981	Sh. Satish Kumar	–		–	–
1982	05.03.1982	Sh. Samar Singh, IAS	–	–	–	–
1982	07.12.1982	Sh. Kailash Sankhla	–	–	–	10
1983	08.02.1983	Sir Robert, British High Commissioner	–	–	–	14
1984	15.01.1984	Dr J. G. Ramesh, FAO Representative	–	–	–	14
1985	26.01.1985	Mr Simon Cauls, Ottawa, Canada	–	–	–	14
1985	26.01.1985	Sh. Arun Saxena, WLW Bikaner	–	–	–	18
1986	26.01.1986	Tarry Lingston Ripley	5	13	–	18
1987	23.04.1987	Mahendra Vyas, Member, State WLA Board	–	–	–	14
1987	22.06.1987	Joanna Vancnise	–	–	–	18
1987	24.09.1987	Divya Bhanu Singh, New Delhi	8	–	–	8
1988	25.12.1988	Dr T. H. Bitsett, Alberta, Canada	–	–	–	7
1989	06.01.1989	Pratap Singh, IFS, Arunachal Pradesh	–	–	–	–
1990		No sighting was recorded		–	–	–
1991		No sighting was recorded	–	–	–	–
1992	23.02.1992	R. S. Bhandari, CCF, Rajasthan	–	–	–	4
1992	20.12.1992	Asad Rahmani	10	7	–	17
1993	25.08.1993	V. N. Singh, IFS, IGNFA	–	–	–	33
1994	05.02.1994	Asad Rahmani	–	–	–	17
1994	23.06.1994	Rajesh Bedi	–	–	–	13
1994	26.11.1994	Ulrick A. Motzfelda, Norway	–	–	–	16
1995	16.10.1995	Dr P. L. Kankane, Dehradun	–	–	–	1
1996	15.02.1996	S. S. Dey, Addl. IGF, GOI	–	–	–	8
1997	09.11.1997	Rahul Bhatnagar, ACF, Udaipur	–	–	–	6
1999	03.03.1999	Justice of Supreme Court, India	–	–	–	16
2000	01.02.2000	A. H. Musavi	–	–	–	14
2001	04.05.2001	Mahendra, Research Scholar, WL Management	–	–	–	25
2002	01.02.2002	U. M. Sahay, CCF, Jodhpur	–	–	–	16
2002	25.04.2002	Rajeev Dasot, DIG, BSF, Jaisalmer	–	–	–	15
2003	25.03.2003	U. M. Sahay, CCF, Jodhpur	–	–	–	8
2004	07.02.2004	Henrick Christen, Denmark	–	–	–	14
2005	05.11.2005	Uday Vora, DCF, MPSRET	–	–	–	9
2006	02.02.2006	Kilain Mallary Olse, Denmark	–	–	–	8
2007	13.08.2007	Madhuri Ramesh	19	7	5	31
2007	Dec 2007	Rajpal Singh and M. K. Vijayavargia				17

Source: Visitors Comment Book, Sudasari

Table 11.4 A fast decline in the IUCN Red data status of Great Indian Bustard

Year	Criteria
1988	Threatened
1994	Endangered
2000	Endangered
2004	Endangered
2008	Endangered
2012	Critically Endangered

till 1978–1979. Public protest against falconry in Rajasthan resulted in the International Symposium on Bustards in 1980 at Jaipur to create awareness among the masses and to make conservation efforts to save these vanishing birds. The decline in the population of Houbara Bustard was mainly due to an indiscriminate hunting by Arab hunters and their falcons in Pakistan and other areas [8], loss and fragmentation of habitat due to land hunger for agricultural and other developmental activities, excessive biotic pressure, loss of breeding environment, slow breeding habit of the bird, improper management of habitat, lack of awareness about the importance of the species in the food chain and food web of the ecosystem and also in establishing equilibrium. Severe loss of population in the past and the continuous decline of the GIB population have brought this bird in the "Threatened" category of IUCN Red Data List in 1988 and "Endangered" category from 1994 onwards (Table 11.4).

Most of the bustard sanctuaries, for example, Karera, Nannaj, Rollapadu, Sonkhaliya and Sorsan (Rajasthan), were private agricultural areas or common grazing lands. Respective state governments owning these sanctuaries overlooked the settlements of legal rights which left their respective state forest departments with virtually no control over the land, making it difficult to protect the habitat or carry out improvements. Habitat alteration by expansion of agriculture or by human settlements, development of roads or canals and spread of industries is still going on, and unless immediate measures are taken, some of the sanctuaries like Ghatigaon are likely to totally lose their bustard populations in the next 2–3 years, whereas Karera, Sorsan and Rannibennur have already lost all their bustards.

The dwindling population of bustards clearly indicate the need for a specific protection strategy. The main threat to the bustard is habitat loss due to conversion of grasslands into agriculture fields, intensive agriculture and fragmentation causing loss of foraging and breeding sites. Use of canal irrigation in the area adjoining the DNP has resulted in to drastic changes in the ecology. In addition, unsustainable developmental pressures, increasing human population, hunting, mining and oil exploration are causing severe threats to the existence of GIB at DNP. Apart from the above, military exercises in the vicinity of bustard areas cause severe disturbance to GIB habitat and other wildlife species. Increasing use of pesticides may also cause deaths of bustards, like the Indian Peafowl, *Pavo cristatus* populations in many western districts of the state. Mismanagement of Bustard protected areas and lack of a clearly defined core area, lack of funding, clear-cut policy and political will are also the issues requiring immediate attention. Lack of awareness among local population regarding the need for conservation of bustard population, ill-equipped protection staff and lack of adequate research are also the reasons of decline in bustard population.

Conservation Requirements for Bustards

Although bustards have been the subject of various types of management interventions over the years, these efforts were extremely inadequate. The legal protection given to these birds could not be enforced due to various factors. Recent studies on the GIB reveal that habitat protection and development of core areas for bustard breeding in a large multiple-use area could help in increasing bustard population. As the bustard lives in marginal agriculture areas, the support from locals is absolutely necessary for their long-term conservation planning. Sanctuaries such as Karera, where the core area could not be protected, showed a sudden decline in the bustard population. The GIB is an ideal species to be protected under the concept of community reserve or conservation reserve as suggested under the recently amended Wildlife (Protection) Act, 1972. There are also possibilities of taking recourse to various provisions of the Environment (Protection) Act, 1986 by restricting certain activities in the bustard habitats without displacing or disturbing the human populations.

Conservation Measures

Government of state of Maharashtra, Karnataka, Andhra Pradesh, Madhya Pradesh, Rajasthan and Gujarat declared such areas as protected areas/sanctuaries where the GIB is commonly found. These are Lesser Florican sanctuaries, Sailana and Sardarpura in Madhya Pradesh and the Great Indian Bustard sanctuaries, Desert National Park in Rajasthan, Velavadar National Park and Naliya Sanctuary in Gujarat, Bustard Sanctuary in Maharashtra, Ghatigaon and Karera in Madhya Pradesh, Rannibennur in Karnataka and Rollapadu in Andhra Pradesh. Sorsan and Sonkhalia have been declared as protected areas for the Great Indian Bustard in Rajasthan where it has been long declared as the "State bird". The Houbara Bustard also regularly visits the DNP and the open areas around Ramgarh and Sultana in the Jaisalmer district of Rajasthan. Despite declaring the vast area of the desert districts as Desert Wildlife Sanctuary, the population of the GIB and Houbara Bustard continues to decline. In order to protect all Indian species of bustards and their grassland habitats, the Government of India should start "Project Bustards" on the lines of Project Tiger, Project Elephant and Project Snow Leopard, with the following objectives:

1. To start a long-term *in situ* and *ex situ* conservation and breeding programme.
2. To identify the bustard areas and declare them as community reserves or conservation reserves or eco-sensitive zones.
3. To initiate discussions on and finalise a national grazing and grassland policy for bustard areas.
4. To obtain necessary financial, management and scientific inputs required for the protection of habitat within and outside the protected areas.
5. To plan and implement landscape conservation strategies for grassland management both within and outside the wildlife reserves, with the consent and involvement of local communities.
6. To involve local communities in and around the identified bustard habitats in the protection of these bustards.
7. To create awareness among the masses; educational material in local languages on grassland ecosystems and bustards for publicity may be produced regarding this rare germ plasm and gene pool material of the desert region.
8. To constitute an interstate co-ordination committee of members from Andhra Pradesh, Madhya Pradesh, Gujarat, Maharashtra, Karnataka and Rajasthan to monitor and review the protection measures, and also to take co-ordinate efforts at state level as well as at national level. This committee will also co-ordinate at the central level to solve the interstate as well as transborder issues with neighbouring countries.

Conservation Strategy

Majority of the GIB population is surviving in DNP of Rajasthan. Keeping in view of the conditions in DNP, the following strategies were in place and/or proposed for the conservation of GIB in Rajasthan:

1. Declaration of Desert National Park as a biosphere reserve
   The present area of the DNP was selected and earmarked for declaration as a biosphere reserve. The Ministry of Environment and Forests, Government of India constituted a working group in 1988 under the chairmanship of Dr K. S. Sankhla to prepare a project document on the Thar Desert Biosphere Reserve with the objective of integrating biological, socio-economic and cultural elements. The working group recommended the upgradation of the DNP as the Thar Desert Biosphere Reserve. The BNHS has also submitted a project report for declaring it the Thar Desert Biosphere Reserve. The Planning Commission of India has also recommended the upgradation of Desert National Park as a Biosphere Reserve in the 11th Five-Year Plan (2007–2012). Unfortunately, a firm decision is yet to be taken by the state government, and this should be pursued to

achieve the objective for which this area was selected and demarcated. This will also help in ameliorating the economic condition of the locals and also cater to their basic needs of fuel wood and fodder.

2. Creating inviolate Core Space
   Seventy-three villages and Dhanis are situated in the DNP with a human population of 62,024 and cattle population of 182,352. As per the estimate of the Executive Engineer PWD, Barmer, about 1,150 vehicles ply every day only in the Barmer part of the DNP. With such a degree of heavy biotic interference, it is very difficult to properly manage and protect the habitat of bustards. It is suggested that if all the 73 villages from the park area cannot be removed in the present political set-up, at least nine villages of Jaisalmer district, namely, Sam, Sagaron ki Basti, Matuon ki Basti, Ganga, Kanoi, Nimba, Mehboob Ka Par, Jamda and Bida, should be shifted immediately to create an inviolate Core Space in the DNP to save the fast vanishing GIB, for which the Sub-divisional Magistrate, Jaisalmer, has proposed adequate compensation in his proposal dated July 7, 1999, sent to the State Government. Similarly, six villages of Barmer districts, namely, Nodiyala, Bachia, Punj Raj ka Par, Modardi, Dhagari and Sagorani around Bandhara village, should also be shifted in the second phase of the project.
3. Transfer of revenue land to Forest Department
   Since a major chunk of the DNP land is revenue land, it should be immediately transferred to the Forest Department for its better management and development.
4. Transfer of forest land to DNP
   The forest land, which is a part of the DNP, currently under possession of the territorial division of Barmer should also be handed over to the DNP management.
5. Change in land use pattern
   With a growing network of canals, the land use pattern in western Rajasthan is also changing very fast. The traditional pasture lands are cleared for agricultural fields. Thus, local vegetation and palatable indigenous grasses are depleting very fast, eradicating and exterminating so many species of rare plants and animals in the process. Hence, it is suggested that pasture lands of Sewan (*Lasiurus sindicus*) should be protected at any cost because these grasslands are the lifeline of western Rajasthan for pastoral communities.
6. Integrated approach for protection
   Since the GIB is reported from so many areas outside the DNP, co-operation of the territorial staff other than the wildlife wing of the Forest Department is a must. They should be directed and trained accordingly, and the progress should be reviewed and monitored regularly for the integrated effort to save these magnificent birds. Similarly, the Border Security Force (BSF) can also play a pivotal role in the protection of GIB, Houbara Bustards and other desert species such as Desert Fox, Mongoose, Golden Jackal, Spiny-tailed Lizard and Monitor Lizards by nabbing the poachers because of their vast and sound information network.

7. Total ban on planting exotic tree species
   Planting of tree species particularly exotics like *Acacia tortilis* should be prohibited with immediate effect because it destroys the grassland. Use of insecticides and pesticides in agriculture is also detrimental to the indigenous flora and fauna of the region.
8. Research on the biology of GIB
   Research in the field of bustards is so far inadequate. In order to conserve the bustards, it is necessary to understand the intricacies of the life cycle of bustard. There is a strong need to investigate the seasonal patterns of local migrations, and their reasons should be fully understood in order to improve management. Habitat choice, current population trends and breeding pattern also need to be clearly documented. Captive breeding in near natural conditions is also required to increase the population.
9. Protection from grazing and fire
   Grasslands naturally attract domestic cattle and other herbivores causing disturbances, while protected and undisturbed grasslands are needed for bustards. The anthropogenic pressures also cause fire in these areas, and so there is a need to create separate grasslands near the bustard habitats in order to reduce competition. Please see Chap. 2 for pictures of GIB.

References

1. Ali S, Ripley RD (1983) Handbook of the Birds of India and Pakistan. Vol. 2. Oxford University Press, Compact Edition; 2001, Delhi, pp 737
2. Sankaran R (1997) The relation between bustard body size and display type. J Bomb Nat Hist Soc 94(2):403–406
3. Rahmani AR (2009) Ravi Sankaran's ornithological contribution. Ind Birds 5(4):121–125
4. Rahmani AR (1989) The Great Indian Bustard: Final Report of Project "Study of ecology of certain endangered species of wildlife and their habitats" funded by the U.S. Fish and Wildlife Service, and sponsored by the Ministry of Environment, Government of India.". Published by Bombay Natural History Society, Bombay, pp 234 (pp 207–211)
5. Sankaran R, Rahmani AR, Ganguli-Lachungpa U (1992) The distribution and status of the Lesser Florican *Sypheotides indica* (J.F.Miller) in the Indian subcontinent. J Bomb Nat. Hist Soc 89(2):156–179
6. Rahmani AR, Narayan G, Rosalind L, Sankaran R, Ganguli U (1991) Status of the Bengal Florican *Houbaropsis bengalensis* in India. J Bomb Nat Hist Soc 88:349–375
7. Goriup PD, Karpowicz ZJ (1985) A review of the past and recent status of the lesser Florican. Bustard Stud 3:161–182
8. Baqri QH, Kankane PL (2001) Deserts: Thar. In: Ecosystems of India; ENVIS. Zool Surv Ind, pp 93–122
9. Dharmakumarsinhji DKS (1950) The Lesser Florican [*Sypheotides indica* (Miller)]: its courtship display, behaviour and habits; 1950. J Bomb Nat Hist Soc 49(2):201–216
10. Sankaran R (1987) The Lesser Florican. Sanctuary Asia VII(1):26–37
11. Sankaran R (1987) Back to Florican country. Hornbill 1:22–25

12. Sankaran R (1991) Some aspects of the breeding behaviour of the Lesser Florican *Sypheotides indica* (J. F. Miller) and the Bengal Florican *Houbaropsis bengalensis* (Gmelin). Ph.D. Dissertation. Unpublished. University of Bombay, Bombay, pp 265
13. Sankaran R (1994) Some aspects of the territorial system in Lesser Florican *Sypheotides indica* (J. F. Miller). J Bomb Nat. Hist Soc 91(2):173–186
14. Sankaran R (1996) Territorial displays of the Bengal Florican. J Bomb Nat Hist Soc 93(2):167–177
15. Sankaran R (1996) Aerial display of the Lesser Florican. J Bomb Nat Hist Soc 93(3):401–410

Chapter 12
Distribution, Status and Conservation of Lesser Florican *Sypheotides indicus* in Rajasthan

Rakesh Vyas and B.K. Sharma

Abstract This chapter aptly describes the ecology, habitat and potential threats to the grassland bird, Lesser Florican, based on surveys conducted by the author during 1994–1996 in various districts of Rajasthan. Lesser Florican belongs to family Otididae of bustards, and all the three endemic bustard species of India are endangered. Eleven per cent of the land area of Rajasthan provides grassland and cropland as the most suitable habitat for this bird. The Lesser Florican is a monsoon visitor to central and southern Rajasthan inhabiting the districts of Nagour, Ajmer, Tonk, Bhilwara and Chittourgarh. In the south–east Rajasthan, it has been reported from Kota, Baran and Jhalawar districts. Sawai Madhopur district has also been shown as an ideal habitat. The post-breeding movement of Lesser Florican is not very well understood. In the name of sport, this graceful bird was shot in large numbers all over its distributional range. Its innocent habit of performing an aerial display to achieve conjugal bliss by attracting future mate has been its undoing despite its protected status. In Rajasthan, the annual cycle of sheep migration from west to east and back takes a heavy toll on the grassland ecosystem. According to Birdlife International, the estimated number of Lesser Florican in India was as low as 2206 in 1994, and since then, very little work has been done to assess its status. This chapter also mentions about grassland management practices in Rajasthan. The protection of government-owned and private grasslands has been recommended by the author to improve the fate of this endangered bird which in turn helps the rural population dependent solely on the domestic livestock for livelihood.

R. Vyas (✉)
Hadouti Naturalists Society, Kota, 204, Washleigh Manor,
Rustam Bagh, Bangaluru 560017, India
e-mail: rakeshvyas53@hotmail.com

B.K. Sharma
Department of Zoology, R.L. Saharia Government P.G. College,
Kaladera (Jaipur), Rajasthan, India
e-mail: drbksharma@hotmail.com

B.K. Sharma et al. (eds.), *Faunal Heritage of Rajasthan, India: General Background and Ecology of Vertebrates*, DOI 10.1007/978-1-4614-0800-0_12,

Introduction

The members of family Otididae, popularly known as bustards, are an ancient group of birds with their earliest fossil records from the Eocene Period, some 40–50 million years back. The group is widely distributed throughout Asia, Africa, Europe and Australia. Bustards are medium to large grassland birds with strong flight punctuated by sustained wing beat. Male bustards are conspicuously coloured and may have a crests, moustache and elongated plumes on the head and neck.

The six species of bustards known to occur in India are the Great Indian Bustard *Ardeotis nigriceps*, the Lesser Florican *Sypheotides indicus*, the Bengal Florican *Houbaropsis bengalensis*, the Great Bustard *Otis tarda*, the Little Bustard *Tetrax tetrax* and Houbara Bustard *Chlamydotis undulata*. The Great Indian Bustard (GIB), Lesser Florican and Bengal Florican are breeding residents, whereas the Houbara is a regular migrant to India, there is no recent record in India of the remaining two species [1, 2]. A subspecies of the Bengal Florican is also found in Vietnam and Cambodia [3]. According to Birdlife International, the estimated number of Lesser Florican in India was as low as 2206 in 1994, and since then, very little has been done in terms of its conservation.

The Lesser Florican is among the smallest and most beautiful bustard of the world (Fig. 12.1). It is a monsoon visitor to central, eastern, south-eastern and

Fig. 12.1 Lesser Florican *Sypheotides indicus* in its natural habitat. *Courtesy: Rakesh Vyas*

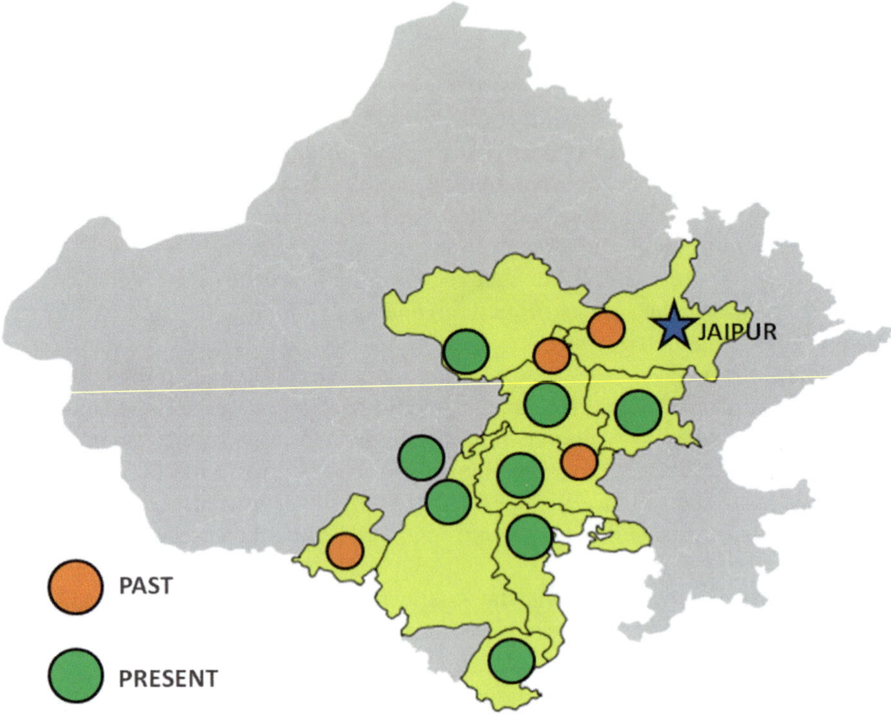

Fig. 12.2 Distribution of Lesser Florican *Sypheotides indicus* in Rajasthan

southern parts of Rajasthan (Fig. 12.2). The birds arrive with the onset of south-west monsoon to breed in the luxuriant grasslands of these areas, most of which are unprotected private farmlands and village grasslands [4, 5]. It breeds in southern Rajasthan, southern and eastern Gujarat and western Madhya Pradesh. The post-breeding movement of the Lesser Florican is not very clear. In the name of sport, this graceful bird was shot in large numbers all over its breeding range all over its distributional range. Its innocent habit of performing an aerial display to achieve the conjugal bliss by attracting its future mate has been its undoing. Even today, in spite of its protected status, it is getting killed during its monsoon sojourn in the grasslands of western and central India. During the surveys conducted in Rajasthan, direct or indirect evidence of its poaching has been found almost everywhere.

The grasslands are the primary habitat for its breeding, but crop fields, margins of the agricultural fields and the undulating degraded pastureland also provide a good alternative habitat. The traditional *charagah*/village *beeds* have become fragmented and are rarely taken care of by the village communities. In Rajasthan, the annual cycle of sheep migration from west to east and back takes a heavy toll of grassland ecosystem. The grasslands are the backbone of rural economy in Rajasthan as the pastoral communities and their livestock depend on them for fodder.

Table 12.1 Past records of Lesser Florican in Rajasthan

Place	Florican details	Date/season	Source
Rajpootana	Common	Jul–Sept	Jerdon [6]
Sambhar Lake	Female	Early monsoon	Adams [7]
Sambhar Lake	Male	Early monsoon	Adams [8]
Rajpootana	Not known	Sept–Oct	Barnes [9]
Rajpootana	Breeding	July–Sept	Barnes [10]
Athun Ganeshpur	Breeding	Aug 1983	Saxena & Meena [11]
Gangwana, Mangaliabas, Ramsar, Bandar, Goyla	Male, female	Monsoon	Saxena & Meena [11]
Sorsan	Female	Monsoon	Bharat Singh, pers. Comm.
Shahpura	2 Males	Monsoon, 1986	Sankaran [12]
Kalsas, Sangamer	7 Males	Monsoon, 1984 and 1992	Sankaran [12]

Rajasthan has a unique distinction of having the maximum per capita number of domestic animals. The number of milch, non-milch and other domestic animals in Rajasthan is 49.2 million as per the *Resource Atlas of Rajasthan* published by the Government of Rajasthan (1994). The permanent pasture and grazing land constitute approximately 3.5% of the total land area beside degraded undulating grass and scrubland, which is about 7.6% of the total land area. Therefore, it can be conveniently construed that 11% of the land area of Rajasthan could provide some habitat to the Lesser Florican, beside patches of grass on the margins of the crop fields and the crop fields themselves.

In the various surveys conducted since 1982, to ascertain the status and distributional range of Lesser Florican in Rajasthan, the bird was personally observed in central and southern Rajasthan, whereas its presence in south–east Rajasthan is based on personal communication with Shri Bharat Singh and records and descriptions by an older generation of naturalists (Table 12.1) [6–12]. It has been recorded in the districts of Nagour, Ajmer, Tonk, Bhilwara and Chittourgarh during the surveys conducted by Late Dr Ravi Sankaran, former Director of Salim Ali Centre for Ornithology and Natural History (SACON) with the author during 1994, 1995 and 1996 [11–14]. In south–east Rajasthan, it was known as "Mal Moradi" and was found in Kota, Baran and Jhalawar districts. Some of the grasslands in Sawai Madhopur district have also been shown as ideal habitat for the Lesser Florican (Table 12.2).

The yearly movement and breeding of Lesser Florican in Rajasthan depends on south-west monsoon, and in the good years, its numbers also swell. The management and protection of government-owned or private grasslands and cropland would ultimately decide the fate of this endangered bird, and the conservation of grassland, in turn, will help the rural population which is dependent on fodder procured from these beeds/*charagah* for their domestic stock. The conservation measures and recommendations for the protection of grasslands are discussed.

Table 12.2 Lesser Florican survey in Rajasthan, 1994, 1995, 1996

District	Place	Habitat	1994/1995	1996
Ajmer	Gangwana	Grassland/cropland	0	10–12
	Alniyawas	Cropland	1	2
	Sonkhalia	Grassland/cropland	4	8
	Dhani Rathodan	Cropland	0	1
Banswara	Mirch Ghati	Grassland	1	4
	Hathiadili	Grassland/cropland	0	5
Bhilwara	Bated	Grassland/cropland	1	0
	Baldharkha	Grassland/cropland	2	5
	Ghasta Beed	Grassland	2	6
	Kalsas	Grassland/cropland	Not visited	4
	Loolas/Mundetha	Grassland/cropland	Not visited	10
	Jamoli	Cropland	5	10
Nagour	Merta City	Cropland	Not visited	2
Pratapgarh	Chiklad	Grassland/cropland	0	10
	Bajrangarh, Gandher, Belara, Kultana, Bilesri, Akaypur	Grassland/cropland	5	20
Pali	Boya	Grassland/cropland	6	10
	Birolia	Grassland/cropland	Not visited	4
	Omkali	Grassland/cropland	Not visited	3
Tonk	Lahen, Lamba, Dinghara	Cropland	Not visited	20

Study Area and Methodology

The surveys in the year 1994, 1995 and 1996 were conducted in Ajmer, Bhilwara, Banswara, Kota, Nagour, Pali, Pratapgarh, Tonk and Sawai Madhopur districts [12, 13, 15]. Information from review of available literature, historical records and personal communication with local wildlifers and Forest Department has been compiled. The general survey was done by using a vehicle, but most of the large and prospective grasslands were covered by walking along the periphery from a safe distance, so as not to disturb the birds. The local forest guards, farmers and knowledgeable people were shown the picture of the bird and were asked about its presence in their area. The prevalent agricultural practices, state of their livestock and pastures were inquired. A study of the status of the grasslands, local management practices and annual grazing regimes was also done.

Results and Discussion

The dispersal and movement of the Lesser Florican is strongly determined by the distribution and quantum of rainfall which also determines its optimum breeding environment [15]. Therefore, the arrival pattern of Lesser Florican to breeding areas is closely related to rainfall. As a result, during the breeding season, it seeks the areas of good rainfall and grass growth. The first two years of survey were bad rainfall years with poor and irregular precipitation in the south–east and southern Rajasthan. In the first year (1994), only Ajmer, Bhilwara, Pratapgarh, Chittourgarh and Tonk districts were covered. The Lesser Florican was seen at Sonkhalia (4), Pratapgarh (5) and Bated and Kalsas (1). The habitat of the Lesser Florican in Rajasthan is of mixed type with grassland and cropland overlapping each other. The Sonkhalia grassland is a protected area for the Great Indian Bustard, but since most of the land is revenue land, crop fields encroach upon the grasslands. The birds are equally distributed in both the habitats. The suitability and safety of the display site for the male is the determining factor for its presence. The ownership of the grasslands is also equally divided among Forest and Revenue Department, *panchayat* and individuals. The earlier studies by (Late) Dr Ravi Sankaran have shown that if the rainy spells are dominated by dry, sunny periods, the birds leave the grassland. The arrival of male(s) preceded that of female(s) by 11 (1986–1987) and 15 days (1985), respectively.

In 1996, the good monsoon had brought regular rainfall, which was fairly distributed all over Rajasthan. Intensive surveys were undertaken in the month of August with many new destinations added as the reported sightings from many places. In Ajmer district, maximum birds were seen at Gagwana and Sonkhalia. The Gagwana grassland is very good with crop fields on the margins, and 12 birds were recorded during that year. At Sonkhalia also, eight birds were counted in two days. Two birds were found in the cropland at Alniyawas and Dhani Rathodan. Baneda *tehsil* in Bhilwara district, the area bordering Madhya Pradesh in Pratapgarh district and grasslands close to Tonk were most productive from the sighting point of view. In the places like Loolas, Mundetha and Jamoli in Bhilwara district, 5–10 birds were seen in the grassland. The Pratapgarh district has some very important Lesser Florican areas requiring immediate protection. The grassland and cropland near the villages Bajrangarh, Akaypur, Gandher, Silarpur, Belara, Kultana and Bilesri support good population of breeding Lesser Florican. In two days, about 20 birds were seen in different areas. In the grassland of Boya in Pali district, 10 birds were recorded, and the nearby areas of Biroliya and Omkali also had few birds. The grasslands near Tonk at Lahen, Lamba and Dinghara supported about 20 birds in 1996 (Table 12.2).

The rainfall in Rajasthan shows considerable interannual variation, and the same intervariation is seen in the growth of grass and height. In such environments, green grass is present only during the wet season, i.e. June to September. It is the rainfall which initiates and determines grass growth in the Lesser Florican habitat, and a well-distributed rainfall bears a stronger influence over growth patterns than the total annual rainfall [15]. An irregular and deficient rainfall in Central India and Rajasthan during 2001–2005 had an adverse impact on the number of Lesser Florican arriving in Sardarpura area of Dhar district of Madhya Pradesh (PM Lad, *pers. com.*).

The movement of Lesser Florican shows a south-westerly migration from their nonbreeding areas. Although, the breeding range covers a vast area, major breeding areas are located in Kutch, Kathiawad, western Madhya Pradesh, north-western Maharashtra and central and south-eastern Rajasthan [8–10] . Historically, the erstwhile area of Rajputana had wide distribution of the bird, right from Sambhar Lake to Pindwara in Sirohi district. The hunters of older and present days had been very much familiar with the bird and had informed about its recent as well as past presence in their areas. Skin of this bird in Rao Madho Singh Museum of Kota was seen; the bird was hunted near Kota in early part of the twentieth century. The Lesser Florican has been reported from the grassland of Sorsan in Baran district (Bharat Singh, *pers. com.*). The maximum number of Lesser Floricans was found in Ajmer, Nagour, Chittourgarh, Pratapgarh, Pali and Banswara districts. There are some past records from Tonk and Bharatpur also. The erstwhile rulers of Jodhpur had often shot 100 floricans in a morning shoot party of four or five guns [15]. The number of Lesser Floricans in the surveys of 1994, 1995 and 1996 did not come anywhere close to the past records.

Recently, a survey to assess the present status and distribution of Lesser Florican in the north-western India, i.e. in Gujarat, Madhya Pradesh and Rajasthan, was carried out in the month of August 2010, which is a part of breeding season of this species, when most of the males display in the grasslands [16]. A total of 84 individual Lesser Floricans (83 males and one female) were sighted in three states of north-western India, which is 65% less than the sightings reported in 1999 by Dr Sankaran [17]. Among the three states, more sightings of Lesser Florican were reported in the state of Gujarat (54) followed by Rajasthan (18) and Madhya Pradesh (12). But in 1999, more sightings of florican were reported in Gujarat (141) followed by Madhya Pradesh (63) and Rajasthan (34). Maximum number of birds (6) were seen in Pratapgarh district, followed by four in Malpura area of Tonk district, Shahpura area of Bhilwara (5) and Saunkhaliya area of Ajmer (3). Out of 20 potential grasslands, 18 were surveyed which include grasslands surveyed during 1999 [17]. Out of the surveyed grasslands, floricans were found in 10 grasslands as against seven in 1999. Floricans were sighted in all surveyed districts, i.e. Bhilwara, Tonk, Ajmer and Pratapgarh. Although, the number is continuously declining in Rajasthan, the distribution range is the same. More than 80% surveyed grasslands (*beeds*) were owned by private owners, and these lands were under serious pressure for farming and agriculture. In Pratapgarh district, a drastic reduction of Lesser Florican was reported even in 2008 [18]. Most of the grasslands in the state are invaded by *Prosopis juliflora* and *Capparis decidua*, especially in Ajmer and Bhilwara districts, which is taken as major conservation threat.

There is an urgent need to look into the present problems of its habitat and efforts to rectify the situation by restoring the grassland ecosystem in the state. Complacency about its hunting should not exist as during the survey conducted by author, they saw the feathers and plumes of a freshly shot florican. So, stricter enforcement of law and awareness campaigns among the hunting communities will also be required to keep the state safe for the Lesser Florican.

The Management Practices of Grasslands in Rajasthan

The government-owned grasslands are mostly controlled by the Department of Forests which protects the grass from grazing during the July–October period and then auctions the grass. The lease is permitted to either cut fodder or graze the grassland.

Table 12.3 Important sites for Lesser Florican conservation in Rajasthan

District	Place	Location	Area
Ajmer	Gangwana	8 km from Ajmer on Jaipur highway	
	Sonkhalia	Near Nasirabad	
Bhilwara	Loolas	Baneda	
	Dabla	Baneda	
	Baldharka	Baneda	
	Jamoli	Off Shahpura–Jahajpur road	
	Dhani Beed	Shahpura forest range	315 ha
	Karamdas	Baneda forest range	c.200 ha
	Nogawa	Baneda forest range	c.200 ha
	Dhikola	Baneda forest range	c.200 ha
Banswara	Mirch Ghatti plantation	Kushalgarh forest range	850 ha
	Jharnia Chhaparia	Banswara forest range	390 ha
Pali	Bali	Bali forest range	c.265 ha
	Guda–Indala	Pali forest range	c.250 ha
Pratapgarh	Lakhia Beed	Pratapgarh forest range	c.100 ha
	Gandher Radi	Pratapgarh forest range	c.250 ha
	Kultana	Pratapgarh	
	Bajrangarh	Pratapgarh	
Sawai Madhopur	Bamanbas	Gangapur	
Tonk	Lahen	Tonk	
	Lamba	Tonk	
	Dinghara	Near Jhirana	

In the years of scarcity, the harvesting of grass is controlled by the department. The privately owned grasslands are protected for their fodder output, and the owner uses it for livestock or leases it out to some cattle graziers. The vast grasslands owned by the princely states had been taken over under the Land Ceiling Act and were converted into cropland after the independence of India as seen in Pali and Bhilwara districts. The management and protection status of grasslands by the State Forest Department is not adequate. The plantation programmes undertaken by the department and introduction of *Prosopis juliflora* have negative impact on grasslands and grass production.

The recommendations are being dealt separately in this chapter, and a list of potential sites in need of immediate protection is also given (Table 12.3). Dr Asad Rafi Rahmani in his editorial in the *Journal of Bombay Natural History Society* writes "most Protected Areas are threatened by all the following or some of them: roads, railways, dams, illicit wood cutting, urban expansion, over-grazing, invasive species, encroachment, unrestricted tourism and mismanagement or plain neglect" [19]. Scientists are now looking into the ecological implications of bird decline and calculating the cost of ecosystem services offered by the bird species as each avian functional group is important [20]. The need of the hour is to promote research, save habitat and manage ecosystems effectively to tide over the loss of biodiversity in general and critically endangered species in particular.

Recommendation

The survival of the bustard group of birds depends upon the protection of healthy grassland habitat. The survival of agro-pastoral communities of Rajasthan also depends upon restoration of flourishing well-maintained and community-managed grasslands. It is to our advantage to protect grasslands, in turn, protect some of the most endangered birds in the world like the Lesser Florican, the Great Indian Bustard and the Bengal Florican [21]. The existing policies and practices of grassland management have not yielded desired results. Therefore, a fresh scientific outlook would be required to tide over the problem of diminishing fodder supply for the livestock and habitat for grassland avifauna. After much deliberation among scientists, wildlife officers and the members of non-governmental organizations and stakeholders, the consensus has emerged and following recommendations were made:

1. Making a national policy on grassland management: It is important to make sustainable management of grassland resources without harming their ecological services, for which a national policy on grassland management is suggested.
2. There should be a total ban on the plantation of trees in reserved grasslands.
3. The non-reserved grasslands in the revenue land or *panchayat* land should be managed by the fodder co-operative in consultation with the State Government Department of Forests and Wildlife. No free grazing should be allowed. The fodder must be harvested and supplied as per the local need.
4. In degraded grasslands, the savannah-type indigenous vegetation may be grown in the clumps and leave large areas in between for production of fodder.
5. The specified sites in Rajasthan (Table 12.3) should be brought under the purview of Wildlife Protection Act, 1972 by declaring them as a conservation or community reserves.
6. The wastelands around the villages should be handed over to the fodder co-operative, to be developed as fodder-producing grasslands.
7. The private grasslands should not be used, and illegal altered land use or encroachments must be dealt strongly.
8. Special conservation education programmes, highlighting the importance of grassland and its unique fauna, should be started in the villages and in the schools.
9. A state-level "florican watch" should be launched with the co-operation of NGOs, stakeholders and the Forest Department officials to monitor the annual arrival of the Lesser Florican and its breeding in the state.

Please also see Chap. 2 for more pictures.

Acknowledgements (RV) I dedicate this note to the memory of my friend and foremost florican expert in India, Dr Ravi Sankaran. He left us for his heavenly abode on 17 January 2009. The note is a product of many hours of discussions with him and long field surveys in Rajasthan about a decade back. He rose to the position of the Director of Salim Ali Centre for Ornithology and Natural History (SACON), Coimbatore, at a very young age. The loss of this young scientist is irreparable, and the best homage to his memory is to carry on his work honestly and earnestly. I am thankful to many naturalists, who have given encouragement and suggestions from time to time. I put on records my gratitude to Shri Bharat Singh, Minister for Panchayati Raj and Rural development, Government of Rajasthan; Late Shri Shantanu Kumar, DGP (retd.); Shri P. M. Lad, CWLW (retd.); Dr A. R. Rahmani, Director, Bombay Natural History Society; and Dr G. S. Bhardwaj, Rajasthan Forest & Wildlife Department. I am especially thankful to Shri P. M. Lad for photographs. I am thankful to the members of Hadoti Naturalists' Society for being a part of surveys and workshops on Lesser Florican at Kota.

References

1. Ali S, Ripley SD (1983) Compact handbook of the birds of India and Pakistan. Oxford University Press, New Delhi, pp 737
2. Osborne PN, Collar N, Goriup PD (1984) Bustard. Dubai Wildlife Research Centre, Dubai, UAE
3. Sankaran R, Rahmani AR, Ganguli-Lachungpa U (1992) The distribution and status of the Lesser Florican *Sypheotides indica* (J. F. Miller) in the Indian subcontinent. J Bomb Nat Hist Soc 89:156–179
4. Dharamkumarsinhji RS (1950) The Lesser Florican (*Sypheotides indica* Miller): its courtship display, behaviour and habits. J Bomb Nat Hist Soc 49:201–216
5. Sankaran R (1997) Habitat use by the Lesser Florican. J Bomb Nat Hist Soc 94(1):40–47
6. Jerdon TC (1864) Birds of India. Vol. III, Calcutta, India
7. Adams RM (1873) Notes on the birds of Sambhar Lake and its vicinity. Stray Feathers 1:393
8. Adams RM (1874) Additional notes on the birds of Sambhar Lake and its vicinity. Stray Feathers 2:337–341
9. Barnes HE (1886) Birds nesting in Rajpootana. J Bomb Nat Hist Soc 1:38–62
10. Barnes HE (1891) Nesting in Western India. J Bomb Nat Hist Soc 6:13–14
11. Saxena VS, Meena BL (1985) Occurrence of Lesser florican in forest plantations in Rajasthan. Bustard Studies 3:183–184
12. Sankaran R (1996) The Status and Conservation of the Lesser florican in Rajasthan. Salim Ali Centre for Ornithology and Natural History, Coimbatore, India
13. Sankaran R (1996) Conservation of Lesser Florican—Background Paper for the Workshop at Kota, Rajasthan
14. Sankaran R (1994) The Status of Lesser florican in 1994, Preliminary Report. Salim Ali Centre for Ornithology and Natural History, Coimbatore, India
15. Sankaran R (1995) A fresh initiative to conserve the Lesser Florican. Orient Bird Club Bull 22:42–44
16. Bhardwaj GS, Sivakumar K, Jhala YV (2011) Status, distribution and conservation perspectives of Lesser Florican in the North-Western India: A Survey Report. Wildlife Institute of India, Dehra Dun, India. p 106
17. Sankaran R (2000) The status of the Lesser Florican *Sypheotides indica* in 1999. Salim Ali Centre for Ornithology and Natural History, Coimbatore in Collaboration with Bombay Natural History Society, Mumbai, India
18. Bhardwaj GS (2010) Status of Lesser Florican *Sypheotides indicus* in Pratapgarh district, Rajasthan, India. Indian Birds 6(1):20–21
19. Rahmani AR (2008) Editorial. J Bomb Nat Hist Soc. 105(1)
20. Cagan HS (2006) Increasing awareness of avian ecological function. Trends Ecol Evol. 8–21
21. Sankaran R, Rahmani AR (1990) Recommendations for Lesser Florican: Conservation, Status and Ecology of Lesser Florican. Final Report, Bombay Natural History Society, Bombay. 117–123

Chapter 13
Status and Distribution of Raptors in Rajasthan

Rishad Naoroji and Harkirat Singh Sangha

Abstract This chapter presents an overview of the status and distribution of raptors in the state of Rajasthan which has the largest concentration of these birds in the Indian Subcontinent. More than 50 raptor species have been reported by the authors. Data regarding their population from Bharatpur district during 1985–1989, Ranthambhore National Park in 1990 and other long-term surveys in the desert since 1998 onwards are still on going and have been mentioned. A composite picture has emerged over the years, and authors feel that this chapter will provide baseline data applicable for conserving the various species of birds of prey and their habitat in Rajasthan. Range maps for each species throughout Rajasthan are included. The status and distribution of all the resident and migratory species including vagrant species have been discussed at length. The decline of raptors particularly *Gyps* vulture has also been discussed. Conservation threats like increase in human and livestock population due mainly to the development of Indira Gandhi Nahar Pariyojna (IGNP) have been outlined. This chapter highlights that IGNP has caused drastic environmental changes on account of the exploitation of the raptor habitat in the semi-arid areas and Aravallis, and by the growing human population.

R. Naoroji
Godrej Bhavan, 5th Floor, 4A Home Street, Fort, Mumbai 400001, Maharashtra, India
e-mail: rishad@godrej.com

H.S. Sangha (✉)
B-27, Gautam Marg, Hanuman Nagar, Jaipur, Rajasthan, India
e-mail: harkirat.sangha@gmail.com

B.K. Sharma et al. (eds.), *Faunal Heritage of Rajasthan, India: General Background and Ecology of Vertebrates*, DOI 10.1007/978-1-4614-0800-0_13,

Methods

The authors have studied raptors throughout Rajasthan during different seasons. Data on raptors was collected at Bharatpur from 1985 to 1989, Ranthambhore in 1990 and long-term almost-unbroken open width surveys in the Thar Desert from 1998 onwards (both authors) and still continuing. The data will be later quantified and published elsewhere. Extensive data were collected on wintering raptors at Bharatpur (Rishad Naoroji) over three seasons while studying the breeding biology of the Pallas's Fish-eagle. One summer was spent at Ranthambhore, studying the breeding biology of the Bonelli's Eagle and creating a raptor checklist. A composite picture has emerged over the years, and this chapter will provide baseline data applicable for conserving the various species of birds of prey through habitat conservation in Rajasthan. Range maps for each species throughout Rajasthan are included.

Most famous of the wetlands is the Keoladeo National Park at Bharatpur, a World Heritage site which, during winter, supports large concentrations of migrant raptor species recorded within its 29 sq. km area, mainly 9 sq. km of marsh. The well-known wildlife areas in Rajasthan are Keoladeo National Park; Bharatpur, Ranthambhore and Sariska Tiger Reserves, and the Desert National Park. Both desert and semi-arid zones do not exhibit any great endemism, but support high densities of wintering migratory raptors. Forest species such as the Crested Serpent-eagle and Changeable Hawk-eagle can be observed in the Aravallis, mainly the protected parks of Ranthambhore and Sariska, while a few adults and wide ranging juveniles have been observed exploiting the abundant food source at Keoladeo National Park. In raptor species richness, both zones occupy a central position among the country's bio-geographic zones [1]. Out of a total of 50 raptor species recorded in Rajasthan, 34 (including residents, passage migrants and purely wintering species) have been recorded in the desert. Information on flyways and numbers of migratory raptors through Rajasthan is poorly understood, and migration-linked studies are required if we are to protect wintering raptors.

Threats to Raptor Habitat

The desert and semi-arid areas are among the largest concentrations of wintering raptors in the country. Increase in human density due to the development of the Indira Gandhi Nahar Project has resulted in large-scale ecological changes. Traditional crops are being replaced by cash crops which require more irrigation. An intricate network of canals has raised humidity levels and converted large tracts of desert to cultivation. Misuse of irrigation facilities have resulted in water logging and increased salinity along the length of the canal. The changed habitats (plantations, cash crops and seepage wetlands) attract increasing numbers of woodland and wetland birds, while desert and semi-desert bird populations are decreasing along the canal network. Furthermore, due to available of irrigation facilities, settlers from elsewhere have bought large swaths of land from local people and commenced large-scale commercial farming.

The increase in human settlements and livestock population has led to displacement of graziers to non-command areas, exerting even greater pressure on the overgrazed countryside. Grasslands, a key habitat for wildlife in the desert- have therefore drastically reduced. The natural features of the desert in some areas have already disappeared. There is also uncontrolled propagation of the undesirable exotic *P. juliflora* along the canal. Salination has occurred in some parts of the command area where irrigation has been attempted. Uncontrolled tourism (including proliferation of "desert lodges") has degraded prime desert habitat around Jaisalmer.

Problem of electrocution of raptors is more acute in the open habitat of the desert and semi-arid zones which lack adequate natural perches. Power lines benefit raptors by providing perches and nesting and roosting sites. Sizable numbers of electrocuted raptors of five species were found in Bikaner district below power poles, mainly Long-legged Buzzard, Tawny Eagle, White-eyed Buzzard and Common Kestrel. Large raptors including vultures and kites are more susceptible to electrocution because their open wings easily span the distance between live wires when alighting or taking off from power poles. This fact needs to be given much consideration as the design of power lines can be modified to provide safe perches [2].

Keoladeo National Park in the semi-arid zone was famous for a wide variety of up to 46 species of mainly wintering and also resident raptors attracted by the huge concentrations of wintering waterfowl. However, since water from *Ajan Bund* (dam) was diverted, these concentrations have drastically decreased.

Species List

Osprey *Pandion haliaetus* (Linnaeus, 1758)

Race *P.h. haliaetus* (Linnaeus, 1758)

Winter migrant. Distributed mainly east of 72° north longitude. Mainly found along rivers like Chambal, Luni, Banas, and Kalisindh and in the vicinity of waterbodies such as Dhebar Lake, Shahpura Lake and other smaller wetlands throughout the state (Fig. 13.1). Birds on passage could probably be flying over western Rajasthan to winter in the Rann of Kutch. With the advent of the Indira Gandhi Nahar Canal, it is possible that Ospreys may stop over or winter along the length of the canal.

Temporary lakes created by large depressions among the sand-dunes filled through seepage via the Rajasthan Canal have created reed-swamp vegetation. Filled with fish, these transient water bodies attract waterfowl and waders which in turn attract marsh harriers and provide adequate stopovers for wintering Ospreys.

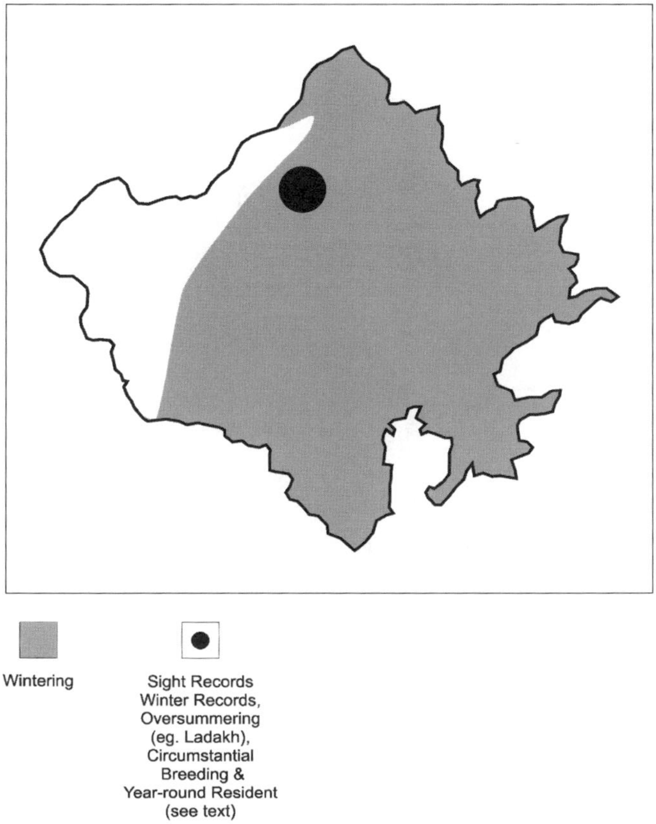

Fig. 13.1 Distribution of Osprey *Pandion haliaetus* (Linnaeus, 1758) in Rajasthan

Oriental Honey Buzzard *Pernis ptilorhynchus* (Temminck, 1821)

Races: *P.p. ruficollis* (Lesson, 1830)
P.p. orientalis (Taczanowski, 1891)

Resident of the semi-arid zone east of the desert where groves of trees are present. The species has been observed along the Indira Gandhi Nahar Canal where irrigation facilities have resulted in plantations and seepage wetlands attracting other woodland species, commonly found in the Aravallis and shows preference for wooded plains, groves of trees and gardens, cultivated landscapes, and rural and urban human habitation such as Bikaner and the Phalodi area (Fig. 13.2). The migratory subspecies *orientalis* also winters in Rajasthan, but not readily distinguishable from the resident subspecies.

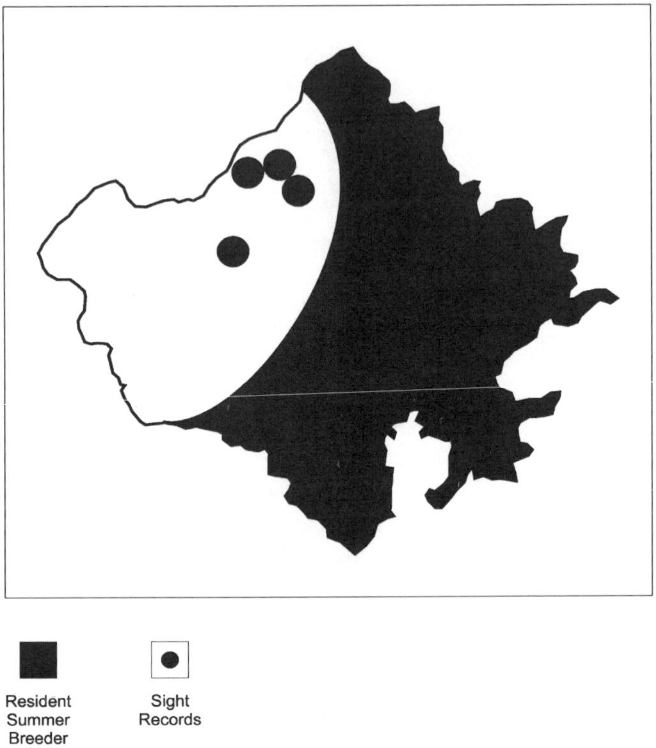

Fig. 13.2 Distribution of Oriental-honey Buzzard *Pernis ptilorhynchus* in Rajasthan

Black-winged Kite *Elanus caeruleus* (Desfontaines, 1789)

Race: *E.c. vociferus* (Latham, 1790)

A common species resident throughout Rajasthan. It prefers open habitat like open woodland, thorny forest, grasslands, cultivated areas near villages, fresh water marshes, scrub and grassland. It is most commonly observed in the semi-arid zone followed by the desert (Fig. 13.3).

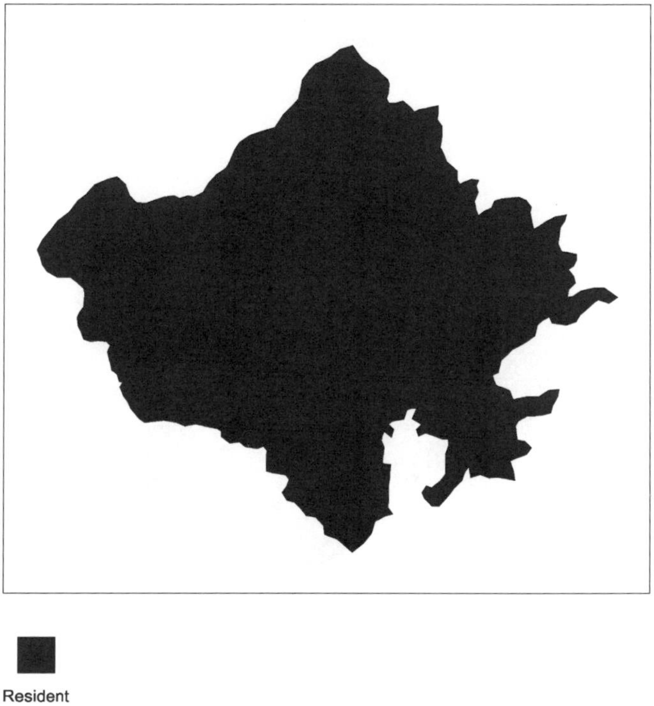

Fig. 13.3 Distribution of Black-winged Kite *Elanus caeruleus* (Desfontaines, 1789) in Rajasthan

Black Kite *Milvus migrans* (Boddaert, 1783)

Races: *M.m. migrans* (Boddaert, 1783)
M. m. govinda (Sykes, 1832)
M. m. lineatus (J.E. Gray, 1831)

Common raptor. The most adaptable and numerous raptor in the world benefiting greatly by its close association with man and his environs. The nominate subspecies *govinda* is found throughout Rajasthan, inhabiting the desert in the vicinity of human habitation except along the western extremity of the border with Pakistan and the Shahgarh Bulge. The Black-eared Kite *Milvus migrans lineatus* is a winter visitor throughout the state including Shahgarh Bulge and the western border, especially near-nomadic settlements (Fig. 13.4). With the "greening" of the desert and increased human habitation, the resident subspecies *govinda* is also increasing its range westwards. The nominate migratory race *M. M. migrans* is known to winter in the westernmost parts of Gujarat and Kutch and likely winters in the Thar Desert. Commensal with humans, this gregarious species is found in large densities around cities, towns and villages throughout Rajasthan.

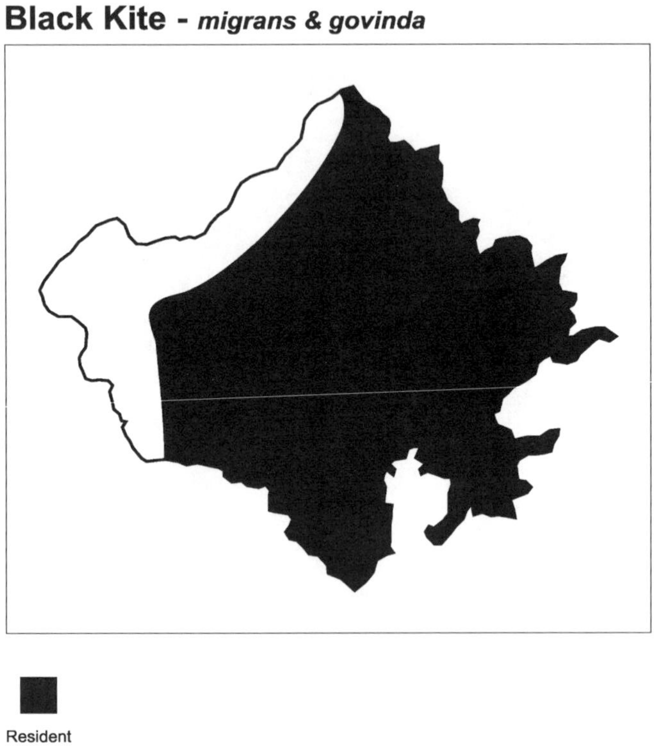

Fig. 13.4 Distribution of Black Kite *Milvus migrans* (Boddaert, 1783) in Rajasthan

Brahminy Kite *Haliastur indus* (Boddaert, 1783)

Race: *H.i. indus* (Boddaert, 1783)

A local winter migrant restricted to southernmost Rajasthan. Mostly resident in the rest of the country. One breeding record from Udaipur (Manoj Kulshreshtha, personal communication). Occasionally seen in Jaipur at Jal Mahal (Fig. 13.5).

Pallas's Fish-eagle *Haliaeetus leucoryphus* (Pallas, 1771)

Race Monotypic

A rare breeding migrant mostly absent from Rajasthan except in the easternmost part of the semi-arid zone bordering Madhya Pradesh and Uttar Pradesh. A few sight records exist from the Chambal River; one pair observed in 1981–1982

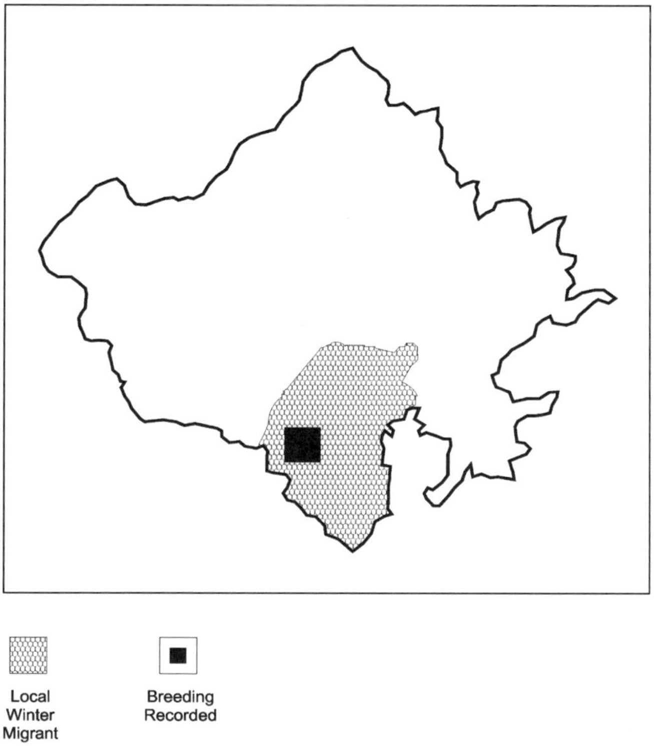

Fig. 13.5 Distribution of Brahminy Kite *Haliastur indus* (Boddaert, 1783) in Rajasthan

winter in Ranthambhore National Park. A sight record of a juvenile photographed from the Chambal River (confirmed by RKN). Two pairs bred over many years at Keoladeo National Park, Bharatpur. For study purpose, the Park is divided in various blocks. The dominant D Block pair had a much larger territory than the L Block pair which did not breed as successfully as the D Block pair. A few years before the water was diverted from the park, the L Block pair disappeared (Fig. 13.6).

IUCN status: Vulnerable C1 [3]

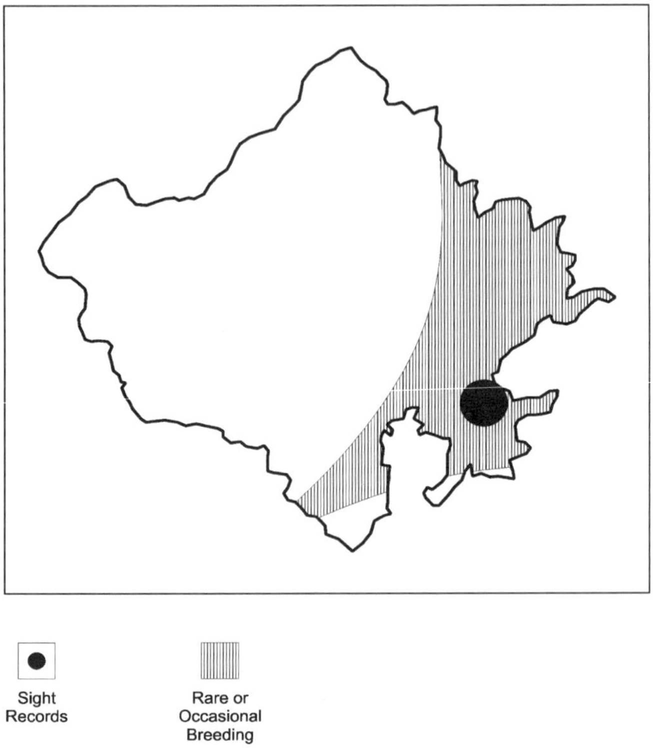

Fig. 13.6 Distribution of Pallas's Fish-eagle *Haliaeetus leucoryphus* (Pallas, 1771) in Rajasthan

White-tailed Eagle *Haliaeetus albicilla* (Linnaeus, 1758)

Race Monotypic

A rare wintering species to eastern Rajasthan where waterbodies exist with repeated sight records from Keoladeo National Park, Bharatpur (Fig. 13.7).

IUCN Status: Least Concern

Lesser Fish-eagle *Ichthyophaga humilis* (Müller & Schlegel, 1841)

Race: *I.h. plumbea* (Jerdon, 1871)

Rare winter vagrant. Vagrants confirmed from Sariska and Keoladeo National Park, Bharatpur [4] (Fig. 13.8). This species displays a marked preference for swift flowing rivers and streams in the Himalayan foothills as breeding resident.

IUCN status: Near Threatened—A1b, c, e; A2b, c, e; C1; C2a [3]

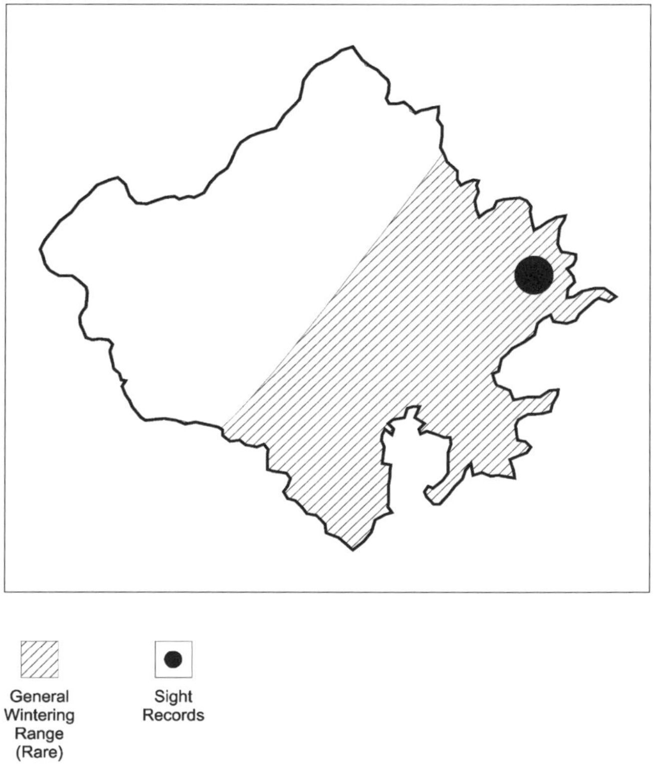

Fig. 13.7 Distribution of White-tailed Eagle *Haliaeetus albicilla* (Linnaeus, 1758) in Rajasthan

Grey-headed Fish-eagle *Ichthyophaga ichthyaetus* (Horsfield, 1821)

Race Monotypic

Vagrant. Rajasthan is not within its normal range, but there are two sight records from Bharatpur (RKN) and Ranthambhore Tiger Reserve (Manoj Kulshreshtha and G. Viswanatha Reddy, personal communication). Within its preferred breeding range, the species has a preference for low-lying moist-deciduous or semievergreen-forested country in the vicinity of perennial waterbodies (Fig. 13.9).

IUCN status: Near Threatened—A1b, c; A2b, c; C1 [3]

Egyptian Vulture *Neophron percnopterus* (Linnaeus, 1758)

Races: *N.p. percnopterus* (Linnaeus, 1758)

N.p. ginginianus (Latham, 1790)

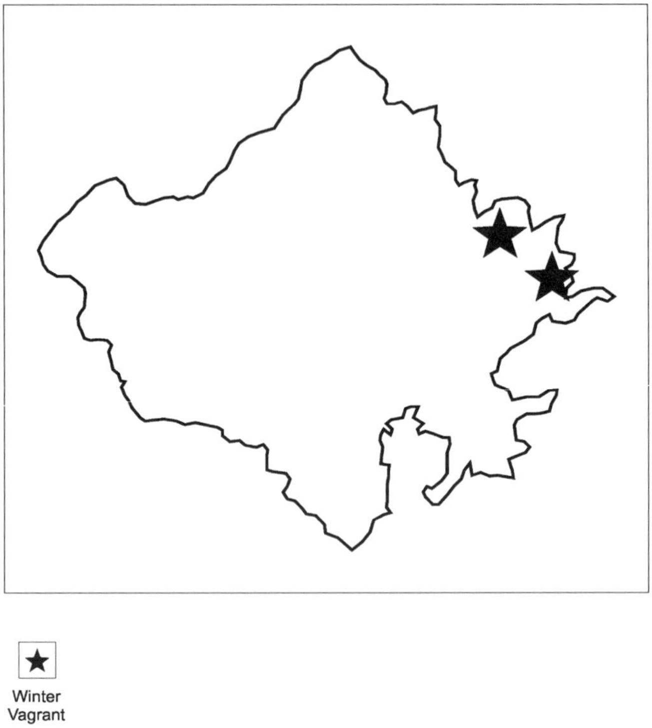

Fig. 13.8 Distribution of Lesser Fish-eagle *Ichthyophaga humilis* (Müller & Schlegel, 1841) in Rajasthan

Resident. Both races are locally migratory and resident throughout Rajasthan. Frequently observed in dry open country such as the desert and semi-arid zones, and agricultural areas (Figs. 13.10, 13.51, 13.52, 13.53, and 13.54). More information is required as to the local movements and status of *ginginianus*. At least 200 Egyptian Vultures were observed near Jodhpur in December 2007 [5].

IUCN Status: Endangered A2bcde+3bcde+4bcde

Indian White-rumped Vulture *Gyps bengalensis* (Gmelin, 1788)

Race Monotypic

Rare resident. Once the commonest raptor in the world; abundant even in the desert and semi-arid zones. Used to be observed earlier at high density around human dwellings and carcass dumps. Now seriously threatened due to widespread use of diclofenac for veterinarian application for cattle. Twenty-five White-backed Vultures were observed near Jodhpur in December (Fig. 13.11) [5].

IUCN status: Critically Endangered—A1c, e; A2c, e [3]

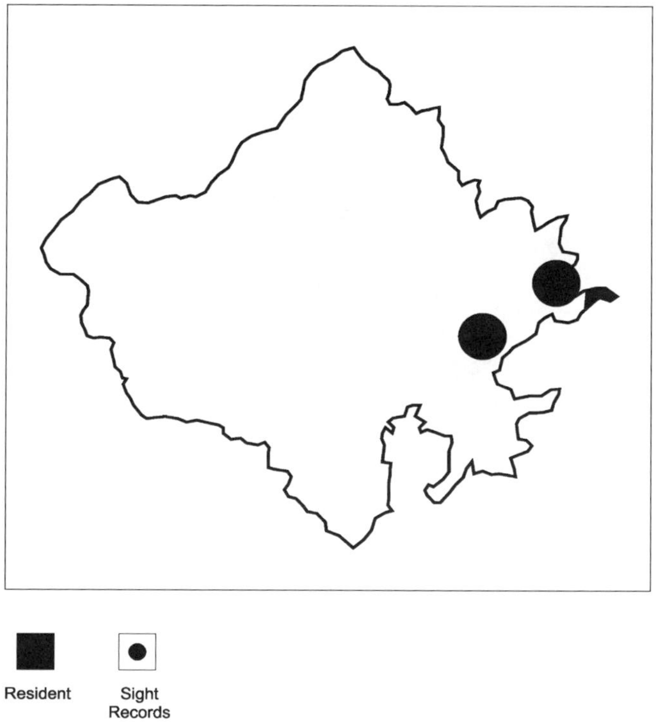

Fig. 13.9 Distribution of Grey-headed Fish-eagle *Ichthyophaga ichthyaetus* (Horsfield, 1821) in Rajasthan

Indian Vulture *Gyps indicus* (Scopoli, 1786)

Race Monotypic

Uncommon to rare resident prove to local migration. In the extreme western part of Rajasthan, mainly a forager. Like all vultures, travels huge distances in search of carrion. Though primarily a cliff nester, occasionally does breed on trees. Located breeding on the *Prosopis cineraria* trees in Nagour district [6] and on a *Khejadi* tree at Tal Chhapar in Churu district (Harkirat Singh Sangha (HSS) or and Harsh Vardhan, personal communication) and at Kodamdesar near Bikaner (authors). Decline first observed from Keoladeo National Park when counts of birds fell from 812 in 1985–1986 to a maximum of only 25 in 1998–1999 [7]. Recently recorded breeding on electricity pylon at Sri Dungargarh in Bikaner district [8] (Fig. 13.12).

IUCN status: Critically Endangered A2bce+4bce [3]

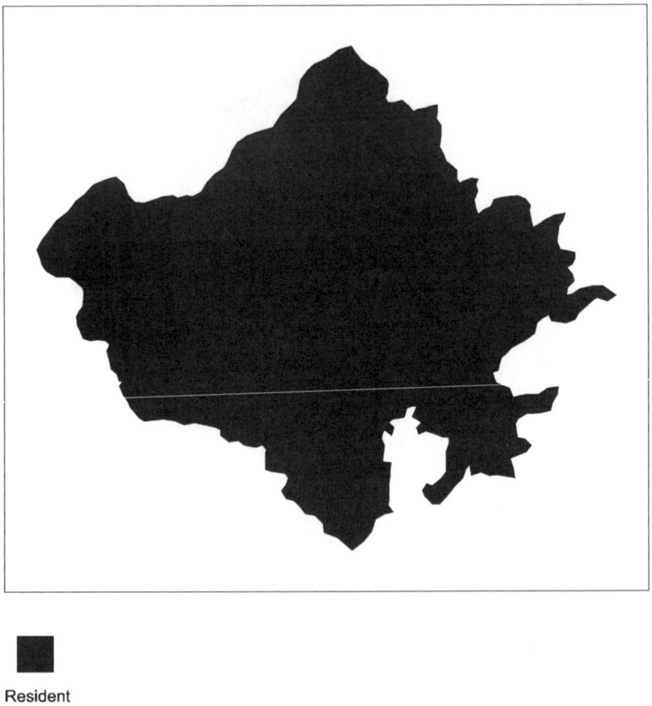

Fig. 13.10 Distribution of Egyptian Vulture *Neophron percnopterus* (Linnaeus, 1758) in Rajasthan

Slender-billed Vulture *Gyps tenuirostris* (G.R. Gray, 1844)

Race Monotypic

Rare vagrant. Four individuals observed at Keoladeo National Park, Bharatpur (John Schmitt, personal communiation), in Naoroji [1] (Fig. 13.13).

IUCN status: Critically Endangered A2ce+4ce [3]

Himalayan Vulture *Gyps himalayensis* (Hume, 1869)

Race Monotypic

Wintering migrant. Compared to earlier years, increasingly large numbers of Himalayan Griffon are observed wintering at carcass dumps throughout Rajasthan; mainly juveniles and immature (Fig. 13.14). Not uncommon during winter.

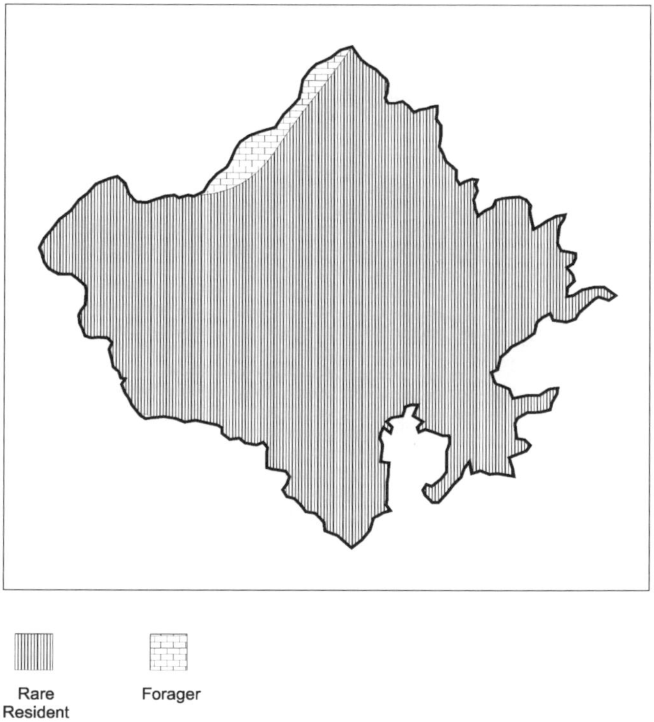

Fig. 13.11 Distribution of Indian White-rumped Vulture *Gyps bengalensis* (Gmelin, 1788) in Rajasthan

Griffon Vulture *Gyps fulvus* (Hablizl, 1783)

Race: *G. f. fulvescens* (Hume, 1869)

Extralimital winter migrant in the desert zone of western Rajasthan where seen in good concentrations at carcass dumps. A rare winter visitor in eastern Rajasthan. Formerly 2–3 individuals were regularly observed wintering at Keoladeo National Park, Bharatpur (Fig. 13.15).

IUCN status: Least Concerned

Cinereous Vulture *Aegypius monachus* (Linnaeus, 1766)

Race Monotypic

Winter Migrant. Common in the desert zone graduating to rarer eastwards in the Semi-arid zone (Fig. 13.16).

IUCN status: Near Threatened—C1 [3]

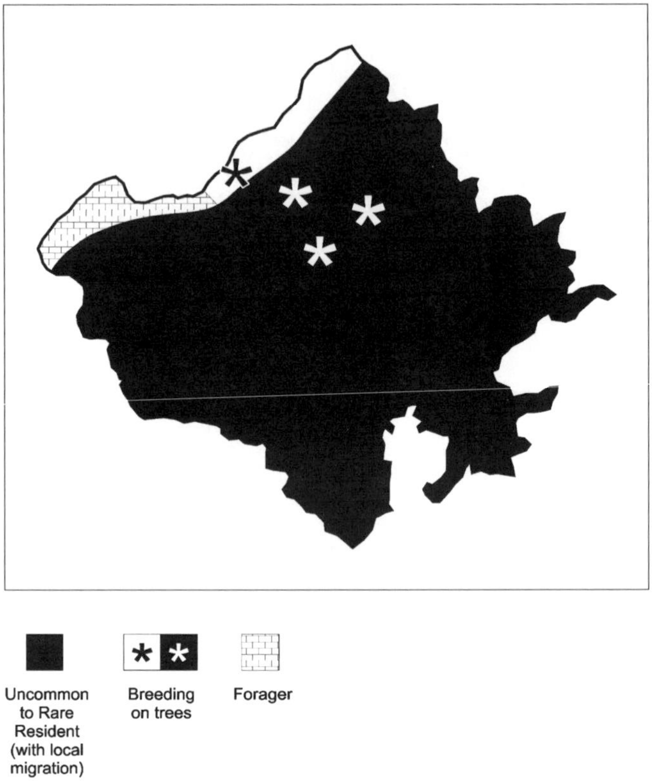

Fig. 13.12 Distribution of Indian Vulture *Gyps indicus* (Scopoli, 1786) in Rajasthan

Red-headed Vulture *Sarcogyps calvus* (Scopoli, 1786)

Race Monotypic

Widespread resident throughout Rajasthan transitioning to uncommon along the border. Not generally a communal species and singles or pairs are usually seen. Mostly a scavenger but also observed taking stranded terrapin and occasionally fish in the drying marsh at Bharatpur in summer (Fig. 13.17).

IUCN status: Near Threatened—A1a, b, c, d, e; A2a, b, c, d, e [3]

Short-toed Snake Eagle *Circaetus gallicus* (Gmelin, 1788)

Race Monotypic

Widespread common resident in Rajasthan becoming progressively rarer towards the Pakistan border where appears to be a transitional visitor. During winter resident

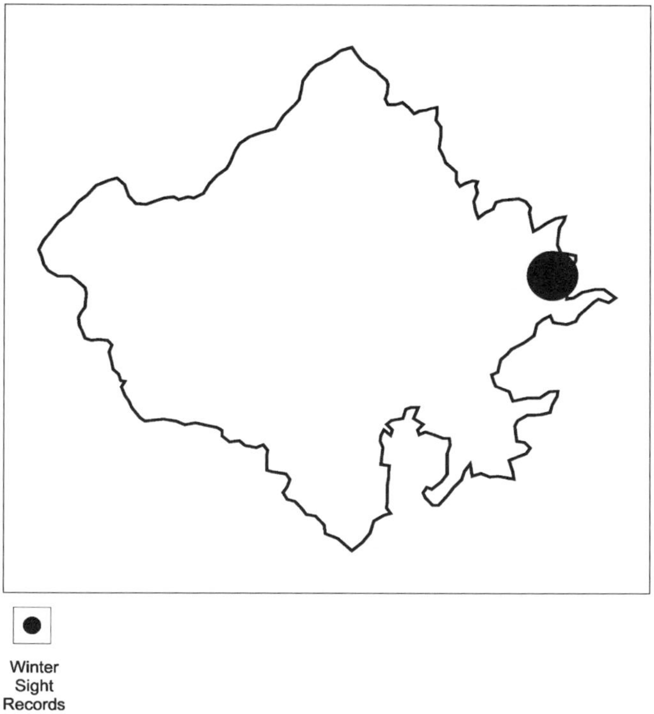

Fig. 13.13 Distribution of Slender-billed Vulture *Gyps tenuirostris* (G.R. Gray, 1844) in Rajasthan

population probably augmented to some extent by migrants from Turkestan and Mongolia. Marked preference for open country including desert and semi-arid zones. At Keoladeo National Park, observed taking watersnakes and hunting in open woodland and grassland within the park (Fig. 13.18).

Crested Serpent-eagle *Spilornis cheela* (Latham, 1790)

Races: *S. c. cheela* (Latham, 1790)

Common resident in forested areas of eastern Rajasthan and Mt. Abu (Fig. 13.19).

Western Marsh Harrier *Circus aeruginosus* (Linnaeus, 1758)

Race: *C.a. aeruginosus* (Linnaeus, 1758)

Migratory. Mainly observed from west central Rajasthan eastwards wherever marsh, waterbodies and bordering grassland exist. Winters from late August with

Fig. 13.14 Distribution of Himalayan Vulture *Gyps himalayensis* (Hume, 1869) in Rajasthan

numbers tapering till end of April. Once a common species at Keoladeo National Park, Bharatpur, when the marsh was active; a total of 14 communal roosts located with eight in Keoladeo National Park and six in adjoining areas within a variety of habitats [9]. One specimen was obtained by Dr. King at Nakki Lake [10] (Fig. 13.20).

Northern Harrier *Circus cyaneus* (Linnaeus, 1758)

Race: *C. c. cyaneus* (Linnaeus, 1758)

Migrant; 2–3 records from Koladahar grassland, Keoladeo National Park, Bharatpur (Fig. 13.21).

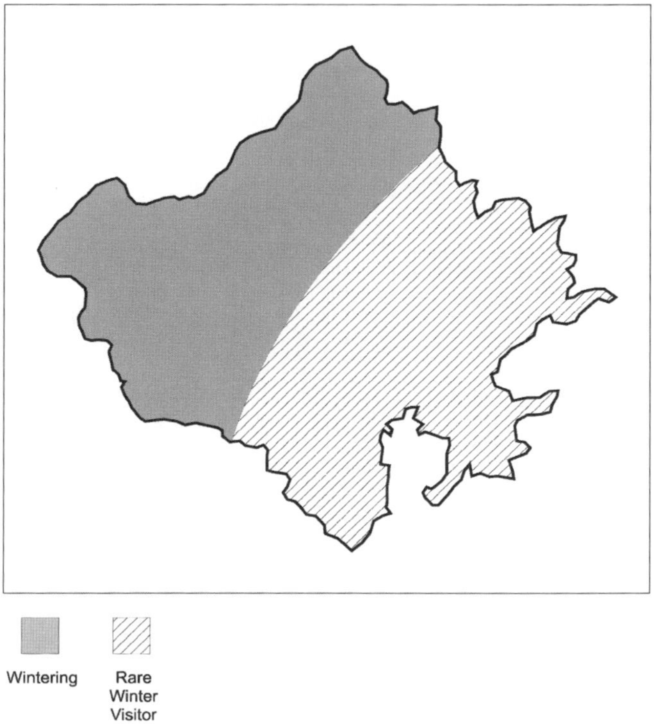

Fig. 13.15 Distribution of Vulture Griffon *Gyps fulvus* (Hablizl, 1783) in Rajasthan

Pallid Harrier *Circus macrourus* (Gmelin, 1770)

Race Monotypic

Widespread migrant throughout Rajasthan (Fig. 13.22). Less common than the Montagu's.

IUCN status: Near Threatened—A1c, d, e; A2c, d, e [3]

Pied Harrier *Circus melanoleucos* (Pennant, 1769)

Race Monotypic

Migrant. Uncommon in Rajasthan (Fig. 13.23). Two to three birds regularly observed wintering in the grasslands of Keoladeo National Park, Bharatpur, (RKN) and one sight record near Kota [11].

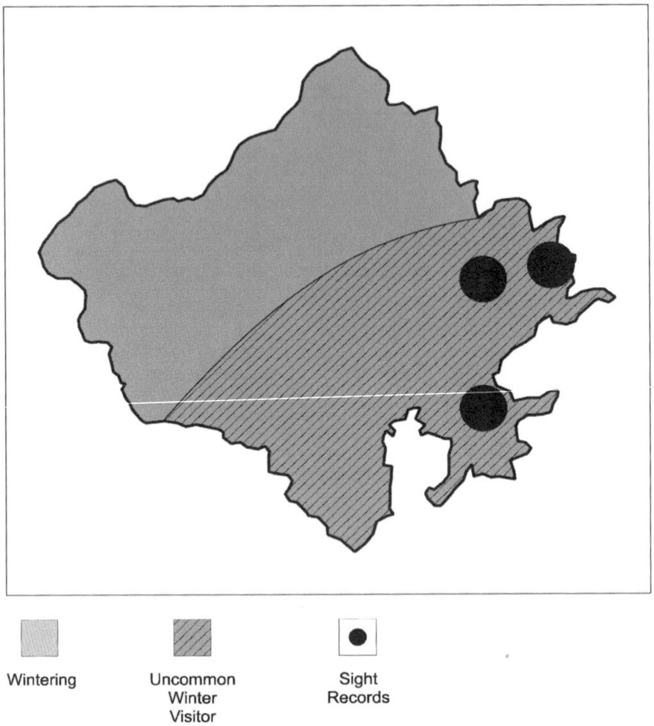

Fig. 13.16 Distribution of Cinereous Vulture *Aegypius monachus* (Linnaeus, 1766) in Rajasthan

Montagu's Harrier *Circus pygargus* (Linnaeus, 1758)

Race Monotypic

Migrant. The commonest migrant harrier, winter visitor and mainly passage migrant. Wintering birds exploit a variety of habitats such as open grassland, bare hilly tracts, cultivated areas, the dry semi-desert and scrub country (Fig. 13.24). Habitat preference include open grassland, marshy, boggy and moist ground.

Shikra *Accipiter badius* (Gmelin, 1788)

Races: *A. b. cenchroides* (Severtsov, 1873)
A. b. dussumieri (Temminck, 1824)

Common. This species has adapted well to cultivated areas and tree groves around habitation in vicinity of Bikaner and Jaisalmer. It is rare in the western

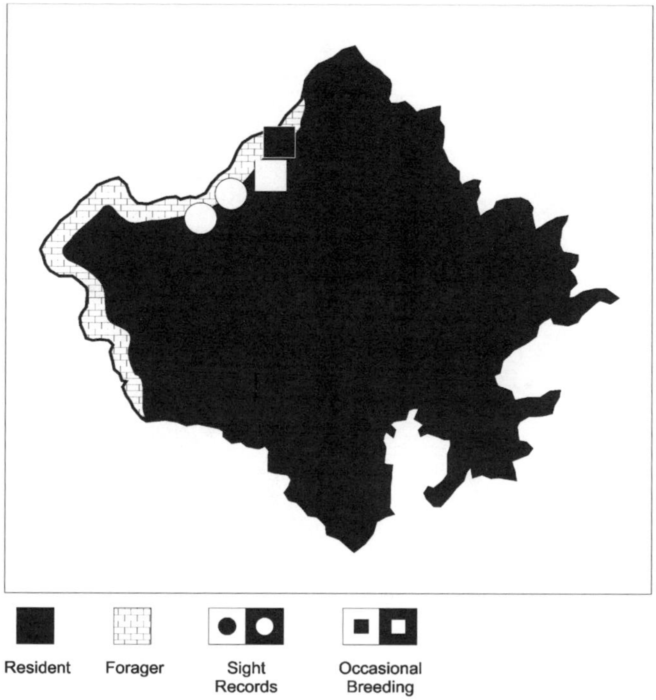

Fig. 13.17 Distribution of Red-headed Vulture *Sarcogyps calvus* (Scopoli, 1786) in Rajasthan

extremity of the desert; more frequent in the semi-arid areas of the state. The population of the resident race *dussumieri* is augmented by the migrant Central Asian race *cenchroides*. However, it is extremely difficult to distinguish the two races in the field. Adapts readily to different habitats such as open wetland, dry deciduous and thorn (*Acacia*) forests, hills, cultivated plains, village and temple groves and gardens within towns and cities. Feeds on a wide variety of birds, reptiles, insects and mammals and recorded attacking a foot-long Checkered Keelback at Bharatpur (Fig. 13.25). Also feeds on carrion.

Besra *Accipiter virgatus* (Temminck, 1822)

Races: *A. v. affinis* (Hodgson, 1836)

Rare migrant. Mostly, a forest species with three sight records from Rajasthan (Fig. 13.26). Adult male ringed at Bharatpur.

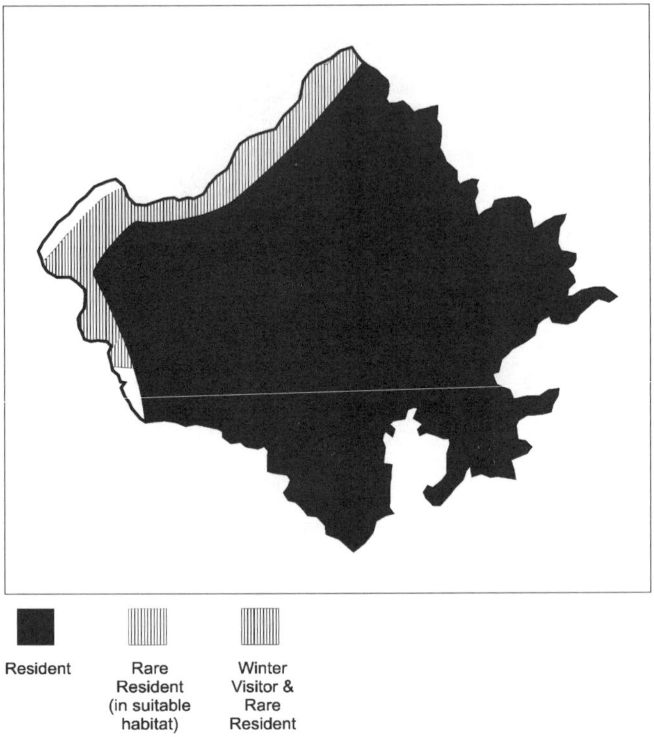

Fig. 13.18 Distribution of Short-toed Eagle *Circaetus gallicus* (Gmelin, 1788) in Rajasthan

Eurasian Sparrowhawk *Accipiter nisus* (Linnaeus, 1758)

Races: *A. n. nisosimilis* (Tickell, 1833)
A.n. *melaschistos* (Hume, 1869)

Winter visitor. Race *nisosimilis* is a rare winter visitor to the desert and commonly winters from central to eastern Rajasthan. It is more commonly observed during return spring migration. Up to six birds regularly seen at Tal Chhapar. The authors saw two individuals at Ghotaru in February/March 2008 (Fig. 13.27). Wintering range of race *melaschistos* was not determined due to confusion in the field with *nisosimilis*.

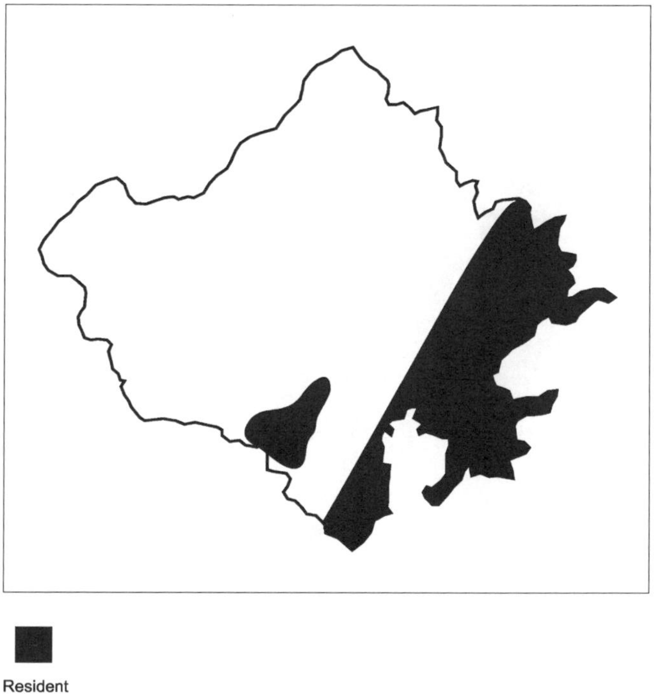

Fig. 13.19 Distribution of Crested Serpent-eagle *Spilornis cheela* (Latham, 1790) in Rajasthan

Northern Goshawk *Accipiter gentilis* (Linnaeus, 1758)

Race: *A. g. schvedowi* (Menzbier, 1882)

Rare winter visitor and passage migrant. Individual records from Keoladeo National Park, Bharatpur, one sighting at Gajner Wildlife Sanctuary, Bikaner district; Revasa, Sikar district; Ranthambhore Tiger Reserve, Sawai Madhopur, and Tal Chhapar Wildlife Sanctuary (Fig. 13.28).

White-eyed Buzzard *Butastur teesa* (Franklin, 1831)

Race Monotypic

Common. Resident in Rajasthan. Displays preference for semi-arid and desert zones. Locally common or uncommon in different areas, its movement is governed

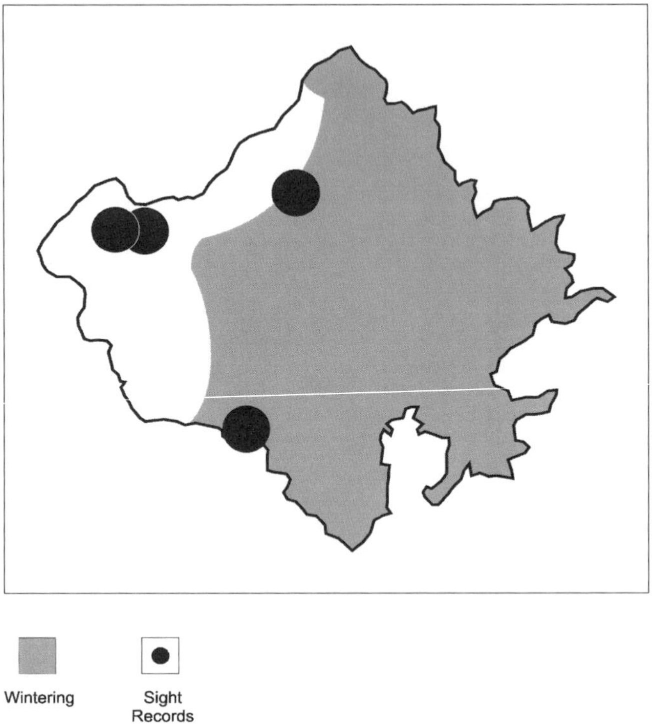

Fig. 13.20 Distribution of Western Marsh Harrier *Circus aeruginosus* (Linnaeus, 1758) in Rajasthan

by cyclic fluctuations in prey populations; as many as a hundred birds were seen near Sariska Tiger Reserve until the food supply was exhausted [12]. Up to 300 mostly juveniles were observed at Tal Chhapar in September 2008 (Fig. 13.29). Monsoon congregation varies yearly at Tal Chhapar depending on the rainfall and availability of prey, viz. locusts and grasshoppers [13].

Common Buzzard *Buteo buteo* (Linnaeus, 1758)

Races: *B. b. vulpinus* (Gloger, 1833)

B. b. japonicus (Temminck & Schlegel, 1845)

Distribution throughout Rajasthan indeterminate as intraspecific variation is poorly understood. The subspecies *vulpinus* is mainly a passage migrant to the semi-arid areas of Rajasthan (Fig. 13.30). One individual, an adult of race *japonicus*, was recorded near Bikaner [1].

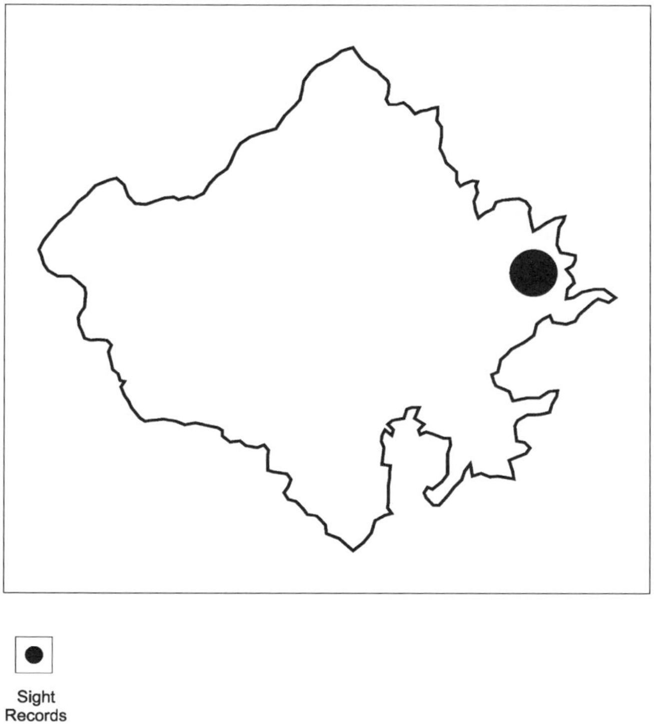

Fig. 13.21 Distribution of Northern Harrier *Circus cyaneus* (Linnaeus, 1758) in Rajasthan

Long-legged Buzzard *Buteo rufinus* (Cretzschmar, 1827)

Race *B.r. rufinus* (Cretzschmar, 1827)

Common winter migrant. Frequenting mainly the desert zone and to a lesser extent, the semi-arid zone. Observed frequently along the western border with Pakistan (Fig. 13.31). Mostly observed perching on harvested haystacks, fences, bushes, hedges, sand-dunes, mounds and prominently on the crown of trees such as *Acacia nilotica* on the lookout for prey.

Black Eagle *Ictinaetus malayensis* (Temminck, 1822)

Race: *I. m. perniger* (Hodgson, 1836)

Uncommon. Rare vagrant. Only three records: two from Ranthambhore Tiger Reserve [12, 14] and a sight record from Darrah Sanctuary, Kota (HSS) (Fig. 13.32).

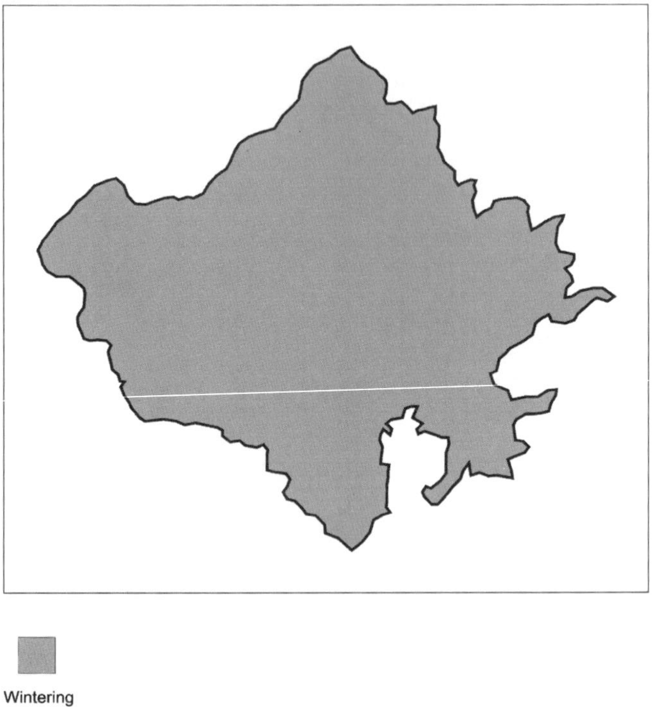

Fig. 13.22 Distribution of Pallid Harrier *Circus macrourus* (Gmelin, 1770) in Rajasthan

Lesser Spotted Eagle *Aquila pomarina* (Lesson, 1831)

Race Monotypic

Rare to uncommon resident in Rajasthan. Absent from the desert zone but observed in the semi-arid zone. Breeds regularly at Keoladeo National Park, Bharatpur, and Tal Chappar National Park, and sighted occasionally in Ranthambhore Tiger Reserve (Fig. 13.33).

Greater Spotted Eagle *Aquila clanga* (Pallas, 1811)

Race Monotypic

Mainly a migrant to the semi-arid zone. Frequents open-wooded tracts and village groves in the vicinity of waterbodies, marshes and river systems. Two sight records from the desert. One at Desert National Park (Steve Madge, personal communication) and the other from Bikaner carcass dump (RKN). One unsuccessful breeding attempt was observed at Keoladeo National Park, Bharatpur [15]

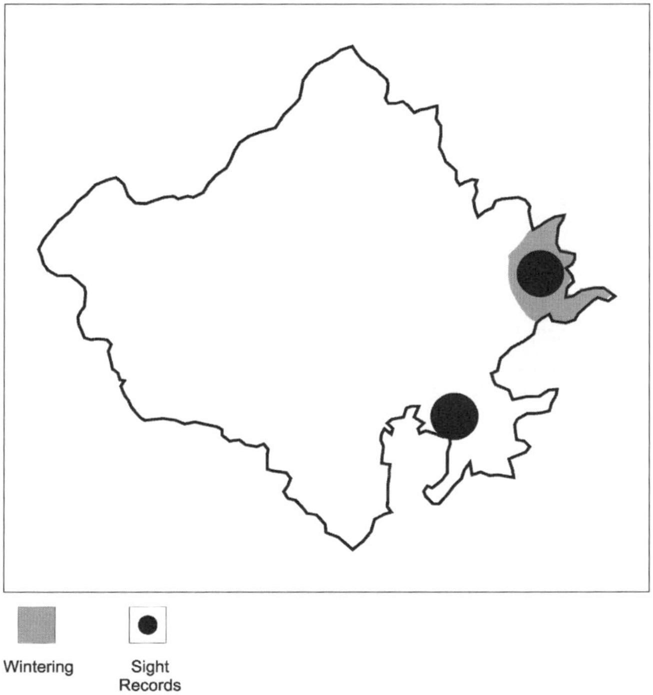

Fig. 13.23 Distribution of Pied Harrier *Circus melanoleucos* (Pennant, 1769) in Rajasthan

(Fig. 13.34). Bharatpur marsh probably supported the largest concentration of Greater Spotted Eagle in the country, attracted no doubt by the large concentration of migrant waterfowl and marauding Painted Stork, Grey Heron, egret and open-billed breeding colonies [16].

Global status: Vulnerable C1 [3]

Tawny Eagle *Aquila rapax* (Temminck, 1828)

Race: *A. r. vindhiana* (Franklin, 1831)

Resident throughout Rajasthan. Has preference for dry, arid country. Commonest in the desert and western extremity of the semi-arid zone (Fig. 13.35, 13.56, 13.57).

Fig. 13.24 Distribution of Montagu's Harrier *Circus pygargus* (Linnaeus, 1758) in Rajasthan

Steppe Eagle *Aquila nipalensis* (Hodgson, 1833)

Race: *A. n. nipalensis* (Hodgson, 1833)

A winter migrant to Rajasthan. Most commonly seen in the eastern half of the desert and semi-arid zone, congregating in large numbers at carcass dumps. In the western extremity of the desert, the species winters to some extent but is mainly a passage migrant. Large numbers of Steppe Eagle of varying age-classes can be seen at the Bikaner carcass dump and also at Bharatpur when the marshes were active (Fig. 13.36, 13.58, 13.59) [16].

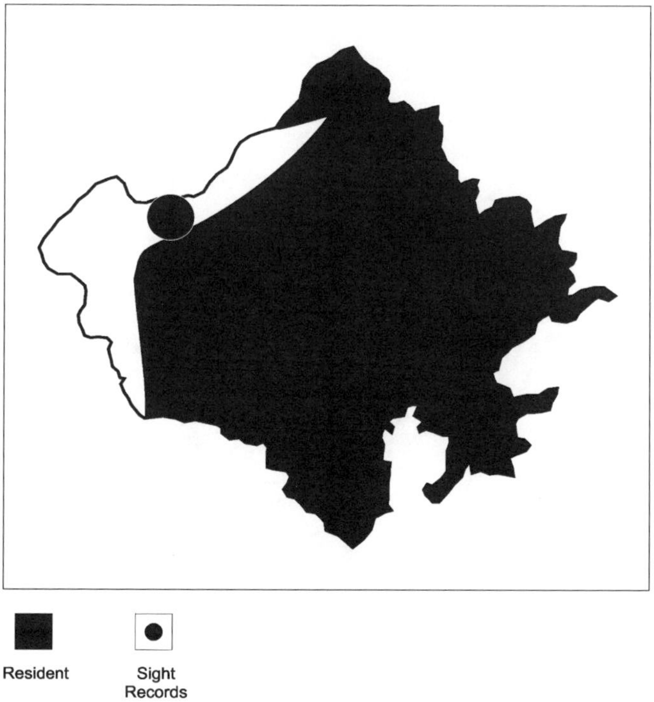

Fig. 13.25 Distribution of Shikra *Accipiter badius* (Gmelin, 1788) in Rajasthan

Eastern Imperial Eagle *Aquila heliaca* (Savigny, 1809)

Race Monotypic

Winters primarily along the eastern section of the desert zone and throughout the semi-arid zone. Occasionally sighted in the desert. Seen regularly at Gajner, district Bikaner and at Bharatpur where attracted by the large number of waterfowl. In November, a few immatures and subadults are observed in Bikaner and Jaisalmer districts (Fig. 13.37, 13.60, 13.61).

IUCN status: Vulnerable—C1 [3]

Bonelli's Eagle *Aquila fasciatus* (Vieillot, 1822)

Race: *H. f. fasciatus* (Vieillot, 1822)

Resident breeder in eastern Rajasthan along the Aravallis. Common in Ranthambhore National Park and Chambal ravines and relatively common in

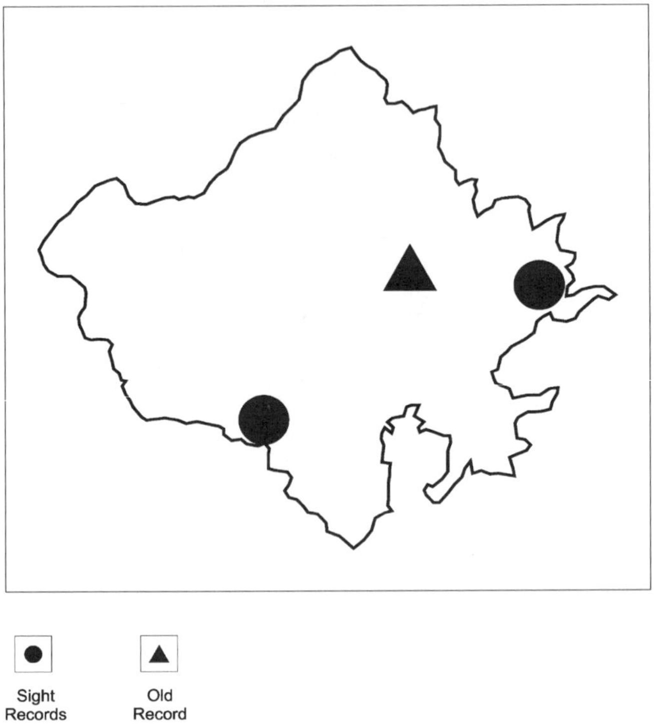

Fig. 13.26 Distribution of Besra *Accipiter virgatus* (Temminck, 1822) in Rajasthan

Mt. Abu and Sariska Tiger Reserve. Wintering individuals normally seen in the Rajasthan desert are mostly juvenile, immature and subadults which range over vast area. Also known to breed on rocky outcrops in the Thar Desert. Juveniles were observed twice in Fossil Park, Jaisalmer (authors). Winter visitor to Keoladeo National Park, Bharatpur, attracted by the abundant availability of waterfowl (Fig. 13.38).

Booted Eagle *Hieraaetus pennatus* (Gmelin, 1788)

Race Monotypic

Winter visitor to Rajasthan except the western extremity of the Thar Desert where occasionally observed (Fig. 13.39).

Changeable Hawk-eagle *Nisaetus cirrhatus* (Gmelin, 1788)

Race: *S. c. cirrhatus* (Gmelin, 1788)

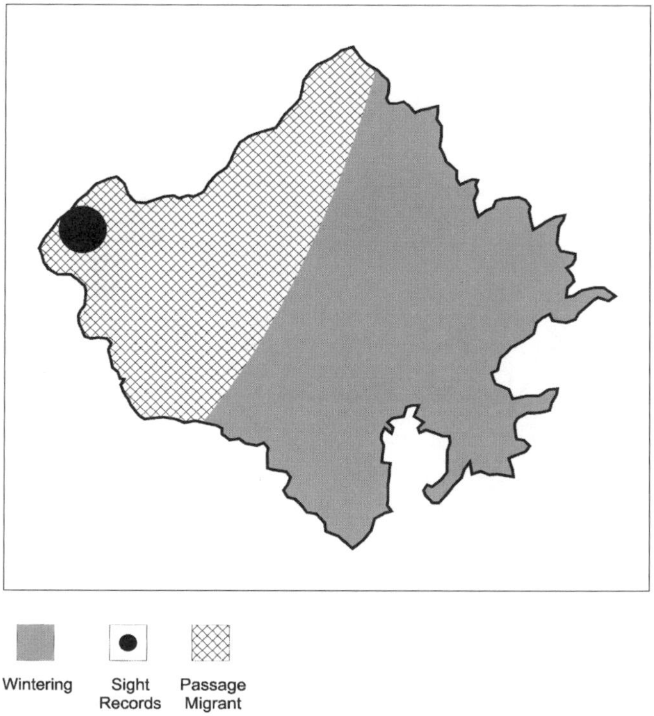

Fig. 13.27 Distribution of Eurasian Sparrowhawk *Accipiter nisus* (Linnaeus, 1758) in Rajasthan

Resident. Subspecies *cirrhatus* ranges throughout eastern Rajasthan due north of Mt. Abu, eastwards to the Uttar Pradesh boundary and south to the Gujarat border. Juveniles occasionally observed at Keoladeo National Park (Fig. 13.40).

Lesser Kestrel *Falco naumanni* (Fleischer, 1818)

Race Monotypic

Mainly a passage migrant through eastern Rajasthan. One sight record from Jaisalmer [17] and Keoladeo National Park, Bharatpur [17, 18]. There is a recent record from Tal Chhapar, Churu district (S. S. Poonia, personal communication) (Fig. 13.41).

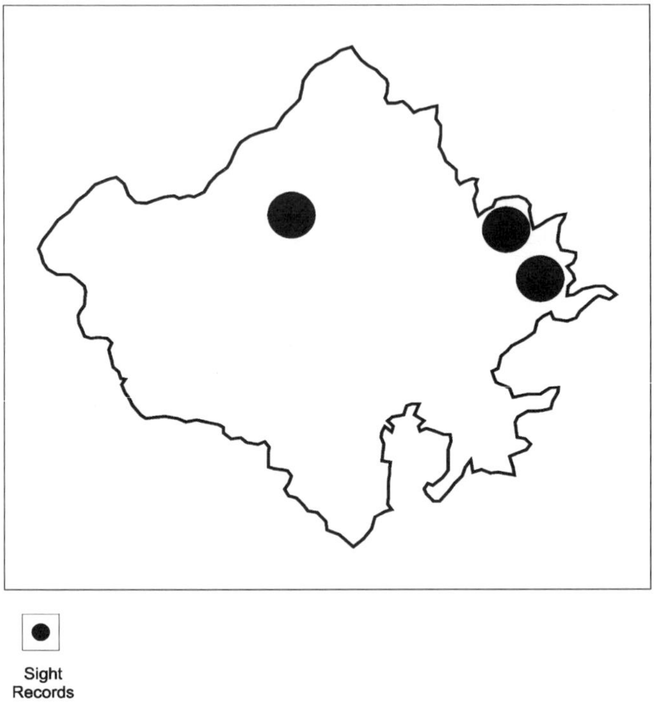

Fig. 13.28 Distribution of Northern Goshawk *Accipiter gentilis* (Linnaeus, 1758) in Rajasthan

Common Kestrel *Falco tinnunculus* (Linnaeus, 1758)

Races: *F. t. tinnunculus* (Linnaeus, 1758)
F. t. interstinctus (McClelland, 1840)

Common winter migrant throughout the desert and semi-arid zones of Rajasthan (Fig. 13.42).

Red-headed Falcon *Falco chicquera* (Daudin)

Race *F.c. chicquera*, Daudin, 1800

Resident throughout Rajasthan except the westernmost section of the desert bordering Pakistan where local winter migrant winters (Fig. 13.43). Has a preference for open country even in the vicinity of human habitation. Avoids forest. Not very conspicuous in the field as often overlooked due to its small size and crepuscular habits and its penchant for perching out of sight in foliage.

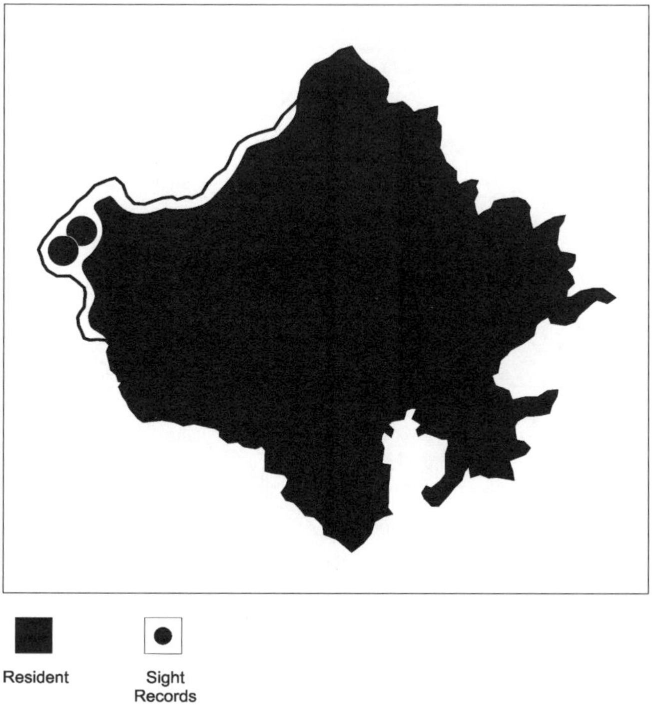

Fig. 13.29 Distribution of White-eyed Buzzard *Butastur teesa* (Franklin, 1831) in Rajasthan

Amur Falcon *Falco amurensis* (Radde, 1863)

Race Monotypic

Passage migrant. One sighting from Rajasthan. At least five birds were observed on 3 October 2003, at Phulwari Sanctuary in Dhedmariya forest range, southwest of Udaipur, a first record for Rajasthan (Fig. 13.44) [19].

Merlin *Falco columbarius* (Linnaeus, 1758)

Races: *F. c. insignis* (Clark, 1907)

F. c. pallidus (Sushkin, 1900)

Rare winter visitor to Rajasthan and often overlooked due to its diminutive size. A few scattered records from Tal Chhapar, Churu district, and Kanod, Jaisalmer district (Fig. 13.45) [20].

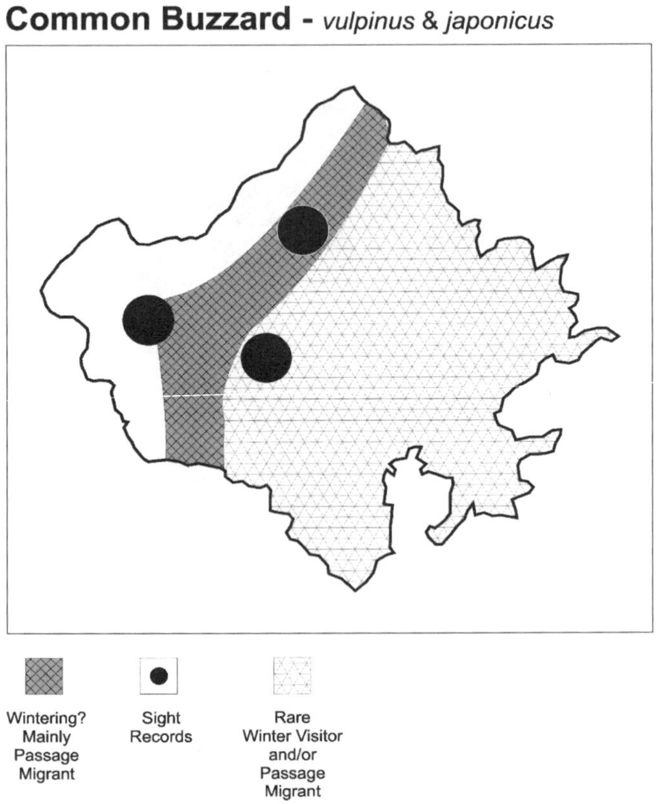

Fig. 13.30 Distribution of Common Buzzard *Buteo buteo* (Linnaeus, 1758) in Rajasthan

Eurasian Hobby *Falco subbuteo* (Linnaeus, 1758)

Race *F.s. subbuteo* Linnaeus, 1758

Winter visitor throughout Rajasthan except the western extremity along the India/Pakistan border and Shahgarh Bulge (Fig. 13.46).

Oriental Hobby *Falco severus* (Horsfield, 1821)

Race Monotypic

Passage migrant to Rajasthan from the eastern margins of the desert zone throughout the semi-arid zone. Sight records from Keoladeo National Park, Bharatpur, and Dholpur and Chittourgarh districts. Also Silised and Kankwari in Sariska Tiger Reserve, November 1986 [21]. Observed in Kota and upstream towards Jawahar Sagar (Fig. 13.47) [22].

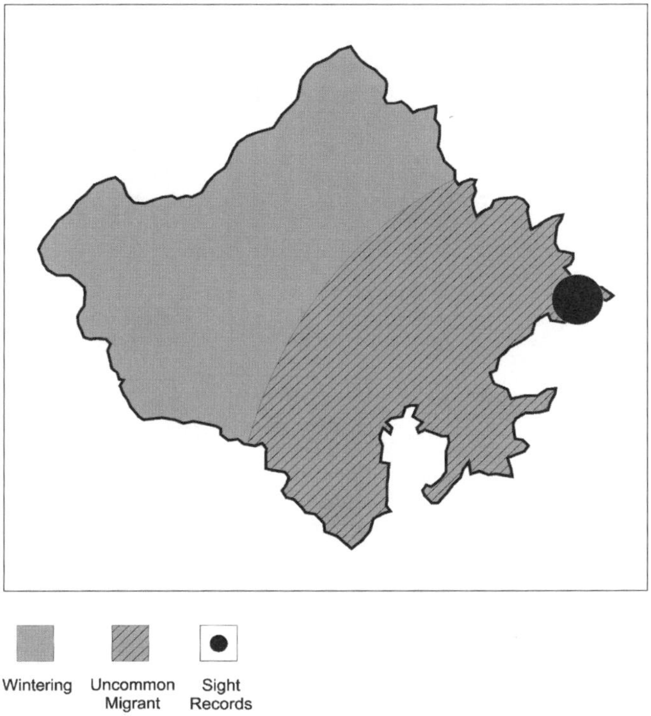

Fig. 13.31 Distribution of Long-legged Buzzard *Buteo rufinus* (Cretzschmar, 1827) in Rajasthan

Laggar Falcon *Falco jugger* (J.E. Gray, 1834)

Race Monotypic

Resident and common throughout the desert and western semi-arid zones. A rare resident in eastern Rajasthan (Fig. 13.48, 13.62, 13.63, 13.64).

Saker Falcon *Falco cherrug* (J.E. Gray, 1834)

Races: *F. c. cherrug* (J.E. Gray, 1834)

F. c. milvipes (Jerdon, 1871)

Rare migrant to Rajasthan. Less frequently observed along the western margins of the desert. Overall food spectrum not known but in the desert subsists mainly on Spiny-tailed Lizard, Desert Gerbil *Meriones hurrianae* and other small mammals and birds such as sandgrouse (*Pterocles* spp.) (Fig. 13.49).

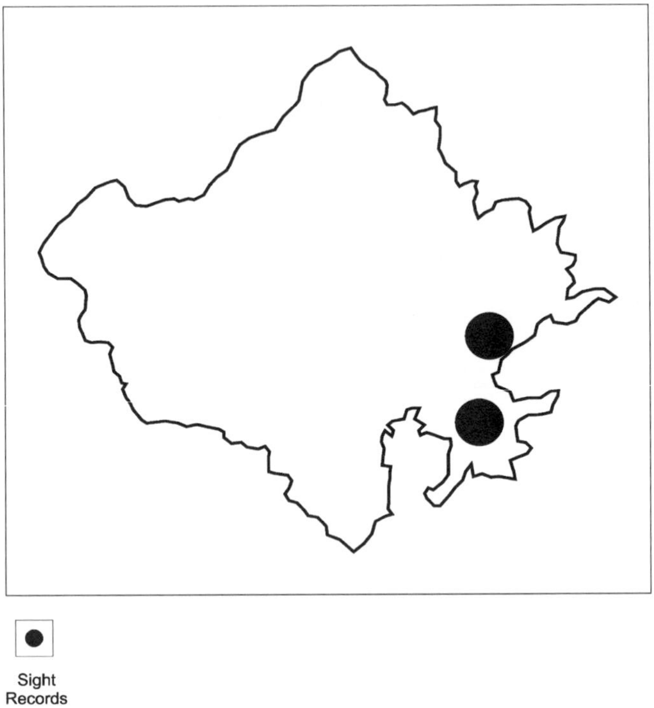

Fig. 13.32 Distribution of Black Eagle *Ictinaetus malayensis* (Temminck, 1822) in Rajasthan

Peregrine Falcon *Falco peregrinus* (Tunstall, 1771)

Races: *F. p. calidus* (Latham, 1790)
F. p. babylonicus (P.L. Sclater, 1861)
F. p. peregrinator (Sundevall, 1837)

F. p. calidus is a winter visitor throughout the semi-arid zone and eastern section of the desert. Usually observed near wetlands and Rajasthan Canal (Fig. 13.50). Observed from the ramparts of Jaisalmer fort hunting Rock Pigeons *Columba livia* (authors).

Red-naped Shaheen *F. p. babylonicus.* Also called Barbary Falcon *F. pelegrinoides*

Rare winter visitor to Rajasthan except the western extremity of the desert. Favours rocky barren hills in open country and desert (Fig. 13.51).

Shaheen *F. p. peregrinator*

Resident in eastern Rajasthan along rocky outcrops and Aravallis to the Rajasthan/Gujarat border. Recent records from Harshnath, Sikar district; Ranthambhore National Park [23] and Mt. Abu (Fig. 13.52) [24].

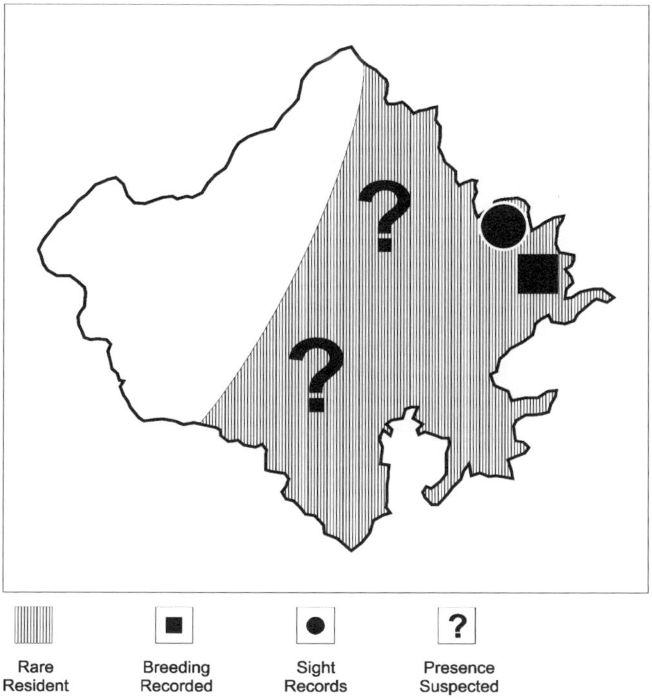

Fig. 13.33 Distribution of Lesser Spotted Eagle *Aquila hastata* (Lesson, 1831) in Rajasthan

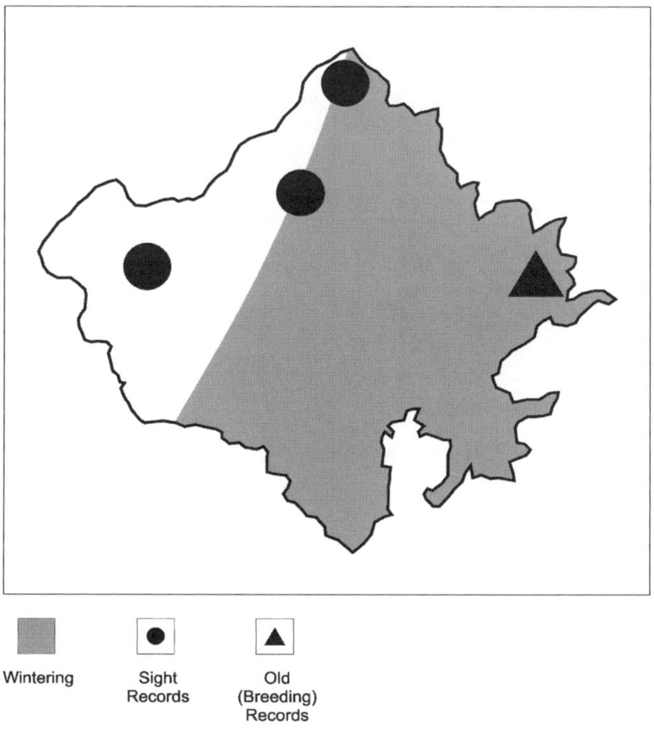

Fig. 13.34 Distribution of Greater Spotted Eagle *Aquila clanga* (Pallas, 1811) in Rajasthan

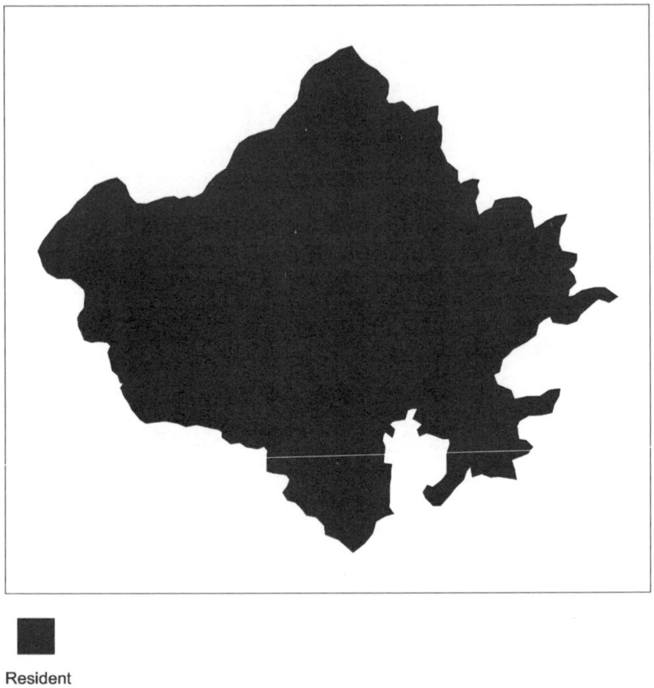

Fig. 13.35 Distribution of Tawny Eagle *Aquila rapax* (Temminck, 1828) in Rajasthan

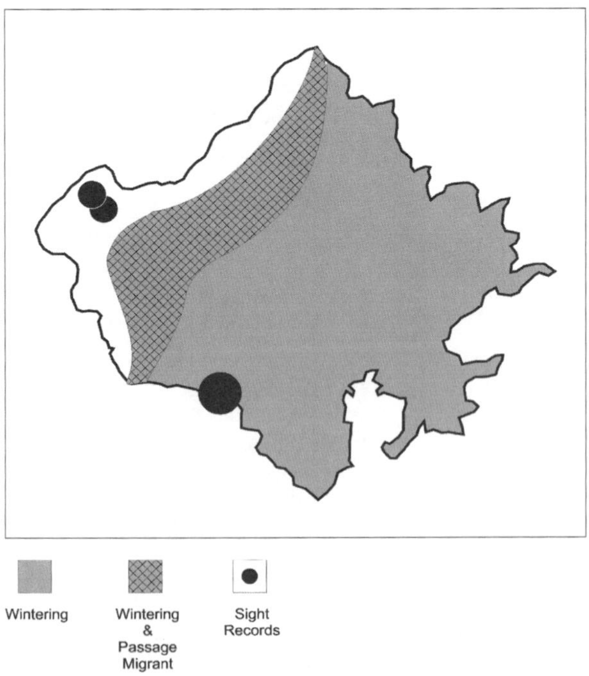

Fig. 13.36 Distribution of Steppe Eagle *Aquila nipalensis* (Hodgson, 1833) in Rajasthan

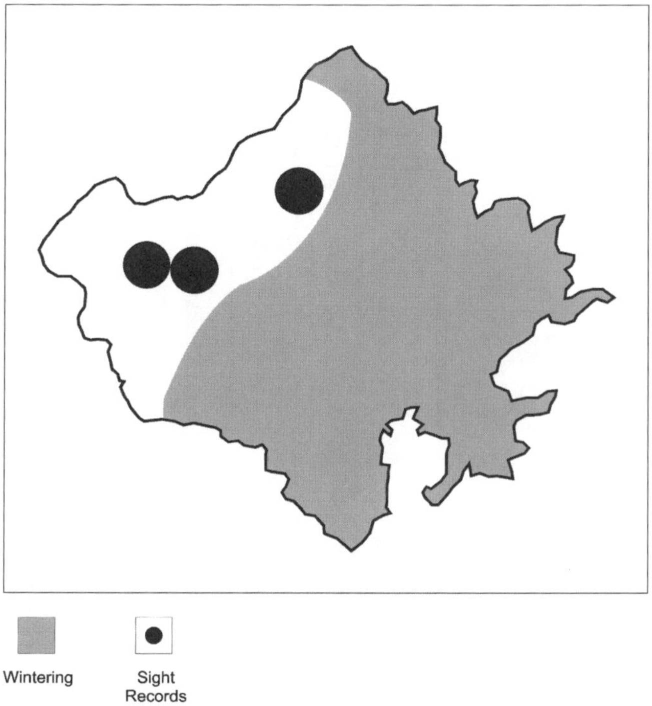

Fig. 13.37 Distribution of Eastern Imperial Eagle *Aquila heliaca* (Savigny, 1809) in Rajasthan

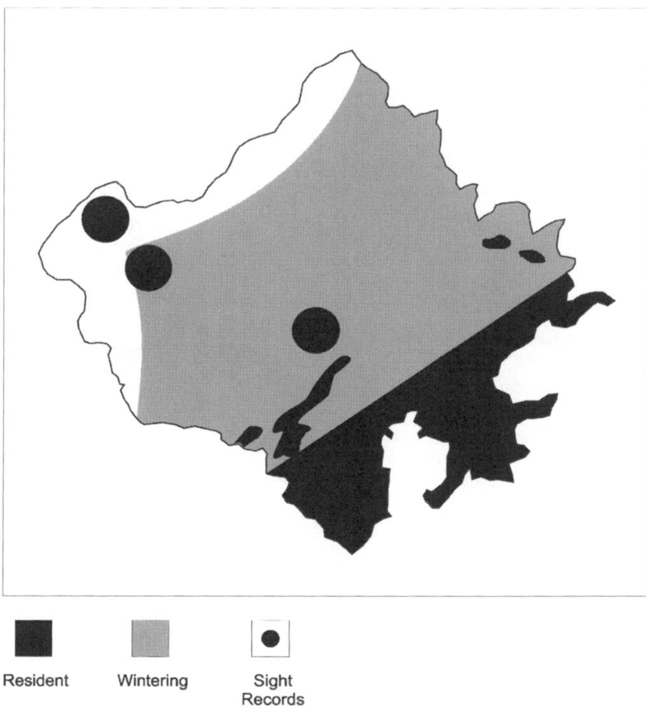

Fig. 13.38 Distribution of Bonelli's Eagle *Aquila fasciatus* (Vieillot, 1822) in Rajasthan

Fig. 13.39 Distribution of Booted Eagle *Hieraaetus pennatus* (Gmelin, 1788) in Rajasthan

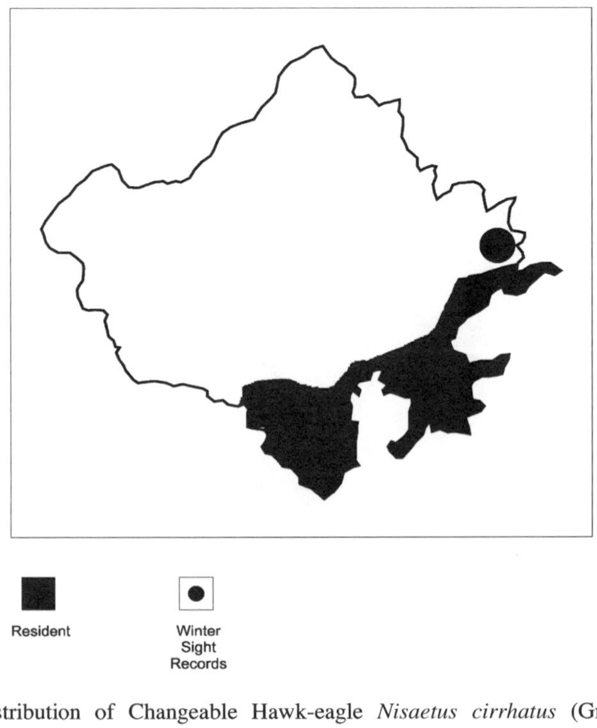

Fig. 13.40 Distribution of Changeable Hawk-eagle *Nisaetus cirrhatus* (Gmelin, 1788) in Rajasthan

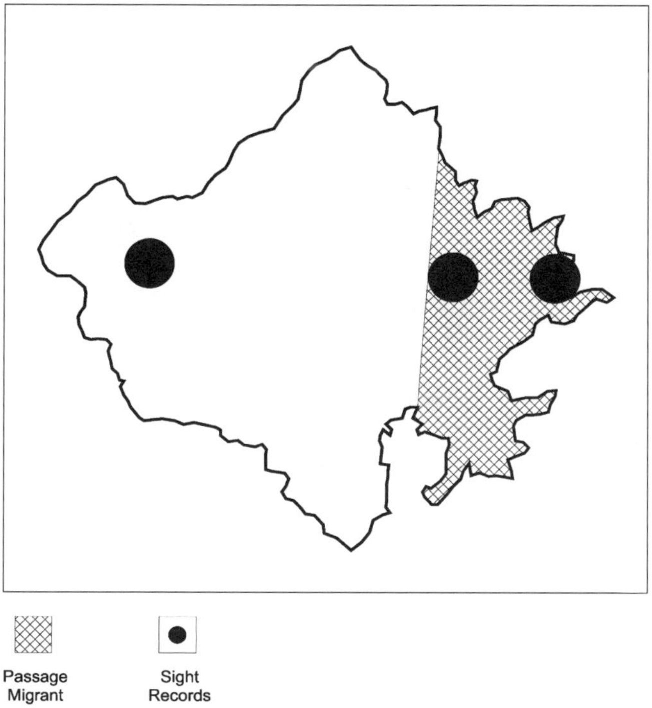

Fig. 13.41 Distribution of Lesser Kestrel *Falco naumanni* (Fleischer, 1818) in Rajasthan

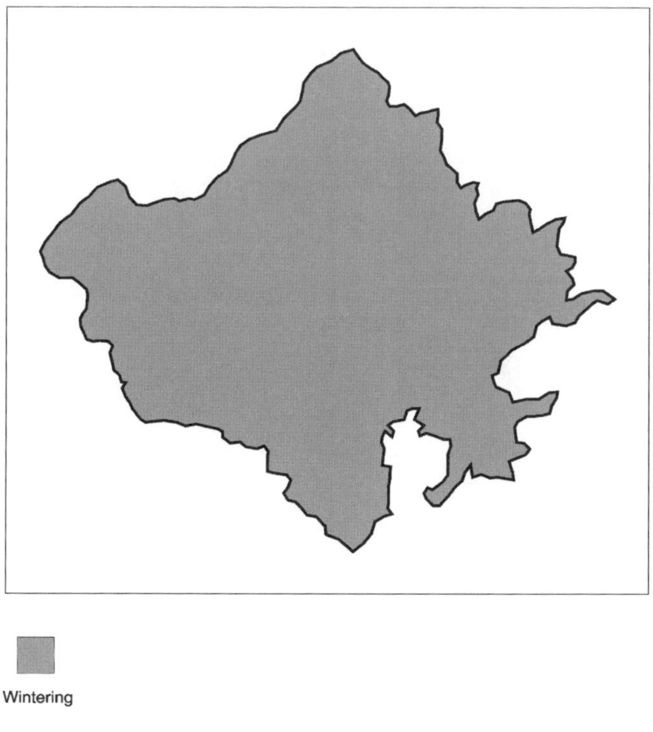

Fig. 13.42 Distribution of Common Kestrel *Falco tinnunculus* (Linnaeus, 1758) in Rajasthan

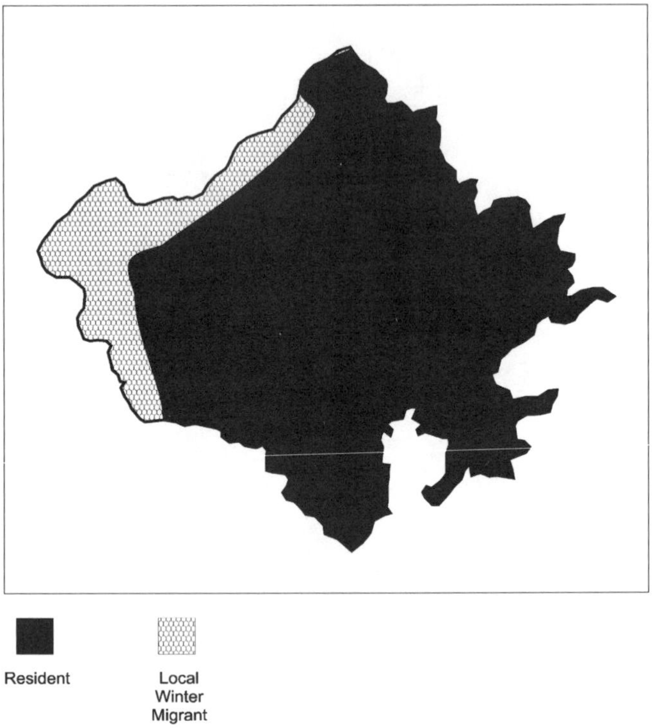

Fig. 13.43 Distribution of Red-headed Falcon *Falco chicquera* (Daudin, 1800) in Rajasthan

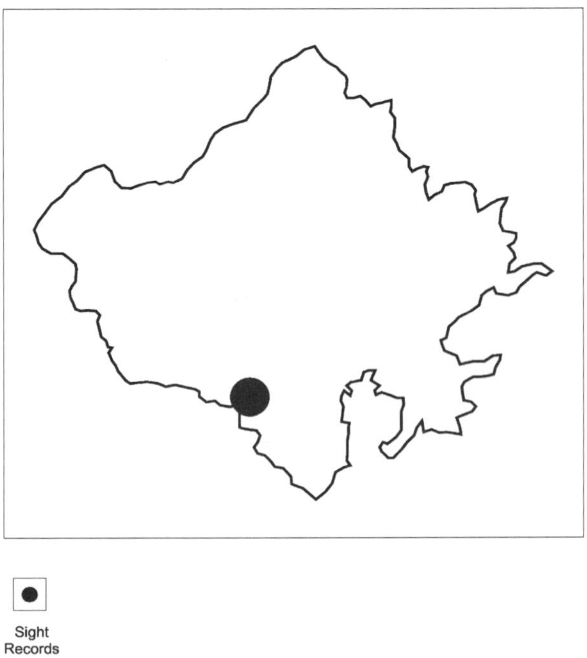

Fig. 13.44 Distribution of Amur Falcon *Falco amurensis* (Radde, 1863) in Rajasthan

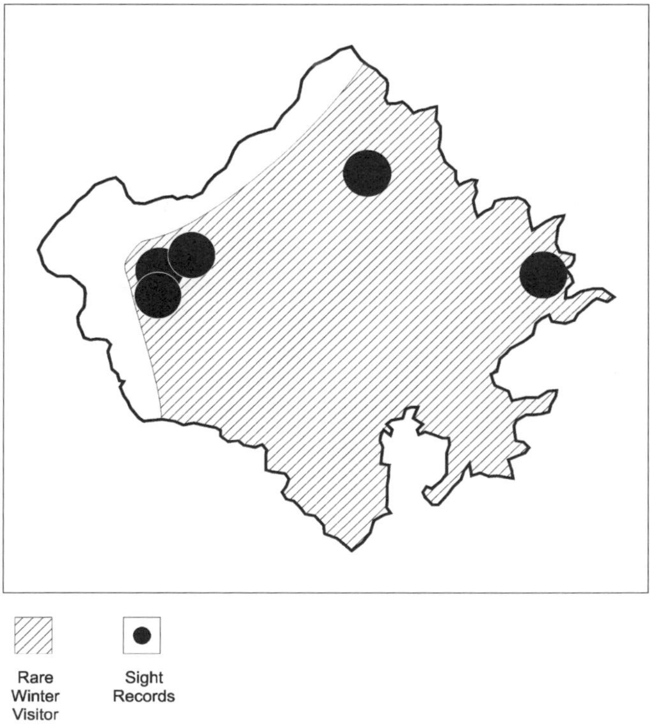

Fig. 13.45 Distribution of Merlin *Falco columbarius* (Linnaeus, 1758) in Rajasthan

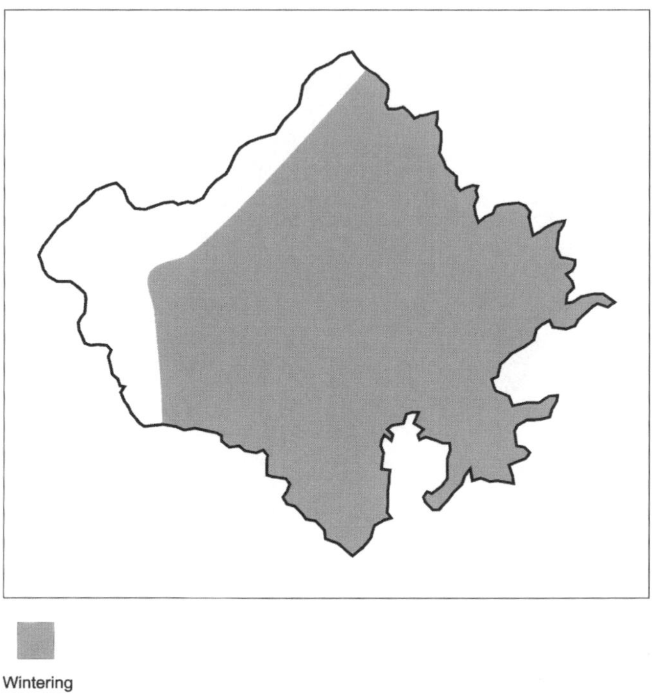

Fig. 13.46 Distribution of Eurasian Hobby *Falco subbuteo* (Linnaeus, 1758) in Rajasthan

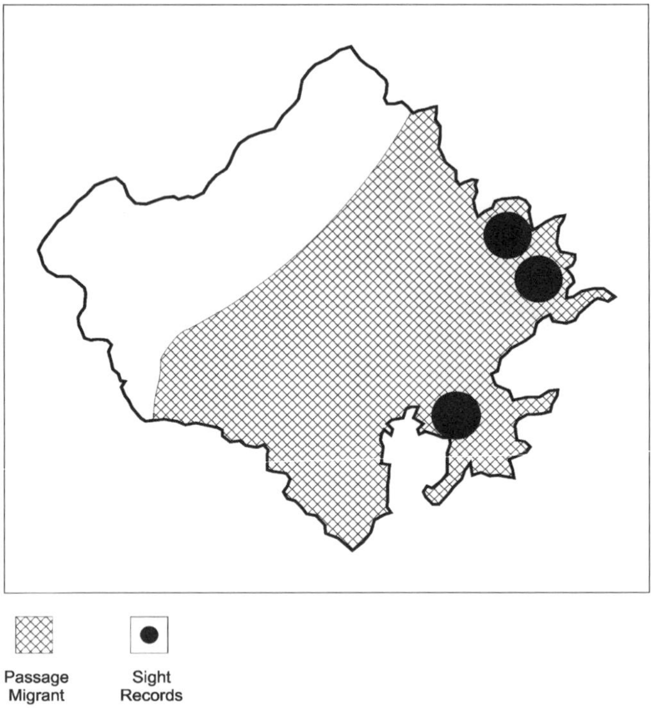

Fig. 13.47 Distribution of Oriental Hobby *Falco severus* (Horsfield, 1821) in Rajasthan

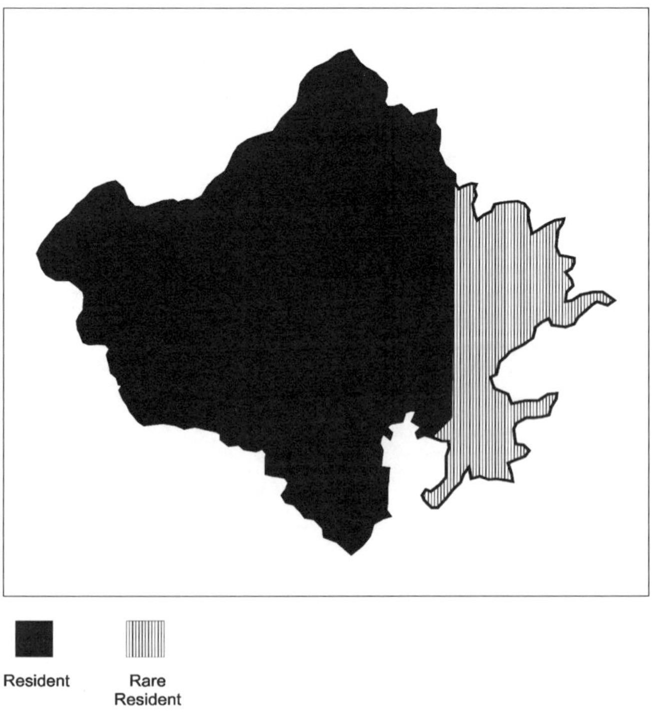

Fig. 13.48 Distribution of Laggar Falcon *Falco jugger* (J.E. Gray, 1834) in Rajasthan

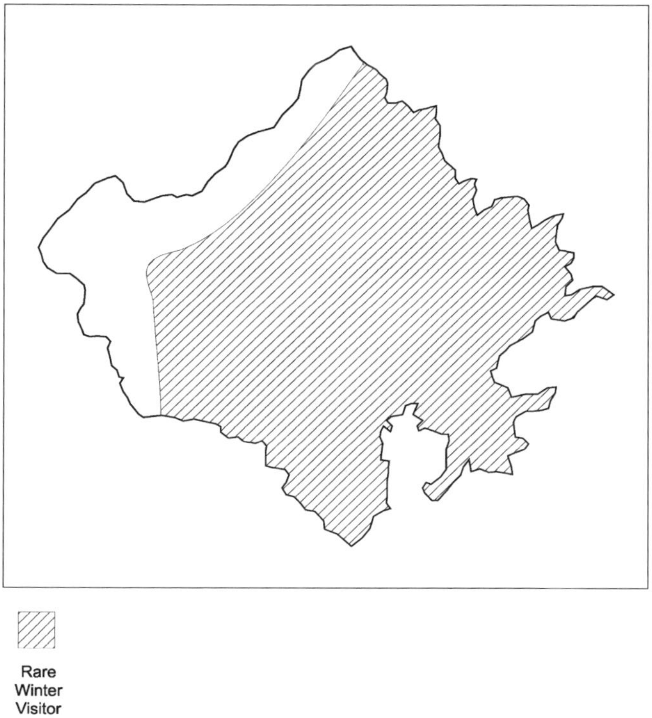

Fig. 13.49 Distribution of Saker Falcon *Falco cherrug* (J.E. Gray, 1834) in Rajasthan

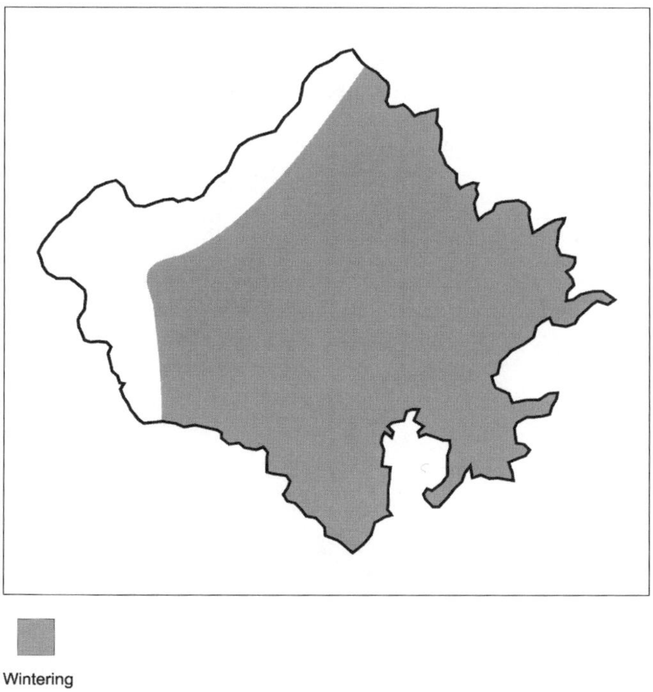

Fig. 13.50 Distribution of Peregrine Falcon *Falco peregrinus* (Tunstall, 1771) in Rajasthan

Fig. 13.51 Distribution of Red-naped Shaheen *F. p. babylonicus* in Rajasthan

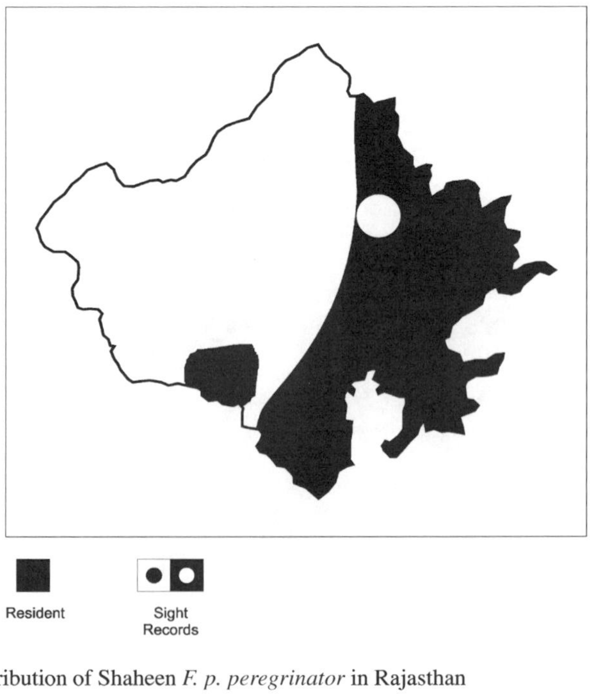

Fig. 13.52 Distribution of Shaheen *F. p. peregrinator* in Rajasthan

Fig. 13.53 Egyptian Vultures *Neophron percnopterus*. Adults. Bikaner carcass dump. Rajasthan Desert (18th November 2007)

Fig. 13.54 Egyptian Vulture *Neophron percnopterus*. Subadult. Advanced 3rd to 4th plumage. Bikaner carcass dump. Rajasthan Desert (06th January 2008)

Fig. 13.55 Egyptian Vulture *Neophron percnopterus*. Near adult (5th plumage type) with some retained greyish underwing-coverts. Dungargarh to Bikaner railway crossing. Thar Desert (26th November 2010)

Fig. 13.56 Egyptian Vulture *Neophron percnopterus*. Adult. Nominate race percnopterus. *Jor Beed*, Bikaner carcass dump. Thar Desert (07th December 2007)

Fig. 13.57 Short-toed Snake Eagle *Circaetus gallicus*. Adult. Dist. Jaisalmer. Thar Desert (08th January, 2008)

Fig. 13.58 Tawny Eagle *Aquila rapax*. Subadult. Pithla to Akal. Thar Desert (29th November 2010)

Fig. 13.59 Tawny Eagle *Aquila rapax*. Pale adult. Pithla to Akal. Thar Desert (29th November 2010)

Fig. 13.60 Steppe Eagle *Aquila nipalensis*. Advanced second plumage 2nd to 3rd cal year. Chinu to Bap. Dist Jaisalmer. Thar Desert (17th December 2007)

Fig. 13.61 Steppe Eagle *Aquila nipalensis*. Third plumage. Dungargarh to Bikaner railway crossing. Thar Desert (26th November 2010)

Fig. 13.62 Eastern Imperial Eagle *Aquila heliaca*. Fifth plumage-type individual. Thar Desert (28th November 2010)

Fig. 13.63 Laggar Falcon *Falco jugger*. Subadult. Ramgarh to Jaisalmer. The Thar Desert (28th November 2010)

Fig. 13.64 Laggar Falcon *Falco jugger*. Immature carrying most likely Laughing Dove *Streptopelia senegalensis*. Bikaner carcass dump. The Thar Desert (29th November 2010). *Photo Courtesy: Rishad Naoroji* (Figs. 13.53–13.64)

Acknowledgments At Keoladeo National Park, Bharatpur, 1985–1989 (RKN) thanks to former Chief Wildlife Wardens Mr. O.P. Mathur, Mr. V.D. Sharma, and Mr. R.G. Soni for readily providing facilities for research; former Deputy Chief Wildlife Wardens, Rajan Mathur, K.L. Saini, A.S. Brar, Shruti Sharma and Mr. R.N. Mehrotra, the Principal Chief Conservator of Forests and Chief Wildlife Warden, Jaipur. Bholu Abrar Khan shared his intimate knowledge of the park. The staff at Keoladeo, primarily Sohanlal, Amarnath, Prakash Chand Sain, Habib and Iqbal Khan, Suresh and Shiv Singh, were extremely helpful. At Ranthambhore, Fateh Singh Rathore, former Field Director, generously provided all facilities. The authors acknowledge Shantanu Kumar I.G. BSF who provided all facilities along border areas in Bikaner and Jaisalmer districts. Mr. K.S. Jangpangi, D.I.G. BSF, Bikaner district, helped out with logistics along the border; Brig. (Dr.) B.K. Khanna, SM, VSM, D.I.G. BSF, Jaisalmer district, and Mr. D. V. Saraswat, D.I.G. BSF, Jaisalmer district.

References

1. Naoroji RK (2006) Birds of prey of the Indian Subcontinent. Christopher Helm, London. pp 692
2. Olendorff RR, Miller AD, Lehman RN (1981) Suggested practices for raptor protection on power lines—the state of the art in 1981. Raptor Res Rep 4:111
3. BirdLife International (2001) Threatened birds of Asia: the BirdLife International Red Data Book. BirdLife International, Cambridge
4. Robson C (2004) From the field. India. BirdingASIA 1:79–80
5. Robson C (2007) From the field. India. BirdingASIA 7:92–96
6. Kulshreshtha M (2001) Long-billed Vulture *Gyps indicus indicus* nesting on trees in the Thar Desert, Rajasthan. J Bomb Nat Hist Soc 98(3):446–450
7. Prakash V (1999) Status of vultures in Keoladeo National Park, Bharatput, Rajasthan, with special reference to population crash in *Gyps* species. J Bomb Nat Hist Soc 96(3):365–378
8. Sangha HS (in press) First record of the Long-billed Vulture *Gyps indicus* nesting on electricity pylon near Sri Dungargarh, Bikaner district, Rajasthan. BirdingASIA 15:74–75
9. Verma A, Prakash V (2007) Winter roost habitat use by Eurasian Marsh Harriers Circus aeruginosus in and around Keoladeo National Park, Bharatpur, Rajasthan, India. Forktail 23:17–21
10. Butler EA (1875) Notes on the Avifauna of Mount Aboo and Northern Guzerat. Stray Feathers 3(6):437–500
11. Vyas R (1992) Pied Harrier *Circus melanoleucos* (Pennant) in South-East Rajasthan. J Bomb Nat Hist Soc 89(2):248
12. Samant JS, Prakash V, Naoroji RK (1995) Ecology and behaviour of resident raptors with special reference to endangered species, Final Technical Report 1990–93. Bombay Natural History Society, Bombay
13. Sangha HS, Bhardwaj GS, Poonia SS (in press) Huge concentrations of the White-eyed Buzzard *Butatur teesa* at Tal Chhapar Wildlife Sanctuary, Churu district, Rajasthan. Indian Birds 7(5):139–140
14. Dharmakumarsinhji KS (1985) The Black Eagle *Ictinaetus malayensis* Temm. and Laug at Sawai Madhopur, (Rajasthan). J Bomb Nat Hist Soc 82(3):655
15. Prakash V (1988) Greater Spotted Eagle (*Aquila clanga*) breeding in Keoladeo National Park, Bharatpur. J Bomb Nat Hist Soc 85(2):418
16. Naoroji RK (1990) Predation by Aquila Eagles on nestling Storks and Herons in Keoladeo National Park, Bharatpur. J Bomb Nat Hist Soc 87(1):37
17. Turin R, Heegaard M, Priemé A (1987) Northern part of the Indian subcontinent 87. Unpublished birdwatching report
18. Holman D (1987) Northern India, 17th February–5th March 1987. Unpublished

19. Sangha HS, Devarshi D, Sharma SK (2007) The Amur Falcon Falco Amurensis Radde in Phulwari Wildlife Sanctuary—first record for Rajasthan. J Bomb Nat Hist Soc 103(1):98–99
20. Sangha HS (2002) A supplementary note on the avifauna of the Thar Desert (Rajasthan). J Bomb Nat Hist Soc 99(1):120–126
21. Sankar K, Mohan D, Pandey S (2006) Birds of Sariska Tiger Reserve, Rajasthan, India. Forktail 8:133–141
22. Kumar S (1993) Additions to the check list of Kota. Newslett Birdwatch 33(2):38
23. Sangha HS (2006) Birds of Mount Abu Wildlife Sanctuary, Rajasthan, India. Indian Birds 2(2):26–32
24. Dharmakumarsinhji KS (1955) Birds of Saurashtra. Compiled HS Sangha, Dil Bahar, Saurashtra

Chapter 14
White-naped Tit *Parus nuchalis*: A Vulnerable Species in Rajasthan

J.K. Tiwari, Devendra Bhardwaj, and B.K. Sharma

Abstract This chapter is entirely based on the status survey of White-naped Tit *Parus nuchalis* carried out in the year 1995 in seven districts of Rajasthan, namely, Pali, Jodhpur, Jalore, Sirohi, Ajmer, Jaipur and Nagour, and Kutch, Palanpur (Banaskantha district) and Taranga hills (Mehsana district) of Gujarat. Second survey of the above-mentioned places done in 2005–2006 included important sanctuaries of the area. This chapter presents a comparison of the data obtained from these two surveys. In Rajasthan, key areas for the White-naped Tit were near Beawar, Bar, Sendra and Amarpura to Ajmer which bore patches of tropical thorn forests in 1995. In many places where the Tit was reported and seen during the 1995 survey, the habitat of the endemic tit were wiped out when the authors revisited the area in 2005–2006. Sambhar Salt Lake is also one area where the bird has not been sighted since the last five years. However, the White-naped Tit was observed for the first time at Jamwa Ramgarh Wildlife Sanctuary and Nahargarh Biological Park in Jaipur district in the year 2010. Interestingly, *Bishnoi*-dominated areas of western Rajasthan near Jodhpur also showed no sighting of the bird. Author's observation on the status of the White-naped Tit indicates that the survival of this endemic species is dependent on the tropical thorn forests and protection of dead and decaying *Acacia* trees. The main cause of deforestation in Gujarat and Rajasthan are illegal charcoal making, gathering of fuelwood and invasion of exotic plant *Prosopis juliflora* into the tit habitat. Nest-site competition is also another important factor as the tits are secondary hole

J.K. Tiwari (✉)
Centre for Desert and Ocean Village, Moti Virani, Taluka-Nakhtrana, Kutch, Gujarat, India
e-mail: cedoindia@yahoo.com

D. Bhardwaj
Department of Forests and Wildlife, Government of Rajasthan, Jaipur, India
e-mail: devendra_bhardwaj@yahoo.com

B.K. Sharma
Department of Zoology, R.L. Saharia Government P.G. College,
Kaladera (Jaipur), Rajasthan, India
e-mail: drbksharma@hotmail.com

B.K. Sharma et al. (eds.), *Faunal Heritage of Rajasthan, India: Ecology and Conservation of Vertebrates*, DOI 10.1007/978-1-4614-0800-0_14,

nesters; they use abandoned nests of other birds. Cutting of the nest-bearing trees also badly affects the population of tits. Conservation of thorn forests is strongly suggested as the only way out to protect and revive this beautiful bird.

Introduction

The White-naped Tit, *Parus nuchalis*, commonly known as White-winged Tit, White-winged Black Tit or Pied Tit is popularly called as *Kabri Ramchakri* in Gujarati dialect. It is endemic to India with a disjunction and restricted geographic range. It is considered as Vulnerable (Version 3.1) by the IUCN and BirdLife International (2012). The bird has glossy black upper parts with large white flight feathers and prominent nuchal spot. Female is slightly less glossy black. Some individuals have washed pale yellowish parts instead of white. Juvenile is dull black with lesser white in wings and tail (Fig. 14.1). Sizes of the bird include a length of 108–111 mm, 28–30 mm head and 50–58 mm tail [1]. It is a species of the tropical dry thorn scrub forest characterized by *Acacia leucophloea*, *A. nilotica*, *A. senegal*, *Prosopis cineraria*, *Ziziphus jujuba*, *Capparis aphylla*, *Salvadora oleoides*, *S. persica* and *Grewia tenax* (Fig. 14.2). Its patchy distribution may be due to fragmentation of scrub forest habitat by human interference. The White-naped Tit is seen in a pair or sometimes a family part of 3–5 birds in number and exhibits a shy nature.

Fig. 14.1 A White-naped Tit *Parus nuchalis* perching on the branch. *Courtesy: J. K. Tiwari*

Fig. 14.2 White-naped Tit *Parus nuchalis* sitting on *Acacia* tree branch. *Courtesy: J. K. Tiwari*

Table 14.1 Sightings of the White-naped Tit in Gujarat and Rajasthan during the 1995–1995 survey

Date	Place	Nos.	Sighted by
7 October 1996	Sendra Reserve forest	2	J.K. Tiwari
14 October 1996	Kishangarh	2	J.K. Tiwari
7 April 1994	Ravli Todgarh	1	Harkirat Sangha
14 May 1995	Nasirabad	2	Harkirat Sangha
20 January 1996	Near Ramsar, Ajmer	1	Harkirat Sangha

It can be detected by the musical whistling note. Its calls are described as *tee-whi-whi* or *see pit-pit-pit-pit* which differs from that of the Great Tit.

According to Collar [2], 12 specimens were seen in Maroth near Sambhar Lake area. Tiwari [3] has also taken observations during surveys on White-naped Tit in Gujarat and Rajasthan. Club 300 Foundation for Bird Protection, Sweden, kindly provided fund to study the status and distribution of White-naped Tit in Rajasthan and some parts of Gujarat. In 1996, the lead author carried out status surveys in Kutch and with S.N. Varu in Palanpur (Banaskantha district), Taranga hills (Mehsana district) and Gujarat (Table 14.1). Similar surveys were also carried out by the first author in the seven districts of Rajasthan as mentioned above (Table 14.2). A restricted range of distribution is reported from the Nellore area of Andhra Pradesh (A.P.) and near Bangalore and Mysore in southern India [4]. The Biligiriranga Hills area had a relict population of the White-winged Tit as reported by Ali [5] and Uttangi [6].

Table 14.2 Sight records of *Parus nuchalis* from Ajmer District, Rajasthan, in the 1995–1996 survey

Date	No.	Place	Sighted by
14 May 1994	1	Ravli Todgarh, Ajmer	Harkirat Sangha
7 April 1994	2	Nasirabad, Ajmer	Harkirat Sangha
20 January 1996	1	Ramsar, Ajmer	Harkirat Sangha
20 May 1996	1	Balaram, North Gujarat	J.K. Tiwari & S.N. Varu
11 Sep 1996	2	Jethi	J.K. Tiwari
4 Oct 1996	2	Sundha Mata, Jalore	J.K. Tiwari
7 Oct 1996	2	Bar, Pali district	J.K. Tiwari
9 Oct 1996	2	Maroth, Nagour district	J.K. Tiwari
9 Oct 1996	2	Panchota Hill, Nagour district	J.K. Tiwari
9 Oct 1996	2	Sambhar Salt Works Jaipur district	J.K. Tiwari
9 Oct 1996	2	Kanota, Nasia Jaipur	J.K. Tiwari
14 Oct 1996	2	Kishangarh, Ajmer district	J.K. Tiwari

Table 14.3 White-naped Tit seen in the 2005 survey

Date	Place	No.
8 May 2005	Desuri ki Nal	1
10 May 2005	Sajjangarh Wildlife Sanctuary	1
10 May 2005	Jaisamand forest	2
12 May 2005	Bassi Wildlife Sanctuary	2
18 May 2005	Bar and Sendra area	2
9 November 2005	Jalore Sundha Mata Hill	2
11 November 2005	Nasia old fort Jaipur	2

The second survey by the lead author was carried out in May 2005 and November 2005 covering Sirohi, Udaipur, Chittourgarh, Ajmer, Pali and Jaipur districts (Table 14.3) of Rajasthan. The White-naped Tits were seen for the first time in forest areas of Sajjangarh Wildlife Sanctuary, Jaisamand Wildlife Sanctuary and Bassi Wildlife Sanctuary near Chittourgarh. During the 2005 survey, the *Acacia leucophloea* forest was found to be fragmented by several small to large villages and croplands with the exotic mesquite (*Prosopis juliflora*) forest. The habitat destruction (cutting, lopping, burning and clearing seen in many areas) was excessive in Maroth (near Sambhar) and Sambhar Salt Works reserved forest. No tits were seen in these areas. The habitat was totally cleared by local people and there were hardly any trees left. This area was rich in wildlife during the 1996 survey but the entire habitat was cleared by villagers when revisited in 2005. *Parus major*, Grey Tit and *Parus nuchalis* can coexist in the same patch of forest, and they are not mutually exclusive at least in some parts of India like Sambhar Lake in Rajasthan, Balaram Ambaji Wildlife Sanctuary and in Biligiriranga Hills WLS in Karnataka [7].

The White-naped Tit is seen in a flock of 2–6. The flock remains in contact by their musical call notes and whistles. They are shy and agile in nature and constantly move from branch to branch on *Acacia leucophloea*, *A. senegal*, and *A. nilotica* in search of insects. The tits roost in hollows of trees or overhanging bark of dead and decaying *Acacia* trees. The White-naped Tits show strong affinity to their roosting sites. They tend to roost at the same site for months, if not disturbed. The Brahminy Starling *Sturnus pagodarum* and Chestnut-shouldered Petronia *Petronia xanthocolis* compete with the White-naped Tits for nest and roost sites. The nesting season of White-naped Tit is July to September in Kutch [8–10]. They use the abandoned nest hollows of the Yellow-fronted Pied Woodpecker. The status and nesting and roosting behaviour of the species were studied in Kutch district when the lead author was working with the Bombay Natural History Society on the Bird Migration study project and the Grassland Ecology Project, funded by the US Fish and Wildlife Service during 1990–1995.

Survey Results

Mount Abu Wildlife Sanctuary (Area 328 km²)

Mount Abu (altitude 1,219 m) is a hill oasis in the Thar or Great Indian Desert. The sanctuary comprises of an old mountain range. The ecology of this hill is unique as it recieves a rainfall of nearly 1,500 mm in otherwise low rainfall area, so we see many faunal and floral elements that are generally not found in the arid Thar Desert. The Abu Hill Range was thoroughly searched (stay period one year) for Paridae. The Great Tit *Parus major* and Black-lored Tit *Parus xanthogenys* were seen. It was interesting to note the altitudinal distribution of the Genus *Parus*. The Great Tit occurs at the height of 1,219 m and above, the Black-lored Tit at relatively lower height reaches up to 600–1,300 m and the Black-naped Tit occurs at the lowest height, not 400 m above sea level. The Black-lored Tit was found to be a common resident of Mount Abu Aravalli Range. The White-naped Tit was seen at Balaram and Ambaji Hill Range in 1999, but not during the present survey. Balaram is 55 km from Mount Abu. The other species endemic to India is Green Avadavat *Amandava formosa* seen frequently (28 birds at Oriya; 55 birds at Sunrise Valley; 20 birds near Dilwara, Teachers Training School; 8 birds near Ganesh point and 12 birds near Sunset Point).

Jessore Sloth Bear Sanctuary and Balaram Ambaji Forest Survey

The Jessore Sloth Bear Sanctuary was first surveyed with S.N. Varu and was resurveyed on 13 March 2005 with S.N. Varu and D. Meghani. The area is a dry deciduous forest, with Semul *Bombax ceiba* and *Butea monosperma* and no *Lantana camara*. Commonly seen birds were Chestnut-shouldered Petronia *Petronia xanthocollis*, Rufous Treepie *Dendrocitta vagabunda*, Rosy-ringed Parakeet *Psittacula krameri*, Brown-headed Barbet *Megalaima zeylanica* and Little Green Bee-eater *Merops orientalis*. Jessore Sloth Bear Sanctuary is a prime forest for the Sloth Bears *Melursus ursinus*.

Balaram and Ambaji Forest Survey

The forest range of Balaram Ambaji is a part of Ambaji Wildlife Sanctuary. White-naped Tit was not seen during the present survey. In May 1999, the authors had reported the White-naped Tits for the first time and the Great Tits from the same area. Kher or *Acacia leucophloea* along with *Acacia nilotica* is seen abundantly in this forest. Balaram is a temple forest; the areas near the Jethi River are ideal locations for the sighting of White-naped Tit (two were seen in 1999).

Ranakpur Forest (Malgarh and Sumer Forest Survey), 7–8 May 2005

Ranakpur is a temple forest (famous Jain temple lies in the lower hill ranges of Aravallis) with semi-deciduous and tropical thorn forest. White-naped Tit was not seen during the survey. Due to the presence of Jain temple, this forest is safe from destruction. One Indian Eagle-Owl *Bubo bengalensis* was seen near the Shilpi guest house. Grey Junglefowl *Gallus sonneratti* and Red Spurfowl *Galloperdix spadicea* come to feed the grains along with hundreds of Grey Partridge *Francolinus pondicerianus* and Indian Peafowl *Pavo cristatus* at the Shakti Mata temple in Ranakpur in the wee hours. The birds seen commonly at Ranakpur are Ashy-crowned Sparrow-lark *Eremopterix griseus*, Spotted Dove *Stigmatopelia chinensis*, Indian Grey Hornbill *Ocyceros birostris*, White-bellied Drongo *Dicrurus caerulescens*, Paradise Flycatcher *Terpsiphone paradisi*, Oriental Magpie-Robin *Copsychus saularis*, Greater Coucal *Centropus sinensis*, Plum-headed Parakeet *Psittacula cyanocephala* and Black-rumped Flameback *Dinopium benghalense*. Mammalian fauna such as Ruddy Mongoose *Herpestes smithii*, Nilgai *Boselaphus tragocamelus*, Chital *Axis axis*, Indian Hare *Lepus nigricollis* and Hanuman Langur *Semnopithecus entellus* were also seen in the area.

Kumbhalgarh Wildlife Sanctuary and Environ Survey, 8 May 2005

Ranakpur, Syra, Malaki Chowki, and Jhalaki Chowki areas from Ranakpur to Kumbhalgarh were scanned thoroughly for the *Parus nuchalis*, other tit species and Green Avadavat. The White-naped Tit was not seen in this tract of forest during the survey period. However, the Great Tit was seen commonly. The forest tract from Ranakpur to Kumbhalgarh is sparse detached hilly dry deciduous and thorny at the base of hills. Brahmno ka Kalvana, Diyan, Dhikoda, Kitavato ka Guda villages were also searched. Red-rumped Swallow *Hirundo daurica*, Brown Rockchat *Cercomela fusca*, Eurasian Golden Oriole *Oriolus oriolus*, Oriental Honey Buzzard *Pernis ptilorhyncus*, Purple Sunbird *Nectarinia asiatica* and Chestnut-shouldered Petronia *Petronia xanthocollis* were found to inhabit the above-mentioned villages. There is a big roost (around 5,600) of the Indian Flying Fox *Pteropus giganteus* at Bhanpura (15 km from Kumbhalgarh) on seven mango trees. The bats were safe from villagers. No poaching or killing of the bats was seen.

Kumbhalgarh Wildlife Sanctuary, 8 May 2005

White-naped Tit was not seen in this forest. Kumbhalgarh WLS is situated in the most rugged hills of Aravalli Range and adjoins the borders of Pali, Udaipur and Rajsamand districts of Rajasthan. *Anogeissus pendula* forest is dominant. This area has no *Lantana camara* although it is a problem in the entire Aravalli Range. For example, in Mount Abu, the forest fires on the frost-bitten dry *Lantana* resulted in a heavy loss to the forest (an average 1,000 ha is burnt per year).

Desuri ki Nal and Sumer ki Nal Survey

The forest type is *tropical Acacia leucophloea* and *Acacia nilotica* dominated thorn scrub. At Sumer ki Nal (Naal means a riverine forest), one White-naped Tit and two Great Tits were seen. Desuri ki Nal and Sumer ki Nal are two contiguous riverine forests located 4.5 km apart in the Pali District of Rajasthan. Reza Tehsin had seen White-naped Tit in Sumer ki Nal area on 26 June 2004.

Sajjangarh Wildlife Sanctuary Survey, 10 May 2005

This pristine forest is an example of efforts of the Forest Department to protect the forest near a thickly populated urban area. The boundary wall around the Sajjangarh WLS is effective for the protection of the forest. A thick *Acacia leucophloea* and *Acacia nilotica* forest is seen at the lower reaches of the hills. *Anogeissus pendula* and *Stercularia urens* trees grow profusely in this area at higher reaches of the hills. Four Great Tits and one White-naped Tit were seen at the Sajjangarh WLS. The White-naped Tit was seen at a water hole half a kilometre away from the gate.

Naathdwara, Haldi Ki Ghati, 10 May 2005

This entire tract is devoid of a contiguous thorn forest. However, clusters of *Acacia nilotica* trees can be seen in the agriculture fields. No tits were seen in this forest.

Haldi Ghati Forest

At Khamnor village, only a good patch (5 ha) of *Acacia leucophloea* thorn forest was seen, but no tits were seen on this patch. On Udaipur to Nathdwara road, 8 km from Udaipur, a good thorn forest can be seen from the roadside. No tits were seen in this forest.

Jaisamand Wildlife Sanctuary Survey, 10 May 2005

Two White-naped Tits were seen in this forest. A 50 ha forest of *Acacia leucophloea* is present near 18.5 km of Jaisamand Lake. Jaisamand Sanctuary is situated 50 km south of Udaipur. An adult Jungle Cat *Felis chaus* was found as a roadkill on the way to Jaisamand. Human interference was evident in the entire hill range which runs parallel to the road from Udaipur to Jaisamand. Hanuman Langurs are seen commonly in troops along the road.

Sitamata Wildlife Sanctuary, 11 and 12 May 2005

No White-naped Tit but Great Tit was seen in this area. Sitamata Sanctuary spreads over Aravalli and Vindhyachal Mountain Ranges and is the only forest region where teak tree, *Tectona grandis*, of building value is found. The Sitamata Wildlife Sanctuary is home to Indian Large Brown Flying Squirrel *Petaurista philippensis*.

Bassi Wildlife Sanctuary, 12 May 2005

On Chittourgarh to Kota area in Rajasthan, lies the Bassi WLS area, (13, 805 ha) notified in 1988. In this forest, two White-naped Tits and a Grey Tit were seen. White-naped Tits were feeding on some insect from the pods of an *Acacia* tree. *Lannea grandis, Aegle marmelos, Acacia catechu and Acacia nilotica* trees were dominating. Two young ones of Hyaena were seen in a den in the Bassi WLS area. Four Chinkara *Gazella bennettii* and one Booted Eagle *Hieraaetus pennatus* were also sighted in the area.

Chittourgarh to Sambhar Salt Lake, 13 May 2005

Broken hilly areas with scattered thorn scrub were seen. From Chittourgarh to Ajmer, 5-km-long hill runs parallel to the road (20 km away from Chittourgarh) which has a good thorn forest. Near Nasirabad, *Acacia leucophloea* thorn forest is present. *Parus nuchalis* has been sighted in this forest. No tit has been sighted earlier in this forest.

Sambhar Survey, 13 May 2005

Sambhar is the largest inland salt lake of India. Adam [11] had seen White-naped Tit from the Maroth and Sambhar areas. This habitat was revisited after 10 years by the lead author. In 1995, the forest was in good shape and two White-naped Tits were seen, but in May 2005, the forest condition was degraded with clear signs of lopping, deforestation and cattle grazing. No tit was seen in the habitat.

Maroth Survey

This is a historical site and a forest patch where RM Adam [9] had seen the White-naped Tit and the Grey Tit. The lead author had reconfirmed coexistence of these two species from the same patch of thorn forest near the Maroth village in 1995. But the survey in May 2005 resulted in no tit; in fact, the complete patch of thorn forest was cleared by villagers. A place where once a thick forest of *A. nilotica* and *A. leucophloea* existed, is presently occupied by sand-dunes with stumps of cleared forest.

Sambhar Lake City, Jhapok and Shakambri Mata Area Survey, 14 May 2005

Sambhar Salt Lake and surrounding environs ecologically resemble that of Rann of Kutch. Jhapok village is in the middle of the Salt lake. Strong possibility of an occurrence of White-naped Tit is there in Jhapok village. *Acacia* groves are in good condition in this area. However, the present search did not result in sighting of the White-naped Tit.

Sendra Forest and Bar Survey, 17–18 May 2005

On Pali-Ajmer border forest of Rajasthan, thorn forest occurs but the invasion of weed *Prosopis juliflora* was seen in the entire range. Beawar (a big town and cement industry area) is the place where one must stay to survey these hill ranges. Beawar to Sendra is 12 km and Sendra to Bar forest is 15 km. Pakshidham (feeding ground of birds) is a spot where 500 Rosy-ringed Parakeets come to feed with at least 20 Plum-headed Parakeets. Two White-naped Tits were seen in this area where 50 ha good *Acacia leucophloea* forest exists. One nest of an Indian Vulture *Gyps indicus* Vulture and two White-rumped Vultures *Gyps bengalensis* were seen soaring.

Other forest areas in the vicinity are Paluna *Beed*, Taragarh *Beed* and mixed thorn forest. Between Phulera and Jaipur at a place called Dhindaa, 50 ha thorn forest and near Asalpur-Jobner, a 10 ha thorn forest was scanned for bird life. There are chances of *Parus nuchalis* in this forest but the authors could not locate it.

Survey of Western Rajasthan, 5 November to 15 November 2005

The survey was conducted in the forest areas of Bar, Sendra, Pali, Beawar, Punaghar ki Bhakri, Vijaynadi, Rohat, Guda Vishnoia, Vishnoi ki Dhani, Jodhpur, Sanchore and Jalore in Rajasthan and Vav, Tharad and Sooegam area in Gujarat state. The western Rajasthan is a part of Great Indian Desert (Thar Desert). *Acacia* forest cover is good in *Bishnoi*-dominated areas. The survey failed to find tit in this desert tract of Rajasthan. However, the first survey resulted in the sighting of *Parus nuchalis* from Bar, Sendra and Beawar area. The tit is not reported from other areas before. The *Bishnoi*-dominated areas were searched thoroughly for wildlife. At least 11,000 Chinkaras *Gazella bennetti* were found in the district of Pali and Jodhpur; Blackbuck also survive in this belt in good numbers. Rare trees like *Tecomella undulata* are seen in the agricultural fields of *Bishnoi* farmers. In Jalore and Sanchore areas, at least 60,000 Rosy Pastors *Sturnus roseus* and 4,000 Demoiselle Crane *Anthropoides virgo* were seen. Vav and Tharad area are potential sites for White-naped Tit; although the *Acacia* plantation along the roadside was searched, no tit was seen. Sooegam is the area from where the White-naped Tit was reported by Adam [11] some 120 years back; Currently the area is devoid of this bird. Only the exotic weed *Prosopis juliflora* grows in place of *Acacia leucophloea* forest.

Jalore District, 9 November 2005

Two White-naped Tits were seen near Sundha Mata Hill.

Jaipur District, 11 November 2005

Two White-naped Tits were seen in an *Acacia nilotica* and *A. leucophloea* mixed forest at Nasia (Old fort) near Kanota, 11 km away from Jaipur city. The species was reported from Jaipur district, by (late) Shantanu Kumar Singh, former Director Survey in 2010 General Police, Government of Rajasthan in 1993 (pers. comm.).

First Record of White-Naped Tit *Parus nuchalis* at Jamwa Ramgarh WLS, Jaipur, and Nahargarh Biological Park, Jaipur, in 2010

White-naped Tit has not been reported from Jamwa Ramgarh and Nahargarh Biological Park till 2010. Mr. Tej Kumar Sharma, posted at Nahargarh Biological Park, claimed that these areas are the habitats of the White-naped Tit being dry thorn-scrub forests, particularly dominated by *Acacia* species. The main tree species of Jamwa Ramgarh Sanctuary and Nahargarh Biological Park are *Acacia senegal, A. leucopholia, A. nilotica tomentosa, A. tortilis, Anogeissus pendula, Butea monosperma, Capparis decidua, Ehertia laevis, Holoptelea integrifolia, Ziziphus mauritiana,* etc., making the best habitat for the White-naped Tit. Survey was conducted at four main areas for White-naped Tit considering the habitat in February 2010 and October 2010.

Nahargarh Biological Park

Mr. Tej Kumar Sharma, a forester, was continuously observing the behaviour of White-naped Tit in the Nahargarh Biological Park from Feb 2010. There are two main spots, i.e. *Surra Ki Baori* and Deer Enclosure. The major trees observed by him for the perching of birds were *Acacia catechu, A. senegal, Anogeissus pendula, Prosopis cineraria* and bushes like *Rhus mysorensis*. The birds were seen flying around the water hole or sitting and drinking water with the Great Tit. On 7 July 2010, two birds were seen together, otherwise always reported solitary (Tables 14.4 and 14.5).

Table 14.4 Sight records of White-naped Tit from the Sura Ki Baori area, Jaipur, in 2010

Dates	Nos. sighted
28 March 2010, 17 April 201, 25 May 2010, 16 August 2010, 24 November 2010, 10 January 2010 and 2 March 2011	1

Table 14.5 Sight records of White-naped Tit from the Deer Enclosure area, Jaipur, 2010

Dates	Nos. sighted
22 February 2010, 15 March 2010, 6 April 2010, 16 June 2010, 22 June 2010, 12 July 2010, 29 August 2010, 12 September 2010, 28 September 2010, 7 October 2010, 9 November 2010, 28 November 2010, 5 December 2010, 16 December 2010, 26 December 2010, 15 February 2011 and 22 February 2011	1
25 July 2010	2

Jaitpur Khichi Forest Area Near Achrol, 28 September 2010

This is a ravenous area located in the Aravalli foothills. *Acacia tortilis* is seen well established with existing flora. Only one bird was sighted on *Acacia senegal* tree on 28 September 2010.

Dantala Forest Block, 6 Oct 2010

This is also a ravenous area in Aravalli foothills. *Acacia tortilis* flourished here with existing flora of dry deciduous thorny forest like *Acacia senegal*, *Dichrostachy cinerea*, *Anogeissus pendula* and *Rhus mysorensis*. On 6 Oct 2010, only one bird was sighted by the second author on *Dichrostachy cinerea*.

Godiana Forest Block, 1 and 6 Oct 2010 and 15 December 2010

A ravine area near Bangagna River had *A. nilotica*. Now, *Acacia tortilis* is well established with existing flora of dry deciduous thorny forest like *Acacia senegal*, *Dichrostachy cinerea*, *Anogeissus pendula and Rhus mysorensis*. Only one bird was sighted on *Dichrostachy cinerea*, *Prosopis cineraria*, *Holoptelea integrifolia*, and *Acacia leucophloea* branches at different times on 6 and 10 October and 15 December 2010.

Conservation

Observation on the status of White-naped Tit in Kutch, north Gujarat, Rajasthan and BR Hills indicate that the survival of this endemic species is dependent on the conservation of the tropical thorn forest and protection of dead and decaying *Acacia* trees [10]. The main cause of deforestation in Gujarat and Rajasthan are illegal charcoal making, gathering of fuelwood and invasion of exotic plants into the White-naped Tit habitat.

Measures to Protect the White-Naped Tit and Its Habitat

Strict protection should be given to areas where *Acacia*-dominated habitat occurs, for example, Sundha Mata near Jalore, Bar and Sendra near Beawar and Sajjangarh and Jaisamand near Udaipur. Emphasis should be given on the forestation programmes of *Acacia nilotica* and *Acacia senegal*.

Acknowledgements The authors are thankful to Club 300 Foundation for Bird Protection for providing fund to carry out the survey and to the forest departments of Gujarat and Rajasthan states. Thanks are due to Dr. Henri K Lind, Kiran Chavda, Shailesh Patel, Anil Mathur, RFO Mount Abu, DFO Mount Abu, RFO Bassi Lalit Singh Rathore, Digvijay Gupta DFO Chittourgarh, owner of Hotel Kumbhal castle, Apna Hotel, Shilpi guest house staff, RFO Bhagvansingh for all the help and encouragement. Authors thank Kevin Vang and Wojciech Debrowika for sharing their pictures of *Parus nuchalis*. Thanks are also due to S.N. Varu for accompanying us in the survey of Jessore Sloth Bear Sanctuary and to Mr. Tej Kumar Sharma for his keen observations of White-naped Tit in the Nahargarh Biological Park, Jaipur, Rajasthan.

References

1. Rasmussen PC, Anderton JC (2005) Birds of South Asia. The Ripley Guide vol. 1 and 22. Smithsonian Institution and Lynx Edicions, Washington DC and Barcelona, pp 16
2. Collar NJ, Andreev AV, Chan S, Crosby MJ, Subramanya S, Tobias JA (eds) (2001) Threatened Birds of Asia. BirdLife International, pp 2415–2416
3. Tiwari JK (2001) Status and distribution of the White-naped Tit *Parus nuchalis* in Gujarat and Rajasthan. J Bomb Nat Hist Soc 98(1):26–30
4. Lott Eric J, Lott Christine (1999) On the occurrence of White-naped Tit *Parus nuchalis* in southern India. Forktail 15:93–94
5. Ali S (1978) The Book of Indian Birds. Bombay Natural History Society and Oxford University Press, Mumbai. 13th edn. Revised 2002, pp 380
6. Uttangi JC (1995) A rare occurrence of the globally threatened White-naped Tit *Parus nuchalis*, in areas of Dharwad, Karnataka, India. Newslet Birdwatch 35(6):114–115
7. Tiwari JK (1999) Status of the Pied Tit *Parus nuchalis* in South India. Newslet Birdwatch 39(2):36
8. Hussain SA, Tiwari JK (1992) Status and distribution of White-winged Black Tit in Kachchh, Gujarat India. Bird Conserv Internat 2:115–122
9. Tiwari JK, Rahmani AR (1997) The current status and biology of the White-naped Tit *Parus nuchalis* in Kutch, Gujarat, India. Forktail 12:79–85
10. Hussain SA (1997) The White-winged Tit. Newslet Birdwatch 36(1):18–19
11. Adam RM (1873) Notes on the birds of the Sambhar Lake and its vicinity. Stray Feathers 1:361–404

Part IV
Faunal Ecology: An Insight – The Mammal Conglomerate

Chapter 15
Non-human Primates of Rajasthan

L.S. Rajpurohit, A.K. Chhangani, G. Sharma, S.M. Mohnot, and B.K. Sharma

Abstract The chapter describes socioecology of the Rhesus Monkey *Macaca mulatta* and Hanuman Langur or Northern Plains Gray Langur *Semnopithecus entellus* found in Rajasthan. The forest-dwelling Rhesus Monkey has successfully invaded human habitats, even up to the extent of creating menace in many towns and cities of Rajasthan. On the contrary, Hanuman Langur is a bit shy than Rhesus Monkey and large populations of these langurs exist in forests of Sariska Tiger Reserve, Ranthambhore National Park and in all the wildlife sanctuaries of Rajasthan, other than Desert National Park and Tal Chhaper Wildlife Sanctuary. Although Hanuman Langurs have also made their way to human settlements of the state, they have not caused much havoc. Habitat depletion has been described as a major cause of encroachment to human habitations like villages, towns or cities. Since both the monkey species have been well-studied in nature by the authors, details of group organization, feeding and foraging, reproductive behaviour, social communication, morphology, sociobiology and ecology are thoroughly described in this chapter. In addition, infanticide in Hanuman Langurs has been mentioned as an interesting

L.S. Rajpurohit (✉) • G. Sharma
Department of Zoology, Jai Narain Vyas University, Jodhpur, Rajasthan 342001, India
e-mail: drlsrajpurohit@yahoo.in

A.K. Chhangani
Department of Environmental Science, Maharaja Ganga Singh University, Bikaner, Rajasthan 334001, India
e-mail: chhanganiak@yahoo.com

S.M. Mohnot
The School of Desert Sciences, 109, Nehru Park, Jodhpur, India

B.K. Sharma
Department of Zoology, R.L. Saharia Government P.G. College, Kaladera (Jaipur), Rajasthan, India
e-mail: drbksharma@hotmail.com

B.K. Sharma et al. (eds.), *Faunal Heritage of Rajasthan, India: General Background and Ecology of Vertebrates*, DOI 10.1007/978-1-4614-0800-0_15,

behaviour. Use of these monkeys in a variety of biomedical researches in the laboratories of the state had received great opposition from NGOs working for the prevention of cruelty towards animals who finally succeeded in stopping this practice.

Introduction

Non-human primates include apes, monkeys, lemurs and langurs are vocal, group-living and easily seen animals more popular than other mammals. The study of non-human primates not only has a close bearing on the understanding of human, social and psychological problems but is also important in medicine (human and veterinary) and agriculture (protecting crops from non-human primate pests). Biologically, they provide an understanding of the morphological, physiological and even behavioural aspects of human evolution, especially of early man before the advent of agriculture [1]. These mammals are also good indicators of the health of an ecosystem, a property that can be used for the appropriate conservation planning of the given ecosystem.

The first living members of the order primates date back to approximately 55 million years ago (mya) at the beginning of the Eocene epoch. The end of the Miocene and the beginning of the Pliocene epoch (5 mya) is marked by the major radiation of Old World monkeys in Africa and Eurasia [2].

Linnaeus listed primates as the first and highest order of class Mammalia and gave four genera: Homo (man), Simia (monkeys and apes), Lemur (lemurs and lorises) and Vespertilo (bats). However, bats were later removed from the order primates [3]. After this first attempt to classify primates, there have been many changes in the taxonomy of primates, the latest being the one by Groves [4]. Still, some more changes are in the offing, especially concerning the taxonomy of the Asian primates.

Non-human primates are very similar to humans and the principal reason for this similarity is simple: humans are primates. Human and non-human primates have many characteristics in common, i.e. tool use, long-lasting social relationships and complex communication system. By studying and learning about non-human primates we may learn more about ourselves [5]. Human and non-human primates also share physiological characteristics like the organization, neuroanatomical studies of the brain, basic biological phenomena such as reproduction, treatment of diseases such as AIDS and the development of drugs, treatment and vaccines for the promotion of better health for human beings. In absence of other inventory information, the species richness of non-primates at a locality is a useful indicator of the probable richness of the primate mammals [6].

Non-human primates are naturally distributed on five continents out of the seven in 92 countries (except in Australia and Antarctica), inhabiting mostly four major biogeographical tropical and subtropical regions (23° N 23° S), (i.e. South and Central America, Africa, Madagascar and Asia), making up for a total of 63 genera having more than 620 species/subspecies (Table 15.1).

In India, 16 species of non-human primates are found (Table 15.2) and three of them are considered as commensal in villages, towns, cities, temple sites, parks

Table 15.1 Global distribution of primates

Region	Genera (no.)	Species/subspecies (no.)
South America	16	204
Africa	20	190
Madagascar	14	50
Asia	13	176
About 92 countries	63	620

Table 15.2 List of Indian non-human primates

Family	Subfamily	Common name	Scientific name
Lorisidae	Lorinae	Slender Loris	*Loris tardigradus*
		Slow Loris	*Nycticbus coucang*
Cercopithecidae	Cercopithecinae	Lion-Tailed Macaque	*Macaca silenus*
		Pig-Tailed Macaque	*Macaca nemestrina*
		Bonnet Macaque	*Macaca radiata*
		Assamese Macaque	*Macaca assamensis*
		Crab-Eating Macaque	*Macaca fascularis*
		Rhesus Macaque	*Macaca mulatta*
		Stump-Tailed Macaque	*Macaca arctoides*
		Arunachal Macaque	*Macaca munzala*
	Colobinae	Hanuman Langur	*Semnopithecus entellus*
		Nilgiri Langur	*Trachypithecus johnii*
		Golden Langur	*Trachypithecus geei*
		Phayres Langur	*Trachypithecus phayrei*
		Capped Langur	*Trachypithecus pileatus*
Hylobatidae		Hoolock Gibbon	*Bunopithecus hoolock*

gardens, orchards, etc. They are Rhesus Monkey (*Macaca mulatta*), Hanuman or Northern Plains Gray Langur (*Semnopithecus entellus*) and Bonnet Macaque (*Macaca radiata*). Rhesus Monkeys and Hanuman Langur are present in northern, central and north eastern parts of India including Rajasthan. Bonnet Monkey is found below River Godawari in the whole of Peninsular India. Both the non-human primate species found in Rajasthan (i.e. Hanuman Langur and Rhesus Monkey) are well studied during last half of the century [1, 7–9]. This chapter covers the socio-ecological aspect, such as the details of group organizations, food-feeding and foraging, reproductive behaviour and various social behaviour of Hanuman Langur and Rhesus Monkey found in Rajasthan.

Hanuman Langur or Northern Plains Gray Langur—*Semnopithecus entellus* (Dufresne [13]; Blanford [14])

Systematic Position
Class: Mammalia (Mammals)
Subclass: Eutheria (Placental mammals)
Sub-order: Anthropoidea (Simians)
Infraorder: Catarrhini (Old World monkeys)

Superfamily: Cercopithecoidea
Family: Cercopithecidae
Sub-family: Colobinae (Leaf monkeys)
Genus: *Semnopithecus*
Species: *entellus*
Subspecies: *entellus* (and 15 other subspecies in South Asia)
Other common names: English—Grey langur, Common langur
Local names: Hindi—Langur, Kumaun—Gooni; Rajasthan—Languria; Gujarati—Vandar; Marathi—Wanar; Maka, Makur, Wanga; Kannada—Mushya, Mustya, Tamil—Korungoo, Vella Monthi; Telugu—Kumdamuchu; Sinhalese—Wanderoo, Vandhura, Kondevandhura; Burmese—Meeauk
Scientific synonyms: *Presbytis entellus*, *Simia entellus*

Taxonomic Note

Gray [10] described family Cercopithecidae and its subfamily Cercopithecinae and Jerdon [10] mentioned subfamily Colobinae. However, Hill [11] and Groves [12] divided family Cercopithecidae into two separate families Cercopithecinae and Colobinae. Initially Dufresne [13, 14] described the monkey as *Semnopithecus entellus* by keeping *Semnopithecus* as a separate Genus [15] but later, Szalay and Delson [16] considered it as a subgenus of *Presbytis*. Groves [12] again separated *Semnopithecus* from *Presbytis* (Fig. 15.1).

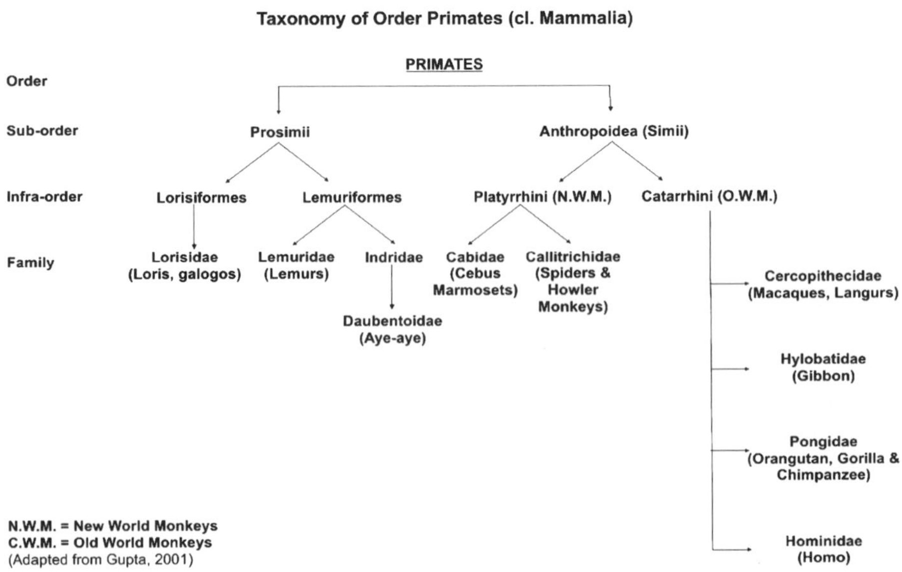

Fig. 15.1 Taxonomy of order primate (Class—Mammalia)

Morphology

Hanuman Langur is a large, black-faced, grey-bodied animal with long limbs, well-developed sex skin in adult males but does not show any cyclic changes. Tail is longer than head and body, differing in carriage and shows geographical and subspecies variations [17]. Adult has hair behind the brow radiating from a frontal whorl hair rising into a crest or tuft. Eyebrows are well developed. At the birth, infants have black coloured coat which changes to normal grey in 4–5 months while face, palms and soles are pinkish-white which changes to black in 2–3 weeks [1, 18–20]. Langurs exhibit clear-cut sexual dimorphism. On an average, an adult male weighs around 18.0 kg and an adult female weighs around 12.0 kg [21]. Head and body length is 58–65 cm in males and 52–57 cm in females. The tail length is 85–106 cm in males and 78–90 cm in females [1, 22–25] (Fig. 15.2).

Fig. 15.2 Hanuman Langur *Semnopithecus entellus* female with an infant. *Courtesy*: *Goutam Sharma*

Ecological Distribution

The Northern Plains Gray Langur or Hanuman Langur (*Semnopithecus entellus*) is the most adaptable south Asian colobine, found in India (except the northeastern part), Pakistan, Nepal, Bhutan, Bangladesh, South Tibet and Sri Lanka. Since they live in a wide range of diversified habitats differing in ecological conditions, from 3,660 m altitude in the Himalayas and peninsular forests to semiarid woodlands, in villages and towns, and on cultivated land [1, 14, 17, 18, 24, 26–32], Hanuman Langur should be considered as one of the most promising primate models for building and testing socioecological hypothesis concerning phylogenetic adaptations and modificational adaptability.

Sixteen subspecies of *Semnopithecus entellus* are recognized and all are found in South Asia [1, 23, 25, 33, 34]

1. *S.e. archetes*. Southern India—Dharwad, Bellary, Karava.
2. *S.e. achilles*. Sikkim and Nepal at high altitudes.
3. *S.e. aeneas*. Southern India.
4. *S.e. ajax*. Himachal Pradesh, Pakistan and Parts of J&K.
5. *S.e. anchises*. Madhya Pradesh and the Eastern Ghats.
6. *S.e. dussumieri*. Southern India: Malabar.
7. *S.e. elissa*. Southern India.
8. *S.e. entellus*. Northern India: Bengal to Gujarat and Rajasthan.
9. *S.e. hypoleucos*. Southern India: Kerala.
10. *S.e. iulus*. Southern India: Karnataka.
11. *S.e. lania*. Southern Tibet.
12. *S.e. priam*. Southern India: Tamil Nadu.
13. *S.e. priamelus*. Southern India: Kerala.
14. *S.e. schistaceus*. Uttar Pradesh, Uttarakhand and Nepal.
15. *S.e. shanicus*. Northern States.
16. *S.e. thersites*. Southern India, Sri Lanka.

Groves [4] provides five more subspecies in addition to the above mentioned subspecies, which are as follows: *S.e. albepes*, *S.e. hector*, *S.e. nipalensis*, *S.e. pallipes* and *S.e. petrophilus* found in South Asia. However, their geographical distribution is not well marked.

Legal Status—The Hanuman Langur is included in Schedule II of Indian Wildlife (Protection) Amenment Act, 2006 in Appendix I of CITES and in USA as ESA-Endangered.

Reproduction

Primate females in general begin reproducing between 1 and 13 years of age, considering the gemut from Mouse Lemur to Chimpanzee and human being. It is worth while to mention that they are slower to mature than other mammals of equal body weight. The majority of monkeys and even the larger lemurs do not bear offspring before 3 years of age [35].

In the last four decades intensive field studies on natural populations of Hanuman Langurs have been carried out in India, Nepal and Sri Lanka (Table 15.3). Langur females seem to be very anthropomorphic to environmental fluctuations as found in several field studies.

The estimated age at which a langur reaches maturity is 2.5–3.0 years in females and 6–8 years in males [1, 17, 18, 21, 54–56]. No regular and well-defined change occurs in the external genitalia during oestrus and menstrual periods [17]. A female langur in oestrus may initiate copulation and she solicits adult males by dropping her tail, shaking her head and beating the ground with her hands [1, 17, 18, 55–57]. Regional differences in this type of reproductive behaviour may occur. In multimale bisexual groups, only the most dominant male copulates, other subordinate or young adult males also have consort relations when more than one female are in oestrus simultaneously. The copulating females may be attacked by other females of the group [18]. The gestation period is about 200 days and average cycling length is 24 days [21, 39, 55, 58]. The sex-ratio at birth (M–F) is 1.1:1.0 [30, 54]. The range of inter-birth interval (IBI) in langurs at Dharwar noted by Jay [17] and Sugiyama [18] is 20–25 months, at Abu 15–30 months [39] and at Jodhpur, 8–22 months [30] with an average of 15.3 months [55]. Rajpurohit *et al.* [59] reported a wide range of inter-birth interval between 7.0 and 76.5 months (average 16.8 months, n:112). Infant loss under the age of 4.1 months influences the IBI and after the birth of next surviving infant, the IBI is significantly longer. Abortions and still birth may reduce the IBI [56, 58–60].

As a rule, a single young is born, but twin birth also occurs occasionally [30, 59]. However, triplet and quadruple birth have also been reported from Jodhpur [53, 61] (Fig. 15.3).

The different age–sex categories and their descriptions are given in the Table 15.4 [54]. The categories for males are elaborated on the basis of Moore [46].

Langurs breed round the year at Kankori [17], Orcha [62], Jodhpur [59], Kumbhalgarh [53] and Polnnaruwa [63], while at other sites like Rajaji National Park [64], Gir Forests [37] and Dharwar [18], they breed only during some months of the year (Table 15.3). Dodsworth [27] noted that in the western Himalayas, the breeding season extends throughout the warmer weather, when the females are seen carrying infants. In northern India, births are concentrated around the hot and dry months (March–May). Infants born at this time begin to supplement their milk diet with leafy food within 2–3 months (July–September), when plenty of green food is available. At Jodhpur [59], births occur round the year but the maximum births take place during the first quarter (January–March) and minimum during the last quarter (October–December).

The reproductive success of high-ranking females is found significantly higher than that of low-ranking female. The long-term study of Jodhpur [7, 48, 56], and Mt. Abu [39] monkeys shows that younger females are more fertile than older ones. The long-term data of Jodhpur langurs [8, 56, 64] and Berkeley colony [65] illustrate that females may survive for more than 30 years. Male life span in langurs are extremely variable, they may live for about 22–25 years [56].

Table 15.3 Comparative climatic and demographic data of langur study sites in India, Sri Lanka and Bangladesh

Study site	Mean annual rainfall	Months with >50 mm rain (N)	Mean annual temp.	Mean troop size (N)	Month with birth occur (N)	Mean adult female/troop	Birth/adult female/year	One male bisexual troops	Source (s)
Bhimtal	1,308	9	13.9	23.0	2	11.7	0.23	33	Boggess [43]
Dharwar	1,492	7	24.0	15.5	12	8.0	0.20	77	Sugiyama [18]
Gir Forest	674	4	26.8	30.4		13.6	0.31	40	Starin [40], Rahaman [37]
Jaipur	648	4	25.1	52.1	12	24.6	0.51	84	Mathur and Manohar [49]
Jodhpur	370	3	27.0	38.5	12	20.5	0.54	100	Mohnot [30], Mohnot *et al.* [7], Sommer and Rajpurohit [48], Rajpurohit and Sommer [54], Rajpurohit *et al.* [8]
Junbesi	2,500			12.0	5	3.4	0.11	33	Boggess [43], Curtin [38], Bishop [41]
Jossore				15.8		7.6		40	Khan [42]
Kanha	1,600	4	23.3	21.7	5	9.4	0.43	87	Newton [47], Kankane [44]
Kaukori	992	4	25.9	54.0	2	19.0	0.73	0	Jay [17]
Kumbhalgarh	725	3	26.0	38.8	5	17.3		100	Chhangani [53]
Melemchi	1,395	6	17.7	32.0	4	8.0	0.75	0	Bishop [41]
Mt. Abu	1,800	4		21.0	12	8.3	0.65	88	Hardy [39], Bishop [41]
Orcha	2,000			19.0	12	6.0	0.72	0	Jay [17]
Polonnaruwa	1,727	10	28.0	25.0	12	8.8	0.22	27	Muckenhim [36]
Rajaji	1,200		20.1	46.3	5	13.5		25	Laws and Laws [45]
Ramnagar	1,778	6	23.4	28.7	5	7.6	0.29	24	Borries [52]
Ranthambhore	900	4		45.7	8	21.7	0.32	67	More [46]
Raipur	1,388	5	26.9	29.1		15.7	0.35	57	Sugiyama [18]
Sariska	824	4	25.0	37.7		17.6	0.51	100	Ross and Srivastava [51]
Simla	1,800	7	13.6	29.1	3	14.8	0.51	42	Ross *et al.* [50]
Singur	1,582	6	27.0	12.0		5.0	0.50	100	Oppenheimer [31]
Shivpuri	1,209	4	25.0	21.2		13.9		90	Kankane [44]
Wilpattu	1,329	6	27.7	25.1	12	11.1	0.62	29	Muckenhim [36]

Fig. 15.3 Quadruplet birth in Hanuman Langur *Semnopithecus entellus* (considered as the first case in the world). *Courtesy: Dr. A.K. Chhangani*

Table 15.4 *Semnopithecus entellus*: age classification/categorization

Age class	Female	Male
Infant I	'Black coat: from birth to the age of about 5 months until completion of fur colour change to gray	
Infant II	'White coat: after completion of fur colour change to completion of weaning, i.e. from about 5 to about 15 months of age	
Juvenile	15 months to the onset of menarche, i.e. about 2.5 years of age	15 months to about 4 years testes internal in younger and descended in older juveniles, glans of penis generally not visible
Subadult	No subadult category in females	About 4–6 years of age; glans usually visible, ischial pads still undeveloped, canine teeth not yet fully erupted, capable of copulatory behaviour
Young adult	From regular cycling to birth of first infant; however, not older than 4 years	About 6–7/8 years of age; glans always visible, ischial pads small and usually pale, canine teeth fully erupted, teeth unworn, not yet full size
Adult	From birth to first infant, but at least from 4 years onwards	From 7 to 8 years onwards; full size; ischial pads pink, often puffy
Old adult	About 20 years onward	About 15/16 years onwards

Ecology and Behaviour

Food and Feeding: According to their diet components, non-human primates are commonly known as frugivorous, folivorous, insectivorous or omnivorous. Like other colobines, langurs are usually classified as folivorous, but study at Jodhpur suggests besides leaves, langurs utilize fruits, flowers, bark, gum, insect larvae and even sand [30, 57, 66]. In an extremely arid environment, Langurs utilize the maximum number (about 210) of plant species ever reported for colobine monkeys and the most preferred ones are: *Acacia senegal*, *Prosopis juliflora*, *P. cineraria*, *Ficus benghalensis*, *F. religiosa*, *Salvadora persica*, *Zizyphus numularia*, *Acacia nilotica* and *Anogeissus pendula*. Moreover, locals provide a variety of fruits, vegetables, baked wheat and millet cakes to the langurs. The natural staple food also varies considerably from group to group involving orchards, gardens and cultivated fields.

Langurs have sacculated stomach specialized for the digestion of leaves and other plant parts [67]. Their diet consists of about 65% fruits and 35% leaves and flowers buds. Like other members of Colobinae, the common langur exhibits a number of anatomical and behavioural adaptations associated with leaf-eating [68] which include dental and gut specializations (e.g. high pH, microbial flora, etc.). Langurs at Gir forest consume 36–41 species [37, 40], at Singur 68 species [31], at Dharwar 37 plant species [69] and at Polonnarnwa 43–47 species [63, 70] of plants. Mohnot [30] reported 84 plant species consumed by Jodhpur Langurs (Table 15.3). The Hanuman langur eats repulsive and foul-smelling latex-bearing plants also, such as the Aak (*Calotropis procera*) which is avoided by the most of animals [30] and *C. gigantean* in Sri Lanka [63]. Langurs also lick stones, hard grounds and soil from termite mounds regularly, or even break the pieces of hard soil, to fulfill the requirement of salts and minerals like calcium, magnesium and trace elements. Ripley [63] reported that mortar from old buildings and soil from termite mounds are eaten by langurs in Sri Lanka. Roonwal and Mohnot [1] found Langurs eating soil, ashes and bone pieces from cremation grounds at Jodhpur. Srivastava [57] provided details analysis of soil licks (geophagy) at Jodhpur which strongly supports the view that the Langurs are selective in their choice of soils. There are many reports revealing plant exudates as an important category of the natural diet of non-human primates. Gum may have more importance as a mineral source than as a source of carbohydrate. Langurs around Jodhpur exploit five plant species for gum [30, 57] as compared to ten in Kanha National Park [47].

Insects are not generally considered to be an important component of langur diet [71]. Yoshiba [72] has noted langurs eating caterpillars and insects galls from the leaves of *Terminolia temendora* at Dharwar. Rahman [37] observed a case of langur eating bird's egg while surveying the langurs of the Gir forest. Moore [46] reported five episodes of insect predation at Mt. Abu. Newton [47] observed langurs devoting 2.8% of feeding time to insectivory at Kanha. Srivastava [57] reported that insects may form a part of their diet. Earlier studies around Jodhpur found langurs being vegetarian [30, 66]. The common langur drinks water regularly from pools, streams, lakes, hollow tree trunks, directly from overhead tanks in the human habitations while living near villages/towns/temples [30, 56]. Langurs feed actively in the morning and evening. Temple groups are usually bolder and not afraid of human beings, showing greater scattering of group members within the home range than normally found in the wild groups [73].

Social Organization: The reproductive units are bisexual, many females (*Harem*) with a single adult breeding male. Multimale troops are very rare but sometimes, a multi-male situation emerges. With a few exceptions, females remain for the life in their native troop. Males emigrate usually as juveniles to unisexual all-male bands whose home range can be as large as 20 km while the bisexual troop occupies the home range of about 0.5–1.5 km^2 [56]. The Langur population at Jodhpur has been recorded having 97.8% single births and 2.2% twin births, but quadruplets (in February 2000) and triplet birth (in 2006) are also reported [61, 74]. Females in the late pregnancy are sluggish, do not participate in group activities, prefer isolation and spend most of the time sitting peacefully. The process of parturition is simple and the location is never fixed. Generally, it takes place in the morning hours. Mother licks the black fur of the infant, along with occasional inspection, watching and hugging. All the while, the infant is kept gently within the folds of her arms and legs and close to the ventrum. The mild screams by the new born baby are followed by slow movements of its head and limbs. The screaming stops when it sucks the nipple.

Clinging: A newborn cannot cling tightly to its mother's belly (ventrum) during the first 2–3 days and needs frequent support from the mother [30]. Jay [17], however, believes that a newborn can cling tightly to the mother's ventrum within a few hours after birth. Clinging is more frequent in black-coat infants, less so among white-coat infants and is absent among the weaned juveniles.

Lactation and Weaning: Lactation lasts for 10–12 months, or a little longer, the period varying from individual to individual. Feeding on the mother's breast is frequent among the black-coat infants up to about 4 months of age and then decreases gradually. However, they may frequently resort to a nipple-hold without actual suckling. Occasionally, yearlings hold the nipple side by side with the feeding newborns (Table 15.4). Thus, clinging and nipple-hold are important devices by which young infants are fed and protected and the older infants are relieved of fear and stress [20, 30, 56] Weaning occurs when the infants are about 10–12 months old. From the beginning to the end, it is a process of stress for the infant. The mother during this period becomes too harsh and develops great differences towards her infant. By this time, the infants are completely weaned, fully independent and enter into the juvenile phase. It is observed that male infants are weaned more quickly than females [20, 56, 75].

Infant Transfer: In the first few weeks, the exclusive rearing and responsibility of infants by their mothers is greatly relaxed by the process of infant transfer. In this process, mothers allow other group members to handle the young ones. The handler/allomother is very affectionate at an initial stage but slowly or soon becomes indifferent and, as a result, the infant starts screaming till it is passed on to another handler or is retrieved by the mother. The infant in contact with several group members learns the technique of growing up. In addition to such short-term foster care, longer foster mothering was also observed [30]. In such case, an infant was raised by non-mother after the death of its own mother. Adult male, however, lacks this behaviour except that it protects the group as a whole at the time of danger—a feature common in all the habitats studied by various authors [56, 58, 61].

Maternal behaviour towards dead infants: Adult females continue to show maternal behaviour towards their dead infants for several days. Around Jodhpur, females other than the mother, were not found to take any interest in dead infants. Rajpurohit [56, 76] revealed that the older females carry the corpses of their infants for longer period than the younger ones. As proposed by Zuckerman [77] and Mohnot [30] the langurs do not recognize the phenomenon of death.

Social Communication

Grooming: It is an important process of tactile communication in which frequent interactions occur either between two or more individuals of the same or of different age and sex. Normally, black-coat and white-coat infants of both sexes never groom one another. The most frequent grooming occurs between adult females. Sugiyama [18] and Mohnot [30] observed a high frequency of grooming among females above 4 years of age and in mother grooming their infants. They further stated that grooming is only a circumstantial evidence and not a conclusive proof of the relationship between two individuals. In bisexual groups, the dominant male is often groomed by one to four females at a time. It, in turn, never observed grooming any one, not even the oestrous females. Self-grooming occurs for short duration, while mutual grooming is rare and occurs only for very short duration and ultimately turns into reciprocal grooming. Developmentally, grooming appears to be a substitute for nursing and acts as a pacifier.

Auditory Communication: One of the common forms of auditory communications is the resounding whoop associated with a preparation for movements, finalization of sitting arrangements, etc. It is given at any time in the day but usually in the mornings, evenings and when different groups spot one another. There are several other vocalizations, which the Langurs make. Many of them were heard around Jodhpur repeatedly. Mohnot [30] has concluded that eight Langurs have an extensive repertoire of vocalizations consisting of at least 15–17 kinds [78]. Ripley [79] has obtained a full set of recorded vocalizations for the Ceylon subspecies, *Presbytis entellus thersites*.

Play: A 2-week old infant starts showing signs of playful tendency which becomes pronounced and spectacular as the infant grows. A variety of play patterns have been observed among infants, juveniles and subadults. It appears that the contacts of infants and juveniles with adult members of the group, and among themselves, ensure smooth and rapid socialization of the young ones. The playful tendencies continue to exist even in adulthood and the mothers occasionally play with infants. Play behaviour is also an important way of acquiring skills [17, 18, 63].

Sub-grouping: The formation of temporary subgrouping by juveniles or by juveniles and subadults is probably influenced by some factors, such as, lack of interest in adult activity or out of fear, or both. Among adult males, temporary separation is probably induced by sexual urge, demand for food and antagonism to avoid direct competition [30]. Separate subgroup formation by an ousted resident and juvenile males (presumed sons/kin) has been reported by Rajpurohit [80].

Origin of All-Male Bands: The weaned male infant (now termed as male juvenile) joins the all-male group nearby when repeatedly forced by the new resident/dominant male of their group to desert. Due to close mother–infant ties, male juveniles resist the threats of the dominant male in the initial stages. However, repeated threats, chasing, air bites and injury caused by the dominant male frighten them. They avoid the leader and start living on the periphery of the group independently [30, 56, 81]. After leading about 4–6 weeks of very stressful life, these juveniles join the neighbouring all-male bands. All-male bands on their part receive the juveniles without any antagonism. Within a short period of a few weeks, these juveniles are fully integrated in the all-male band and become its permanent constituents [81].

Dominance Structure: There is a marked dominance hierarchy in all-male bands and particularly among adult males [56, 82] which becomes more pronounced in specific situations. Most of the dominance interactions are non-violent. Among adult males, a functional rank order is maintained on the basis of several dominance interactions like procurement of food, leading the group, production of whoops and warning signals, invading bisexual groups, getting oestrous females, etc [30, 56, 57]. However, according to Sugiyama [18], all-male groups in Dharwar (South India) are less rigid and poorly organized and are without a functional rank order. Such ranking order can be noticed in bisexual troops [57, 66].

Infanticide in Langurs

Social changes are common in all types of groups of Hanuman Langurs. The important social changes are: the replacement of the resident male of a unimale bisexual troop, division of a unimale bisexual troop into two unimale bisexual troops and the sudden death or disappearances of several members of a group. Some changes of a permanent nature like births, deaths and disappearance in bisexual troops and death and disappearance in all-male bands do alter the social structure of a group.

Replacement of Resident Males: A large number of cases of leadership changes are observed in bisexual groups around Jodhpur in the last 40 years [21, 30, 56–58, 60, 61, 75, 82, 83]. Of these, in more than 40% cases of resident replacement, neither infant killing nor any apparent evidence to suspect the killings was found. Of the remaining 60% cases, either direct evidence of infant killing or suspected infant killings and associated behaviour of the aggressive males were found. The first comprehensive details of infant killing were provided by Mohnot, [84] Hrdy [85], Sugiyama [18], Makwana [83] and Agaramoorthy and Mohnot [58]. Recently, Rajpurohit *et al.* [86] have reported a sudden resident replacement and infanticide in Jodhpur langurs.

The infanticide in Hanuman Langurs seems to be a common phenomenon and is not inevitable as found in field observations in the last four decades. In more than half (i.e. ca 60%) of the cases of resident replacement infant killing occurs, while no killings occur in the remaining cases. Infanticide has become the basis of new concept of sociobiology in non-human primates. Infanticide in langurs may be considered as a male reproductive strategy exhibited by male potentiality, supremacy or terror strategy to rival males and to the troop members [86]. But the issue remained yet unresolved. More information of the significance of this spectacular phenomenon can be obtained by studying the blood profile and genetic makeup via DNA fingerprinting.

Rhesus Monkey—*Macaca mulatta* (Zimmermann [87])

Systematic Position
Class: Mammalia (Mammals)
Subclass: Eutheria (Placental mammals)
Order: Primates
Suborder: Anthropoidea (Simians)
Infra-order: Catarrhini (Old World monkeys)
Superfamily: Cercopithecoidea
Family: Cercopithecidae
Subfamily: Cercopithecinae (Macaques)
Genus: *Macaca*
Species: *mulatta*
Subspecies: *mulatta* (and 3 others)
Local names: Hindi: Bandar, Kashmiri: Punj, Ponj, Wander, Bengali: Market, Burmese: Myouk sat
Scientific synonyms: *Cercopithecus mulatta*, *Macacus rhesus*
Taxonomic note and Distribution: *Cercopithecus mulatta* [87], *Macacus rhesus* [14], *Macaca mulatta* [23, 25, 33, 88]
Four subspecies are recognized and found in the South Asia and South-east Asia and the south of the Godavari River in India. Prakash [29] provided distribution details in Rajasthan. This species does not occur in and around Jodhpur; however, three to four animals were seen.
M. m. mulatta. Nepal; Bhutan; India (Assam and Northern and Peninsular India); Thailand, Laos; Cambodia; Vietnam; Southern China.
M.m. mcmohoni. Northeastern Afghanistan; Pakistan, about 1,100 m altitude.
M.m. vertita. Tibet
M.m. villosa. Northern India (Southern Kashmir, Upper Panjab and Kumaun Hills).
Fooden [89] stated that *M. fascicularis* is a subspecies of *M. mulatta*, but Hill [11] has suggested that it is an independent species.

Fig. 15.4 A Rhesus Monkey *Macaca mulatta* female with young one. *Courtesy: Goutam Sharma*

Morphology

The Rhesus Monkey *Macaca mulatta* is found in the North and the Central India. However, its distribution in North India is discontinuous [62]. The Aravalli Mountain Range is the eastern limit while they are not found in the Thar Desert.

Rhesus Monkey is medium sized animal with a rather short tail. Crown hair grow back from the brows, without a part, whorl or cap. The face is light pink, flesh coloured or reddish. The upper back is olive, the rump and base of the tail is orange red. The hind quarters are bright red in adults and flesh-coloured to light red in juveniles (Fig. 15.4).

Head and body length of males is 48.5–63.5 cm and of females 47.0–53.1 cm. Tail length of males is 20.3–30.5 cm and of females, 19.0–28.5 cm. The body weight of males is 5.6–10.9 kg and of females is 3.0–10.7 kg. Generally, males are heavier than females: 100:69 [25, 90]. According to Krishnan [62], forest-dwelling animals are generally smaller.

Ecology, Sociobiology and Behaviour

The Rhesus Monkey, the most common monkey in the South Asia, has played an important role in the ecology, culture and traditions of India. The close interaction of these macaques with the people of India forms perhaps the most intense relationship between human and non-human primates anywhere in the world. This species can be easily tamed and taught various tricks, especially when young, but is never fully domesticated. The older individuals get rather vicious and have a tendency to bite. It is the common performing monkey of northern India and monkey-man is very popular among children. It is not regarded as sacred by the Hindus, in the sense in which Hanuman Langur (*Semnopithecus entellus*) is, but it is tolerated. Because of easy handling under laboratory conditions, the *Macaca mulatta* is also of great biomedical importance as its disease spectrum is very similar to that of human beings.

Macaca mulatta lives in a wide variety of habitats including cities, villages, farms, forests and mountains. It is also found in a semiarid environment in Rajasthan and in mangrove swamps in the Sundarbans (Bengal). In several towns in Uttar Pradesh, such as Ayodhya, and north-east towns of Rajasthan such as Jaipur, Alwar and Bharatpur. Rhesus Monkey is a permanent resident of railway platforms, often snatching food from unaware passengers. In the sub-Himalayas (in Uttrakhand, Himachal Pradesh and northern Uttar Pradesh), it is found at various altitudes, from about 500 m to about 1,500 m. In northeastern (Assam), it inhabits the margins of forests but does not enter dense forests [18].

The Rhesus Monkey is diurnal and less arboreal than several other macaque species. It attacks cultivated fields, orchards and gardens. Both young and old ones are good swimmers. They are largely vegetarian and their diet includes leaves, flowers, fruits, berries and seeds of many species of plants, grass and grains and algae from ponds. It also eats insects and spiders. It is not known to eat small birds, lizards, or similar small animals, but in the Sundarbans it eats crabs. It frequently eats soil (from termite mounds) in small quantities [14, 62, 91–93], and mushrooms during mid-monsoon (July and August) [94]. The leaves and stems of some climbers and shrubs are only consumed when other food is not obtainable. Mukherjee [95] noted that stagnant water from roadside ditches or small ponds is drunk by them two or three times a day.

Southwick, Beg and Siddiqi [96] observed the daily routine of a temple population in Aligarh in northeastern India. Each group has a characteristic night-resting location, rooftops or trees in the temple courtyards. The movement begins early in the morning. They slowly and randomly move out of their roosting clusters within 15–20 min, parts of groups would begin definite progression or directed movements and by about 8.00, the groups are in the position of feeding themselves either on the trees or the food given by temple devotee/local people. By about 9.30, adults assume resting and grooming positions, while the infants and juveniles play. Groups sometimes remain there for the whole day. In the early afternoon, from 4.30 to 6.30, groups return to their roosting/lodging places. By the sunset, most individuals are in their typical lodging positions. After a certain amount of aggressive interactions between individuals within a group, they settle down in their final sleeping clusters.

Group Home Range, Size and Composition

Each group has a home range varying from 1 to 16 km^2. The groups are generally agonistic to each other; subordinate groups usually avoid the dominant ones but intergroup fighting occasionally occurs. Forest groups normally spend the night on the foraging trees where they feed themselves at the end of the day. Most groups travel from one part of the range to another during foraging. During these movements the leading animal may be an adult female, an adult male or sometimes even a subadult male, positioned in the last.

The size of bisexual group may vary from small to large (of about 8- 180 individuals). Most of the groups are multimale; unimale groups are relatively uncommon. Occasionally, solitary males are found. The all-male groups appear to be rare [1]. Prakash [97] observed that groups consisted of about 50–100 individuals. A group of 100 consisted of 40 adults and 60 subadults, juveniles and infants. The group may split into a number of groups after reaching a critical maximum size, each headed by a leader and sometimes by a number of subleaders.

So far as is known, sex ratio at birth is equal. In adults the prevalence of females over males is apparently due to higher mortality rates of juvenile males, earlier adult maturation of females, possibly greater longevity of females and the solitary behaviour of some males [96]. The males ranking second and third in a group are extremely peripheral, often separated from the rest of the group by a distance of as much as 200 m, after the alpha male shifted to another group, these two males moved much closer to the group centre and maintained this position even after the alpha rank is filled [98].

Reproduction and Reproductive Behaviour

The breeding in wild populations of Rhesus Monkey is seasonal. According to Prakash, [29, 97] births occur in the summer and autumn in Rajasthan,. In northern India, Southwick *et al.* [96] and Lancaster and Lee [99] recorded births from March to June, with a few in the September. Lindburg [98] noted that mating begins in September, reaching its peak during the month of November, and is also common in the first half of December. Births mostly occurred in April and May, with a few in March. Krishnan [62] reported that breeding in southern India was not limited to a particular season. About two months before the onset of the mating season, males show an increase in redness of the sex skin; the colour is brightest at the peak of the mating season. On the other hand, when females reach adolescence, the skin of the perineum, tail, thighs, abdomen and back looks greatly swollen, turgid and convoluted, and pink or red coloured. The swelling gradually disappears in older females. In most females, mating is continued to several successive days, followed by longer periods of no mating.

The mating behaviour in this species has been studied by several researchers. Mounting and presenting occur in connection with copulation. Mounting between adult males is almost absent and adult females rarely mount other individuals [98]. Presenting occurs most frequently when an animal signals its intention to mount by grasping the body of the intended mountee. Copulation may proceed by little or no display. Typically, 5–25 mounts occur in 10–20 min, each lasting three to four minutes and consisting of several thrusts [96]. During mounts leading to ejaculation, the male frequently bare its teeth and issue a high-pitched vocalization. Consort pairs are seen in all months except March. In general, peak mating activity occurs in October. The consort period lasts from a few hours to a few days. Dominant males form consort relationship more readily than subordinate males. Lindburg [98] reported that during the period of receptivity, a female entered into a consort association with one or several successive males. The consort pair is characterized by long sessions of reciprocal grooming, frequent copulations and considerable restlessness and excitability. Consort pairs are found well outside the group area, sometimes as far away as several hundred metres; this separation reduced interference during mating. A male may start following a receptive female, or a female may approach to a male at the group periphery and stay closer to it. Male usually responds to the female's approaches. The highest ranking females mate primarily with the dominant males, but they also mate with other males of the group. This situation may be modified, depending on the availability of females. Thus, if only one receptive female is available and she is of low rank, the dominant male may mate with her.

In Rhesus Monkey, as a rule, only one young one is born per year. Twins are rare, only about one in 700–800 consists of twins. Young males reach sexual maturity at three and a half years, while young females at two and a half years. Most females begin breeding at the age of three years and the mating activity reaches at the peak at seven years of age [100–103]. The rank of a female in the dominance hierarchy has no effect on mating activity. The oestrous cycles lasted about 28 days. On an average three males associate with a female during oestrus. The male sexual activity is generally correlated with dominance rank but not with age. The highest ranking male is the most sexually active and the only male to form many exclusive consort relationships with individual females. Lower ranking males are progressively less active as per their rank.

Copulations begin soon after the females come into oestrus and solicit the attention of males. Parous females without infant tend to mate and give birth earlier than the other. The dominance status of the males is related to the age of the females with which they mated, the more dominant males tending to mate with older females. There is also a direct correlation between the ages of the mating partners, the age of the female partner increasing with the age of the male. Oestrus, fertile mating and births are seasonally synchronized [104].

Interactions Within the Group

Interactions within a group have been examined by Southwick *et al.* [96], and they suggest the following:

Male–male: The relationship between adult males varies from peaceful and even cooperative to highly agonistic. Fairly sharp dominance hierarchies exist, as revealed in natural and forced encounters (e.g. when some artificial food is placed at equal distances between two adult males).

Male–female: In this category, the most important relationship is the sexual consort relationship. Other types of male–female interactions include grooming (usually males groomed by adult females). Occasionally males attack females, particularly in the early stages of oestrus prior to the formation of consort relationships.

Male–infant and male–juvenile: Such relationships, sometimes termed paternal, and are generally neutral or agonistic. Adult males are often seen being hostile towards infants and juveniles at the feeding time.

Juveniles often groom adult males; the reverse situation occurs only rarely. Mukherjee [95] observed a dominant male not only allowing infant to play with and climb on his back but also retrieved them when they are much troubled by juveniles.

Female–female: Adult females commonly feed, rest and groom in close association with one another, certain females showing closer bonds than others. This relationship is generally peaceful. Pattern of spatial distribution and feeding priorities suggest the existence of some dominance hierarchies among females. Agonistic and aggressive behaviour often occur just before and just after oestrous peaks, when sexual excitement is highest.

Female–infant: The aunt–infant category refers to the relationship of females other than the mother to infants. Lindburg [98] noted that large juveniles and adult females often exhibit an attraction to a young infant and interact with it in many ways. They may try to touch it, pull its extremities or merely sit nearby and watch it. The mother gently tries to move away the infant. The mother–infant relationship is the most intimate and long-lasting of all pair relationship.

Mother–juvenile: This relationship includes the weaning process. Weaning, though gradual, is disruptive, it is completed when the next infant is born. The juvenile, however, retains some association with the mother, thus, forming a triple relationship (mother–new infant–juvenile).

Infant–infant: Infants play, chase, jump, wrestle, mount and explore together. Such play-groups usually consist of two or three individuals. This relationship is quite positive and increases in intensity with time.

Juvenile–juvenile: It is more intense and positive than the infant–infant relationship and includes play, exploration, feeding and grooming.

Juvenile–infant: Juveniles behave as an adult like roles towards younger infants and sometimes give maternal care to motherless infants.

In *Macaca mulatta*, weaning begins in the third or fourth month. Its earliest manifestation is a simple termination of nipple contact, the mother pushing the infant away. By this time, the infant has learned to supplement the mother's milk extensively with solid food. By five months, the infant commonly undergoes weaning tantrums, while the mother cuffs, bites or simply avoids the infant. Early periods of resistance by the mother to the infant approaches are generally short (i.e. in minutes) but by 10–12 months, lasting about an hour and simultaneously, the nursing periods become progressively shorter. The weaning process continues until shortly, before the birth of the next infant. Trivers [105] discussed the theoretical aspects of weaning as a parent–offspring conflict.

Prakash [97] observed a female carrying dead infant and stated that it was difficult to dislodge it from the mother. Mukherjee [95] also recorded that a female rhesus macaque carried a dead infant for several days, and when she finally abandoned it, it was only a dried up body. Such observations demonstrate the strength and persistence of the maternal-infant bond.

Social communication is achieved by the usual means, including vocalization, body postures, and expressions. When alarmed, adults and juveniles of either sex give a shrill warning bark; "coo-coo" is produced by some individuals while feeding. This call seems to be understood by the spotted deer which respond by coming to eat the leaves and fruit dropped by the macaques. "Kech-kech" is a shrill repetitive cry when attacked or threatened. An angry macaque threatening others in the group uses the sound "hough-hough" [91]. Adult females near an observer sometimes give a series of staccato vocalizations as long as they remain in sight of the intruder. Lindburg [98] recorded specific vocalizations, including alarm calls (shrill bark), call of aggression (pant-threat series of rapid barks), responses to threat or aggression (growl, bark, screech, and scream), sounds of surprise or defeat (squeak), vocalization to maintain contact or locate lost animals, calls male by infants (gecker, girming) and calls made by females to infants of other females (chortle). Branch-shaking displays were staged during intergroup encounters, ranging in intensity from very minor bouncing on tree to rapid climbing to treetops and vigorous shaking of branches. Most displays are indulged by adult males, fewer by females, and still fewer by juveniles. A pucker face—characterized by protruded lips, lowered eyelids, slightly raised tail and lip smacking—is a reaction of a subadult or a large juvenile male to intense interactions with high-ranking adults. An adult male may stalk a sexually receptive female in a still-legged gait with a pucker face and lip smacking and the female is generally frightened and flees. A mating female swings an arm back towards the male between mounts in a copulation series.

In *Macaca mulatta*, grooming occurs in any part of the day but generally increases towards midday when the macaques are at rest. The greatest amount of grooming is done by adult females, which groom other females and their own offspring. They more commonly groom females close to them in social rank. Grooming among adult males is rare; leader males groom only adult females. As a rule, adult males receive far more grooming than they give. Juveniles participate in more grooming activities with their mothers and siblings than with other group members [98]. Grooming seems to have a relaxing effect on the participants and in addition, serves a variety of other functions. Individuals solicit grooming and also groom without being solicited. Following a fight or any conflict, grooming sometimes eases an aggressor. Grooming is affected by sexual activity. At midcycle the extent the female grooms a male is at the minimum, and the extent the male grooms a females is at the maximum. Fluctuations in grooming also depend upon change in social status [106].

Nuisance to People

Basically, the problem of monkey nuisance lies in their attempts to procure food and space in human habitats. This, in turn, is a fall out of destruction of forests, their natural habitat. With shrinking forests, changed microhabitat, decreased availability of food and water, and decreased human tolerance to increasing number of monkeys. There is not only a conflict between humans and monkeys, but also a mutual hostility. Attracted by food, water and cover, monkey groups invade croplands, settlements and often destroy property, gardens, household furnishing and parked cars. They are over-abundant in temples where devotees feed them out of religious sentiments, in hospital premises and schools. In general, rhesus macaques are not shy as langurs. Angered and irate humans resort to hitting monkeys with stones because of their destructive activities. In response, monkeys become over-aggressive. They threaten people with snarls, snatch food boxes, spectacles and handbags and very frequently bite human beings. Although there are no country-wide systematic surveys on people bitten and harassed by monkeys, the instances of monkey bites are increasing in recent times particularly in big cities like New Delhi, Agra, Jaipur, etc., [107]. Roughly around 100 people are injured by monkey bites every day in the country. The increased number of commensal monkeys not only threatens the welfare of monkeys but also poses a greater problem to public health. The studies have indicated that both langurs and Rhesus Monkeys are naturally infected by simian retroviruses, SRV-6 [108, 109]. We have data that demonstrate unrecognized lentiviral infection of wild Rhesus Monkeys and langurs with implications for public health since bidirectional transmission of pathogens and zoonotic infections pose health risks for both simians and humans [110, 111].

Recommendations

The management of wild primate population and those living as commensals in villages, towns and cities need separate management plan. The management plan will need data on habitat viability, demography and pattern of resource utilization. This is a matter of serious concern that wild primate populations especially the macaque species are relinquishing their permanent forest abodes and are migrating towards human habitations. It seems these urban monkeys, which are often diseased are surviving at the cost of human and livestock health with certain degree of conflicts and competition since these populations cannot be supported in urban areas because of limited carrying capacity of urban habitats. Our suggestions and recommendations are as follows:

1. Vulnerable areas with monkey-menace need a thorough assessment of habitat, vegetation and available biomass in the area with the help of satellite images.
2. Assessment of the forested areas in terms of ground information of primate food, vegetation, biomass and pre-predator relationship to came out with the carrying capacity of that area as well as satellite information for type of forest and ground cover assessment.
3. Exact species wise status of primates, habitat wise and ecology wise in the state is very urgent. This will help to know the monkey population and assess the exact horticulture, agriculture and property damage, which is not available till date scientifically.
4. To enforce complete ban on all activities leading to habitat degradation in the potential non-human primate habitats.

5. The necessary guidelines to be incorporated in the working plan for the areas rich in primate habitat and populations to ensure the correspondence between developmental activities and primate species conservation.
6. To manage man-monkey conflicts, cassation of feeding by people, translocation and fertility control are suggested. Please see Chap. 2 for pictures.

Acknowledgements We thank Dr. S.M. Mohnot, Former Head, Department of Zoology, J.N.V. University and Chairman, Primate Research Centre, Jodhpur for his inputs and guidance. Bulk of the information presented here was collected under a cooperative research programme of University Grants Commission (UGC). Thanks are due to Prof. G.R. Jakher, Vice – Chancellor, and Prof. M.M. Saxena, Head, Department of Environmental Science, Maharaja Ganga Singh University, Bikaner and Prof. D. Mohan, Former Head, Department of Zoology, J.N.V. University, Jodhpur for the logistic support.

References

1. Roonwal L, Mohnot SM (1977) Primates of south Asia: Ecology, sociobiology and Behaviour. Mass Harvard Univ. Press, Cambridge, p 421
2. Fleage JG, Read KE (1999) Phylogenetic and temporal perspectives on primate ecology. In: Fleagle JG, Janson C, Reed KE (eds) Primate Communities. Cambridge Univ. Press, Cambridge, UK, pp 92–115
3. Mivart ST (1973) On *lepilemur* and *cheirogaleus* and on the zoological rank of the lemuroidea. Proc Zool Soc Lond 2:484–510
4. Groves C (2001) Primate Taxonomy. Smithsonian Inst, Press, USA, pp 350
5. Gupta AK (2001) Non-human primates. Envis, Wildlife and protected area. Wildlife institute of India, Dehra Dun, India, pp 1–25
6. Emmons LH (1999) Of mice and monkeys: primates as predators of mammal community richness. In: Janson G, Reed KE, Fleagle JG (eds) Primate Communities. Cambridge Univ. Press, Cambridge, UK, pp 171–188
7. Mohnot SM, Gadgil M, Makawana SC (1981) The dynamics of the Hanuman langurs population of Jodhpur, Rajasthan, India. Primates 22:182–191
8. Rajpurohit LS, Chhangani AK, Mohnot SM (2006) Population dynamics of Hanuman langur, *Semnopithecus entellus* around Jodhpur (India) during 1995–2001. Proc Nat Acad Sci B76(2):141–147
9. Immam I, Malik I (2006) Rhesus monkey, *Macaca mulatta*, problem in India and their management. In: Sridhara S (ed) Vertebrate Pests in Agriculture- The Indian Scenario. Scientific Publisher, Jodhpur, India, pp 453–476
10. Gray JE. London Med Repos. 1821; 15:297
11. Hill WC (1972) Transfer effects of cue-related movement reversal in discrimination reversal learning with rhesus monkeys. J Comp Physiol Psycho 70:184–189
12. Groves CP (1995) Order Primates. In: Wilson DE, Reader DM (eds) Mammal species of the world, a taxonomic and geographic reference, 2nd edn. Smithsonian Inst Press, Washington, DC, pp 243–277
13. Dufresne P (1797) Sur une nouvelle espece de singe, par c. Dufresne (Description d'une nouvelle espece de guenon, sous le nom d'entelle.). Bulletin de Societes d' Philomathique (Paris) 1(7):49
14. Blanford WT (1888–1891) The fauna of British India including Burma and Ceylon: Mammalia. Taylor and Francis: London, pp 617
15. Desmarest AG (1822) Mammalogie, ou description des espèces de mammifères. Part 2 and suppl. Vve Agasse, Paris

16. Szalay FS, Delson E (1979) Evolutionary History of the Primates. Academic, New York, pp 580
17. Jay PC (1965) The common langurs of North India. Ch. 7. In: Holt R (ed) Primate Behaviour. Winston, New York, pp 32–119
18. Sugiyama Y (1965) On the social change of Hanuman langurs (*Presbytis entellus*) in their natural conditions. Primates 6:381–418
19. McCann C (1933) Observations on some of the Indian langurs. J Bomb Nat His Soc 35:616–628
20. Rajpurohit LS, Mohnot SM (1991) Process of weaning in Hanuman langurs *Presbytis entellus* around Jodhpur. Primates 32:213–218
21. Sommer V (1985) Weibliche und mannliche reprodubetionsstrategien der Hanuman Languren (*Presbytis entellus*) von Jodhpur, Rajasthan/India. Ph.D. Dissertation, Georg-August Univ., Goettingen
22. Phillips WWA (1935) Manual of the Mammals of Ceylon. Ceylon J Sci. Dulau & Co., London, pp 30–36
23. Pocock RI (1939) Mammalia Fauna of British India Series, Primates, Carnivora (in part), vol 1, 2nd edn. Taylor and Francis, London, pp 71
24. Oboussier H, Maydell GA V (1960) Zur Kenntnis von *presbutis entellus* (Dufresne, 1797). Ergebnisse der Deutchen Indien-Expedition 1955–57. Leitung G.A. Frhrv Maydell Zoologische Anzeiger (Leipzig) 164:141–154
25. Napier JR, Napier PH (eds) (1967) A Hand book of living primates: morphology, ecology and behaviour of non-human primates. Academic, New York, pp 405
26. Relay KV (1913) Bombay Natural History Society's Mammalian Survey of India. Report No. 9. J Bomb Nat. Hist Soc 22:283–295
27. Dodshworth PTL (1914) Notes on some mammals found in the Shimla district. The Shimla hills estate and Kalka and adjacent country. J Bomb Nat Hist Soc 22:726–749
28. Hingston RWG (1920) A Naturalist in Himalaya. H. F. & G, Witherby, London
29. Prakash I (1960) Breeding of mammals in the Rajasthan desert, India. J Mammalo 41:386–389
30. Mohnot SM.: Ecology and behavior of the common Indian langurs, *Presbytis entellus*. unpublished Ph.D. Thesis, University of Jodhpur, Jodhpur, India (1974)
31. Oppenheimer JR (1977) *Presbytis entellus*, the Hanuman langur. In: Rainier HSH, Brourne GH (eds) Primate Conservation. Academic, New York, pp 469–512
32. Vogal C (1977) Ecology and sociology of *Presbytis entellus*. In: Prasad MRN, Anand Kumar TC (eds) Use of Non-human Primates in Biomedical Research. Indian National Science Academy, New Delhi, pp 24–45
33. Ellerman JR, Morrison Scott TCS (1951) Checklist of pale arctic and Indian Mammals, 1758 to 1946. British Museum, London
34. Roberts TJ (ed) (1977) The Mammals of Pakistan. Ernest Benn Ltd, London & Tonbridge, pp 87–89
35. Jolly A (1985) The Evolution of Primate Behaviour, 2nd edn. Macmillan, New York
36. Muckenhirn NA.: Leaf-eaters and their predators in Ceylon: ecological role of gray langurs, *Presbytis entellus* and Leopards. Ph.D. thesis, University of Maryland (1972)
37. Rahaman H (1973) The langurs of Gir Sanctuary (Gujarat). A Preliminary Survey. J Bomb Nat Hist Soc 70:294–314
38. Curtin RA.: The Socio-ecology of the common langur (*Presbytis entellus*) in the Nepal Himalaya, Unpublished Ph.D. thesis, University of California, Berkeley, (1975)
39. Hardy SB (1977) The langurs of Abu. Harvard Univeristy Press, Cambridge
40. Starin ED.: A Preliminary Study of the Gir Forest Langur, B.A. Thesis, Friends World College, Huntington, New York (1973)
41. Bishop H (1979) Himalayan langurs: Temperate Colobines. J Human Evol 8:251–281
42. Khan MAR (1984) Ecology and Conservation of Common Langur (*Presbytis entellus*) in Bangladesh. In: Roonwal ML, Mohnot SM, Rathore NS (eds.), Current Primate Researches Jodhpur: India, pp 33–31

43. Boggess J (1980) Intermale relations and troop male membership changes in langurs (*Presbytis entellus*) in Nepal. Int J Primatol 1:233–274
44. Kankane PL (1984) Studies on the Hanuman langur, *P. entellus* at the Madhav National Park, Shivpuri, (Madhya Pradesh, India). In: Roonwal ML, Mohnot SM, Rathore NS (eds) Current Primate Researches. Jodhpur University, Jodhpur, India, pp 23–31
45. Laws JW, Laws J (1984) Social interaction among adult male langurs (*Presbytis entellus*) at Rajaji Wildlife Sanctuary. Int J Primatol 5:31–50
46. Moore J (1985) Insectivory by grey langurs [J]. J Bomb Nat Hist Soc 82(1):38–44
47. Newton PN (1985) The ecology and social organization of Hanuman langurs (*Presbytis entellus* Dufresne, 1797) in Kanha Tiger Reserve. Central Indian Highlands. Primat Eye 26:24
48. Sommer V, Rajpurohit LS (1989) Male reproductive success in harem troops in Hanuman langurs (*Presbytis entellus*). Int J Primatol 10:293–317
49. Mathur R, Lobo A (1988) Density estimates of monkeys of Jaipur, India. Primate Rep 19:35–42
50. Ross C, Srivastava A, Pirta RS (1993) Human influences on the population density of Hanuman langurs (*Presbytis entellus*) and rhesus macaques (*Macaca mulatta*) in Shimla, India. Biolog Conserv 65(2):159–163
51. Ross C, Srivastava A (1994) Factors influencing the population density of the Hanuman langur (*Presbytis entellus*) in Sariska Tiger Reserve. Primates 35(3):361–367
52. Borries C (1997) Infanticide in seasonally breeding multimale groups of Hanuman langurs (*Presbytis entellus*) in Ramnagar (South Nepal). Behav Ecol Sociobiol 41:139–150
53. Chhangani AK.: Ecobehavioural diversity of langurs *Presbytis entellus* living in different ecosystems. Ph.D. Thesis, JNV Univ., Jodhpur, Jodhpur, India (2000)
54. Rajpurohit LS, Sommer V (1991) Sex differences in mortality among langurs (*Presbytis entellus*) of Jodhpur, Rajasthan. Folia Primatol 56:17–27
55. Winkler P, Loch H, Vogel C (1984) Life history of Hanuman langurs (*Presbytis entellus*). Reproductive parameters, infant mortality and troop development. Folia Primatol 43:1–23
56. Rajpurohit LS.: Male Social Organization in Hanuman langur *Presbytis entellus*. Ph.D. Thesis, Univ. of Jodhpur, Jodhpur, India (1987)
57. Srivastava A.: Feeding Ecology and Behaviour of Hanuman langur, *Presbytis entellus*. Ph.D. Thesis, Univ. of Jodhpur, Jodhpur, India (1989)
58. Agoramoorthy G, Mohnot SM (1988) Infanticide and juvenilicide in Hanuman Langur (*Presbytis entellus*) around Jodhpur, India. Hum Evol 3:279–296
59. Rajpurohit LS, Srivastava A, Mohnot SM (1994) Birth Dynamics in Hanuman langur *Presbytis entellus* of Jodhpur, India. J Biosci 19(3):315–324
60. Rajpurohit LS, Chhangani AK, Rajpurohit RS, Mohnot SM (2003) Observation of a sudden resident male replacement in a unimale bisexual troop of Hanuman langurs, *Semnopithecus entellus*, around Jodhpur (India). Folia Primatol 74:85–87
61. Sharma G.: Study on the paternal behaviour in Hanuman langur (*Semnopithecus entellus*). Ph.D. Thesis, JNV. University, Jodhpur (2007)
62. Krishanan M (1972) An ecological survey of the larger mammals of peninsular India. Part 1. J Bomb Nat Hist Soc 68:503–555
63. Ripley S.: The Ecology and Social Behaviour of the Cylon grey langur *Presbytis entellus thersites*. Unpublished Ph.D. Thesis, Univ. of California, Berkeley (1965)
64. Prater SH (1965) The Book of Indian Animals. Bombay Natural History Society, Oxford University Press, Mumbai, pp 483
65. Dolhinow P (1979) A behaviour repertoire for the Indian langur monkey (*Presbytis entellus*). Primates 19(3):449–472
66. Winkler P.: Zur Oko-Ethologie freilebender Hanuman Languren (*Presbytis entellus entellus* Dufresne, 1797) in Jodhpur (Rajasthan) India, Ph.D. Thesis, Univ. of Goettingen (1981)
67. Amerasinghe FP, van Cuylenberg BWB, Hladik CM (1971) Comparative histology of the alimentary tract of Ceylon primates in correlation with the diet. Ceylon J Biol Sci 9:75–87

68. Bauchop T (1978) The significance of micro-organisms in the stomach of non-human primates. World Rev Nutr Diet 32:198–212
69. Yoshiba K (1967) An ecological study of Hanuman langurs, *Presbytis entellus*. Primates 8:127–154
70. Hladik CM, Hladik A (1972) Disponsibilities alimentarin et domains vitaux des primate a Ceylon. Terre Vie 26:149–215
71. Hladik CM (1977) A comparative study of the feeding strategies of two sympatric species of leaf monkeys: *Presbytis senex* and *Presbytis entellus*. In: Clutton-Broch TH (ed) Primate Ecology: studies of feeding and ranging behaviour in lemurs. Monkeys and apes. Academic, London, pp 323–353
72. Yoshiba K (1968) Local and intertroop variability in ecology and social behaviour of common Indian langurs. In: Jay PC (ed) Primates Studies in Adaptation and Variability, Holt. Rinehart & Winston, New York, pp 217–242
73. Tiwari KK, Mukherjer RP (1973) Studies on social behavior in the common langur (*Presbytis entellus*) around Ramtek near Nagpur. Proc Ind Sci Cong 60(4):157
74. Mohnot SM, Chhangani AK, Little K. Birth of quadruplets in Hanuman langur (*Semnopithecus entellus*): A World Record. In: Primates in the New Millennium. Abst. IPS Cong, Adelaide, South Australia. 2001; p 146
75. Rajpurohit RS.: Study on conflicts and reconciliation in Hanuman langur, *Semnopithecus entellus entellus* (Dufresne, 1797). Ph.D. Thesis, JNV University, Jodhpur, India (2004)
76. Rajpurohit LS (1997) Why do mothers carry the corpses of their infants in Hanuman langurs, *Presbytis entellus*. J Natcon 9:183–193
77. Zukerman S (1932) The social life of monkeys and apes. Routledge, London, pp 511
78. Bhaker NR.: Role of vocal communication in sociobiology of Hanuman langur, *Semnnopithecus entellus* around Jodhpur (India) Ph.D. Thesis, JNV University, Jodhpur (2001)
79. Ripley S (1980) Infanticide in langurs and man: adaptive advantage or social pathology? In: Cohen NN, Malpass S, Klein HG (eds) Bio-social mechanism of population regulation. Yale Uni. Press, New Haven, Conn, pp 349–390
80. Rajpurohit LS (1991) Resident male replacement, formation of new male band and paternal behaviour in *Presbytis entellus*. Folia Primatol 57:159–164
81. Rajpurohit LS, Sommer V (1993) Juvenile male emigration from natal one-male troop in Hanuman Langurs. In: Pereira ME, Fairbant LA (eds) Juvenile Primates: Life History. Development and Behaviour. Oxford Univ Press, New York, pp 86–103
82. Rajpurohit DS.: Study the dominance hierarchy and its role in social organization in Hanuman langur, *Semnopithecus entellus entellus* (Dufresne, 1797). Ph.D. Thesis, JNV University, Jodhpur (2005)
83. Makwana SC (1979) Infanticide and social change in two groups of Hanuman Langur, *Semnopithecus entellus* at Jodhpur. Primates 20(2):293–300
84. Mohnot SM (1971) Some aspects of social change and infant-killing in Hanuman Langur *Presbytis entellus* (Primates: Cercopithecidae) in western India. Mammalia 35(2):175–198
85. Hrdy SB (1974) Male-male competition and infanticide among the langurs (*Presbytis entellus*) of Abu, Rajasthan. Folia Primatol 22:19–58
86. Rajpurohit LS, Chhangani AK, Rajpurohit RS, Bhaker NR, Rajpurohit DS, Sharma G (2008) Recent observation on resident male change followed by infanticide in Hanuman langurs (*Semnopithecus entellus*) around Jodhpur. Primate Rep 75:33–40
87. Zimmermann EAW von. Geographische Geschichte des Menschen, und der allgemein verbreiteten vierfussigen Thiere, nebst einer hieher gehorigen zoologischen Weltcharte II. Quadrupeden. Leipzig: Weygandschen Buchhandlung; 1780
88. Khajuria H (1954) Catalogue of mammals in the Indian Museum (Zoological Survey of India, Kolkata). Part 3. Primates: Colobidae. Rec Indian Museum (Delhi) 52:195–220
89. Fooden J (1976) Provisional classification and key to living species of macaques (Primates: Macaca). Folia Primatol 25:225–236
90. Schultz AH (1969) The Life of Primates. Weidenfeld and Nicolson, London, pp 178
91. Mandal AK (1964) The behavior of the rhesus monkey (*Macaca mulatta Zimmerman*) in the Sunderbans. J Bengal Nat Hist Soc 33:153–165

92. Mukherjee RP, Gupta S (1965) Habits of the rhesus macaque, *Macaca mulatta* (Zimmerman) in the Sunderbans, 24 Parganas, West Bengal. J Bomb Nat Hist Soc 62:145–146
93. Puget A (1971) Observations sur le macaque rhesus, *Macaca mulatta* (Zimmerman, 1780), en Afghanistan. Mammalia (Paris) 35:199–203
94. Roonwal ML (1956) Macaque monkey eating mushrooms. J Bomb Nat Hist Soc 54:171
95. Mukherjee RP (1969) A field study on the behavior of two roadside groups of rhesus macaque, *Macaca mulatta* (Zimmerman) in northern Uttar Pradesh. J Bomb Nat Hist Soc 66:47–56
96. Southwick CH, Beg MA, Siddiqi MR (1965) Rhesus Monkeys in North India. In: De Vore I (ed) Primate Behaviour: Field studies of Monkeys and apes. Holt, Renehart and Winston, New York, pp 111–159
97. Prakash I (1962) Group organization, sexual behavior and breeding season of certain Indian Monkeys. Japanese J Ecol 12:83–86
98. Lindburg DG (1971) The rhesus monkey in North India: an ecological and behavioral study. In: Rosenblum LA (ed) Primate Behavior: developments in field and laboratory research. Academic Press, New York, p 106
99. Lancaster JB, Lee RB (1965) The annual reproductive cycle in monkeys and apes. In: Rosenblum LA (ed) Primate Behavior: Field studies of monkeys and apes. Holt Rinehart and Winston, New York, pp 486–514
100. Carpenter CR. Societies of monkeys and apes. Biological Symposium. 8: 177–204. Reprinted In: Southwick CH (ed.) Primate Social Behavior. 1963. pp 24–51. Von Nostrand: Princeton (1942a)
101. Altmann SA (1962) A field study of sociobiology of rhesus monkeys, *Macaca mulatta*. Ann New York Acad Sci 102:338–435
102. Kaufmann JH (1965) A three-year study of mating behavior in a free-ranging band of rhesus monkeys. Ecology 46:500–512
103. Sade DS (1968) Inhibition of son-mother mating among free-ranging rhesus monkeys. Sci Psychoanal 12:18–38
104. Loy J (1972) Synchronization of estrus among free-ranging *Macaca mulatta*. Abstract Book 4th Internatnl Cong Primatol, Portland, Oregon, p 47
105. Trivers RL (1974) Parent-offspring conflict. Am Zool 14:249–264
106. Rowell TE (1963) Behaviour and female reproductive cycles of rhesus macaques. J Reprod Fert 6:193–203
107. Malik I (2001) Monkey menace-who is responsible? In: Gupta AK (ed) Non Human Primates of India. Envis Bull (Wildlife and protected area). Wildlife Institute of India, Dehra Dun
108. Nandi JS, Tikute SK, Chhangani AK, Potdar VA, Walimbe A, Mishra MT, Ashtekar RA, Kumari J, Walimbe A, Mohnot SM (2003) Natural infection by simian retrovirus-6 (SRV-6) in Hanuman langurs (*Semnopithecus entellus*) from two different geographical regions of India. Virology 311(1):192–201
109. Nandi JS, Dooren SV, Chhangani AK, Mohnot SM (2006) New Simian ß Retroviruses from Rhesus monkeys (*Macaca mulatta*) and langurs (*Semnopithecus entellus*) from Rajasthan, India. Virus Genes 33:107–116
110. Jayashree SN, Chhangani AK, Mohnot SM (2011) Novel Simian Foamy Virus infection of wild Indian rhesus macaque (*Macaca mulatta*). Retrovirology 8(Suppl 1):215
111. Jayashree SN, Chhangani AK, Mohnot SM, Felipe DG (2011) Unique lentivirus infecting feral simians from forests of Rajasthan, India. Retrovirology 8(Suppl 1):212

Chapter 16
Status of Tiger in Rajasthan

Gobind Sagar Bhardwaj and B.K. Sharma

Abstract The Ranthambhore Tiger Reserve is the westernmost limit of the Bengal Tiger, *Panthera tigris tigris*, concentrated in the 300 km^2 core area of the reserve; however, the source population is facing tremendous pressure from the adjoining human settlements. In addition, the dispersal of transient tigers from the source population to other degraded and less protected areas is a serious threat to its conservation. These areas are, in fact, sinks to the source population as the scarcity of prey base leads to lifting of livestock from the surrounding areas resulting in man-animal conflict. Thus a stage is set for further confrontation often leading to retaliatory poisoning or killing of these transients. Though officially banned, poaching of tiger for skin and other body parts for onward trade to countries like China continues to be a serious problem. Apart from monitoring of such transients through telemetry, other serious protection measures in adjoining areas of the reserve should be considered. The possibilities of shifting such transients to other protected areas based on the experiences gained from the recent tiger reintroduction at Sariska Tiger Reserve should be scientifically worked out.

Introduction

We all have come across stories about tigers and the ferocity, horror and fear associated with this creature, considered as enemy of mankind. *The Jungle Book* of Rudyard Kipling and *Songs of Experience* by William Blake depict the tiger as

G.S. Bhardwaj (✉)
Department of Forests and Wildlife, Government of Rajasthan, Jaipur, Rajasthan, India
e-mail: gsbifs@yahoo.co.in

B.K. Sharma
Department of Zoology, R.L. Saharia Government P.G. College,
Kaladera (Jaipur), Rajasthan, India
e-mail: drbksharma@hotmail.com

B.K. Sharma et al. (eds.), *Faunal Heritage of Rajasthan, India: General Background and Ecology of Vertebrates*, DOI 10.1007/978-1-4614-0800-0_16,

dangerous. On account of their might and majestic beauty, the tiger holds a special status in Hindu mythology since time immemorial. The Hindu Lord Shiva has been shown wearing tiger skin or meditating on it, while his better half, Goddess Parvati, in her incarnation as *Maa Durga*, is shown riding a tiger. Even in Tibetan culture, the wall paintings of old monasteries depict monks vanquishing tigers with chains and shackles. In fact, the tiger was known to man even around 5,000 years ago during the Indus Valley Civilization. Interestingly, a seal found at Mohenjo-daro depicts a man sitting on a tree and angrily addressing a tiger waiting for him below [1].

Likewise, J. Inglis [2] in 1892 described the tiger as *the embodiment of devilish cruelty of hate and savagery*. On the other hand, in the Asian culture, tiger is considered to be a symbol of strength and royal power and was even used as an “executioner” in some courts. Overpowering or killing this beast was considered a valorous act; that is why most paintings and portraits of the nineteenth and first half of the twentieth centuries preserved in the castles and forts of erstwhile kings depict the winner (*shikari*) posing pompously along with the dead body of the looser (tiger). Until the 1980s, the Indian cinema witnessed the audience applauding the moment the hero killed or overpowered a tiger. Flaunting tiger skins and trophies in the drawing rooms is still considered a status symbol, especially in higher sections of the society not only in India but in majority of affluent countries. In fact, the natural history of tiger was studied predominantly along the sight of a rifle [3]. For old Indian *shikaris* (hunters), the proud narration of tall stories of hunting tigers and other animals and exhibiting their trophies in their *havelis* or castles was a matter of pomp and vanity. Its lustrous ochre skin has always lured its slaughters, apart from its bones and body parts which are used in aphrodisiac Chinese medicines. (Late) Kailash Sankhala—the person who painstakingly initiated the Project Tiger in India and worked for their conservation throughout his lifetime in Rajasthan—was once quoted as saying that the tourism industry in the 1950s and 1960s consumed more than 3,000 tigers as trophies [4].

The Saga of Massacre

Early literature on tiger is filled with stories of its hunting and killing, as narrated by *shikaris*, kings, british army officers, civil servants, princes and landlords, reflecting the number of individual tigers they bagged. For example, Gordon-Cumming [5] shot 73 tigers in just one district along the Narmada River in 1863 and 1864 and 10 tigers in just five days along the River Tapti; Forsyth [6] shot 21 tigers in 31 days in Uttar Pradesh; George V and his party shot 39 tigers in 11 days in Nepal in 1911–1912 [7]; William Rice [8] shot or wounded 158 tigers, including 31 cubs, in Rajasthan in a short span of four years between 1850 and 1854; The Maharaja of Nepal and his guests shot 433 tigers between 1933 and 1940 [7]; Colonel Nightingale shot over 300 tigers in the former Hyderabad State [9]; and The Maharaja of Udaipur shot at least 1,000 tigers during his lifetime [10]. Valmik Thapar [11] in his book *The Last Tiger*, says, *The hunting records of Indian princes far outstripped those of the British…Shikar travel agencies had flourished and strong will and measures were needed to curb the excess of this so-called sport*...the author had the horrific experience of seeing some archival footage of the 1960s, from the area of Kota, quite close to Ranthambhore. A series of tiger shoots were organised by the Maharaja's party in broad daylight using buffalo bait where tigers were picked off like flies. Early 1960 witnessed the total extinction of the tiger in many of its natural habitats, just like in 2004–2005 at Sariska. The author opined the so-called sportsmen of those times were much worse than today's poachers" [11]. It is an irony that a total of 1,074 tigers were shot on licence from government in the forests of Rajasthan state between 1929 and 1939 [12]. Only the so-called noblemen could hunt them, and interestingly, they even established extensive game reserves to assure them adequate supply of tigers for hunting. This mass killing of tigers actually resulted in sharp decline in the tiger population in India. By the turn of the twenty-first century, there were about 40,000 tigers in India [13]. Schaller [3] in 1967 stated, "Legal and illegal killing has been the major cause in the decline of tigers throughout the forested areas. Males, females and young are destroyed indiscriminately although many states now have regulations which prohibit the shooting of tigresses when accompanied by cubs" [3]. Few inventions that ushered the world towards modernization were like a curse for the wildlife. While the invention of firearms made it easier to aim at the animals, the advent of cars and jeep in the first half of the twentieth century made the jungles easily approachable and accessible which earlier took days of marching on camels and horses. Every jeep that moved had one or more guns in it, and practically every animal seen, which presents a fair chance of being killed, was fired at [14]. In 1935, it was observed that India is richer than Africa in the variety of wildlife, but is certainly poorer in number, though Africa is one of the most heavily shot countries and has been attracting sportsmen and poachers of all kinds from all over the world for a number of years, African wildlife is being systematically protected. India is in utter confusion with inadequacy of legislations and unwillingness on the part of subordinate government officials and public servants to enforce the existing laws and order [15]. Although this observation was written more than 70 years ago, it still stands as a mirror with changing reflections.

There are records to prove that, prior to ban on tiger shooting, it was a custom for forest officers to shoot tigers before their promotion. The first and last tiger shot by (Late) Kailash Sankhala transformed him into a great tiger saver. The then Inspector General of Forests, MD Chaturvedi, a celebrated tiger hunter during his time, used to tell trainee forest officer, *Shoot your first tiger in your first year* [4]. These were just some of the traditions of the bygone era that too have contributed to the decline of tiger.

Voice for Concern

Earlier, there were very few people who raised their voice for tiger conservation. It was Captain J. Forsyth, settlement officer and deputy commissioner of Nimar, then in the central provinces, who was the first to express concern at the losses in tiger population in 1872 [4]. In 1900, Russel recorded his objection to the reward being offered for every tiger killed in Mysore [4]. Much later in 1969, in her speech at the International Union for Conservation of Nature (IUCN) the then, Prime Minister of India, Indira Gandhi said, *We need foreign exchange, but not at the cost of the life and liberty of some of the most beautiful inhabitants of this continent.* It clearly reflected concern of a great leader of a great nation towards wildlife conservation. Subsequently in 1970, tiger shooting was banned and the parliament enacted the Wildlife (Protection) Act in 1972. In the year 1973, the "Project Tiger" was launched, with the declaration of nine protected areas as tiger reserves. Two areas, namely, Ranthambhore and Sariska, represented the state of Rajasthan. Since then, time has changed. Earlier, more people were happy to see a dead tiger; now, people want to see tiger on a kill instead. Today, most people enjoy seeing this animal in the wild and love to shoot it with a camera. Earlier, there were no supporters for tigers, but today the situation is different. Now, tigers, like celebrities, are models for umpteen commercial campaigns of different consumer products. The tiger is nowadays considered as the top subject of conservation. The government, NGOs and other voluntary organisations are trying their level best to protect and conserve this wonderful creature, sadly now on the verge of extinction. The print and electronic media, especially the National Geographic, Discovery Channel, Animal Planet and Doordarshan, have played a great role in creating mass awareness, leading to sensitisation among society and creating myriad fan followers of tiger, an icon of the Indian jungles.

The King of Felines

Tiger, the flagship species, occupies the top position in the ecological pyramid of a forest ecosystem. It is the largest existing member of order Felidae of class mammalia. Globally, eight subspecies of tigers are described and generally accepted [16], although some very recent work in molecular biology casts doubt on their validity [17]. This includes the *Panthera tigris tigris*, also known as Bengal Tiger, which is found throughout India. The other seven are *P. t. altaica* (Siberian/Amur Tiger), *P. t. amoyensis* (Chinese Tiger), *P. t. balica* (Bali Tiger), *P. t. corbetti* (Indo-China Tiger), *P. t. sondaica* (Javan Tiger), *P. t. sumatrae* (Sumatran Tiger) and *P. t. virgata* (Caspian Tiger). Out of these, the Bali, Javan and Caspian Tigers became extinct in the twentieth century. Once distributed throughout central and southern Asia and even up to eastern Turkey, tigers now survive only in scattered populations from India to southeast Asia, Sumatra, China and far east areas of Russia. This beautiful carnivore is considered to have actually originated in east Asia and is one of the most familiar cats with its distinctive reddish-orange or ochre coat with black vertical stripes on the body area and in the form of rings on the tail (Fig. 16.1). The forelimbs are well muscular with retractile claws. The male being larger than female, the average size varies from 8 feet 9 in. to 10 feet 3 in. [14], while the underside or the belly portion of the body is whitish in colour.

Fig. 16.1 Bengal Tiger *Panthera tigris tigris*: the king of Ranthambhore National Park, Sawai Madhopur (*Courtesy: Dr. Gobind Sagar Bhardwaj*)

The tiger is well adapted for hunting large prey. In Ranthambhore, Sambar dominates the menu of tigers, contributing around 50% of the total kills (Fig. 16.2). During three years of observations, the author observed that out of 36 kills, 17 were of Sambar (47.22%), 11 of domestic livestock (30.55%), 5 of Nilgai (13.88%) and three of Chital (8.33%) [18]. The short yet heavily muscular forelimbs of a tiger with sharp and retractile claws work like lethal weapons to bring down large prey. Tiger is generally solitary in nature, except when in association with the opposite sex or with female accompanying with the cubs. A tigress with cubs is always busy searching, stalking and killing prey for her growing offspring because she has the responsibility of not only feeding them but also of transforming them into skilful and successful predators. A cat's hunting behaviour comprises a five-stage act that includes searching for prey, detecting the prey, orienting and approaching the prey, capturing and killing the prey and finally eating it [19].

Scent marking by spraying, claw markings and calls are the usual means of communication among the tigers. The scent markings help in delineating the territories of different tigers and also help tigers in knowing the reproductive status of the opposite sexes. On a number of occasions, the author has seen tigers spraying urine while ambling on the forest road. The odour of this spray is like ammonia.

Fig. 16.2 A Ranthambhore tigress with the kill (*Courtesy: Dr. Gobind Sagar Bhardwaj*)

The author has often felt it smelled like *mahua* (*Madhuca indica*) flowers. Purring, poking, grunting, meowing, woofing, moaning and roaring are different vocal calls of tigers observed in different times.

Being polyestrous in behaviour [20, 21], the tigers do not have any fixed mating season [14, 22]. They mate with amazing frequency, mating over 50 times a day [23]. In October 2003, the author observed mating tigers for around 90 min, and they mated six times. Though, he could not record the exact duration of the first copulation, but the subsequent ones lasted 12, 13, 12, 9 and 14 s, respectively [18]. Gestation period is very short and varies from 103 to 110 days [23]. Endowed with a very high reproductive potential, a tigress can produce five cubs in a litter and 4–5 litters in her entire reproductive life span. The author keenly observed the tigress of Rajbag area of Ranthambhore who has produced and reared four known litters of nine cubs successfully [18]. Seidensticker also noticed one exceptional female tiger in Nepal's Royal Chitwan National Park, who lived for 15 years, 10 years of which in the same territory. She produced five known litters of cubs, first one in 1975 and the last in 1985. So, she actually reproduced, on an average, about every second year for 10 years. Eleven of her 16 young ones independently survived [19]. Tigress generally keeps her cubs in secluded and secure areas like rock caves or overhangs [24, 25] and even amidst grasslands in *terrai* area. After remaining in the secluded areas for 4–8 weeks [26–28], the young cubs come out with their

Fig. 16.3 A tigress seen kindling a young one at the Ranthambhore National Park, Sawai Madhopur (*Courtesy: Dr. Gobind Sagar Bhardwaj*)

mother. Mortality of the cubs is very high especially during their dispersal period when they venture into new areas for establishing their territories. Dispersal is the permanent movement an individual makes from its birth site to the place where it reproduces or would have reproduced if it survived and found a mate [29]. Naturally, juveniles and subadults are the dispersers. Often, cubs between the ages of 18 and 24 months, who look almost like adult tigers, disperse [23] (Fig. 16.3). They are called transient floaters and keep on looking for vacant territories (Fig. 16.4). When they do not get any space, they move out to venture into new areas called "frustrated dispersal" [30]. Consequently, the chances of their survival diminish outside the protected area as most of them succumb to poisoning or snares laid by the skin mafia or sometimes become man and/or livestock killers and are eventually either shot down or transported to a zoo. The very recent killing of four people by a young male transient that strayed out of the forests in the Pilibhit district in Uttar Pradesh has compelled the government to order a shoot at sight [31]. Similarly, there are two famous records of straying young tigers from Ranthambhore. The young tiger who died in a train accident in 2003 on the rail track passing through the Mukundra Hills National Park in Kota district of Rajasthan was one of the male cubs (popularly known as Broken Tail) of the first litter of the tigress of Rajbag area of Ranthambhore NP. Another example is of the transient male cub of Berda tigress that strayed out of Ranthambhore NP and killed a villager of Mai Kalan in February 2005. Similarly, a forest range officer was seriously injured when a tiger that had strayed outside in Bhoori Pahadi village adjoining the national park area attacked him.

Fig. 16.4 Two sub-adult cubs enjoying the daylight at Ranthambhore National Park, Sawai Madhopur (*Courtesy: Dr. Gobind Sagar Bhardwaj*)

In a true sense, it is only the tigress that has fixed territories. The male who keeps on looking for different females travels long distances in the territories of tigresses thus, the males have territories which are elastic and overlapping in nature. The movement of tigers outside the national park is also a common behaviour, as just after separation from their family, they disperse looking for new territories or for mates.

Status of Tiger in Rajasthan

During the last three decades, Ranthambhore and Sariska Tiger Reserves remained in news whether it is for tiger sightings, visits of important people including head of states or extermination of tigers from Sariska area and their subsequent reintroduction. Located in the 4B biotic province of the semiarid biogeographic zone [32], Ranthambhore Tiger Reserve is the westernmost limit of the Bengal Tiger. Prior to 2005, Sariska was also a part of the same biotic province which boasted to be the home of a natural population of tigers in Rajasthan. Sariska Tiger Reserve remained a source of attraction for tiger sightings especially between the 1960s and 1970s. Unfortunately, in the year 2004, the last official tiger sighting was reported, and subsequently, the news of total extermination of tigers from Sariska shook conservationists around the world. After independence, a number of other sanctuaries of the state had already witnessed local extinction of this wonderful striped creature. Be it Kailadevi, Ramgarh Vishdhari, Kumbhalgarh or Sitamata Wildlife Sanctuaries, the forests appear like haunted vacant homes of the big cat. On account of relatively

Fig. 16.5 Three cubs at a waterbody inside Ranthambhore National Park, Sawai Madhopur (*Courtesy: Dr. Gobind Sagar Bhardwaj*)

better degree of protection and the legal status as a national park, Ranthambhore is the last hope for the dwindling population of tigers in this part of the world. Owing to its excellent tiger sighting, Ranthambhore still leads among the best wildlife tourist destinations in India. It's unique setting and deciduous forest with scanty vegetation provides ideal conditions for tiger sighting (Fig. 16.5).The high density of tigers, habituated to tourist vehicles, is confined to a relatively small area as compared to equally high-density areas like Corbett Tiger Reserve, Kanha Tiger Reserve and Nagarhole National Park. Moreover, the middle-storey vegetation is also scanty, except for the presence of *Grewia flavescens* bushes and other associated species. A good road network in the park area, especially in the valleys and other low-lying areas, further enhances the chances of sighting the tiger. Due to their continuous interaction with tourist vehicles in Ranthambhore, tigers perhaps have changed their hunting and stalking behaviour. As a result, one can see a tiger or tigress chasing and stalking its prey in broad daylight (Fig. 16.6). In fact, some of the most famous movies on tiger hunts have been made here in broad daylight especially during winters. This may be due to the fact that the protection status in the area is considerably better and in cool weather, the tigers can be active at midday too. Otherwise, tigers hunt primarily at night, between dusk and dawn, a time when the wild hoofed animals are most active [3]. In fact, in Ranthambhore, some of the tigers, especially inhabiting the tourism zone, often use vehicles as a cover for stalking their prey.

Spread in an area of around 1,395 km^2, Ranthambhore Tiger Reserve has a core area (Ranthambhore National Park) of 282 km^2, while the rest of the area, the buffer zone, consists of Kailadevi Sanctuary in Karouli district and Sawai Mansingh Sanctuary in Sawai Madhopur district. Earlier *Shikarkhana* department of the princely states of Jaipur and Karouli managed the forests of this area. It was in the

Fig. 16.6 A tigress just before attacking the prey at Ranthambhore National Park, Sawi Madhopur (*Courtesy: Dr. Gobind Sagar Bhardwaj*)

year 1955 that some of the area was notified as Sawai Madhopur Game Sanctuary and declared as Tiger Reserve in the year 1973 when Project Tiger was launched. Subsequently, Ranthambhore National Park was notified in the year 1980, and later, in the years 1983 and 1984, Kailadevi and Sawai Mansingh Sanctuaries were, respectively, notified. It is the River Banas that divides the Ranthambhore Tiger Reserve into two parts—the Kailadevi Sanctuary on the northern side and the Ranthambhore National Park on the southern side—and then confluences with the River Chambal. Some of the areas of River Banas act as permanent water sources for wild animals of the reserve, especially during summer. There is no perennial river inside the park. The solitary Bakola stream manages to feed the park up to the month of April and eventually succumbs to the scorching heat of May–June. The only perennial sources of water in the park are six ponds, namely, Padam *talab*, Rajbag, Milik *talab*, Lahpur *talab*, Gilaisagar and Mansarovar. In the year 2003, they got completely dried up. All other streams and *nullahs* passing through the park are seasonal and solely at the mercy of the rains.

The forest type of this area is tropical dry deciduous and tropical thorn forest with *dhonk* (*A. pendula*) as the dominant species [33]. Having an excellent capability to coppice, the leaves of this tree are highly palatable not only to wild ungulates but also to the livestock of the adjoining villages. During winters, one can often observe Chital (Spotted deer) feeding on dry fallen leaves of this tree. Apart from this, species of *Khair* (*Acacia catechu*), *Dhak* (*Butea monosperma*), *Salar* (*Boswellia serrata*) and *Karaya* (*Sterculia urens*) are also present. Middle-storey vegetation is sparse, and that is why the visibility range is much higher in Ranthambhore as compared to other tiger reserves, thus, making it the first choice as a wildlife tourism destination. In addition, *Grewia flavescens* is important vegetation which acts as an excellent cover for the tiger to take shelter. During the late winter and the entire summer, the forest wears a dry and brownish look, but the first monsoon shower does wonders and within a week, the fresh sprouting leaves of *dhonk* trees magically turn the lifeless brownish jungle into a lush green youthful forest. Hundreds of saplings of ground vegetation sprout, indicating the rich floral diversity of the area. So far, more than 400 species of plants have been recorded in the Ranthambhore National Park, mostly angiosperms. Pteridophytes are represented by only a few species, whereas, there are a number of *mycetozoa*. Countless species of herbs adorn this wild heritage site.

Other cats of Ranthambhore like Leopard (*Panthera pardus*) claim to hold the same position as the tiger. Then, there is Caracal (*Caracal caracal*) and Jungle Cat (*Felis chaus*). A Rusty-spotted Cat (*Prionailurus rubiginosus*) was for the first time reported in 2005 when some vehicle killed it on the Ranthambhore road. The family Ursidae is represented by the Sloth Bear (*Melursus ursinus*). Among ungulates, Nilgai (*Boselaphus tragocamelus*) and Chinkara (*Gazella bennetti*) are the antelopes. Sambar (*Cervus unicolor*) and Chital (*Axis axis*) are two varieties of deer present all over the park. Other mammals include Hanuman Langur (*Semnopithecus entellus*), Striped Hyaena (*Hyaena hyaena*, L.), Golden Jackal (*Canis aureus*), Indian Crested Porcupine (*Hystrix indica*), civet and badgers. One Wild Dog (*Cuon alpinus*) was also commonly seen in Lakarda, Berda and Anantpura area. In April 2006, the same was seen in Raipur area near the Singhdwar gate. As this area is not known for Wild Dogs this lone animal must be a strayed individual, who probably sneaked into Rajasthan from the adjoining Madhya Pradesh state. Crocodiles, Monitor lizard, turtles and a variety of snakes represent the easily sighted reptiles in this area. More than 300 species of birds including migratory guests during winters [34] are also reported from the park.

Apart from other protection and management activities, monitoring of tigers in Ranthambhore is one of the most important activities being done by the park authorities. As a part of their monitoring exercise, the frontline field staffers of the park remain busy in tracking tigers using the pugmark technique. It was in the year 2005, with the help of Wildlife Institute of India, modern camera trap technique came into use for monitoring the tigers. In the early days, tracking tigers for hunting was a tradition among Indian hunters, and this flourished under royal patronage [4]. It was in the 1930s that the characteristics of pugmarks were published [27, 35]. Population estimation is considered to be an important part of tiger monitoring. The pugmark-based expert system was developed by S. R. Choudhary in the early 1970s and later improved upon by H.S. Panwar in 1979 and again by V. B. Sawarkar in 1987. In 2001, Sharma [36] used the technique to differentiate individual tigers based on various pugmark measurements and recommended the use of 11 potential variables for an effective way of identifying individual tigers. The traditional pugmark analysis method is based on recording several distinguishable morphological features of the pugmarks which helps in distinguishing individual tigers. Though this technique is low cost, it is highly specialised and requires lots of expertise, which our frontline staff sometimes lack. We all know that tigers travel long distances in their territories, often passing through different terrain and different soil types. This difference of soil and terrain often has an adverse effect on the plaster cast of the pugmark, and it is very difficult to analyse these plaster casts having slight differences and reach any substantive conclusion.

After the total extermination of tigers from Sariska and the alleged news of missing tigers from Ranthambhore, the Government of Rajasthan set up a high-powered state-empowered committee in the year 2005. In its report entitled "*Securing the Future- The Report of the State Empowered Committee on Forests and Wildlife Management*", the committee also pointed out such drawbacks of the pugmark count method. It said the "*pugmark method is highly dependent on the quality of plaster cast obtained initially, which in turn is a function of range of variables (terrain, skills, integrity, etc.) and is also prone to a number of subjective assessments during the analytical phase. It is, therefore, recommended that this method be used only for ascertaining habitat occupancy and not arriving at precise numbers as is presently the case.*"

Due to these shortcomings of pugmark analysis method, the author was on the way of adopting a new monitoring system in Ranthambhore based entirely on the difference in their stripe patterns by direct observations of various tigers. The stripe pattern of every individual tiger is unique, and one can readily identify individual tigers through the stripes on their bodies [3, 37–39]. As the in charge of Ranthambhore Tiger Reserve for 38 months, the author successfully located tigers and their families for at least 209 times. An SLR camera and a pair of binoculars were his simple tools for identifying different individual tigers based on their stripe patterns. All in all, the author sighted 34 different tigers including cubs and one dead tigress. All these individual tigers were sighted on different occasions; it does not reflect the tiger count at any particular time. In his book, "*Wild Tigers of Ranthambhore*", Thapar stated: "We have recorded over several generations, more than 125 tigers with our cameras" [40]. After this, there were reports of not less than 50 cubs in Ranthambhore till the end of 2007. It means that there were a total of not less than 175 (125+50) different tigers at different intervals of time. So far, from 1975 to 2007, in 32 years, around a dozen bodies of tigers (source of this information from management plan of Ranthambhore TR and staff) had been recovered, while one died in the Kota train accident, so we were left with 163 tigers. As per the census figure of year 2007, the total count of tigers in Ranthambhore is 32. This means that 131 tigers have gone missing from the Ranthambhore National Park in the last 32 years, with an average of four tigers every year. It seems to be alarming, but obviously, these might have died due to poaching, natural death, territorial fights or old age.

Today, Ranthambhore is the only breeding ground of wild tigers in the entire western India. The park is also considered to be the maternity ward for the entire reserve, producing cubs almost every year, rearing and nourishing them to adulthood and finally bidding goodbye to the dispersing tigers, usually going to the adjacent areas. Some experts believe that these adjacent areas act as sinks to the source populations. The author has observed many juvenile tigers or mature cubs straying outside the national park just after separation from their mother either on account of being expelled by the resident dominant tigers or in search of females or due to prey scarcity. However, there is no scarcity of prey in Ranthambhore. Thus, tigers are straying out to areas where the prey base is very less and often kill livestock, leading to man-animal conflict. In most of the cases, the affected villager will never approach the forest department for a meagre compensation; instead, he approaches some *Mogiya* (a traditional hunter community) for lifting that tiger. It is very difficult to manually trace the trails of tigers, particularly those who stray from the national park. Once they go out, the chances of their return are dim, as the degree of protection outside the national park is almost negligible. We do not know how many tigers have been lost due to this. We have already lost tigers from all sanctuaries in Rajasthan, except Ranthambhore. The disappearances of tigers from Ramgarh Vishdhari Sanctuary and Sariska Sanctuary Tiger Reserve are recent tragedies. This monster of extermination is gradually advancing towards the last remaining tiger population of Ranthambhore too. Luckily, the tiger is still a growing population—from December 2005 to December 2006, more than a dozen cubs have been reported in Ranthambhore National Park, and the population of tigers has increased. The concern and efforts of the forest department are clearly visible in this flourishing population of tigers.

The tiger population recovered in just two years after its dip in 2005. This resiliency is really noteworthy if we compare it with the population dip in the years 1992 when the crashed population of tigers took years to recuperate.

The increase in tiger population at Ranthambhore encouraged the government to think about the translocation of some tigers to Sariska which got totally devoid of them. In 3 years, from 2008 to 2010, five young tigers from Ranthambhore National Park, two males and three females, were relocated to the forests of Sariska Tiger Reserve. This reintroduction experiment conducted by the state's forest department with technical assistance from scientists of the Wildlife Institute of India, Dehra Dun, became feasible due to good population of tigers in Ranthambhore. As stated earlier, due to favourable conditions and better security, Ranthambhore is a breeding haven for tigers, but being a small area, it cannot accommodate the grown-up cubs as they reach adulthood. Thus, the transient tigers are compelled to leave the park in search of new areas and mate, and outside the secure park, they may become victim of poachers or die untimely due to other reasons. The experiment of translocation done for Sariska is a good start for saving such transients or surplus individuals. This should be tried in other forest areas too. Apart from translocation, their proper scientific monitoring is also necessary. The best and effective way to monitor tigers in this modern era of technology is through satellite tracking. Tracking tigers and studying their behaviour only inside the tiger reserve by any other monitoring method is not sufficient. The need of the hour is that we have to keep an eye on straying tigers and dispersing floaters going beyond the forest boundaries, which is only possible through satellite tracking. Apart from studying its vital activities and its movement range, it will also be a great tool in apprehending the poachers.

The other urgent need in tiger conservation is the development of green corridors connecting all national parks and adjoining sanctuaries. The tigers of Ranthambhore move not only to the Kailadevi Sanctuary but also to the Mukundra Hills National Park through Sawai Mansingh and Ramgarh Sanctuaries. If green linkages having sufficient prey base, water bodies and cover and a good degree of protection interconnect these forest areas, so many floaters can be saved from being poached. There is clearly a great need for political will, so as to upgrade the status of these adjoining sanctuaries to national parks and simultaneously develop the entire connecting landscape where our tigers can move freely and safely.

References

1. Burton R (1933) The book of the tiger. Natraj Publishers, Dehra Dun, India. Reprint 1989. pp 177
2. Inglis J (1892) Tent life in tiger land and sport and work on the Nepal Frontier. William Clowes & Sons, London
3. Schaller George B (1967) The Deer and the Tiger. A study of wildlife in India. University of Chicago Press, Chicago, IL. pp 384
4. Sankhla Kailash (1978) Tiger. The story of Indian Tiger. Rupa, Calcutta, India. pp 220
5. Gordon-Cumming R (1872) Wild men and wild beast: Scene in Camp and Jungle (1829–1908). Edmonston & Douglas MD, London
6. Forsyth W (1911) The number of cubs in tigress litter. J Bomb Nat Hist Soc 20(4):1148
7. Smythies E (1942) Big game shooting in Nepal. Thaker, Spink & Co, Calcutta. pp 174
8. Rice W (1857) Tiger shooting in India. Smith, Elder & Co, London. pp 219
9. Perry R (1964) The world of the tiger. Cassell, London. pp 263
10. Singh K (1959) The tiger of Rajasthan. London. Smith, J.L.D. 1978. Smithsonian Tiger Ecology Project Report No. 13. Smithsonian Institution, Washington, DC
11. Thapar Valmik (2006) The last Tiger. Oxford University Press, New Delhi. pp 290
12. Prater S (1940) The number of tigers shot in reserved forest in India and Burma during the year 1937–1938. J Bomb Nat Hist Soc 41(4):881–89
13. Thapar Valmik (1989) The secret life of Tiger. Rodale Press, Emmaus, PA. pp 160
14. Brander A (1923) Wild animals in Central India. Edward Arnold & Co, London. pp 296
15. Corbett J, Morris RC, Jafry HA (eds) (1935) Indian Wildlife. Official Organ of the All India Conference on the Preservation of Wildlife 1(1) July 1936
16. Ellerman J, Morrison-Scott T (1951) Check list of Palaearctic and Indian mammals, 1758 to 1946 British Museum, London
17. Jackson P (1992) International specialists discuss China's threatened cats. Cat News. Bougy-Villars, Switzerland, vol 17. pp 9–10
18. Bhardwaj GS (2008) Tracking Tigers in Ranthambhore. Aureole Publishing, New Delhi. pp156
19. Seidensticker J, Susan Lumpkin (2004) Cats in question: the Smithsonian answer book, Smithsonian Books, Washington DC. pp 304
20. Asdell Sydeny A (1964) Patterns of mammalian reproduction. Comstock Publishing Associates, Ithaca. p 670
21. Crandall L (1964) The management of wild animals in captivity. University of Chicago Press, Chicago. pp 639
22. Inverarity J (1888) Unscientific notes on the tiger. J Bomb Nat Hist Soc 9(4):143–154
23. Karanth K Ullas (2001) The way of the tiger: National History and conservation of the endangered big cat. Voyageur Press, Stillwater, Minnesota. p 132
24. Anderson K (1961) The call of man eater. George Allen & Unwin, London. pp 285
25. Baikov N (1936) Big game hunting in Manchuria. Hutchinson & Co., London. pp 285
26. Novikov G (1962) Carnivorous mammals of the fauna of the USSR. Israel program for Scientific Translations, Washington, DC
27. Brander A (1930) Tiger tracks. J Bom Nat Hist Soc 33:972
28. Ognev S (1962) Mammals of USSR and adjacent countries. Vol. III: Carnivora. Israel program for scientific Translation, Washington, DC
29. Howard E (1960) Innate and environmental dispersal of the individual vertebrates. Am Mid Nat 63:152–161
30. Personnel communication with Mel Sunquist, Tiger expert from Florida, USA (2005)
31. Times of India, New Delhi, 17 Jan 2009
32. Rodger A, Panwar HS (1988) Planning a Wildlife Protected Area Network in India. Wildlife Institute of India, Dehradun, India. pp 339
33. Champion HG, Seth SK (1968) A revised survey of forest types of India. GOI Press, New Delhi, India. pp 404

34. Reddy GV, Tyagi RK, Bhatnagar D (2002). Management plan of Ranthambhore, Report submitted to the department of Forests, Government of Rajasthan
35. Champion FW (1929) Tiger tracks. J Bom Nat Hist Soc 33:284–287
36. Sharma S (2001) Evaluation of pugmark census technique. M.Sc. thesis submitted to Saurashtra University, Rajkot, Gujarat, India
37. McDougal C (1977) The face of the tiger. Rivington Books, London. pp 180
38. Karanth KU (1995) Estimating tiger *Panthera tigris* populations from camera-trapping data using capture recapture models. Biol Conserv 71:333–338
39. Franklin N, Bastoni Sriyanto, Siswomartono D, Manansang J, Tilson R (1999) Last of the Indonesian Tigers: A census for concern. In: Seidensticker J, Christie S, Jackson P (eds) Riding the Tiger: Tiger conservation in a human dominated landscape. Cambridge University Press, Cambridge
40. Thapar Valmik, Rathod FS (2000) Wild Tigers of Ranthambhore. Oxford University Press, Delhi. pp 171

Chapter 17
Status, Distribution and Conservation of Leopard *Panthera pardus fusca* in Rajasthan

Krishnendu Mondal, Shilpi Gupta, K. Sankar, and Qamar Qureshi

Abstract This chapter describes the present status and distribution of the second important big cat, the Leopard, which is an Endangered animal as per Appendix-1 of CITES and Schedule I of the Indian Wildlife (Protection) Amendment Act, 2006. Worldwide, eight subspecies of Leopard are identified based on DNA studies. The Indian subspecies, *Panthera pardus fusca*, is distributed all over the country, being absent only in the arid desert and above the timber line in the Himalayas. The latest available figures based on pugmark census the status, population and conservation issues of Leopard in 16 protected areas are discussed. Leopard is one of the least studied species in Rajasthan. The text discusses the feeding behaviour, adaptability for diverse diet including domestic livestock, man–leopard conflict and ability to withstand anthropogenic pressure and decline in its natural prey-base. Habitat destruction, loss of wild prey, poaching for skins, bones and claws and poisoning carcasses of livestock killed by leopard are significant threats to this species. Under these circumstances, the need for basic research on Leopard ecology, for example, movement, range, feeding ecology, habitat utilisation and man–animal conflict, have been stressed upon for the conservation of this magnificent animal in Rajasthan.

Introduction

The Leopard is found in almost every kind of habitat, from the rainforests of the tropics to desert and temperate regions [1] and throughout sub-Saharan Africa and southern Asia with scattered populations in China and North Africa [2–4]. Worldwide, there are eight subspecies of Leopard recognised based on DNA studies [5]. The Indian

K. Mondal • S. Gupta • K. Sankar (✉) • Q. Qureshi
Wildlife Institute of India, Chandrabani, P.O. Box 18, Dehradun, India
e-mail: krishtigris@yahoo.co.in; guptashilp@gmail.com; ksankar62@yahoo.com; qnq@wii.gov.in

B.K. Sharma et al. (eds.), *Faunal Heritage of Rajasthan, India: General Background and Ecology of Vertebrates*, DOI 10.1007/978-1-4614-0800-0_17,

Fig. 17.1 A young Leopard *Panthera pardus fusca* resting on a Peepal (*Ficus reliogiosa*) tree during the day hours at Ranthambhore Tiger Reserve, Sawai Madhopur, Rajasthan

race, *Panthera pardus fusca*, is distributed all over the country, being absent only in the arid deserts and above the timber line in the Himalayas [6]. Although it is most common among the big cats, Myers [7] recommended that it remains on Appendix 1 of CITES because extensive hunting had depressed their population in several parts of Africa. In India, it figures in Schedule I of the Indian Wildlife (Protection) Amendment Act, 2006 [8]. It has been feared that Leopard population will decline across most of its range because of habitat destruction and poaching [9]. In India, too, habitat destruction, loss of wild prey, poaching for skins, bones and claws and poisoning carcasses of livestock killed by Leopards are significant threats to the animal [10]. The reduction in tiger populations also meant that increasing poaching pressure may be brought upon on the Leopard to meet the demands of the skin and bone trade. The conflict originating due to loss of livestock and people is one of the major causes of Leopard persecution. Leopards are involved in conflicts with people due to their large home ranges, adaptability for diverse diet including domestic livestock, ability to withstand anthropogenic pressure and decline in natural prey base.

The Leopard is the most adaptable and widely distributed among all the big cats [3, 10] (Fig. 17.1). This species is known for its use of habitat edges and its ability to live in close to human habitation [11]. Leopard shows plasticity in changing behaviour as condition changes [4]. Leopard's ability to feed on a broad spectrum of prey makes it the most successful predator among big cats and its size gives the ability to feed on a variety of prey species ranging in size from a young buffalo to the smallest rodent [3, 4, 11–17]. They are also known as ecological generalists.

Detailed studies undertaken on Leopard in Asia are in Wapato National Park, Sri Lanka [12]; dry tropical forest of Thailand [15]; Chit wan National Park, Nepal [11]; Sanjay Gandhi National Park, India [18]; in Peninsular Malaysia in Mara Posh, Kuala Teenage and Kuala Koh [19]; in Satpura-Bori, Madhya Pradesh [20] and in Pauri Garhwal, Uttarakand [21].

The presence of Leopard is reported from 16 protected areas PAs that include two national parks (NP) and 14 wildlife sanctuaries (WLS) in Rajasthan. These are Ranthambhore and Sariska NP, Darrah, Bassi, Bhensrodgarh, Jaisamand, Jamwa Ramgarh, Jawahar Sagar, Kailadevi, Kumbhalgarh, Mount Abu, Nahargarh, Phulwari Ki Nal, Sajjangarh, Sitamata and Todgarh-Raoli WLSs. There are no recent records of leopard's presence in Keoladeo NP, Bandh Baratha, Ramgarh Vishdhari and Shergarh WLSs. No scientific information is available on Leopards in Rajasthan except from Sariska Tiger Reserve. The status, population and conservation issues of Leopard in 16 PAs of Rajasthan have been discussed based on pugmark census and waterhole census conducted in different protected areas by the state forest department [22].

Darrah Wildlife Sanctuary

The Darrah Wildlife Sanctuary spreads over an area of 274.41 km^2. The forest is of dry deciduous mixed type [23] and the terrain is hilly. Sixteen Leopards were reported by the forest department in 2002 which declined to 7 in 2007 (density 2.6/100 km^2) [22]. Hyaena, Grey Wolf, Golden Jackal and Indian Fox were the other carnivores reported. Except for Common Langur the other wild prey species population was reported to be low (Wild Boar 1.13/km^2, Nilgai 0.55/km^2, Chinkara 0.29/km^2, Chital 0.41 km^2 and Sambar 0.15/km^2). The Sanctuary has 19 villages inside and 82 villages around the periphery.

Bassi Wildlife Sanctuary

This is located in Chittourgarh district in the hilly areas of Aravalli and Vindhyan Ranges. The total area is 138.69 km^2. The area supports dry deciduous mixed forests [23] having mainly *Anogeissus pendula*, *Acacia catechu*, *Diospyros melanoxylon*, *Boswellia serrata*, *Lannea coromandelica* and *Zizyphus mauritiana*. The Leopard is the main predator and Carnivores like Hyaena, Wolf, Jackal and Jungle Cat are found here. The population of Leopard was 13 in 2004 but declined to 8 in 2007 (density 5.77/100 km^2) which rose to 10 in 2010. The wild prey species densities are reported to be low (Nilgai 1.70/km^2, Chinkara 1.00/km^2, Wild Boar 0.63/km^2 and Chital 0.12 km^2) [22]. There are 12 villages located inside the Sanctuary.

Bhainsrodgarh Wildlife Sanctuary

This is located in Chittourgarh district and covers an area of 229.14 km^2. The forest types found are *Anogeissus* dominated forest and miscellaneous forest. The reported Leopard density in 2007 was 5.7/100 km^2 and the wild prey species are very low (Wild Boar 1.80/km^2, Nilgai 0.34/km^2 and Chinkara 0.10/km^2) [22]. Hyaena, Golden Jackal and Jungle Cat are the other carnivores found in the Sanctuary. There are 21 villages located inside the Sanctuary. In all, 09 Leopards were counted in the WS in the year 2010 by Wildlife Census of forest department.

Jaisamand Wildlife Sanctuary

Jaisamand WS covers an area of 52.34 km^2. As per the classification the sanctuary falls under tropical dry deciduous forests and tropical dry mixed deciduous forest [23]. The density of Leopard is reported to be 13.4/100 km^2. Hyaena, jackal, fox and Jungle Cat are the other carnivores reported in the sanctuary [22]. Except for Common Langur the other wild prey species density is low in this Sanctuary (Hanuman Langur 11.00/km^2, Wild Boar 2.00/km^2, Chinkara 1.50/km^2, Nilgai 0.50 km^2 and Chital 0.20 km^2). There are 32 villages located in and around the sanctuary. Nine Leopards were counted in Wildlife Census, 2010.

Jamwa Ramgarh Wildlife Sanctuary

Jamwa Ramgarh WS covers an area of 300 km^2. The forest types are tropical dry deciduous and tropical secondary scrub forest [23]. The dominant vegetation is *Anogeissus pendula* mixed with *Acacia senegal*, *Boswellia serrata* and *Lannea grandis*. Eleven Leopards were reported in 2007 with a density of 3.7/100 km^2 which declined to only 06 in 2010 Wildlife Census. Except for Nilgai (2.26/km^2) and Common Langur (2.32/km^2) other wild prey species densities are reported to be low (Wild Boar 0.26/km^2, Chital and Sambar 0.07/km^2). The sanctuary also supports a number of medium and lesser carnivores (hyaena, jackal, wolf, fox and Jungle Cat) [22].

Jawahar Sagar Wildlife Sanctuary

This Sanctuary covers an area of 153.41 km^2. The forest area falls under northern tropical dry-decidous mixed forest [23]. In the year 2002, eight and in 2007 only two Leopards were reported [22] which increased to 06 in 2010. One Bengal Tiger was also reported during 2003. Hyaena, jackal, wolf and fox are the other carnivores found in the sanctuary [22]. Except for Common Langur (3.16/km^2), the other wild prey species are found in low densities (Nilgai 0.33/km^2, Chinkara 0.31/km^2, Wild Boar 0.28/km^2, Chital 0.06 km^2 and Sambar 0.02 km^2).

Kailadevi Wildlife Sanctuary

Kailadevi Wildlife Sanctuary covers an area of 676.4 km^2. The forest types found are *Anogeissus* dominated forest, *Anogeissus* mixed forest, *Acacia* mixed forest and riverine forest. Presence of six tigers is reported in this sanctuary by the forest department [22]. In total, 36 adult Leopards and five cubs were reported in 2004. The abundance of wild herbivores is very low (Nilgai 0.85/km^2, Chinkara 0.50/km^2, Wild Boar 0.29/km^2, Hanuman Langur 0.15/km^2, Chital and Sambar 0.01/km^2). Hyaena, jackal, wolf, fox and Jungle Cat are the other carnivores reported in the Sanctuary [22]. In all, 09 Leopards were counted during Wildlife Survey of state forest department.

Kumbhalgarh Wildlife Sanctuary

Kumbhalgarh Wildlife Sanctuary covers an area of 608.57 km^2. As per the Champion and Seth's classification [23], the forest tract of this sanctuary falls under tropical dry deciduous forests with *Anogeissus pendula* as the dominant vegetation. This sanctuary holds the highest Leopard population in the state ($n = 89$) with the density of 14.3/100 km^2. Chital (0.42/km^2), Sambar (0.23/km^2), Nilgai (1.89/km^2), Chinkara (0.03/km^2), Wild Boar (0.41/km^2) and Common Langur (9.17/km^2) are available as wild prey species. Hyaena, jackal, wolf, fox and Jungle Cat are the other carnivores reported in the park [22]. There are 24 villages located inside the sanctuary whereas 138 villages are on the periphery.

Mount Abu Wildlife Sanctuary

Mount Abu Wildlife Sanctuary holds an area of 288.84 km^2 and is located on the western most limits of Aravalli. It is dominated by subtropical evergreen forest [23]. A total number of 35 Leopards were reported in 2007 with a density of 30.9/100 km^2. The available wild prey species are Langur (8.40/km^2), Nilgai (2.11/km^2), Wild Boar (1.55/km^2) and Sambar (0.15/km^2). There are 14 villages located inside the sanctuary and 30 villages on the periphery.

Nahargarh Wildlife Sanctuary

Nahargarh Wildlife Sanctuary covers an area of 50 km^2. Nahargarh Biological Park is being developed in the sanctuary area. This biological park spreads over 7.2 km^2

within the sanctuary. The vegetation of the region is represented by tropical dry deciduous and tropical thorn forests [23]. The forest types are *Anogeissus* dominated forest, *Anogeissus-Boswellia* mixed forest, *Acacia* mixed forest and *Butea* mixed forest. Presence of 13 Leopards was reported in this sanctuary along with jackal, hyaena, wolf, fox and Jungle Cat as other carnivores. Nilgai (8.1/km^2) and Common Langur (12.4/km^2) are the wild prey base found [22].

Phulwari ki Nal Wildlife Sanctuary

Phulwari Ki Nal Sanctuary (23°0′ to 24°0′ N; 73°0′ to 74°0′ E) covers an area of 492.68 km^2. As per the Champion and Seth's classification [23], the forest tract falls under tropical dry deciduous forests. Twelve Leopards were reported during 2007 census. The estimated density of Leopard is 2.4/100 km^2. The other carnivore species found are hyaena, jackal, fox and Jungle Cat. The wild prey species found are Chital, Wild Boar and Common Langur. A very low density of Chital 0.2/km^2 was reported in 2004 census. There are 134 villages located inside the Sanctuary whereas 53 villages are on the periphery.

Ranathambhore Tiger Reserve

The Ranthambhore Tiger Reserve (1,334 km^2) is located in eastern Rajasthan where the Aravalli Hill Range and the Vindhyan Plateau meet. The Ranthambhore National Park (392 km^2) is the core zone of the Tiger Reserve. In 1992, Kailadevi Sanctuary having area of 674 km^2 of protected forest, Sawai Mansingh Sanctuary with an area of 127 km^2 and Kualji (the then, Close Area) of 7.58 km^2 were added to the Reserve. The forest is northern tropical dry deciduous type [23], with *Anogessius pendula* as the dominant tree species followed by *Acacia leucophloea*, *Lannia coromondelica*, *Butea monosperma* and *Zizyphus mauritiana*. Ranathambore reported the presence of 29 Leopards in 2007. Chital was the most abundant wild prey available (31/km^2) followed by Common Langur (21.7/km^2), Sambar (17.1/km^2), Nilgai (11.3/km^2), Wild Boar (9.7/km^2) and Chinkara (5.6/km^2) [24]. A high density of large prey in parts of Ranthambhore makes the dry forests in western India a potential site for long-term conservation planning for tigers and Leopards because large ungulates can attain high densities even in successional and disturbed forests [25]. There are 332 villages within a 5 km radius of the reserve, four of which are inside the core area. Consequently, human pressure is high, with >143,000 heads of livestock depending on it. There are >3,000 heads of livestock residing in the core area itself. The livestock population has been increasing in the region by 2% annually [26]. The people–park interface is being addressed by eco-development projects initiated in the Tiger Reserve [27].

Sajjangarh Wildlife Sanctuary

Sajjangarh Wildlife Sanctuary covers an area of 5.19 km^2. As per the Champion and Seth's classification [23] the forest tract falls under tropical dry deciduous type. Presence of five Leopards was reported in this sanctuary in 2007 along with hyaena, jackal, fox and Jungle Cat as other carnivores. The wild prey species found are Chital (2.89/km^2), Sambar (2.31/km^2), Nilgai (4.43/km^2), Wild Boar (1.54/km^2) and Common Langur (24.08/km^2) [22].

Sariska Tiger Reserve

Sariska Tiger Reserve is situated in the semi-arid biogeographic zone of India [28]. The total area of the tiger reserve is 881 km^2, out of which core zone comprises 497 km^2, and buffer area 384 km^2. Sariska core zone is in three isolated pockets: Core I (273.8 km^2), II (126.5 km^2) and III (97.5 km^2). The status of Core I has been notified as a national park in 1982 [29]. The vegetation of Sariska corresponds to northern tropical dry deciduous forests and northern tropical thorn forest [23]. *Anogeissus pendula* is the dominant tree species covering over 35% area of the forest [29]. Two studies were conducted on Leopard population estimation using camera traps from a smaller study area (68–91 km^2) in Sariska [18, 30]. Both the studies reported a high Leopard population in the trapped area (13–16 individuals/km^2), whereas a recent study which was conducted in a large area and over a longer period estimated the Leopard population [31] to be 7 individuals/km^2. The analysis of 66 scats of Leopard showed that the Leopard preyed largely on Chital and Sambar. Considering the overall diet pattern of Leopard, Chital (27.3%) and Sambar (22.7%) constituted the major portion as the diet of Leopard followed by rodents (16.1%), birds (12.1%), cattle (10.6%), Nilgai (7.6%), Rufous-tailed Hare (6.1%) and Common Langur (6.1%), while goats (3%), Porcupine (1.5%) and plant material (1.5%) occurred in low percentage. The results obtained were compared with the previous study conducted in the same study area when tigers were present [17]. The previous study showed that rodent remains were present in Leopard scats in very high proportion (45.6%). Chital and Sambar remains were found to be 20.8% and 20%, respectively, in the Leopard scats. The higher percentage of rodents may be attributed to high rodent availability in the study area [17]. It is evident that chital and sambar have now become an important prey base for Leopard in Sariska. The Chital and Sambar were, indeed, an important prey base for tigers in Sariska before they were exterminated [32]. The previous study did not record any livestock remains in the Leopard diet. However, a recent study [31] reported that Leopards fed on cattle (10.6%) and goats (3%). There are 32 villages found inside the Tiger Reserve. The estimated overall livestock population for the National Park was 9,933 which comprised 2,643 buffaloes, 896 cows, 6,160 goats and 234 sheep. The livestock depredation by Leopard may lead to confrontation with local people.

Sitamata Wildlife Sanctuary

Sitamata Wildlife Sanctuary is situated in the southeast region of the Rajasthan, where three mountain ranges, namely, the Aravalli, the Vindhayas and the Malva Plateau, meet. The total area of the sanctuary is 422.94 km^2. As per the Champion and Seth's classification [23], the forest tract falls under tropical dry deciduous forest with *Tectona grandis* as dominant tree species. Presence of 43 Leopards is reported in this sanctuary with the density of 10.2/100 km^2. Hyaena, hyaena, jackal, fox and Jungle Cat are the other carnivores found. Nilgai, Wild Boar, Common Langur and Feral Cattle are the prey species of Leopard [22]. The densities of Nilgai and Wild Boar are 0.9/km^2 and 0.4/km^2, respectively. There are 13 villages situated inside the Sanctuary.

Todgarh Raoli Wildlife Sanctuary

Todgarh Raoli Wildlife Sanctuary covers an area of 495.27 km^2. The forest area falls under dry tropical forests as per Champion and Seth's classification [23]. The reported density of Leopard in this sanctuary is 4.2/100 km^2. The population of Leopard was 51 in 2004, which declined to 21 in 2007. The other carnivores found are hyaena, jackal, fox and wolf and Jungle Cat. The reported densities [22] of prey species are Common Langur (5.78/km^2), Nilgai (2.25/km^2), Wild Boar (0.42/km^2), Sambar (0.09/km^2) and Chinkara (0.02/km^2). Twenty-seven villages are located in and around the Sanctuary.

Conservation of Leopard

The presence of Leopard is reported from 16 protected areas in Rajasthan with an estimated total population of 527 individuals by the Forest Department based on pug mark census (Wild Animal Census, 2010). 408 Leopards inside PAs and 119 Leopards outside PAs have been found. The largest number of Leopard was reported from Kumbhalgarh WLS (n=89) followed by Sitamata WLS (n=43), Sariska (n=38), Mount Abu WS (n=35), Ranthambhore (n=29) and Todgarh Raoli WLS (n=21). These PAs have better chances of survival of Leopard as they hold large areas with adequate wild prey base. However, all PAs in the state showed decline in Leopard population over the years.

Dietary studies have found that Leopards feed on a range of prey. In the Himalayas, Leopards have been known to prey upon Goral (*Naemorhedus goral*) (Mukherjee and Mishra *pers. com.*). The Shivalik Hills of Rajaji National Park analysis of scats showed that Leopards feed on Chital, Sambar, Barking Deer, Goral and livestock (Johnsingh *pers. com.*). In Sariska, a large proportion of Leopard scats collected and analysed for the period 1988–1990 contained rodents [17]. The leopards in the Mundanthurai Plateau preyed mainly on Sambar [33] while Johnsingh [14] found that in Bandipur most of the Leopard kills were of Chital. Ravi Chellam [34] found that of the 200 leopard scats analysed from Gir, 40% contained chital remains while langur remains were found in 25% of the scats. Near Mumbai, Leopards living near urban areas survive to a large extent on domestic dogs and rodents [18]. Thus, from the available literature, it is evident that Leopard feeds on a range of prey items such as rodents, primates (Hanuman Langur) and large herbivores (livestock, Chital, Sambar).

PAs in Rajasthan with <100 km^2 area and having <10 Leopards may not sustain Leopard population in the long run due to heavy biotic interference. In these PAs, the availability of wild prey species is found to be low, resulting in Leopard predating largely on domestic livestock (cattle and goats) and dogs, thus leading to human–leopard conflicts. Further, rapid decline in forest resources (fuel wood, fodder) in these PAs is reported due to heavy dependency by local people. The increasing human population, changing land use practices, soaring demands from urban population and more recently fast-expanding economic activity have started straining the delicate balance on which Leopard survives. Leopards, being large territorial animals, require large spaces while in small and isolated protected areas, they frequently venture out and come in direct conflict with local people and experience high mortality. Fragmentation and loss of habitat and its quality usually result in loss of prey availability. Inadequate prey base affects Leopard's reproductive success and its survival. All the PAs in Rajasthan where Leopard is present are more or less isolated. More recently poaching for commercial use is a serious threat to Leopard population. No research has been conducted on Leopard in Rajasthan, except in Sariska. More scientific information is required on Leopard population, habitat use, prey availability, food habits and human–leopard conflict for effective management of the species. Please see Chap. 2 for more pictures.

Acknowledgements We thank the Director and Dean, Wildlife Institute of India, for extending support to work in Sariska and the Rajasthan Forest Department under the "Leopard Ecology Project".

References

1. Kitchener A (1991) The natural history of the wild cats. Christopher Helm, A & C Black, London, pp 280
2. Scott J (1988) The Leopard's tale. Jonathan Scott. Elm Tree Books, London, p 192
3. Bailey TN (1993) The African leopard, Ecology and behaviour of a solitary felid. Columbia University Press, New York, pp 429
4. Daniel JC (1996) The leopard in India—a natural history. Natraj Publishers, Dehra Dun, India, pp 228
5. Miththapala S, Seidensticker J, Phillips LG, Fernando SBU, Smallwood JA (1989) Identification of individual leopards (*Panthera pardus*) using spot pattern variation. J Zool Lond 218: 527–536
6. Prater SH (1980) The book of Indian animals. Bombay Natural History Society, Bombay, Oxford University Press, Mumbai, India, pp 483

7. Myers N (1976) The leopard, *Panthera pardus* in Africa. IUCN monograph no. 5. pp 62
8. Anonymous (1993) The Wildlife (Protection) Act 1972. Natraj Publishers, Dehradun, India
9. Martin RB, De Meulanaer T (1988) Survey of the status of the leopard (*Panthera pardus*) in Sub-Saharan Africa. Secretariat on the Convention on International Trade in Endangered Species of Wild Fauna and Flora, Lausanne, Switzerland
10. Nowell K, Jackson P (1996) Wild Cats: status survey and Conservation Plan. Gland, Switzerland. IUCN/SSC Cat Specialist Group; p 12–16
11. Seidensticker JC, Sunquist ME, McDougal CW (1990) Leopards living at the edge of Royal Chitawan National Park, Nepal. In: Daniel JC, Seraro JS (eds) Conservation in developing countries: problems and prospects. Bombay Natural History Society, Oxford University Press, India, pp 415–423
12. Eisenberg JF, Lockhart M (1972) An ecological reconnaissance of Wilpattu National Park, Ceylon. Smithsonian Contrib Zool 101:1–118
13. Santiapillai C, Chambers MR, Ishwaran N (1982) The leopard *Panthera pardus fusca* (Meyer, 1794) in the Ruhuna National Park, Sri Lanka and observations relevant to its conservation. Biol Cons 23:5–14
14. Johnsingh AJT (1983) Large mammalian predators in Bandipur. J Bomb Nat Hist Soc 80:1–57
15. Rabinowitz AR (1989) The density and behaviour of large cats in a dry tropical forest in Huai Kha Khaeng Wildlife Sanctuary, Thailand. Nat Hist Soc Bull Siam Soc 37(2):235–251
16. Karanth KU, Sunquist ME (1995) Prey selection by tiger, leopard and dhole in tropical forests. J Anim Ecol 64:439–450
17. Sankar K, Johnsingh AJT (2002) Food habits of tiger (*Panthera tigris*) and leopard (*Panthera pardus*) in Sariska Tiger Reserve, Rajasthan, India, as shown by scat analysis. Mammalia 66(2):285–289
18. Edgaonkar A, Chellam R (1998) A preliminary study on the ecology of the leopard (*Pathera pardus fusca*) in the Sanjay Gandhi National Park, Maharashtra. RR-98/002. Wildlife Institute of India, Dehra Dun, p 33
19. Kawanishi K, Sunquist ME (2004) Conservation status of tigers in a primary rainforest of Peninsular Malaysia. Bio Conserv 120:329–344
20. Qureshi Q, Edgaonkar A (2006) Ecology of leopard in Satpura-Bori conservation area, Madhya Pradesh. Interim Report. Wildlife Institute of India, Dehra Dun
21. Goyal SP, Chauhan D (2007) Status and ecology of leopard in Pauri Garhwal: Ranging patterns and reproductive biology of leopard (*Panthera pardus*) in Pauri Garhwal. Comprehensive Report 2000–2006. Wildlife Institute of India, Dehra Dun, p 48
22. Anonymous (2007) A Wildlife Census report on wild animals in National Parks and Sanctuaries, Rajasthan. Unpublished report. Rajasthan Forest Department, Jaipur, India
23. Champion HG, Seth SK (1968) A revised survey of the forest types of India. Manager of Publications, Govt. of Indian Press, New Delhi, pp 402
24. Bagchi S, Goyal SP, Sankar K (2003) Prey abundance and prey selection by tigers (*Panthera tigris*) in a semi-arid, dry deciduous forest in western India. J Zoo Lond 260:285–290
25. Karanth KU, Sunquist ME (1992) Population structure, density and biomass of large herbivores in the tropical forests of Nagarhole, India. J Trop Ecol 8:21–35
26. Jain P (2001) Project Tiger Status Report, Project Tiger, MoEF, New Delhi, Govt., of India
27. MacKinnon K, Mishra H, Mott J (1999) Reconciling the needs of conservation and local communities: global environment facility support for tiger conservation in India. In: Siedensticker J, Christie S, Jackson P (eds) Riding the tiger: tiger conservation in human dominated landscapes. Cambridge University Press, Cambridge, UK, pp 307–315
28. Rogders WA, Panwar HS (1988) Planning a Wildlife Protected Area Network in India. Wildlife Institute of India, Dehra Dun, Uttrakhand, India, pp 339
29. Sankar K (1994) Ecology of three large sympatric herbivores (chital, sambar, nilgai) with special reference to the reserve management in Sariska Tiger Reserve, Rajasthan. PhD Thesis. University of Rajasthan. Jaipur, India

30. Chauhan DS, Harihar A, Goyal SP, Qureshi Q, Lal P, Mathur VB (2005) Estimating leopard population using camera traps in Sariska Tiger Reserve, Wildlife Institute of India, Dehradun, India, p 23
31. Sankar K, Qamar Qureshi, Krishnendu Mondal, Worah D, Srivastava T, Gupta S et al. (2008) Ecological studies in Sariska Tiger Reserve, Rajasthan. Final Report. Wildlife Institute of India, Dehra Dun, p 145
32. Sankar K, Goyal SP, Qureshi Q (2005) Assessment of status of tiger (*Panthera tigris*) in Sariska Tiger Reserve. A report submitted to the Project Tiger Directorate. Wildlife Institute of India, Dehradun, p 39
33. Sathyakumar S (1992) Food habits of leopard (*Panthera pardus*) on Mundanthurai plateau, Tamil Nadu, India. Tiger Paper 21:8–9
34. Chellam R (1993) Ecology of the Asiatic lion *Panthera leo persica*. PhD Thesis, Saurashtra University. Rajkot, Gujarat, India

Chapter 18
Small Cats of Rajasthan

Shomita Mukherjee

Abstract This chapter describes the least thought of but important felines "the small Cats." Out of the 36 species of cats found worldwide, India is home to 15 species of cats, the highest any country has and hence, is truly the "land of cats." Despite this, apart from the large cats (lion, tiger, leopard and Snow Leopard), practically nothing is known of the smaller species. In some cases, such as the Fishing Cat and the Rusty-spotted Cat, to date, even their distributions are not clearly documented. The Rusty-spotted Cat was first recorded from Rajasthan only in the 1990s. This hiked up Rajasthan's score of cats to seven, including 05 small cats, namely Fishing Cat, Caracal, Jungle Cat, Wild Cat/Indian Desert Cat, and the Rusty-spotted Cat. In this chapter, the author discuses whatever information is available on these cats till date, including distributions at various scales, i.e., global, Indian, and in Rajasthan along with the feeding behavior, habitat choice, the time of activity, competition, reproduction, and taxonomy. The author has also mentioned the various techniques used for the study of cats and the potential for future work on this family. Finally, conservation issues related to small cats in Rajasthan have been looked into and ways to address these issues have been suggested.

Introduction

India's geographic location is fascinating from a bio-geographic perspective. Species from various major bio-geographic realms (Ethiopian, Palaearctic, and Oriental) have been able to colonize the region from different directions [1].

S. Mukherjee (✉)
Sálim Ali Centre for Ornithology and Natural History (SACON),
Coimbatore (Tamil Nadu), India

National Centre for Biological Sciences, Laboratory-3, GKVK Campus,
Bellary Road, Bengaluru (Karnataka), India
e-mail: shomitam@gmail.com

B.K. Sharma et al. (eds.), *Faunal Heritage of Rajasthan, India: General Background and Ecology of Vertebrates*, DOI 10.1007/978-1-4614-0800-0_18,

Additionally, a vast array of habitats allows species to exploit newer avenues wherever possible. A great example of this is reflected in the family Felidae. Fifteen of the 36 species of cats that occur worldwide are found in India, the highest for any country [2]. This is despite the loss of the Cheetah (*Acinonyx jubatus*) in middle of the last century, a grim reminder of how a successful and widely distributed species can be driven to extinction in decades by various anthropogenic actions. A recent example from Rajasthan is the extinction of tigers from Sariska Tiger Reserve. With a population of a billion, and still expanding, these are matters of concern especially when dealing with species that are relatively rare, such as felids because, protected areas within the country cover a tiny fraction of the geographical area, perhaps a higher population of small cats exists outside, in human dominated landscapes.

Rajasthan is home to seven species of cats. For some, like the Caracal (*Caracal caracal*) and the Wild Cat/Desert Cat (*Felis silvestris ornata*), India forms the eastern limit of geographical distribution while the Rusty-spotted Cat (*Prionailurus rubiginosus*) is endemic to India and Sri Lanka. According to the IUCN (2012), which assesses global populations of species, the Fishing Cat (*Prionailurus viverrinus*) is Endangered, the Rusty-spotted Cat is Vulnerable C2 a(i), while the Wild Cat, Caracal, and Jungle Cat (*Felis chaus*) are of Least Concern category [3]. The CITES places Indian populations of Caracal and Rusty-spotted Cat in Appendix I, and the Wild Cat, Jungle Cat and Fishing Cat in Appendix II. The Indian Wildlife (Protection) Amendment Act, 2006 [4] places all Indian small cats in Schedule I, except the Jungle Cat which is placed in Schedule II.

The biggest threat to small cats is habitat loss and degradation, with poaching perhaps at the second place [5]. According to the latest (2001) census figures, human densities touch 165 per km^2 in Rajasthan, with a decadal growth rate of 25%, as compared to the national average of 21.34% (http://www.rajrelief.nic.in/humancensus.htm, accessed on 3rd January 2011). Approximately 70% of this population is rural. The livestock density is around 140 per km^2. To most, this would seem like an ecological/conservation disaster, and yet, five small cats along with their array of prey and competitors still persist in this region. Unfortunately, we do not know much about their ability to pull through into the future.

How Does One Study Cats?

Much of our lack of knowledge on cats stems from their very nature. Being rare, secretive, small, solitary and nocturnal, they are extremely difficult to study. Nevertheless, rapid progress in technology has reduced the challenge and obtaining vital information on this group is now possible.

The techniques used for any study are contingent upon the questions asked in that study, which can range from the academic to the applied, and most often it is difficult to tear these apart. Most studies, if appropriately designed, can contribute towards conservation. However, the most popular questions asked in conservation are oriented towards surveys of presence in various habitats and abundances.

Many longer term studies in ecology focus on diet, social systems, spatial distributions, competition, resource distribution, etc. The difficulty arises while designing a study that focuses on rare and elusive species. This is particularly difficult when addressing small cats, since many of them occur together, and indirect evidence such as scat (carnivore fecal matter) and tracks while indicating the presence of a cat, cannot visually be separated to species, since all cats have similar looking scat and tracks. This is where technology plays a pivotal role.

As mentioned earlier, Rajasthan is home to seven cats. These include the Jungle Cat, Caracal, Wild Cat, Rusty-spotted Cat, Fishing Cat, Tiger (*Panthera tigris*), and Leopard (*Panthera pardus*). Thus, the largest (Tiger, avg. body mass = 150 kg) and the smallest (Rusty-spotted Cat, avg. body mass = 1.2 kg) cats in the world are found in the state. Apart from the Jungle Cat, Caracal, and to a very minor extent, the Fishing Cat, no other small cat has been studied in Rajasthan. Until recently (1990s), it was not even known that the Rusty-spotted Cat occurred in Rajasthan!

In this chapter, the author has demonstrated the way small carnivores have been studied in the past and how new technology has influenced carnivore studies currently.

Study Area

The field work in STR was conducted for three years from 1994 to 1996 and data analysis and writing took another two years. I addressed questions related to co-occurrence or sympatry among three small carnivores, Golden Jackal (*Canis aureus*), Jungle Cat, and Caracal. Sariska Tiger Reserve is located between 74°17′ to 76°34′N and 25°5′ to 27°33′ E, in eastern Rajasthan. Encompassing an area of 800 km^2, Sariska has three core areas of which core 1, with an area of 273.8 km^2, forms the proposed National Park. This area functioned as the study site and the intensive study area was approximately 30 km^2. STR is characterized by rugged terrain, valleys and plateaus with an altitudinal variation from 540 to 777 m.

Objectives and Methods

With the objectives of finding whether these three carnivores use habitats and diets differentially, an initial reconnaissance and vegetation characterization was done to categorize various habitats (open scrub, dense scrub, grassland, hill forests, and mosaic). For further categorization, circular plots of 10 m radius were marked, in which the number of trees of various species, number of bushes and ground cover (a visual percentage estimate) were counted.

For identifying habitat differentiation, indirect as well as direct techniques were used. Track counts were used (expressed as percent visitation rates) on track plots spaced 500 m apart along dirt tracks in each habitat, as indirect evidence of habitat use by predators. It was also used as an evidence for presence of prey. Every morning, I would record the species that moved on these track plots through the tracks they left behind and prepare the plot for the next 24 h, 100 such plots scattered around STR in various habitats were collected.

Direct methods involved encounter rates and time of sightings of predators, through vehicle drive counts and on several one km feet transects through various habitats. Similarly, I quantified prey through sightings of ground birds and hare on the same foot transects. For rodent prey, I used Sherman—live traps for capturing murid rodents in order to obtain estimates of relative abundance of various rodent species in the various habitats during summer and winter. I had 200 traps placed in a grid (each trap 100 m from the next) and each trapping session lasting five days and five nights, after which I moved to another habitat. In all, I had a total of 9,500 trap nights.

Since, scats have remains of undigested food; one can study what the predator's diet is. Scats were collected from the various habitats for estimating diet of the three predators. A major problem using scats was in identifying or assigning them to a species. Since all three predators were of similar body size, it was very difficult to differentiate between their scat. Though, the mean diameter of scats of the three predators varied, they were not significantly different. Since most predators have a diet rich in fat, various bile acids secreted to digest fat are excreted in the scat. These bile acids differ in the type and quantity secreted by each species. Using thin layer chromatography (TLC), these can be separated after extraction and the identity of the predator can be obtained. I used TLC to assign scats to respective species in my study. In all, I had more than 1,000 scats collected after three years. Since it was not possible to analyze all the 1,000, given the limited time frame, they were randomly sampled for each habitat. A total of 227 scats were analyzed for diet.

Once the scats were classified, they were teased apart and the various food items were separated and results were presented as percentage of scats having a particular food item. Teeth of rodents came out intact in scat, and from the pattern of dentition, the species of rodent eaten can be identified. Moreover, by counting the number of jaws in the scat, I could estimate the number of rodents eaten per scat. I had measured the body mass of rodents captured in field. A combination of the number of rodents found in each scat, the average body mass of each species of rodent, and a calculation for energetics extrapolated from studies of feeding trials on Bobcats (*Lynx rufus*)and Coyotes (*Canis latrans*) were, then used for further analysis. This helped in determining the energy obtained by eating rodents alone, which was then compared among the three species [6].

Results

Results from habitat characterization showed that grass height was highest in the grassland area (winter=111 cm±7.8; summer 75 cm±7.3) (±SE) followed by dense scrub (winter=25 cm±3.17; summer=53 cm±12.6), hill forest (summer=53 cm±12.6). Open scrub and mosaic had a grass height less than five centimeter in winter and summer. The above ground grass cover (percent) in winter was similar in open scrub, dense scrub and grassland (53–63%) and was significantly higher than in mosaic (27% ± 6.5). In summer, the above ground grass cover was highest in grassland (53% ± 4.04) followed by dense scrub (38% ± 7.42), open scrub (29% ± 4.27), mosaic (24.5% ± 7.36) and hill forest (4.3% ± 3.05). Disturbance from domestic livestock quantified through track counts, showed open scrub and mosaic to be the most disturbed habitats during both seasons.

Potential prey species identified were ground birds (class 1 ≤ 250 g, class 2 = 250–500 g, and class 3 ≥ 1 kg), murid rodents, and hare. Track plot results show that in winter, peafowl abundance was highest in mosaic (70% ± 4.6), partridge in open scrub (68% ± 5.0), and small birds were equally abundant in open scrub, dense scrub, grassland, and mosaic (59–60%). Use of hill forests was significantly ($p < 0.05$) lower than all other habitats for all three classes of birds. In summer, track counts of birds were similar in all habitats. Visitations to plots by hare was highest in the open scrub during both seasons (winter = 22% ± 3.9; summer = 17% ± 4.6). Encounter rates of ground birds on transects showed similar results as track plots for peafowl, which were encountered most often in mosaic (1.4/km ± 0.31), and was significantly higher than in other habitats. There was no significant difference in encounter rates of other prey among the various habitats in winter. In summer, no significant difference was noted among different habitats for any prey.

Trapping success of murid rodents was very low (0.9%). Three murid rodents were captured: *Tatera indica* found in dense scrub and grassland, *Golunda ellioti* in grassland and *Mus platythrix* in open scrub, mosaic, dense scrub, and hill forest. Murid rodents showed an inverse relationship with use of habitat by domestic livestock and were more abundant in dense scrub and grassland.

Overall visitation rates on track plots for Golden Jackal were higher than for Jungle Cat in all habitats except the dense scrub. Track counts showed that in winter, Golden Jackal used open scrub (winter = 32.2% ± 5.5) and mosaic (winter = 22.8% ± 4.0) significantly more than other habitats which had visitation rates between 0.4 and 10.3%. Hill forest had a very low visitation rate (0.4 ± 0.48) and was significantly lower than all habitats. In summer, mosaic and hill forest could not be monitored and there was no significant difference in visitation rate by Golden Jackal in the remaining habitats (4.0–13.5%). Jungle Cat showed greater use of dense scrub in winter (20.4% ± 4.0) than all other habitats which had visitation rates ranging from 3.0 to 8.1%. In summer, though dense scrub (7.7% ± 2.9) had higher visitation than other habitats, the difference was not significant. Despite the efforts, only four sightings of Caracal were obtained: three from open scrub and one from grassland. Track counts also indicated the rarity of this species, as tracks were seen only thrice in the open scrub and once in hill forests. Direct sightings showed the same trend as track counts in winter for Jungle cat ($r = 0.71$, $p = 0.17$) and Golden Jackal ($r = 0.93$, $p < 0.05$). In summer, however, in the case of Jungle Cat, the results from the two methods varied. Both Jungle Cat and Golden Jackal showed high crepuscular activity, but Golden Jackal was cathemeral, whereas the Jungle Cat was crepuscular and nocturnal. Vehicle drive counts were the most reliable way of obtaining sightings of predators.

To sum up the results, through tracks, direct sightings and scat encounters, it was observed that the felids used the scrub more than other habitats, while the Golden Jackal used all other habitats (except the hill forests) equally. Within the scrub, the Jungle Cat used the relatively less disturbed, more vegetated dense scrub, which had a higher rodent abundance, more than the disturbed open scrub.

Scat analysis showed that the three predators were mostly eating small mammals which occurred in 90% of all scats of which 75% contained rodent remains. Birds formed the next important prey and occurred in more than 35% of scats for all three predators. Observations showed that scavenging from large cat kills formed a major source of food for the Golden Jackal.

Although biomass of rodents consumed did not differ among the predators, the energy obtained through rodents alone, differed considerably. The felids obtained more than 50% of their daily metabolisable energy (ME) from rodents alone, while for the Golden Jackal this amounted to 32–45% [6]. Results showed that each Jungle cat in STR ate between three to five rodents per day, and hence approximately 1,095–1,825 per year, while the Caracal ate eight to nine rodents per day, amounting to 2,920–3,285 rodents per year, and the Golden Jackal ate between five to six rodents per day and 1,825–2,190 rodents per year. Together, one Jungle Cat, one Caracal, and one Golden Jackal eat 5,840–7,300 rodents per year. One does not really have to look for more reasons to conserve these species!

Discussion

Evidence suggests that habitat loss is a major problem for species like the Fishing Cat and perhaps the Caracal. However, since no quantified study exists to demonstrate this, it is difficult to recommend conservation actions and policy for this region. A matter of major concern, though, is government schemes of developing wasteland. Although classified by the Ministry of Environment and Forests, the categories of land that are designated as wasteland are appalling, without any ecological thought, concern or knowledge [7–9]. An example of this is the Jaisalmer district, of which, shockingly 70% is classified as "wasteland." Forty-four percent of this falls in one category of wasteland called "sandy areas," which is obviously a natural ecosystem in an arid region (http://dolr.nic.in/WastelandsAtlas2005/Rajasthan.pdf)! On the other hand, the Desert National Park (a Protected Area with the status of a sanctuary) is partly located in Jaisalmer [7]. Hence, the same natural ecosystem, on one hand, is afforded the highest protection (on paper), while on the other, is declared a wasteland!

This can only be rectified if we have more information on the presence of these species in various land regimes and how policies impact their persistence in these land categories. It is very likely that some policies, though not consciously intending to, may benefit small cats, while on the other hand they could also severely impact their persistence. However, it is clear that we need more information to plan future conservation actions and policies. The encouraging bit is that it is very possible now to gather such information.

Until now most techniques used to study small carnivores relied heavily on field oriented work and very little laboratory based work. These were very tedious and most often could only focus on a limited geographical scale due to logistical constraints. Techniques very popular amongst carnivore ecologists have been camera trapping and radio telemetry. Both these, though extremely useful in providing vital information, have their limitations and disadvantages, especially considering more suitable technology available in current times. Camera trapping is logistically difficult and has severe limitations in human dominated landscapes. Theft is common, and moreover, with small cats that do not have strong coat patterns (Caracal, Jungle Cat, Rusty-spotted Cat), the information obtained is limited to presence and at times gender determination. Radio-telemetry is very invasive, since it involves trapping, tranquilizing, and collaring individuals. Furthermore, the individual is burdened with the collar for the rest of its life.

It must be stressed here, however, that in many cases, and in the past with no other better techniques available, this has proved to be an extremely important method in gathering information on species. To this date, there are some questions related to ranging, daily movement, breeding, hunting strategies, interactions with con-specifics and other competitors, among others, that cannot be answered using any other method. At the same time, this technique should be used only when there are no other options, and when the question to be addressed is specific and necessitates the use of this tool.

Throughout the chapter, the importance of new technology and its integration into conservation has been pointed out. Unfortunately, state-of-the-art molecular technology is not very popular in ecological studies in India, but it has tremendous potential and can be an extremely elegant and powerful way to study small cats. Perhaps, amongst most other techniques used on the field to track predators (track counts, direct observations, telemetry) and even thin layer chromatography to assign scats to species which has an error of up to 30% in identification [10], molecular technology can be used more effectively to address the same issues.

When an animal defecates, the cells of the lining of the intestine get sloughed off with it, coating the scat on the outer surface. These cells contain DNA, the blueprint for the making of the individual, linking species and taxa further back on the evolutionary path. By extracting and analyzing various portions of the DNA, a variety of information can be gathered. On the broadest scale, the species can be detected and one can zoom in further to tell the gender, individual, relatedness of that individual to others around (when their DNA is also obtained), and a host of other information that was unthinkable to obtain even a couple of decades ago. This has immense applications in surveys. If scats are picked up along with GPS (Global Positioning System) co-ordinates, species, individuals, genders, etc., can be mapped spatially. With Remote Sensing Imagery and digital maps at varied scales, barriers for species and individual movements, habitat preferences, spacing among competitors and more, can be picked up. Based on these, predictions can be made over landscapes and land regimes for policies that should be implemented. Conservation and landscape genetics studies are not just an idea but have been conducted in many regions for many species [11–17].

What is most fortunate is that the complete cat genome has been deciphered [18]. This has opened up a new world to cat ecologists, since some sequences and molecular markers are now available, or at least easier to design, for all the 36 species of cats in the world [19, 20]. Using this, I am currently studying the Jungle Cat and Leopard Cat (*Prionailurus bengalensis*) across India and looking at how genetic variation is related to spatial distributions. This will enable us to pick out populations that are unique and need immediate conservation attention.

In the future, molecular tools to survey endangered and rare species like the Fishing Cat and Caracal along with other co-existing cats, not only in Rajasthan but in other parts of India over varied land regimes and spatially map their locations for future monitoring are planned.

Recommendations

Though protected areas are a necessity, it is not possible to have them over large continuous landscapes, and many small species such as small cats exist in larger numbers outside PAs. Over the years, difficulty in surveying rare and elusive species has been a reason for lack of initiative in conducting studies on small cats all across India. We now have techniques that can generate the data required over large landscapes.

The biggest threat to small cats is habitat loss and degradation, with poaching perhaps coming in at second place. Habitat loss can be minimized through informed land policies that take into account human welfare and conservation issues. Small cats have demonstrated their value as natural rodent controllers. In many parts of Rajasthan, where the human population is predominantly vegetarian and agrarian, with some support from good land policies and awareness, natural populations of small wild cats can thrive without conflict. Having large bushes within crop fields that provide shelter for the cats, during the day and through the breeding season, is one step towards this.

A review of the current Wasteland Development Program is an absolute prerequisite for conservation planning. Rajasthan with its unique ecosystems like deserts, wetlands, and semi-arid landscapes can pave way for robust conservation planning by integrating scientific research with land policy that other states can follow.

Acknowledgments Most of the work presented here was conducted at the Wildlife Institute of India (WII). The studies have been conducted as the part of my doctoral work at the Sariska Tiger Reserve (STR). I thank everyone at WII for the help and support. I specially thank my guide Dr. A.J.T. Johnsingh and my supervisor Dr. S.P. Goyal for support during my PhD work. I thank the Forest Department of Rajasthan for extending all the support, not only during my PhD work but also for my current work. I thank Dr. Uma Ramakrishnan of the National Centre for Biological Sciences (NCBS), and everyone from laboratory 3 of NCBS for my work on the genetic aspects of small cats.

Epilogue from the Editors The leading English daily of India 'Times of India' (Jaipur edition, 26 June, 2012) has published a report on the front page quoting Mr. RS Shekhawat, Field Director, Sariska Tiger Reserve that, a Rusty-spotted Cat – one of the smallest (35 to 45 centimeters in length with a 15–30 centimeter tail and weighing between 900 gm to 1.6 kg) known wild cats in the world, was recently sighted in the mountain near the Dakkan Chowki area of Sariska Tiger Reserve, Alwar, Rajasthan. According to previous reports, a pair of the cat was also sighted at the Baalquila area of Alwar. Their status is listed as vulnerable by International Union for Conservation of Nature (IUCN). Please see Chap. 2 for the picture of Rusty-spotted Cat (Fig. 2.119).

References

1. Eisenberg JF (1981) The mammalian radiations. Athlone, London
2. Nowell K, Jackson P (eds) (1996) Wild cats, status survey and conservation action plan. IUCN, Gland Switzerland
3. IUCN (2011) IUCN Red List of Threatened Species. Version 2010.4. www.iucnredlist.org. Downloaded on 3 Jan 2011
4. Anonymous (1992) The Indian Wildlife (Protection) Act 1972. Government of India. Natraj Publishers, Dehradun
5. Sharma VD, Sankhala K (1984) Vanishing cats of Rajasthan. In: The plight of cats. Proceedings of the meeting and workshop of the IUCN/SSC Cat Specialist Group, Madhya Pradesh, India, pp 117–135
6. Mukherjee S, Goyal SP, Johnsingh AJT, Leite Pitman MRP (2004) The importance of rodents in the diet of Jungle cat (*Felis chaus*), Caracal (*Caracal caracal*) and Golden jackal (*Canis aureus*) in Sariska Tiger Reserve, Rajasthan, India. J Zool (Lond) 262:405–411
7. Rahmani AR (1989) The uncertain future of the Desert National Park in Rajashtan, India. Environ Conserv 16:237–244
8. Prakash I (2001) Biological invasion and loss of endemic biodiversity in the Thar Desert. Resonance. 76–85
9. Joshi V, Sharma RC, Singh M, Singh H, Sharma K, Sharma Y, Adha S (2005) Entomological studies on malaria in irrigated and non-irrigated areas of Thar desert, Rajasthan, India. J Vect Borne Dis 42:25–29
10. Taber AB, Novaro AJ, Neris N, Colman FH (1997) The food habits of sympatric jaguar and puma in the Paraguayan Chaco. Biotropica 29:204–213
11. Ernest HB, Penedo MCT, May BP, Syvanen M, Boyce WM (2000) Molecular tracking of mountain lions in the Yosemite Valley region in California: genetic analysis using microsatellites and faecal DNA. Mol Ecol 9:433–441
12. Farrell LE, Roman J, Sunquist ME (2000) Dietary separation of sympatric carnivores identified by molecular analysis of scats. Mol Ecol 9:1583–1590
13. Eizirik E, Kim JH, Menotti-Raymond M, Crawshaw PG, O'Brien SJ, Johnson WE (2001) Phylogeography, population history and conservation genetics of jaguars (*Panthera onca*, Mammalia, Felidae). Mol Ecol 10:65–79
14. Manel S, Schwarts MK, Luikart G, Taberlet P (2003) Landscape genetics: combining landscape ecology and population genetics. Trends Ecol Evol 18:189–197
15. Coulon A, Guillot G, Cosson JF, Angibault JMA, Aulagnier S, Cargnelutti B et al (2006) Genetic structure is influenced by landscape features: empirical evidence from a roe deer population. Mol Ecol 15:1669–1679
16. Holderegger R, Wagner HH (2006) A brief guide to landscape genetics. Land Ecol 21:793–796

17. Storfer A, Murphy MA, Evans JS, Goldberg CS, Robinson S, Spear SF et al (2007) Putting the "landscape" in landscape genetics. Heredity 98:121–142
18. O'Brien SJ, Johnson W, Driscoll C, Pontius J, Pecon-Slattery J, Menotti-Raymond M (2008) State of cat genomics. Trends Genet 24:268–279
19. Menotti-Raymond M, David VA, Lyons LA, Schaffer AA, Tomlin JF, Hutton MK, O'Brien SJ (1999) A genetic linkage map of microsatellites in the Domestic Cat (*Felis catus*). Genomics 57:9–23
20. Johnson WE, Eizirik E, Pecon-Slattery J, Murphy WJ, Antunes A, Teeling E, O'Brien SJ (2006) The late Miocene radiation of modern Felidae: a genetic assessment. Science 311:73–77

Chapter 19
Major Canids of Rajasthan

Satish Kumar Sharma and B.K. Sharma

Abstract This chapter presents the status report of Dhole or Wild Dog and a brief account of the Indian Gray Wolf (*Canis lupus pallipes*) belonging to family Canidae in Rajasthan which is considered as Endangered mammal of India. Out of the three subspecies of Wild Dog known in India, the race known from the geographic limits of Rajasthan is *Cuon alpinus dukhunensis or Dhole*. Once upon a time, this species was present in the forests of Ajmer, Alwar, Baran, Chittourgarh, Pali, Pratapgarh, Sawai Madhopur, and Udaipur districts. *Dhole* was reported from Sariska and Ranthambhore National Parks and Shahabad zone during 1980–1990. To and fro movements of wild dog were recorded between forest areas of Sawai Madhopur and Baran districts and the surrounding areas of Rajasthan and Madhya Pradesh states of India. Habitat loss, rapid decline in prey species, and human-animal conflicts are described as the major reasons of the extermination of this species. No confirmed records of its presence in the state during the running decade are available. The Indian wolf was present in various habitats of Barmer, Jodhpur, Jaisalmer, Nagour, Sikar, Jhunjhunu, Churu, Jalore, Pali, Bikaner, Udaipur, Rajsamand, Jaipur, Ajmer, Dungarpur, Alwar, and Chittourgarh districts of the state about forty years ago, but now, they are rare in most of their distribution ranges. At present, Kumbhalgarh Wildlife Sanctuary is one of the best sites in Rajasthan to observe them where they are also breeding successfully. The major conservation threats to this species such as habitat loss, food scarcity, and human-animal conflict have been discussed in the text. The two other canids, namely, Golden Jackal *Canis aureus* and Indian Fox *Vulpes bengalensis*, have also been discussed in brief.

S.K. Sharma (✉)
Sajjangarh Wildlife Sanctuary, Department of Forest and Wildlife,
Government of Rajasthan, Udaipur, Rajasthan, India
e-mail: sksharma56@gmail.com

B.K. Sharma
Department of Zoology, R.L. Saharia Government P.G. College,
Kaladera (Jaipur), Rajasthan, India
e-mail: drbksharma@hotmail.com

B.K. Sharma et al. (eds.), *Faunal Heritage of Rajasthan, India: General Background and Ecology of Vertebrates*, DOI 10.1007/978-1-4614-0800-0_19,

Introduction

The canids are predatory mammals belonging to the family Canidae and are represented by 14 genera and 36 species on the earth. In all, six species are present in India: one species each of wolf, jackal and Wild Dog and three species of foxes [1]. Rajasthan itself has five species, namely, Indian or Gray Wolf (*Canis lupus pallipes*), Golden Jackal *Canis aureus*, Indian or Bengal Fox *Vulpes bengalensis*, and Desert Fox or White-footed Fox *Vulpes vulpes pusilla*. Dhole (Wild Dog) or *Cuon alpinus dukhunensis* has not been sighted since last two decades. Somewhat similar to dogs with bushy tails and five digits, these animals are adapted to cursorial life, and their long canines are prized possessions that have scared humans since time immemorial. The status and distribution of major canids in Rajasthan are given below.

Dhole or *Cuon alpinus dukhunensis*

Dhole (Wild Dog) or *Cuon alpinus dukhunensis* is known by various local names in Rajasthan. A few of the known local names in vogue in the state are *Karu*, *Kona*, and *Kona Karu* in the southern districts such as Udaipur, Chittourgarh, Jojawar, and Desuri (Godwar area of Pali district). The Dhole or Indian Wild Dog belonging to family Canidae, is a social animal which lives and hunts in a pack. This species is distributed in central and eastern Asia from the Altai Mountain and Manchuria, southward through the forest regions of India and the Malay countries [1]. In India, the Wild Dog is present in peninsular, central, northeast India, trans-Himalaya, and Ladakh to Sikkim [2, 3]. Three races are recognized within the limits of India: trans-Himalayan, Himalayan, and Peninsular. The race known from the geographic limits of Rajasthan is *Cuon alpinus dukhunensis* [2, 3].

The species was once present in the forests of Udaipur, Pratapgarh, Chittourgarh, Sawai Madhopur, Pali, Ajmer, Baran, and Alwar districts. A few records of presence of wild dogs in Rajasthan are given in Table 19.1 [4–9] which proves that the distribution range of Wild Dog was quite extensive. The westernmost distribution limit of this species in India was present parallel to the Aravallis in Rajasthan. This species was visible in many pockets of the forest areas of southern Rajasthan till 1970. This species was present from 1980 to 1990 in Sariska, Ranthambhore, and Shahabad zone. To and fro movement of this species was there between forest areas of Rajasthan and Madhya Pradesh in Sawai Madhopur, Baran, and surrounding areas (R.S. Shekhawat, Field Director, Tiger Project, Ranthambhore, Sawai Madhopur, *pers. comm.*). At present, Wild Dogs have been exterminated from Rajasthan, as no confirmed record of its presence has been available in the state for more than a decade. Three Wild Dogs were recorded in the Sariska Tiger Reserve during the summer of 1986 [5]. Its population did not increase in the reserve [6]. Now, no information is available about its occurrence in this tiger reserve. Habitat loss, rapid decline in prey species, and human-animal conflicts are the main reasons for the extermination of wild dogs in Rajasthan. Please see Chap. 3 (Fig. 3.29) for picture.

Table 19.1 Past records of the presence of wild dog or *Dhole* in Rajasthan

S. No.	District	Locality	Comments about presence (with reference)	Present status
1	Ajmer	Forests of Ajmer	Present in the past [4]	Exterminated
2	Alwar	Sariska Tiger Reserve	Three observed on April 23–24, 1986, [3, 5] but population is not increasing [6]	Exterminated
3	Baran	Forests of Shahabad	Present till 1980–1985	Exterminated
4	Chittourgarh	Forests of Chittourgarh	Present during past time [7]	Exterminated
5	Pali	Desuri, Jojawar	Present during past time [7]	Exterminated
6	Pratapgarh	(1) Pratapgarh and Chhoti Sadri	Once abundant [8]	Exterminated
		(2) Dhariyawad	Present until 1970 (Raza Tehsin, *per comm.*)	Exterminated
7	Sawai Madhopur	Ranthambhore	Present in tiger reserve [3, 5]	Exterminated
8	Udaipur	(1) Manasi, Phulwari-Ki-Nal, Mota Magra (All in Phulwari WLS)	Present until 1947–1948 (Rana Manohar Singh Solanki, *per comm.*)	Exterminated
		(2) Sendwada (Gogunda-Ogna road)	Present until 1970 [9]	Exterminated
		(3) Jaisamand	Present until 1973 [9]	Exterminated
		(4) Nal Mokhi, Palewa Ghati	Present until 1970 (Raza Tehsin, *per. comm.*)	Exterminated

Fig. 19.1 Indian Gray Wolf *Canis lupus. pallipes* (*Courtesy: Dr Satish Sharma*)

Records of the Indian Wolf (*Canis lupus pallipes*) in Rajasthan State

The wolf is known by different names in the local dialects in different areas of Rajasthan, such as *Bhedav* in Alwar district, *Nyali* and *Chhali-nar* in Jaipur and Dausa districts, and *Vari* and *Vargda* in Udaipur, Pali, Ajmer, Rajsamand, Chittourgarh, and Dungarpur districts (Fig. 19.1). A pack of wolves is called *Sata-Ravan* in many pockets of Rajasthan. A few decades ago, shepherds in Rajasthan would gift a live goat to the *shikari* that would bring the head of a wolf. People used to fear them and set fire to their dens, even if it contained a litter.

The Indian wolf is a threatened species according to Wildlife (Protection) Amendment Act, 2006 and placed in Schedule I and Appendix I of CITES. IUCN has categorized the species as Least Concerned due to its wide distribution on the globe. The Gray a Wolf of India (*Canis lupus pallipes*) is distributed in districts situated to the west of the Aravallis, in the Thar Desert area of Rajasthan. Their numbers are quite less and dwindling fast. Wolves are also present in many nondesert districts of Rajasthan like Ajmer, Alwar, and Chittourgarh. Wolves are also present in Bichhiwara Forest Range of Dungarpur district (late Phool Singh, RFO, *pers. comm.*, 2006). According to Choudhary [10], the wolf is sighted in Kailadevi WLS in Karouli district. According to Tehsin [11–13], this species is also present in many localities in Udaipur district but is confined to southern Aravalli, near Nathdwara area (now a part of Rajsamand district). According to Prakash [14], the beast is present in Sitamata WLS of Chittourgarh district. Kumbhalgarh Wildlife Sanctuary present at the confluence of Udaipur, Pali, and Rajsamand districts is inhabited by a good number of wolves. Grassy plains dotted with bushes and trees are much liked by the wolves in this sanctuary. They successfully breed in *Joba* forest area, near Ghanerao town. It is one of the best sanctuaries in Rajasthan for viewing wolves.

In 1990, their population was about 252–350 in the whole of the Rajasthan [15]. Wolves have been seen many times from February 1999 to January 2001 in the Nahargarh WLS, situated on the northern outskirts of Jaipur city, covering an area of 52.40 km^2. They prefer sandy tracts of the northern Aravalli foothill zone in this sanctuary. Many dens of wolves are present toward the Khurad village area while wolves are also present toward the Mayla Bagh area of this sanctuary. On December 8, 1999, just after the sunset, a pair of wolves entered an enclosure in Khurad village for lifting a goat, but the crowd encircled the enclosure, blocked all the escape routes and was successful in killing the female wolf. During this operation, a woman was injured by the cornered male, who succeeded in escaping from the enclosure. Wolves are also sighted toward Khania village, which is present in the south-eastern outskirts of Jaipur city. The ravines present in the east of Chulgiri Hills provide suitable habitat to wolves. This species was seen many times in this zone from 1988 to 1992 and 1999 to 2000.

Small populations of wolves have been also sighted in Shergarh, Darrah, Raoli-Todgarh, and Bandh-Baretha WLS. Raoli and Bijaji-ka-Guda are very important localities of the Raoli-Todgarh Sanctuary, where a small population of wolves can be seen due to the availability of suitable habitat and food. A population of three and five wolves was censused in Sitamata and Mt Abu WLS, respectively, during 1999. During migration of sheep herds from Bhilwara district, packs of wolves were seen following the sheep herds (Lalit Singh Ranawat, Asst. Conservator of Forests, *pers. comm.*, 2004, and BP Singh, FES, Bhilwara, 2006). Suitable habitat for wolves is present in many pockets of Bhilwara district. A small population of wolves was seen at Rojhda-ka-Badiya in Kareda *tehsil* of Bhilwara district during 2004. Wolves are also present near Gyangarh, Suliya, and Baddu villages in Mandal *tehsil* of Bhilwara district. Once, a big population was known from Gyangarh area, but now, their number is declining fast (PS Chundawat, ACF, *pers. comm.*, 2003). A few wolves are still present in Bassi WLS of Chittourgarh district (Manoj Parashar, DFO, pers. comm., 2008). A small population of the canid is seen in Chandrai Jod, *tehsil* Ahore, Jalore district. Wolves still dot the Kurabar and Haldighati area of Udaipur district. Scrub jungles of these localities prove to be a suitable habitat for wolves. In this zone, they much depend on goats and fawns of Nilgai and Chinkara. Habitat loss, food scarcity, grassland degradation, and human-animal conflicts are major conservation threats to this species.

Golden Jackal *Canis aureus*

Golden jackal *Canis aureus* is globally Low-Risk and Least Concerned species (IUCN 2012), which was abundant in the past, but its number is now declining fast in Rajasthan. Still, they are more in number than other canids. As per the Wildlife (Protcetion) Amendment Act, 2006 they are categorized as Schedule III species.

The species and its body parts are banned for trade and listed in Appendix III of CITES. They are, in general, found in the semiarid areas and semi-urban forested habitats in Rajasthan.

It is a medium-sized canid with buff gray coat, interspersed with black hair on the back, while undersides and the area around eyes and lips are white. The golden jackal is omnivorous in diet and opportunistic in feeding habit. After the decline of vultures, it plays the role of an important scavenger [16]. They are distributed widely almost in all parts of Rajasthan.

Bengal Fox *Vulpes bengalensis*

Indian fox or Bengal Fox *Vulpes bengalensis* is endemic to the Indian Subcontinent. It is a nocturnal and crepuscular animal. It inhabits semiarid, flat to undulating terrains, scrublands and avoids dense forest. They are restricted to plains and open scrub forests in the Indian Peninsula. They are placed in Schedule III of Indian Wildlife (Protection) Amendment Act, 2006 and Least Concerned species as per IUCN criteria. They have typical vulpine appearance with disproportional ears, black nose, and lips and eyes with dark tear marks [16]. Tail is long and almost half of the body length. The animal is omnivorous and an opportunistic feeder.

In Rajasthan, they are less threatened and are often seen in Desert National Park (Jaisalmer), Sariska (Alwar), Ranthambhore Tiger Reserve (Sawai Madhopur), Jaisamand (Udaipur), and Mount Abu (Sirohi) [17]. There are records of Indian fox at KNP, and one of the coeditors, Dr. Seema Kulshreshtha has seen it in the year 2003 at KNP carrying some food chunk and crossing the entrance road of the park. The animal is sensitive to habitat change and prey availability that is contributing to its declining population in Rajasthan. There is also a limited localized trade for skin, tail, fur, and teeth and claws (for medicinal and charm purpose).

Desert Fox *Vulpes vulpes pusilla*

Desert Fox or White-footed Fox *Vulpes vulpes pusilla* is the smallest and lightest of all the three foxes found in India [1]. It is also known as the White-footed Fox due to its white feet. The animal possesses short grayish fur with rusty-brown hair. It is distributed in the desert of Rajasthan, Gujarat, Punjab, and Madhya Pradesh. It is a rare species. It avoids dense forests. It plays an important role in the control of pest species and helps in the dispersal of seeds. It is placed in Least Concern (LC) category in the IUCN Red Data List (2012) and categorized as Schedule I species in Wildlife (Protection) Amendment Act, 2006. This data-deficient species needs research on its ecology and distribution, since not much is known about it.

Please see Chaps. 2 and 3 for more pictures of the Canids.

References

1. Menon V (2003) A field guide to Indian mammals. Penguin Books India (P) Ltd, pp 200
2. Prater SH (1980) The book of Indian animals. Bombay Natural History Society, Mumbai. Oxford University Press, Chicago, Illinois, USA, pp 384
3. Johansingh AJT (1999) Dhole: dog of the Indian jungle. In: Mahesh R (ed) Indian wildlife, vol II. Oxford University Press, New Delhi, pp 182–190
4. Sharma SK (2007) Study of biodiversity and ethnobiology of Phulwari Wildlife Sanctuary, Udaipur (Rajasthan). PhD. Thesis submitted to Mohanlal Sukhadia University, Udaipur
5. Divyabhanusinh (1986) Notes on Indian Wild dog (*Cuon alpinus*) in Sariska National Park. J Bomb Nat Hist Soc 83(3):654
6. Divyabhanusinh (1988) Interaction between Sambar (*Cervus unicolor*) and Indian wild dogs (*Cuon alpinus*) in Sariska National Park. J Bomb Nat Hist Soc 85(2):410–411
7. Singh KR (1989) The Sariska sanctuary. India Magazine 9:54–63
8. Sehgal KK (1977) Gazetteer of Chittorgarh. Directorate, District Gazetteer, Govt. of Rajasthan, p 457
9. Tehsin R (1986) How wild dog (*Cuon alpinus*) use their chemical weapons in hunting. Tiger Paper 13(2):23
10. Tehsin R (1987) The Wolf (*Canis lupus*) of Mewar Region, Rajasthan. J Bomb Nat Hist Soc 84(2):422–24
11. Choudhary SS (2000) Ranthambhore beyond Tigers. Himanshu Publications, Udaipur and New Delhi, pp 20
12. Tehsin R (1995) Aggressive behavior of a thirsty Leopard *Panthera pardus* (Linn.). J Bom Nat Hist Soc 92(1):112–113
13. Prakash I (1999) Biodiversity rich spots in Rajasthan-Ecological changes. Second Dr. M.L. Bhatia Memorial Lecture 1998, Feb. 4, 1999. Department of Zoology, University of Delhi, p 22
14. Jhala YV, Gile S (1991) Status and conservation of wolves in Gujarat and Rajasthan. Conservat Biol 5(4):68
15. Sharma SK (1999) A preliminary survey of biodiversity of Nahargarh Wildlife Sanctuary, Jaipur, Rajasthan (India). Wildlife Wing, Department of Forest, Rajasthan, p 41
16. Wildlife Census (2010) Forest and Wildlife Department, State Government of Rajasthan, Jaipur, Rajasthan, India
17. Vijayan VS (1983) Vertebrate Fauna of Keoladeo National Park. A Report submitted to Bombay Natural History Society, Mumbai

Chapter 20
The Status and Conservation of Sloth Bear in Rajasthan

Harendra Singh Bargali and B.K. Sharma

Abstract This chapter presents the ecology, behaviour and status of Sloth Bear in Rajasthan. Four species of bear, namely, Sloth Bear (*Melursus ursinus*), Asiatic Black Bear (*Ursus thibetanus*), Himalayan Brown Bear (*Ursus arctos*) and Malayan Sun Bear (*Ursus malayanus*), are found in India. Listed as Vulnerable by the IUCN, Sloth Bear is listed in Appendix I of CITES and is protected under Schedule I of the Indian Wildlife (Protection) Amendment Act, 2006. Sloth Bear is endemic to the Indian Subcontinent and is found in India, Sri Lanka, Nepal and Bhutan. In India, it is distributed from the southern tip of the Western Ghats mountains to the foothills of the Himalayas. The quality of its habitat is determined by the availability and seasonal variations in food, shelter and vegetation cover. Sloth Bears are reported from a few protected areas (PAs) and 13 administrative divisions of the Department of Forests, Government of Rajasthan, located in the Vindhyas and the Aravallis Hills. They are common in PAs such as Kailadevi, Kumbhalgarh, Mount Abu, Sawai Mansingh, Raoli-Todgarh Wildlife Sanctuaries and Ranthambhore National Park. The bears do not occur in the desert margin of the northern and western part of the state. Sloth Bear is a key species of the hills and mountains of the semi-arid region of western India. Conservation threats to the Sloth Bear population such as loss, degradation and fragmentation of habitats and poaching for gall bladder and other body parts have been discussed.

H.S. Bargali (✉)
The Corbett Foundation, P.O. Dhikuli, Ramnagar 244715 Nainital, India
e-mail: hsbargali@gmail.com

B.K. Sharma
Department of Zoology, R.L. Saharia Government P.G. College,
Kaladera (Jaipur), Rajasthan, India
e-mail: drbksharma@hotmail.com

B.K. Sharma et al. (eds.), *Faunal Heritage of Rajasthan, India: General Background and Ecology of Vertebrates*, DOI 10.1007/978-1-4614-0800-0_20,

Fig. 20.1 Sloth Bear *Melursus ursinus* at Aravalli Hills

Introduction

There are eight species of bear worldwide [1] which are the Malayan Sun Bear (*Ursus malayanus*), Sloth Bear (*Melursus ursinus*), Asiatic Black Bear (*Ursus thibetanus*), Brown Bear (*Ursus arctos*), American Black Bear (*Ursus americanus*), Polar Bear (*Ursus maritimus*), Giant Panda (*Ailuropoda melanoleuca*) and Spectacled Bear (*Tremarctos ornatus*). Bears have a wide global distribution and are found in every continent except Africa, Antarctica and Australia [2]. Two species occur in Europe, three in North America, one in South America and six in Asia. The bear occupies a special place by playing an important role in a number of indigenous societies and has been venerated as an object of worship for thousands of years by numerous human cultures around the world [3]. On the flip side, bear body parts have been used in traditional medicines for centuries (China over 5,000 years). Currently, the killing of bears for sports, medicinal products and protection of livestock and crops has led to the decline of the species [3–6].

Sloth Bear

The Sloth Bear is one of the four species of bears found in India. It is a medium-sized bear with a distinctively large shaggy black coat and a broad U-shaped chest blaze [7, 8] (Fig. 20.1). The hairs are especially long around the neck and at the back of the head. It has a dull white muzzle covered with thin short greyish white hairs. The Sloth Bear differs from other species of bears. It has a highly specialised morphological adaptation to feed on insects by having the absence of the first maxillary incisors, raised elongated palate, protrusible mobile lips, nearly naked mobile snout, long slightly curved front claws, long shaggy coat and nostrils which can be closed voluntarily. Adult male generally weighs 80–145 kg, while female weighs about 60–100 kg. Sloth Bears stand 65–85 cm at shoulder and are 140–170 cm long [8, 9].

The Sloth Bear is endemic to the Indian Subcontinent and is found in India, Sri Lanka, Nepal and Bhutan. In India, it ranges from the southern tip of the Western Ghats mountains to the foothills of the Himalayas. The desert region of Rajasthan limits the western distribution. Habitat degradation, diminished food sources, trade of body parts and increased conflicts with humans [7, 10–13] are posing a serious threat to the Sloth Bear population in its entire distribution range. Moreover, attacks on human and crop damage by the Sloth Bear have created fear and animosity among the public in many areas of its range [14]. Sloth Bears are listed under Schedule I of the Indian Wildlife (Protection) Amendment Act, 2006, Appendix I of CITES and as "Vulnerable" by the IUCN Red List of Threatened Species (2012) [15].

Materials and Methods

The Sloth Bear population is distributed in fragmented patches of the Aravalli Hills, especially in southern and eastern region of Rajasthan. The animal also occurs in patchy habitats of the Vindhya Hills in southeast Rajasthan. North-western deserts limit the westernmost distribution of Sloth Bear in the state.

Information on the sloth bear population has been acquired from the census figures reported by the state forest department. The number cannot be derived in a precise manner due to various reasons, but converting the numbers into a range gives an estimate of the trend in the bear population. Practically, an estimate of bear population across the state is a difficult task for any agency; therefore, information given by the concerned forest department has been considered to draw the inference.

Results

From the conservation point of view, there are protected areas supporting bear populations which seems viable at the moment. On the contrary, populations outside protected areas in territorial administrative divisions except for a few divisions are very low. Sloth Bears are reported from more than 10 PAs [16] and 12 administrative divisions outside the PAs [17]. Protected areas like Kailadevi, Kumbhalgarh, Mt. Abu, Ranthambhore, Sawai Mansingh, Todgarh Raoli, Phulwari Ki Nal and Darrah have a good population of Sloth Bear, whereas forests of Jalore, Sirohi and Udaipur support a viable population (Tables 20.1 and 20.2). Please see Chap. 2 for more pictures.

Table 20.1 Detail of Sloth Bear population inside Protected Areas in Rajasthan

S. No.	Protected Area	No. of bears Census 2007	Census 2009
1	Bhainsrodgarh WLS	1–10	1–10
2	Chambal WLS	1–10	1–10
3	Darrah WLS	11–20	11–20
4	Jawahar Sagar WLS	1–10	11–20
5	Kailadevi WLS	51–60	51–60
6	Kesarbagh WLS	1–10	1–10
7	Kumbhalgarh WLS	161–170	181–190
8	Mount Abu WLS	121–130	131–140
9	Phulwari Ki Nal WLS	11–20	21–30
10	Ranthambhore National Park	41–50	51–60
11	Sajjangarh WLS	NR	1–10
12	Sawai Mansingh WLS	21–30	21–30
13	Todgarh Raoli WLS	41–50	41–50
14	Van Vihar WLS	1–10	1–10

NR not reported

Table 20.2 Detail of Sloth Bear population outside Protected Areas in Rajasthan

S. No.	Administrative divisions	No. of bears Census 2007	Census 2009
1	Baran—DFO Baran	1–10	1–10
2	Bhilwara—DCF Bhilwara	1–10	1–10
3	Bundi—DFO Bundi	NR	NR
4	Chittourgarh	NR	1–10
5	Chittourgarh—DFO Chittourgarh	1–10	1–10
6	Dholpur	NR	11–20
7	Dungarpur	NR	1–10
8	Jalore—DCF Jalore	11–20	11–20
9	Karouli—DCF (SC) Karouli	1–10	NR
10	Kota—DFO Kota	1–10	1–10
11	Pali—DCF Pali	1–10	1–10
12	Rajsamand—DCF Rajsamand	1–10	1–10
13	Sirohi—DFO Sirohi	71–80	61–70
14	Udaipur—DCF (South) Udaipur	1–10	NR
15	Udaipur—DCF (North) Udaipur	1–10	NR
16	Udaipur—DCF (Central) Udaipur	11–20	1–10

NR not reported

Recommendations

1. The reproductive rates of bears are the lowest among the terrestrial animals [18]. If detection of population decline takes years, then even after identification, it may be difficult to reverse the trend because of the low productivity of bear population [19]. Therefore, it is recommended that the Sloth Bear population should be monitored periodically across their range in the state.

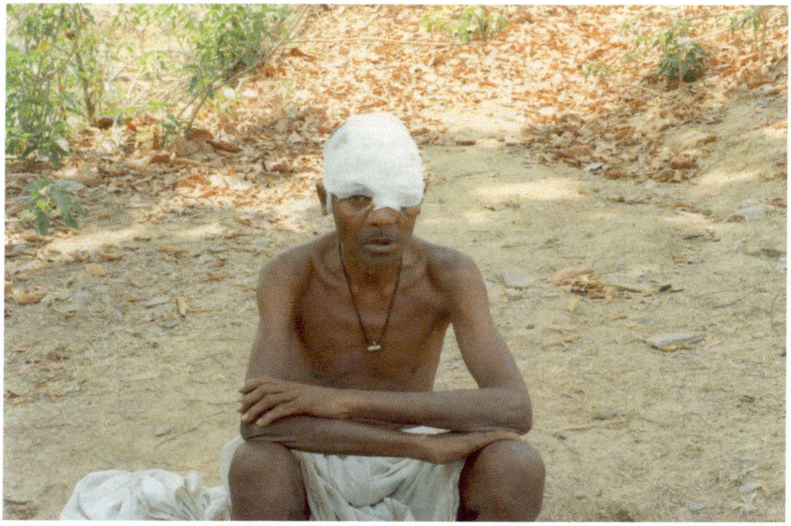

Fig. 20.2 A victim of Sloth Bear attack. Human-bear conflict is a threat to the Sloth Bear conservation

2. Considering the fragmentation of available habitat and decline in Sloth Bear population, it is recommended that important habitats should be quantified and mapped and corridors should be restored.
3. Though large populations are reported to be confined within PAs habitats outside PAs in Jalore, Sirohi, Kota and Udaipur forest divisions of Rajasthan should be managed for long-term conservation of the Sloth Bear.
4. To ensure long-term conservation of Sloth Bear in the state, there is a need to manage crucial corridors between the Rajasthan and forests of the adjoining states like Gujarat and maintain connectivity between populations, wherever possible. Regional conservation plans should be adopted especially in the western Aravallis.
5. There are reports of human-bear conflicts in and around the Mount Abu Sanctuary (Fig. 20.2). A database should be established to understand trends in human-bear conflicts to help in adopting timely measures to mitigate the problem.
6. Excluding a few PAs and forest divisions, population of the Sloth Bear, across its range in fragmented habitats, is small. If issues such as human-bear conflict and poaching are not handled with keen attention, these populations will be vulnerable to eventual extirpation.
7. Education and awareness programmes emphasising on behavioural ecology and the role of the Sloth Bear in maintaining forest ecosystem should be considered time to time. This will help in reducing fear and animosity towards bear, gain community support for anti-poaching endeavours and add to long-term conservation of the species in the state. Awareness programmes could be organised by the state forest department or non-governmental organisations or in collaboration by both.

References

1. Waits L, Paetkau D, Strobeck C (1999) Genetics of the Bears of the World. In: Servheen C, Herrero S, Peyton B (eds) Bears: status survey and conservation action plan. International Union for the Conservation of Nature and Natural Resources, Gland, Switzerland, pp 25–32
2. Nowak RM, Paradiso JL (1983) Walker's mammals of the world, vol II. John Hopkins Univ. Press, Baltimore, p 1362
3. Kemf E, Wilson A, Servheen C (1999) Bears in the Wild. WWF Species Status Report, World Wide Fund for Nature, Gland, Switzerland, pp 44
4. Mills JA, Servheen C (1991) The Asian trade in bears and bear parts. TRAFFIC-USA/ World Wide Fund, Washington DC, USA, pp 113
5. Servheen C, Herrero S, Peyton B (1999) Bears: status survey and conservation action plan. IUCN/SSC Bear and Polar group Specialist Groups, IUCN, Gland, Switzerland and Cambridge, UK, p 309
6. Read BE (1982) Chinese materia medica: animal drugs. Southern Materials Centre, Inc., Taipei, p 281
7. Laurie A, Seidensticker J (1977) Behavioural ecology of the sloth bear (*Melursus ursinus*). J Zool 182:187–204
8. Prater SH (1980) The book of Indian animals. Bombay Natural History Society, Bombay, India and Oxford University Press, Mumbai, pp 483
9. Garshelis DL, Joshi AR, Smith JLD, Rice CG (1999) Sloth bear conservation action plan. In: Servheen C, Herrero S, Peyton B (eds) Bears: status survey and conservation action plan. International Union for the Conservation of Nature and Natural Resources, Gland, Switzerland, pp 225–240
10. Rajpurohit KS, Chauhan NPS (1996) Study of animal damage problems in and around protected areas and managed forest in India. Phase-I: Madhya Pradesh, Bihar and Orissa. Wildlife Institute of India, Dehradun, India
11. Garshelis DL, Joshi AR, Smith JLD (1999) Estimating density and relative abundance of sloth bears. Ursus 11:87–98
12. Bargali HS (2003) Ecology of the problematic sloth bear (Melursus ursinus) and mitigation of human-bear conflicts in North Bilaspur forest division, Madhya Pradesh. PhD Thesis, Saurashtra University, Rajkot, India
13. Bargali HS, Akhtar N, Chauhan NPS (2005) Characteristics of sloth bear attacks and human casualties in North Bilaspur forest division, Chhattisgarh, India. Ursus 16(2):263–267
14. Bargali HS, Akhtar N, Chauhan NPS (2004) Feeding ecology of sloth bears (*Melursus ursinus*) in a disturbed area in central India. Ursus 15(2):212–217
15. Garshelis DL, Ratnayeke S, Chauhan NPS (2010) *Melursus ursinus*. In: IUCN 2010. IUCN red List of Threatened Species. 2008 Version 2010.3. http://www.iucnredlist.org. Downloaded on 07 September 2010
16. Anonymous (2009) State of forest report 2009, Ministry of Environment and Forests, Government of India, Forest Survey of India, Dehradun
17. Wildlife Management Information System (2010) Department of Forests, Government of Rajasthan. http://www.rajforest.nic.in/cwlw/cwlw_main.html. Downloaded on 07 September 2010
18. Bunnell F, Tait D (1981) Population dynamics of bears—implications. In: Smith T, Fowler C (eds) Dynamics of large mammal populations. Wiley, New York, pp 75–98
19. Miller S (1990) Population management of bears in North America. International Conference on Bear Research and Management 8:357–373

Chapter 21
Chiropteran Fauna of Rajasthan: Taxonomy, Distribution and Status

C. Srinivasulu, Bhargavi Srinivasulu, and Y.P. Sinha

Abstract This chapter gives an intensive account of the 25 species of bats recorded so far from Rajasthan. Some species are restricted in distribution, while others are widespread in the areas where suitable habitats and food resources exist. As many as 15 species from the desert region, 17 species from non-desert region and 16 species from the Aravalli Hills have been reported, out of which seven species are common to all regions of Rajasthan. No taxa are endemic to Rajasthan. Taxonomy, distribution and status of each species have been presented separately in the text. Drastic reduction in the rich diversity and extinction of many local species in Bikaner, Jaisalmer and Jodhpur districts have been attributed to the increasing tourism and the conversion and renovation of old *havelis* (big villas) and palaces into hotels and lodges. The recent appearance of fruit bats and increase in their diversity due to the changing ecology of the Thar Desert after the formation of Indira Gandhi Nahar Project (IGNP) have also been described. Implementation of a ban on disturbing bat roosts in old *havelis* and ruins has been suggested to protect their habitat. More research initiatives need to be taken up for well-planned and detailed field studies.

Introduction

Bats are the only group of mammals capable of sustained flight. This unique feat was achieved by long evolutionary history that progressively brought over morphological and physiological adaptations that were needed for nocturnal and arboreal lifestyle.

C. Srinivasulu • B. Srinivasulu (✉)
Wildlife Biology Section, Department of Zoology, University College of Science, Osmania University, Hyderabad 500 007, Andhra Pradesh, India
e-mail: hyd2masawa@gmail.com; bharisrini@gmail.com

Y.P. Sinha
Gangetic Plains Research Station, Zoological Survey of India, Road No. 11D, Rajendra Nagar, Patna 800 016, Bihar, India

B.K. Sharma et al. (eds.), *Faunal Heritage of Rajasthan, India: General Background and Ecology of Vertebrates*, DOI 10.1007/978-1-4614-0800-0_21,

Representing one-fourth of the known mammals, bats are found throughout the world excepting the cold Arctic, the Antarctica and a few oceanic islands [1]. Order Chiroptera includes about 1,117 species of bats world over in rather two unequal suborders—the Megachiroptera (consisting 186 species of Old World fruit bats in one family) and the Microchiroptera (consisting 931 species in 16 families) [2]. As many as 141 species and subspecies of bats are known from India [3].

In this chapter, we summarize the current knowledge of the chiropteran diversity of Rajasthan based on historical and contemporary primary literature and our surveys and studies in various parts of Rajasthan. The works quoted in this chapter are to the best of our present knowledge (cutoff date: 30 November 2010).

Sources of Information

Historical Information (Pre-Independence Sources)

There are very few historical sources concerning bats in Rajasthan. The first account of bats from *Rajputana* (British name for Rajasthan and its surrounding states) dates back to 1857 in the work of R.F. Tomes who provided descriptions of *Scotophilus pachyomus* (presently *Eptesicus serotinus pachyomus*) and *Vespertilio blythi* (presently *Myotis blythi blythi*) collected from Nasirabad, 130 km south of Jaipur in the present-day Ajmer district. Blanford [4] puts on record the presence of *Hesperoptenus tickelli* (Blyth, 1855) from Nasirabad. In the early twentieth century, mammal collections made by C.A. Crump at Mt. Abu under the aegis of the Bombay Natural History Society's Mammal Survey of India, Myanmar and Sri Lanka have added more information on the chiropteran diversity of Rajasthan [5].

Modern Information (Post-Independence Sources)

The first work that appeared post-Independence was that of Garg [6] who studied the skull morphology of insectivorous bats from the Ajmer district. Prakash [7, 8] synthesized the distributional information of bats in the Great Indian Desert. Advani and Sinha commenced their work in the late 1970s and have contributed significantly in this field. Sinha [9, 10] provided the first comprehensive knowledge on distribution, status and ecology of Rajasthan bats. Sinha [11] added Black-bearded Tomb Bat *Taphozous melanopogon* Temminck, 1941 to the existing list. Gaur [12] studied the general ecology of bats of Thar Desert, and Sinha [13] revisited and updated the information on bats of the Thar Desert. Advani's work chiefly related to ectoparasites of bats [14], but he also contributed significantly on ecology of bats [15–20]. Basing on his extensive works, later Advani [21] summarized the distribution and status of bats of Rajasthan.

Earlier to these and during this period, works on other aspects of bat biology and ecology were being sporadically studied in different parts of Rajasthan. Some of the important works include that of Ramaswami and Anand Kumar [22], Anand Kumar [23], Wason [24], Agarwal and Gupta [25], Lall [26], Bhupathy [27], Gupta and Trivedi [28] and Trivedi and Lall [29]. Sharma [30] reported the presence of Painted Bat *Kerivoula picta* in Rajasthan. The only report of injury caused to human due to a bat was reported by Agarwal *et al.* [31], of a case pertaining to corneal injury caused by a flying insectivorous bat in a village near Udaipur.

Post-1990, important works on bats pertain to that of Trivedi [32], Purohit and Senacha [33], Purohit and Senacha [54, 59], Senacha [34, 35], Tak and Dookia [36], Trivedi *et al.* [37], Dookia [38], Dookia and Tak [39], Purohit and Senacha [40, 41], Senacha and Purohit [42], Trivedi and Lall [43, 44], Purohit and Vyas [45], Senacha *et al.* [46], Srinivasulu and Srinivasulu [47] and Purohit *et al.* [48].

Conservation Issues

The major conservation issue is that of habitat loss due to anthropogenic activities. In many areas throughout Rajasthan, bats have been reported to use old derelict mansions (*havelis*), large wells (*baoris*) and other such man-made structures. In the recent past, due to development in tourism sector, many old *havelis* are being converted into lodges and rooms that had been used by bats as roosts were subjected to treatment to remove them. This has led to local extermination of many populations of bats that were known inhabiting such places since long. As many as three species, namely, *Megaderma lyra lyra*, *Hipposideros fulvus pallidus* and *Tadarida aegyptiaca thomasi*, have become locally extinct from Bikaner and Jodhpur [33, 40, 41, 48, 49], and one species, namely, *Taphozous perforatus perforatus*, has become locally extinct from Jaisalmer [40].

Recent reports of increased sightings of fruit bat species from the desert areas of Rajasthan also indicate that the conversion of arid zones for cultivation due to the Indira Gandhi Canal Project has led to changes in the climate and vegetation, leading to many species of bats spreading to the earlier arid regions of Rajasthan [46, 50].

Bats of Rajasthan

As many as 25 species of bats have been reported from Rajasthan, none of them are endemic to the state (Table 21.1 and Fig. 21.1). Though not collected in the recent times, we have included all the species that have been reported from Rajasthan on the basis of authentic information. Few workers (see Topal [51]) have doubted the records of Himalayan elements, such as *Eptesicus serotinus pachyomus*, *Barbastella leucomelas darjilingensis* (now as *Barbastella darjelingensis*) and *Myotis blythi blythi* from Nasirabad. With respect to *Myotis blythi blythi*, Topal [51] suggested that the locality could possibly be Naseerabad in the Himalayas which many subsequent workers were unable to locate.

The species accounts provided below include common name, scientific name with author and year of description and complete type description detail with notes

Table 21.1 Chiropteran diversity of Rajasthan, India

S. No.	Family	Genus	Species	Subspecies
1.	Pteropodidae	*Pteropus*	*giganteus*	*giganteus*
2.		*Rousettus*	*leschenaultii*	*leschenaultii*
3.		*Cynopterus*	*sphinx*	*sphinx*
4.	Emballonuridae	*Taphozous*	*perforatus*	*perforatus*
5.			*longimanus*	*longimanus*
6.			*nudiventris*	*kachhensis*
7.			*melanopogon*	*melanopogon*
8.	Rhinopomatidae	*Rhinopoma*	*hardwickii*	*hardwickii*
9.			*microphyllum*	*kinneari*
10.	Megadermatidae	*Megaderma*	*lyra*	*lyra*
11.	Rhinolophidae	*Rhinolophus*	*lepidus*	*lepidus*
12.	Hipposideridae	*Hipposideros*	*fulvus*	*pallidus*
13.			*lankadiva*	*indus*
14.	Molossidae	*Tadarida*	*aegyptiaca*	*thomasi*
15.		*Chaerephon*	*plicatus*	*plicatus*
16.	Vespertilionidae	*Eptesicus*	*serotinus*	*pachyomus*
17.		*Hesperoptenus*	*tickelli*	–
18.		*Barbastella*	*leucomelas*	*darjelingensis*
19.		*Scotophilus*	*kuhlii*	*kuhlii*
20.			*heathii*	*heathii*
21.		*Pipistrellus*	*tenuis*	*mimus*
22.			*ceylonicus*	*indicus*
23.		*Scotozous*	*dormeri*	*dormeri*
24.		*Myotis*	*blythi*	*blythi*
25.		*Kerivoula*	*picta*	*picta*

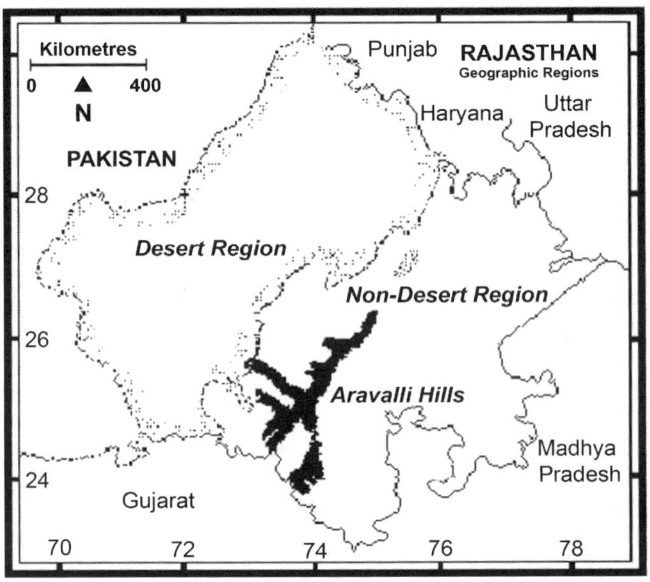

Fig. 21.1 Map of Rajasthan, India, showing three distinct geographic regions

on type locality provided in parenthesis for each species separately. For each entry, synonyms, details of subspecies, distribution range in India, South Asia and Rajasthan and their status, wherever applicable, are provided. It is important to note that excepting Salim Ali's Fruit Bat *Latidens salimalii* and Wroughton's Free-tailed Bat *Otomops wroughtoni* that are listed in Schedule I and other Fruit Bats listed in Schedule V (as vermins), all other bats are not protected under the Indian Wildlife (Protection) Amendment Act, 2006. The Red List categories are following Molur *et al.* [52]. Distribution maps of chiropteran diversity present in Rajasthan are also included. A gazetteer of localities in Rajasthan included in this chapter is provided as Appendix 1.

Species Accounts

Order Chiroptera Blumenbach, 1779
Suborder Megachiroptera Dobson, 1875
Family Pteropodidae Gray, 1821
Subfamily Pteropodinae Gray, 1821
Genus *Pteropus* Brisson, 1762 (Flying Foxes)

Indian Flying Fox *Pteropus giganteus* Brünnich, 1782

1782. *Vespertilio gigantea* Brünnich, *Dyrenes Historie*, 1: 45 (Bengal, India)

Taxonomy

Synonyms: *Vespertilio gigantea* Brünnich, 1782; *Pteropus medius* Temminck, 1825; *Pteropus edwardsi* I. Geoffroy, 1828; *Pteropus leucocephalus* Hodgson, 1835; *Pteropus assamensis* McClelland, 1839; *Pteropus ruvicollis* (misspelt *rubricollis* or *rubicollis*) Ogilby, 1840; *Pteropus kelaarti* Gray, 1871; *Pteropus ariel* Allen, 1908

Subspecies: Three subspecies, namely, *Pteropus giganteus giganteus* (Brünnich, 1782), *Pteropus giganteus leucocephalus* Hodgson, 1835 and *Pteropus giganteus ariel* Allen, 1908, are found in South Asia [3]. Only *Pteropus giganteus giganteus* (Brünnich, 1782) occurs in Rajasthan.

Distribution

South Asia: Distributed throughout India, Bangladesh, Bhutan, Nepal, Pakistan, Sri Lanka and Maldives

Rajasthan (Fig. 21.2): *Pteropus giganteus giganteus* (Brünnich, 1782) has been reported from Jaipur (Jaipur District) [7, 53]; Balsamand (Jodhpur District);

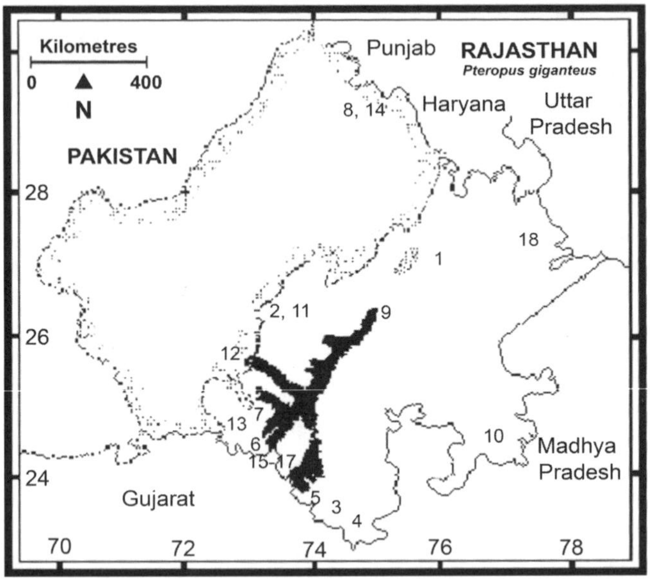

Fig. 21.2 Distribution of *Pteropus giganteus giganteus* in Rajasthan, India. Key: 1—Jaipur, 2 and 11—Balsamand, 3—Dungarpur, 4—Banswara, 5—Parsad, 6—Mt. Abu, 7—Berah, 8 and 14—Naga village, 9—Nasirabad, 10—Jhalawar, 12—Pali, 13—Sirohi, 15—Dewara, 16—Jaswantgarh, 17—Padhuna, 18—Bharatpur

Dungarpur (Dungarpur District); Banswara (Banswara District); Parsad (Udaipur District); Mt. Abu (Sirohi District); Berah (12 km SW of Jawai Dam, Pali District); Naga village (Jhunjhunu District); Nasirabad (Ajmer District); Jhalawar (Jhalawar District) [10]; Jodhpur (Jodhpur District); Pali (Pali District); Sirohi (Sirohi District); Jhunjhunu (Jhunjhunu District) [19, 21]; Dewara, Jaswantgarh, Padhuna (Udaipur District) [35] and Bharatpur (Bharatpur District) [47]. Also see Gaur [12], Purohit and Senacha [33, 54], Senacha [34], Purohit and Vyas [45] and Purohit *et al.* [48].

Status

CITES Appendix II; Schedule V of Indian Wildlife (Protection) Act, 1972 and LC (Least Concern in India) as per Red Data Book.

Genus *Rousettus* Gray, 1821(Rousettes)

Leschenault's Rousette *Rousettus leschenaultii* (Desmarest, 1820)

1820. *Pteropus leschenaultii* Desmarest, *Mammalogie*, in *Encyclop. Méthod.*, 1: 110 (Pondicherry, India)

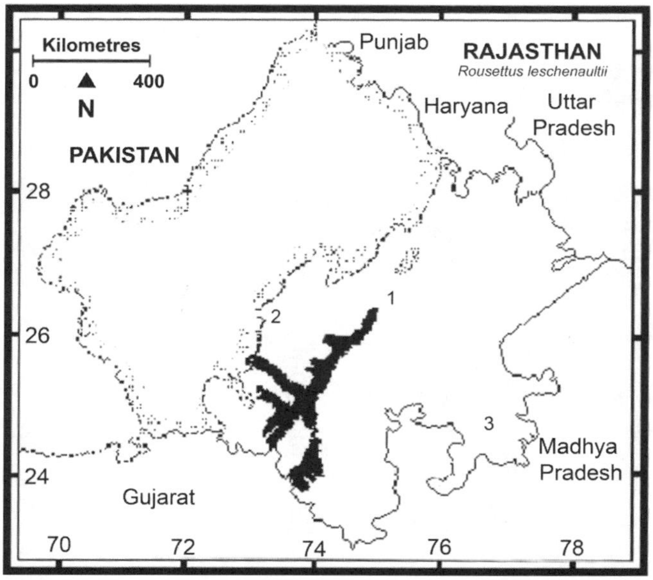

Fig. 21.3 Distribution of *Rousettus leschenaultii leschenaultii* in Rajasthan, India. Key: 1—Nasirabad, 2—Jodhpur, 3—Gagron ka Kila

Taxonomy

Synonyms: *Pteropus leschenaultii* Desmarest, 1820; *Pteropus pyrivorus* Hodgson, 1835; *Cynopterus marginatus* Gray, 1843; *Cynopterus affinis* Gray, 1843; *Pteropus seminudus* Kelaart, 1850; *Eleutherura fusca* Gray, 1870; *Xantharpyia seminuda* Gray, 1870; *Rousettus seminudus* (Gray, 1870); *Cynonycteris infuscata* Peters, 1873

Subspecies: Two subspecies, namely, *Rousettus leschenaultii leschenaultii* (Desmarest, 1820) and *Rousettus leschenaultii seminudus* (Kelaart, 1850), occur in South Asia [3]. Only *Rousettus leschenaultii leschenaultii* (Desmarest, 1820) occurs in Rajasthan.

Distribution

South Asia: Distributed throughout India, Bangladesh, Bhutan, Nepal, Pakistan and Sri Lanka

Rajasthan (Fig. 21.3): *Rousettus leschenaultii leschenaultii* (Desmarest, 1820) has been reported from Nasirabad (Ajmer District) [55, 56], Jodhpur (Jodhpur District) [57, 58] and Gagron ka Kila (Jhalawar District) (Plate 21.1, Image 3) [10, 47]. Also see Gaur [12] and Purohit *et al.* [48].

Plate 21.1 *Image 1*: Gagron ka Kila, Jhalawar, Rajasthan. *Image 2*: Naulakha Kila, Jhalrapatan, Rajasthan. *Image 3*: Colony of Leschenault's Rousette *Rousettus leschenaultii* in one of the chambers of Gagron ka Kila, Jhalawar, Rajasthan. *Image 4*: Mixed colony of Naked-rumped Tomb Bat *Taphozous nudiventris* and Lesser Mouse-tailed Bat *Rhinopoma hardwickii* in one of the chambers of Abelimeni ki Mahal, Jhalrapatan, Rajasthan

Status

Schedule V of Indian Wildlife (Protection) Act, 1972 and LC (Least Concern in India) as per Red Data Book.

Genus *Cynopterus* Cuvier, F., 1824 (Short-nosed fruit bat)

Greater Short-nosed Fruit Bat *Cynopterus sphinx* (Vahl, 1797)

1797. *Vespertilio sphinx* Vahl, *Skr. Nat. Selsk. Copenhagen*, 4(1): 123 (Tranquebar, Madras, India)

Taxonomy

Synonyms: *Vespertilio sphinx* Vahl, 1797; *Vespertilio fibulatus* Vahl, 1797; *Pteropus pusillus* Geoffroy, E., 1803; *Pteropus marginatus* Geoffroy, E., 1810; *Pachysoma*

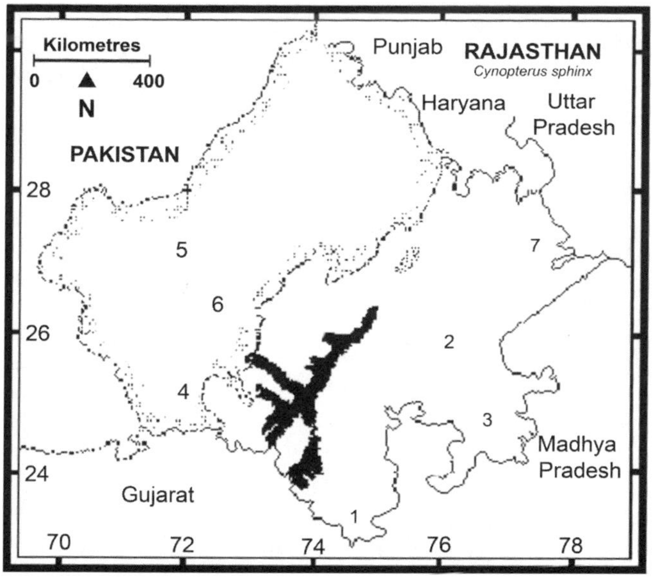

Fig. 21.4 Distribution of *Cynopterus sphinx sphinx* in Rajasthan, India. Key: 1—Banswara, 2—Bundi, 3—Jhalawar, 4—Bhinmal, 5—Lathi, 6—Tinwari, 7—Bharatpur

brevicaudatum Temminck, 1837 (not *Pachysoma brevicaudatum* Geoffroy, I., 1828); *Cynopterus marginatus* var. (*Pachysoma scherzeri*) Zelebor, 1869; *Cynopterus brachyotis scherzeri* Zelebor, 1869; *Cynopterus marginatus* var. *ellioti* Gray, 1870; *Cynopterus angulatus* Miller, 1898; *Cynopterus brachyotis angulatus* Miller, 1898; *Cynopterus sphinx gangeticus* Andersen, 1910

Subspecies: Three subspecies, namely, *Cynopterus sphinx sphinx* (Vahl, 1797), *Cynopterus sphinx scherzeri* Zelebor, 1869 and *Cynopterus sphinx angulatus* Miller, 1898, occur in South Asia [3]. Only *Cynopterus sphinx sphinx* (Vahl, 1797) occurs in Rajasthan.

Distribution

South Asia: Distributed throughout India, Bangladesh, Bhutan, Nepal, Pakistan and Sri Lanka

Rajasthan (Fig. 21.4): *Cynopterus sphinx sphinx* (Vahl, 1797) has been reported from Banswara (Banswara District); Bundi (Bundi District); Jhalawar (Jhalawar District)

[10, 19, 21, 47]; Bhinmal (Jalore District) [38]; Bharatpur (Bharatpur District) [47], Lathi village (Jaisalmer District) and Tinwari village (Jodhpur District) [46]. Also see Gaur [12], Purohit and Senacha [33], Senacha [34], Purohit and Vyas [45] and Purohit *et al.* [48].

Status

Schedule V of Indian Wildlife (Protection) Act, 1972 and LC (Least Concern in India) as per Red List.

Suborder Microchiroptera Dobson, 1875

Family Emballonuridae Gervais, 1855

Genus *Taphozous* E. Geoffroy, 1818 (Tomb Bats)

Egyptian Tomb Bat *Taphozous perforatus* E. Geoffroy, 1818

1818. *Taphozous perforatus* E. Geoffroy, *Descrip. De L'Egypte*, 2: 126 (Kom Ombo, Egypt)

Taxonomy

Synonyms: None

Subspecies: Only the nominate subspecies, namely, *Taphozous perforatus perforatus* E. Geoffroy, 1818, occurs in India.

Distribution

India: Distributed in parts of central and western India, found in Andhra Pradesh, Gujarat, Madhya Pradesh and Rajasthan

South Asia: Other than India, the range of this species also includes Pakistan.

Rajasthan (Fig. 21.5): *Taphozous perforatus perforatus* E. Geoffroy, 1818 has been reported from Jodhpur (Jodhpur District) (Prakash, 1960); Barmer (Barmer District); Dungarpur (Dungarpur District) [10]; Jaisalmer (Jaisalmer District); Jhalawar (Jhalawar District) [21]. Purohit and Senacha [33] point out that this species was not collected during recent years in and around Jaisalmer. Also see Gaur [12], Purohit and Senacha [33], Senacha [34], Purohit and Vyas [45] and Purohit *et al.* [48].

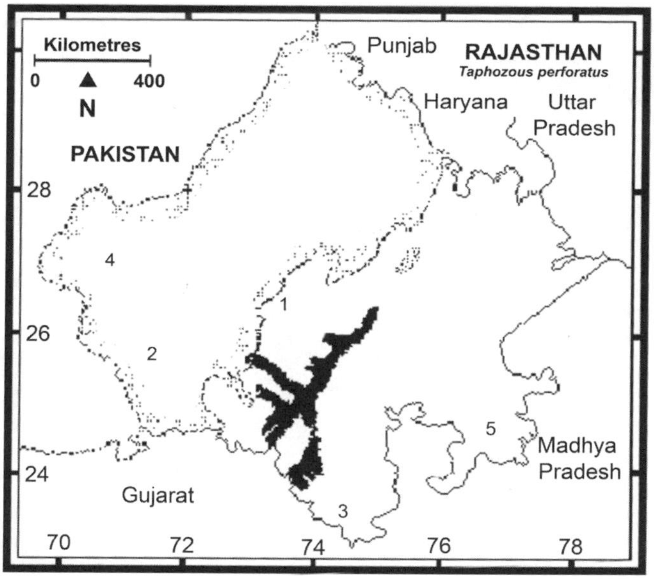

Fig. 21.5 Distribution of *Taphozous perforatus perforatus* in Rajasthan, India. Key: 1—Jodhpur, 2—Barmer, 3—Dungarpur, 4—Jaisalmer, 5—Jhalawar

Status

Red List: LC (Least Concern in India)

Long-winged Tomb Bat *Taphozous longimanus* Hardwicke, 1825

1825. *Taphozous longimanus* Hardwicke, *Trans. Linn. Soc. Lond.*, 14: 525 (Calcutta, India)

Taxonomy

Synonyms: *Taphozous fulvidus* Blyth, 1841; *Taphozous brevicaudus* Blyth, 1841; *Taphozous cantorii* Blyth, 1842

Subspecies: Only the nominate subspecies, namely, *Taphozous longimanus longimanus* Hardwicke, 1825, occurs in India.

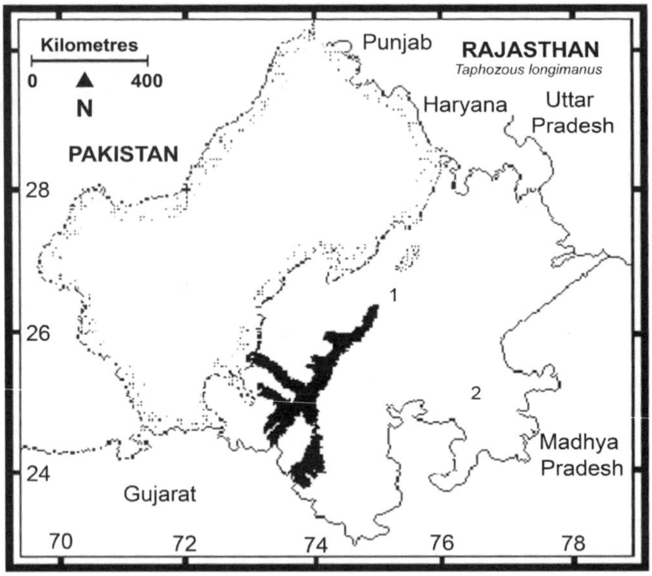

Fig. 21.6 Distribution of *Taphozous longimanus longimanus* in Rajasthan, India. Key: 1—Nasirabad, 2—Kota

Distribution

South Asia: Distributed patchily throughout India, Bangladesh, Nepal and Sri Lanka

Rajasthan (Fig. 21.6): *Taphozous longimanus longimanus* Hardwicke, 1825 has been reported from Nasirabad (Ajmer District); Kota (Kota District) [10].

Status

Red List: LC (Least Concern in India)

Naked-Rumped Tomb Bat *Taphozous nudiventris* Cretzschmar, 1830

1830. *Taphozous nudiventris* Cretzschmar, *In* Rüppell, *Atlas Reise Nordl. Afr. Zool. Saügeth.*, p. 70, Fig. 27b (Giza, Egypt)

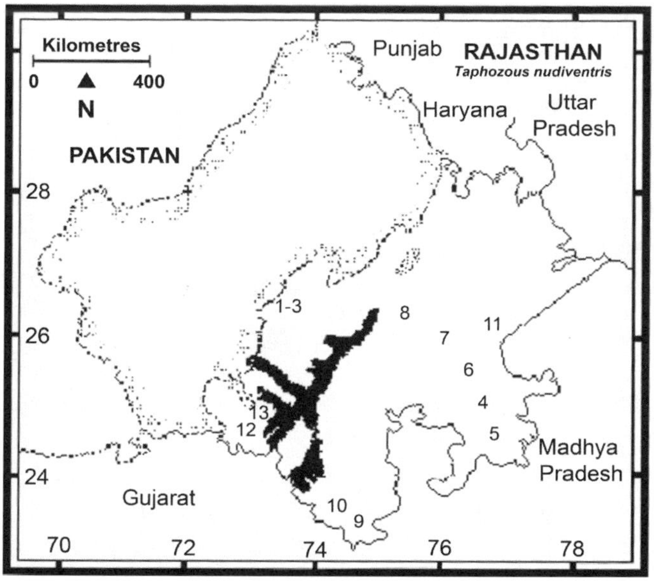

Fig. 21.7 Distribution of *Taphozous nudiventris kachhensis* in Rajasthan, India. Key: 1—Mandore, 2—Balsamand, 3—Bhim Bharak, 4—Kota, 5—Jhalrapatan, 6—Bundi, 7—Tonk, 8—Ajmer, 9—Banswara, 10—Dungarpur, 11—Sawai Madhopur, 12—Sirohi, 13—Pali

Taxonomy

Synonyms: *Taphozous nudiventer* Temminck, 1841; *Taphozous kachhensis* Dobson, 1872

Subspecies: One subspecies, namely, *Taphozous nudiventris kachhensis* Dobson, 1872, occurs in India [3].

Distribution

South Asia: Distributed patchily throughout India, Afghanistan, Bangladesh and Pakistan

Rajasthan (Fig. 21.7): *Taphozous nudiventris kachhensis* (Dobson, 1872) has been reported from Mandore, Balsamand, Bhim Bharak (Jodhpur District); Kota (Kota District); Jhalrapatan (Plate 21.1, Image 4 and Plate 21.2, Image 3) (Jhalawar District), Bundi (Bundi District), Tonk (Tonk District) [10, 47] and districts of

Plate 21.2 *Image 1*: Greater Mouse-tailed Bat *Rhinopoma microphyllum* and Lesser Mouse-tailed Bat *Rhinopoma hardwickii* in a cavern at Bagher, Jhalawar District, Rajasthan. *Image 2*: Close-up of Mouse-tailed Bat *Rhinopoma microphyllum* from Jhalrapatan, Rajasthan. *Image 3*: Close-up of Naked-rumped Tomb Bat *Taphozous nudiventris* at Jhalrapatan, Rajasthan. *Image 4*: Colony of Greater False Vampire Bat *Megaderma lyra* in one of the chambers of Abelimeni ki Mahal, Darrah, Rajasthan

Ajmer, Banswara, Dungarpur, Sawai Madhopur, Sirohi, Pali [15, 17]. Also see Gaur [12], Purohit and Senacha [33], Senacha [34], Purohit and Vyas [45] and Purohit *et al.* [48].

Status

Red List: LC (Least Concern in India)

Black-Bearded Tomb Bat *Taphozous melanopogon* Temminck, 1841

1841. *Taphozous melanopogon* Temminck, *Monogr. Mamm.*, 2: 287 (Bantam, West Java, Indonesia)

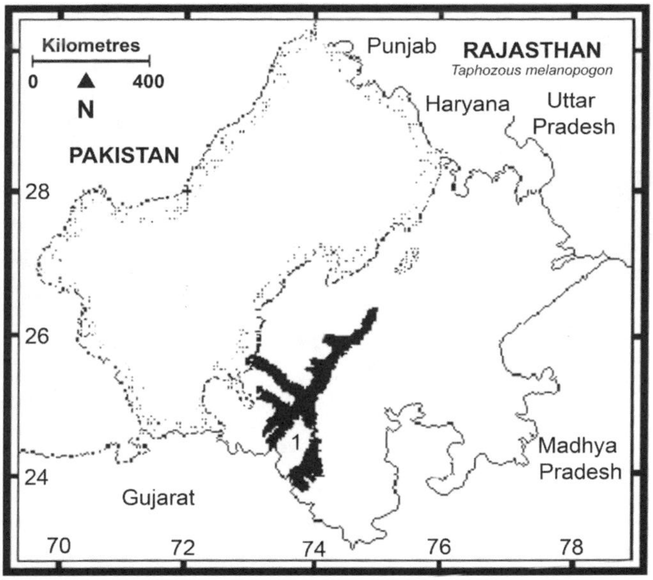

Fig. 21.8 Distribution of *Taphozous melanopogon melanopogon* in Rajasthan, India. Key: 1—Udaipur

Taxonomy

Synonyms: *Taphozous bicolor* Temminck, 1841

Subspecies: Only the nominate subspecies, namely, *Taphozous melanopogon melanopogon* Temminck, 1841, occurs in India [3].

Distribution

South Asia: Distributed patchily throughout India, Bangladesh and Sri Lanka

Rajasthan (Fig. 21.8): *Taphozous melanopogon melanopogon* Temminck, 1841 has been reported from Udaipur (Udaipur District) [10].

Status

Red List: LC (Least Concern in India)

Family Rhinopomatidae Bonaparte, 1838

Genus *Rhinopoma* E. Geoffroy, 1818 (Mouse-tailed Bats)

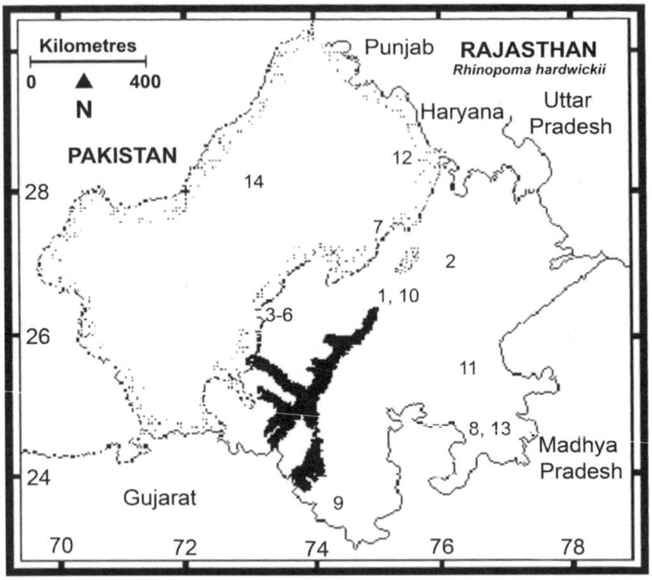

Fig. 21.9 Distribution of *Rhinopoma hardwickii hardwickii* in Rajasthan, India. Key: 1—Nasirabad, 2—Jaipur, 3—Jodhpur, 4—Bhim Bharak, 5—Salawas, 6—Kalyanpur, 7—Solayan, 8—Jhalrapatan, 9—Dungarpur, 10—Ajmer, 11—Bundi, 12—Nangal, 13—Bagher, 14—Bikaner

Lesser Mouse-Tailed Bat *Rhinopoma hardwickii* Gray, 1831

1831. *Rhinopoma hardwickii* Gray, *Zool. Misc.*, 1: 37 (India, restricted to Bengal by Qumsiyeh *et al.*, 1992)

Taxonomy

Synonyms: None

Subspecies: Only the nominate subspecies, namely, *Rhinopoma hardwickii hardwickii* Gray, 1831, occurs in India [3].

Distribution

South Asia: Distributed patchily throughout India, Afghanistan, Bangladesh and Pakistan

Rajasthan (Fig. 21.9): *Rhinopoma hardwickii hardwickii* Gray, 1831 has been reported from Nasirabad (Ajmer District) [58]; Jaipur (Jaipur District) [58]; Jodhpur, Bhim Bharak, Salawas, Kalyanpur (Jodhpur District); Solayan Village, 15 km NE of Kuchaman Road Railway Station (Nagour District); Bagher (Plate 21.2, Image 1), Jhalrapatan (Plate 21.1, Image 4) (Jhalawar District); Dungarpur (Dungarpur District); Ajmer (Ajmer District); Bundi (Bundi District); Nangal Village (Jhunjhunu District); Bikaner (Bikaner District) [3, 10, 59]. Also see Prakash [53], Gaur [12], Purohit and Senacha [33], Senacha [34, 60], Senacha and Purohit [61], Purohit and Vyas [45] and Purohit *et al.* [48].

Status

Red List: LC (Least Concern in India)

Greater Mouse-tailed Bat *Rhinopoma microphyllum* (Brünnich, 1872)

1872. *Vespertilio microphyllus* Brünnich, *Dyrenes Historie*, 1: 50, pl. 6, Fig. 21.1—4 (Giza, Egypt)

Taxonomy

Synonyms: *Vespertilio microphyllus* Brünnich, 1872; *Rhinopoma kinneari* Wroughton, 1912

Subspecies: Two subspecies, namely, *Rhinopoma microphyllum microphyllum* (Brünnich, 1872) and *Rhinopoma microphyllum kinneari* Wroughton, 1912, occur is South Asia [3]. Only *Rhinopoma microphyllum kinneari* Wroughton, 1912 occurs in Rajasthan.

Distribution

India

South Asia: Distributed patchily throughout India, Afghanistan, Bangladesh and Sri Lanka

Rajasthan (Fig. 21.10): *Rhinopoma microphyllum kinneari* Wroughton, 1912 has been reported from Ajmer (Ajmer District) [6]; Udaipur (Udaipur District) [26], Aligarh (Tonk District) [25]; Jodhpur, Mandore, Bhim Bharak, Ransi village (Jodhpur District); Hillocks near Pali on Jodhpur-Pali Road (Pali District); Jhalrapatan (Plate 21.2, Image 2), Gagron ka Kila, Bagher (Plate 21.2, Image 1) (Jhalawar District) [10, 47]; Dungarpur (Dungarpur District); Marot, Shyamgadh, Panchota (Nagour District); Bundi (Bundi District) [10]. Also from the districts of

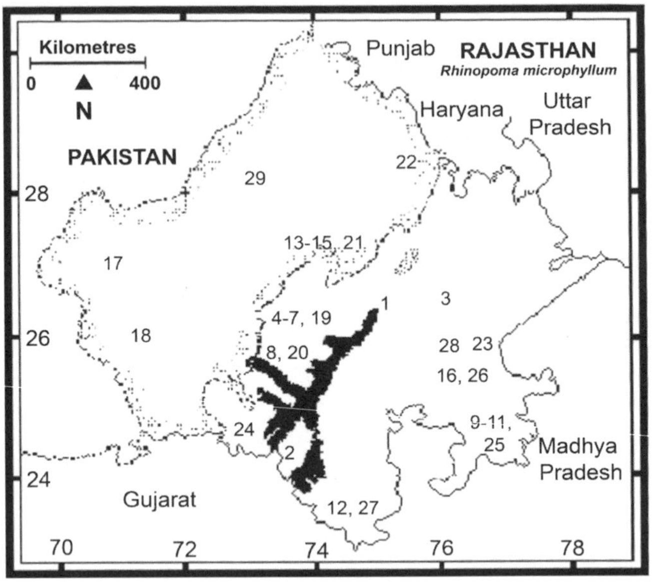

Fig. 21.10 Distribution of *Rhinopoma microphyllum microphyllum* in Rajasthan, India. Key: 1—Ajmer, 2—Udaipur, 3—Aligarh, 4—Jodhpur, 5—Mandore, 6—Bhim Bharak, 7—Ransi, 8—Pali, 9—Jhalrapatan, 10—Gagron ka Kila, 11—Bagher, 12—Dungarpur, 13—Marot, 14—Shyamgadh, 15—Panchota, 16—Bundi, 17—Jaisalmer, 18—Barmer, 19—Jodhpur, 20—Pali, 21—Nagour, 22—Jhunjhunu, 23—Sawai Madhopur, 24—Sirohi, 25—Jhalawar, 26—Bundi, 27—Dungarpur, 28—Tonk, 29—Bikaner

Jaisalmer, Barmer, Jodhpur, Pali, Nagour, Jhunjhunu, Sawai Madhopur, Sirohi, Jhalawar, Bundi, Dungarpur, Tonk [21, 62] and Bikaner [59]. Also see Prakash [53], Gaur [12], Purohit and Kaluram [63], Purohit and Senacha [33], Purohit *et al.* [64], Senacha [34, 60], Purohit and Vyas [45] and Purohit *et al.* [48].

Status

Red List: LC (Least Concern in India)

Family Megadermatidae H. Allen, 1864

Genus *Megaderma* E. Geoffroy, 1810 (False Vampire Bats)

Greater False Vampire Bat *Megaderma lyra* E. Geoffroy, 1810

1810. *Megaderma lyra* E. Geoffroy, *Ann. Mus. Hist. Nat. Paris*, 15: 190 (Madras, India)

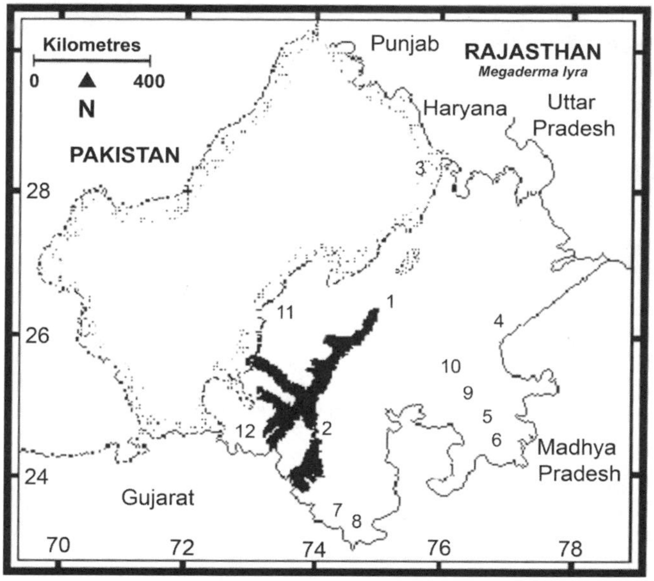

Fig. 21.11 Distribution of *Megaderma lyra lyra* in Rajasthan, India. Key: 1—Ajmer, 2—Udaipur, 3—Nangal, 4—Ranthambhore, 5—Darrah, 6—Jhalrapatan, 7—Dungarpur, 8—Banswara, 9—Bundi, 10—Tonk, 11—Jodhpur, 12—Sirohi

Taxonomy

Synonyms: *Vespertilio (Megaderma) carnatica* Elliot, 1839; *Megaderma spectrum* Wagner, 1844; *Megaderma schistacea* Hodgson, 1847; *Encheira lyra caurina* Andersen and Wroughton, 1907; *Megaderma lyra caurina* (Andersen and Wroughton, 1907)

Subspecies: Only nominate subspecies, namely, *Megaderma lyra lyra* E. Geoffroy, 1810, occurs in India [3].

Distribution

South Asia: Distributed throughout India, Afghanistan, Bangladesh, Nepal, Pakistan and Sri Lanka

Rajasthan (Fig. 21.11): *Megaderma lyra lyra* E. Geoffroy, 1810 has been reported from Ajmer (Ajmer District) [16]; Udaipur (Udaipur District) [28]; Nangal Village (Jhunjhunu District); Ranthambhore (Sawai Madhopur District); Abelimeni ki Mahal, Darrah (Kota District) (Plate 21.2, Image 4); Jhalrapatan (Jhalawar District); Dungarpur (Dungarpur District); Banswara (Banswara District) [10, 47]. Also from the districts of Banswara, Dungarpur, Bundi, Kota, Jhalawar, Tonk, Sawai Madhopur, Jodhpur, Jhunjhunu and Sirohi [21]. See also Prakash [53] and Ramaswami and Ananda Kumar [22]. Gaur [12] reported its decline in Jodhpur, while Purohit and

Senacha [33, 40, 41] and Purohit *et al.* [48] point out that this species has not been recently collected from Jodhpur.

Status

Red List: LC (Least Concern in India)

Family Rhinolophidae Bell, 1836

Genus *Rhinolophus* Lacépède, 1799(Horseshoe Bats)

Blyth's Horseshoe Bat *Rhinolophus lepidus* Blyth, 1844

1844. *Rhinolophus lepidus* Blyth, *J. Asiat. Soc. Bengal*, 13: 486 (Calcutta (uncertain), India)

Taxonomy

Synonyms: *Rhinolophus monticola* Andersen, 1905

Subspecies: Two subspecies, namely, *Rhinolophus lepidus lepidus* Blyth, 1844 and *Rhinolophus lepidus monticola* Andersen, 1905, occur in India [3]. Only *Rhinolophus lepidus lepidus* Blyth, 1844 occurs in Rajasthan.

Distribution

South Asia: Distributed throughout India except the extreme arid north-western India, Afghanistan, Bangladesh, Pakistan and Nepal

Rajasthan (Fig. 21.12): *Rhinolophus lepidus lepidus* Blyth, 1844 has been reported from Bikaner (Bikaner District) [7], Jodhpur (Jodhpur District), Ranthambhore (Sawai Madhopur District) and Sikar Burz 11 km SE of Bundi (Bundi District) [10]. Also see Gaur [12], Purohit and Senacha [33, 59], Senacha [34], Purohit and Vyas [45] and Purohit *et al.* [48].

Status

Red List: LC (Least Concern in India)

Family Hipposideridae Lydekker, 1891

Genus *Hipposideros* Gray, 1831(Leaf-nosed Bats)

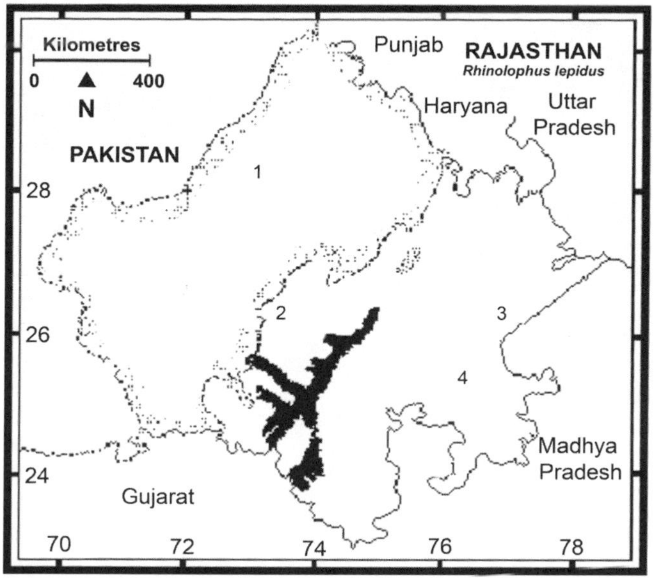

Fig. 21.12 Distribution of *Rhinolophus lepidus lepidus* in Rajasthan, India. Key: 1—Bikaner, 2—Jodhpur, 3—Ranthambhore, 4—Sikar Burz

Fulvus Leaf-Nosed Bat *Hipposideros fulvus* Gray, 1838

1838. *Hipposideros fulvus* Gray, *Mag. Zool. Bot.*, 2: 492 (Dharwar, Karnataka, India)

Taxonomy

Synonyms: *Hipposideros bicolor fulvus* Gray, 1838; =*Rhinolophus fulgens* Elliot, 1839; *Hipposideros murinus* Gray, 1838; *Phyllorhina aurita* Tomes, 1859; *Phyllorhina atra* Fitzinger, 1870; *Hipposideros fulvus pallidus* Andersen, 1918; *Hipposideros bicolor pallidus* Andersen, 1918

Subspecies: Two subspecies, namely, *Hipposideros fulvus fulvus* Gray, 1838 and *Hipposideros fulvus pallidus* Andersen, 1918, occur in India [3]. Only *Hipposideros fulvus pallidus* Andersen, 1918 occurs in Rajasthan.

Distribution

South Asia: Distributed throughout India, Afghanistan, Nepal, Pakistan and Sri Lanka

Rajasthan (Fig. 21.13): *Hipposideros fulvus pallidus* Andersen, 1918 has been reported from Jaipur (Jaipur District) [58, 65], Jodhpur (Jodhpur District) [12, 49], Jhalrapatan (Jhalawar District), Ajmer (Ajmer District) [10] and Bharatpur

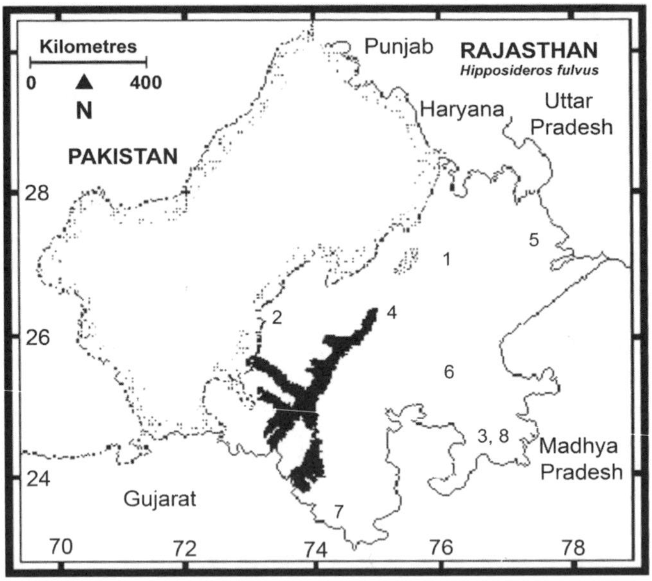

Fig. 21.13 Distribution of *Hipposideros fulvus pallidus* in Rajasthan, India. Key: 1—Jaipur, 2—Jodhpur, 3—Jhalrapatan, 4—Ajmer, 5—Bharatpur, 6—Bundi, 7—Dungarpur, 8—Jhalawar

(Bharatpur District) [27]. It has also been reported from the districts of Bundi, Dungarpur and Jhalawar [18]. Purohit and Senacha [33, 40] and Purohit *et al.* [48] point out that this species has not been recently collected from Jodhpur.

Status

Red List: LC (Least Concern in India)

Indian Leaf-Nosed Bat *Hipposideros lankadiva* Kelaart, 1850

1850. *Hipposideros lankadiva* Kelaart, *J. Sri Lanka Branch Asiat. Soc.*, 2(2): 216 (Kandy, Sri Lanka)

Taxonomy

Synonyms: *Hipposideros indus* Andersen, 1918; *Hipposideros indus mixtus* Andersen, 1918; *Hipposideros lankadiva mixtus* (Andersen, 1918); *Hipposideros*

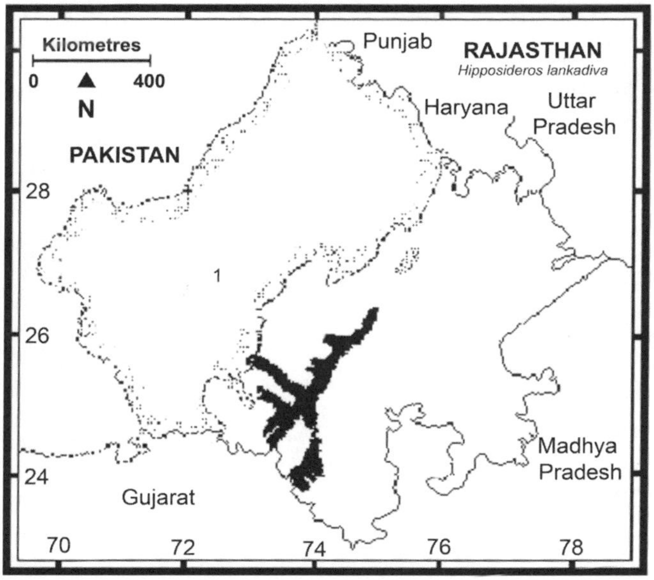

Fig. 21.14 Distribution of *Hipposideros lankadiva indus* in Rajasthan, India. Key: 1—Bhim Bharak

indus unitus Andersen, 1918; *Hipposideros lankadiva unitus* (Andersen, 1918); *Hipposideros schistaceus* Andersen, 1918

Subspecies: One subspecies, namely, *Hipposideros lankadiva indus* Andersen, 1918, occurs in India [3].

Distribution

South Asia: Distributed patchily throughout India, Bangladesh and Sri Lanka

Rajasthan (Fig. 21.14): *Hipposideros lankadiva indus* Andersen, 1918 has been reported from Bhim Bharak (Jodhpur District) [24].

Status

Red List: LC (Least Concern in India)

Family Molossidae Gill, 1872

Genus *Tadarida* Rafinesque, 1814 (Free-tailed Bats)

Egyptian Free-Tailed Bat *Tadarida aegyptiaca* (E. Geoffroy, 1818)

1818. *Nyctinomus aegyptiacus* Geoffroy, E., Discrip. De L'Egypte 2: 128, pl. 2, No. 2 (Giza, Egypt)

Taxonomy

Synonyms: *Nyctinomus aegyptiacus* Geoffroy, E. 1818; =*Dysopes geoffroyi* Temminck, 1826; *Nyctinomus tragata* Dobson, 1874; *Tadarida tragata* (Dobson, 1874); *Tadarida thomasi* Wroughton, 1919; *Tadarida sindica* Wroughton, 1919; *Tadarida gossei* Wroughton, 1919; *Tadarida aegyptiaca gossei* Wroughton, 1919

Subspecies: Three subspecies, namely, *Tadarida aegyptiaca tragatus* Dobson, 1874, *Tadarida aegyptiaca thomasi* Wroughton, 1919 and *Tadarida aegyptiaca sindica* Wroughton, 1919, occur in India [3]. Only *Tadarida aegyptiaca thomasi* Wroughton, 1919 occurs in Rajasthan.

Distribution

South Asia: Distributed patchily throughout India, Afghanistan, Bangladesh, Pakistan and Sri Lanka

Rajasthan (Fig. 21.15): *Tadarida aegyptiaca thomasi* Wroughton, 1919 has been reported from Alwar (Alwar District), Mt. Abu (Sirohi District) [5], Jodhpur (Jodhpur District), Rajgad (Ajmer District), Kota (Kota District), Dungarpur (Dungarpur District), Bundi (Bundi District) [10] and also districts of Jhalawar, Sawai Madhopur, Tonk and Banswara [20, 21]. Purohit and Senacha [33, 40, 41] and Purohit *et al.* [48] point out that this species has not been recently collected from Jodhpur.

Status

Red List: LC (Least Concern in India)

Genus *Chaerephon* Dobson, 1874 (Mastiff Bats)

Wrinkle-lipped Free-tailed Bat *Chaerephon plicatus* (Buchanan, 1800)

1800. *Vespertilio plicatus* Buchanan, *Trans. Linn. Soc. London*, 5: 261, pl. 13 (Puttahaut, Bengal, India)

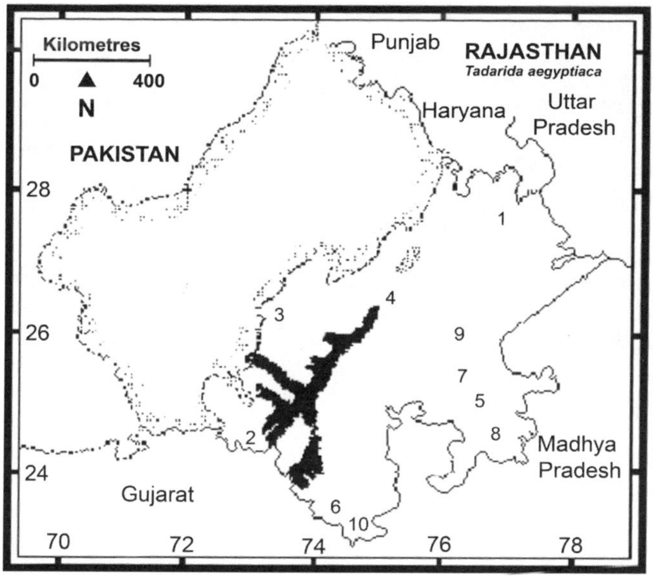

Fig. 21.15 Distribution of *Tadarida aegyptiaca thomasi* in Rajasthan, India. Key: 1—Alwar, 2—Mt. Abu, 3—Jodhpur, 4—Rajgadh, 5—Kota, 6—Dungarpur, 7—Bundi, 8—Jhalawar, 9—Tonk, 10—Banswara

Taxonomy

Synonyms: *Vespertilio plicatus* Buchanan, 1800; *Nyctinomus bengalensis* Desmarest, 1820; *Dysopes murinus* Gray, 1830; *Tadarida plicata insularis* (Phillips, 1932)

Subspecies: One subspecies, namely, *Chaerephon plicatus plicatus* (Buchanan, 1874), occurs in India [3].

Distribution

India

South Asia: Distributed sporadically throughout India and Sri Lanka

Rajasthan (Fig. 21.16): *Chaerephon plicatus plicatus* (Buchanan, 1874) has been reported from Mt. Abu (Sirohi District) [10].

Status

Red List: LC (Least Concern in India)

Family Vespertilionidae Gray, 1821

Genus *Eptesicus* Rafinesque, 1820 (Serotines)

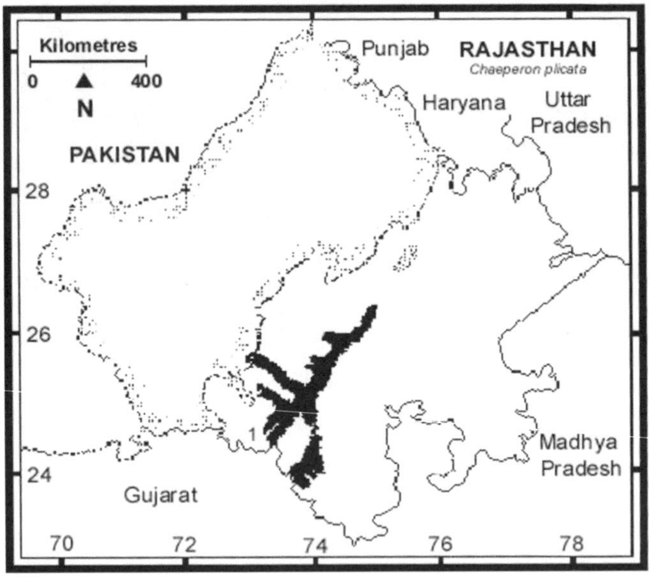

Fig. 21.16 Distribution of *Chaerephon plicatus plicatus* in Rajasthan, India. Key: 1—Mt. Abu

Serotine *Eptesicus serotinus* Schreber, 1774

1774. *Vespertilio serotinus* Schreber, *Die Saugethiere*, 1: 167 (France)

Taxonomy

Synonyms: *Vespertilio serotinus* Schreber, 1774; *Scotophilus pachyomus* Tomes, 1857

Subspecies: One subspecies, namely, *Eptesicus serotinus pachyomus* (Tomes, 1857), occurs in India [3].

Distribution

South Asia: Distributed in the Himalayan tract of India, found in Assam, Jammu and Kashmir and Nagaland. There is also a historic record from Rajasthan, Afghanistan and Pakistan.

Rajasthan (Fig. 21.17): The type locality of *Eptesicus serotinus pachyomus* (Tomes, 1857) is "Rajputana" (present-day Rajasthan), India. The type probably has been collected from Nasirabad (Ajmer District) [10].

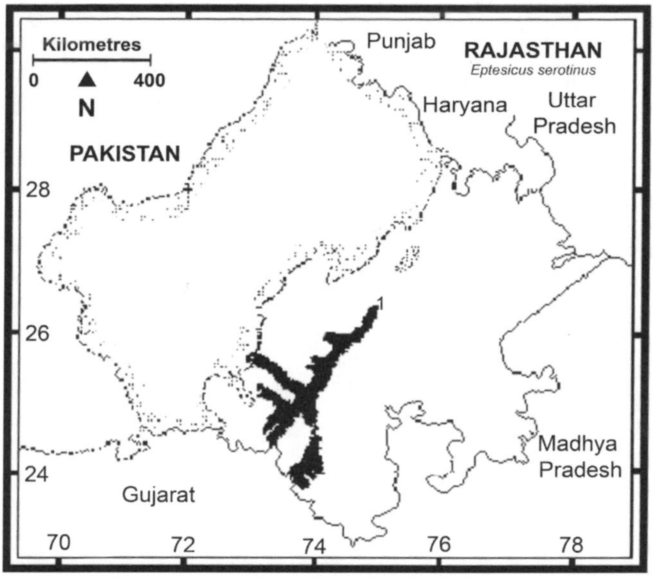

Fig. 21.17 Distribution of *Eptesicus serotinus pachyomus* in Rajasthan, India. Key: 1—Nasirabad

Status

Red List: NT (Near Threatened in India)

Genus *Hesperoptenus* Peters, 1868 (False Serotines)

Tickell's Bat *Hesperoptenus tickelli* (Blyth, 1851)

1851. *Nycticejus tickelli* Blyth, *J. Asiat. Soc. Bengal*, 20: 157 (Chaibasa, Bihar, now in Jharkhand, India)

Taxonomy

Synonyms: *Nycticejus tickelli* Blyth, 1851; *Nycticejus isabellinus* Kelaart, 1850 (*nomen nudum*); =*Nycticejus isabellinus* Horsfield, 1851

Subspecies: No subspecies

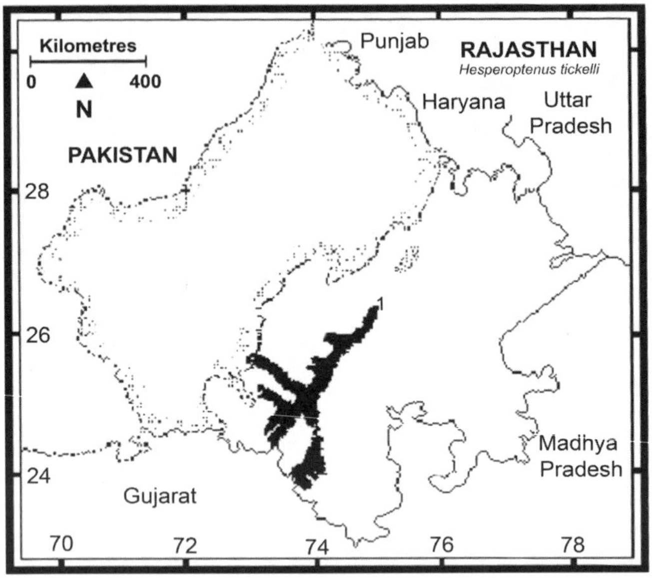

Fig. 21.18 Distribution of *Hesperoptenus tickelli* in Rajasthan, India. Key: 1—Nasirabad

Distribution

South Asia: Distributed patchily in India, Bangladesh, Bhutan, Nepal and Sri Lanka

Rajasthan (Fig. 21.18): *Hesperoptenus tickelli* (Blyth, 1851) has been reported from Nasirabad (Ajmer District) [4, 10].

Status

Red List: LC (Least Concern in India)

Genus *Barbastella* Gray, 1821(Barbastelles)

Large Barbastelle *Barbastella darjelingensis* (Hodgson, 1855)

1855. *Plecotus darjelingensis* Hodgson, In Horsfield, Ann. Mag. nat. Hist., xvi, p. 103

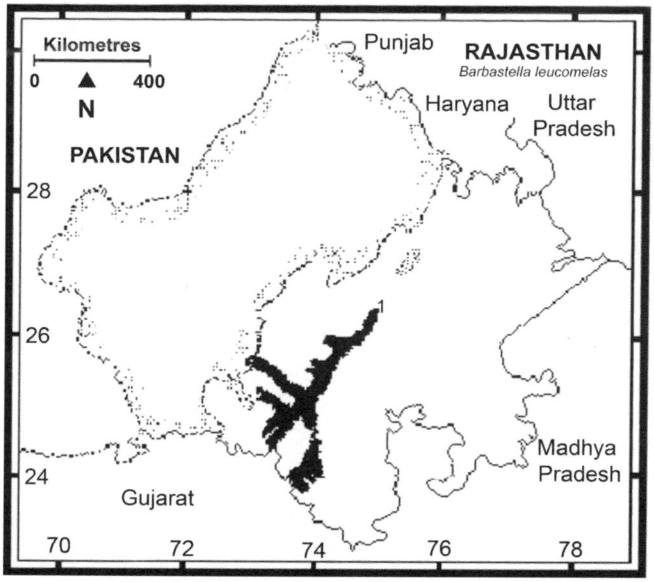

Fig. 21.19 Distribution of *Barbastella leucomelas darjelingensis* in Rajasthan, India. Key: 1—Nasirabad

Taxonomy

Synonyms: *Plecotus darjelingensis* Hodgson, 1855 *in* Horsfield, 1855; =*Synotus darjelingsis* (Hodgson, 1855); *Barbastellus darjelinensis* Dobson, 1875 (sometimes as *dargelinensis*); =*Barbastella blanfordi* Bianchi, 1916

Subspecies: None

Distribution

India: Distributed in the Himalayan tracts of India, found in Assam, Himachal Pradesh, Jammu and Kashmir, Meghalaya, Mizoram, Sikkim, Uttar Pradesh, Uttaranchal and West Bengal. There is also a historic record from Rajasthan.

South Asia: Other than India, the range of this species also includes Afghanistan, Bhutan, Nepal and Pakistan.

Rajasthan (Fig. 21.19): *Barbastella darjelingensis* (Hodgson, 1855 *in* Horsfield, 1855) has been reported from Nasirabad (Ajmer District) [10, 56, 66].

Status

Red List: NT (Near Threatened in India)

Genus *Scotophilus* Leach, 1821 (Yellow House Bats)

Lesser Asiatic Yellow House Bat *Scotophilus kuhlii* Leach, 1821

1821. *Scotophilus kuhlii* Leach, *Trans. Linn. Soc. Lond.*, 13: 71 ("India")

Taxonomy

Synonyms: *Scotophilus fulvus* Gray, 1843; *Scotophilus wroughtoni* Thomas, 1897; *Scotophilus temmincki wroughtoni* (Thomas, 1897); *Scotophilus kuhlii wroughtoni* (Thomas, 1897)

Subspecies: One subspecies, namely, *Scotophilus kuhlii kuhlii* Leach, 1821, occurs in India [3].

Distribution

South Asia: Distributed patchily throughout India, Bangladesh, Pakistan and Sri Lanka

Rajasthan (Fig. 21.20): *Scotophilus kuhlii kuhlii* Leach, 1821 has been reported from Bharatpur (Bharatpur District) [10] and districts of Jodhpur, Jhunjhunu, Alwar, Bharatpur, Ajmer, Sawai Madhopur, Banswara, Dungarpur, Bundi and Kota [21].

Status

Red List: LC (Least Concern in India)

Greater Asiatic Yellow House Bat *Scotophilus heathii* (Horsfield, 1831)

1831. *Nycticejus heathii* Horsfield, *Proc. Zool. Soc. Lond.*, p. 113 (Madras, Tamil Nadu, India)

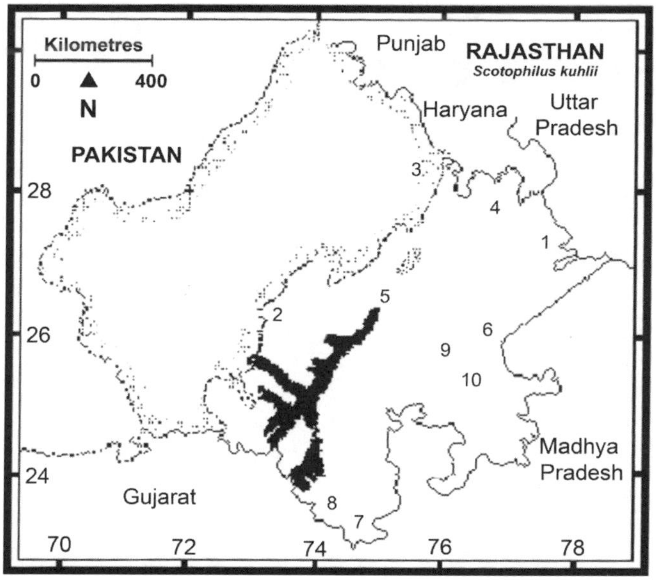

Fig. 21.20 Distribution of *Scotophilus kuhlii kuhlii* in Rajasthan, India. Key: 1—Bharatpur, 2—Jodhpur, 3—Jhunjhunu, 4—Alwar, 5—Ajmer, 6—Sawai Madhopur, 7—Banswara, 8—Dungarpur, 9—Bundi, 10—Kota

Taxonomy

Synonyms: *Nycticejus heathii* Horsfield, 1831; *Vespertilio belangeri* Geoffroy, 1834; *Scotophilus heathi belangeri* (Geoffroy, 1834); *Nycticejus luteus* Blyth, 1851; *Scotophilus flaveolous* Horsfield, 1851

Subspecies: One subspecies, namely, *Scotophilus heathii heathii* (Horsfield, 1831), occurs in India [3].

Distribution

India

South Asia: Distributed throughout India, Afghanistan, Bangladesh, Nepal, Pakistan and Sri Lanka

Rajasthan (Fig. 21.21): *Scotophilus heathii heathii* (Horsfield, 1831) has been reported from Jodhpur (Jodhpur District); Nangal (Jhunjhunu District); Perbeni (Alwar District); Bharatpur (Bharatpur District); Rajgad (Ajmer District); Sawai

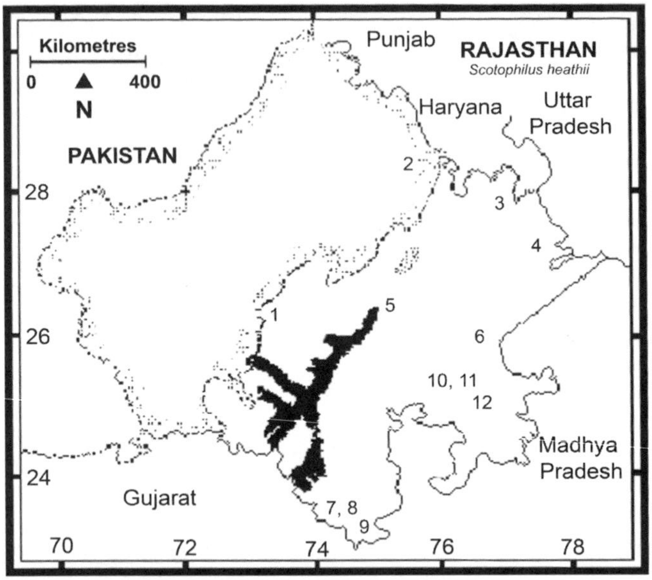

Fig. 21.21 Distribution of *Scotophilus heathii heathii* in Rajasthan, India. Key: 1—Jodhpur, 2—Nangal, 3—Perbeni, 4—Bharatpur, 5—Rajgad, 6—Sawai Madhopur, 7—Dugarpur, 8—Supur, 9—Banswara, 10—Bundi, 11—Sikar Burj, 12—Kota

Madhopur (Sawai Madhopur District); Dungarpur, Surpur (Dungarpur District); Banswara (Banswara District); Bundi, Sikar Burj (Bundi District) [10] and districts of Jodhpur, Jhunjhunu, Alwar, Bharatpur, Ajmer, Sawai Madhopur, Banswara, Dungarpur, Bundi and Kota [21]. Also see Prakash [53], Gaur [12], Purohit and Senacha [33], Senacha [34], Purohit and Vyas [45] and Purohit *et al.* [48].

Status

Red List: LC (Least Concern in India)

Genus *Pipistrellus* Kaup, 1829 (Pipistrelles)

Least Pipistrelle *Pipistrellus tenuis* (Temminck, 1840)

1840. *Vespertilio tenuis* Temminck, *Monogr. Mamm.*, 2: 229 (Sumatra, Indonesia)

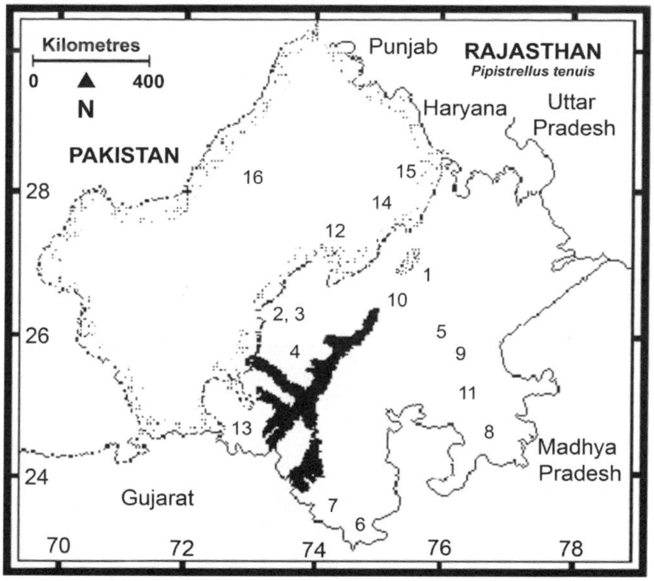

Fig. 21.22 Distribution of *Pipistrellus tenuis mimus* in Rajasthan, India. Key: 1—Jaipur, 2—Jodhpur, 3—Salawas, 4—Pali, 5—Tonk, 6—Banswara, 7—Dungarpur, 8—Jhalawar, 9—Bundi, 10—Ajmer, 11—Kota, 12—Nagour, 13—Sirohi, 14—Sikar, 15—Jhunjhunu, 16—Bikaner

Taxonomy

Synonyms: *Vespertilio tenuis* Temminck, 1840; *Pipistrellus mimus* Wroughton, 1899; *Pipistrellus mimus mimus* Wroughton, 1899; *Pipistrellus principulus* Thomas, 1915; *Pipistrellus mimus glaucillus* Wroughton, 1912; *Pipistrellus mimus principulus* Thomas, 1915

Subspecies: One subspecies, namely, *Pipistrellus tenuis mimus* Wroughton, 1899, occurs in India [3].

Distribution

South Asia: Throughout India, Afghanistan, Bangladesh, Nepal, Pakistan and Sri Lanka

Rajasthan (Fig. 21.22): *Pipistrellus tenuis mimus* Wroughton, 1899 has been reported from Mt. Abu [5]; Jaipur (Jaipur District) [10, 53]; Jodhpur, Salawas (Jodhpur District); Pali (Pali District); Tonk (Tonk District) [10] and districts of Jodhpur, Banswara, Dungarpur, Jhalawar, Tonk, Bundi, Ajmer, Kota [17]; Nagour, Pali, Sirohi, Sikar, Jhunjhunu [21] and Bikaner (Bikaner District) [59]. Also see Gaur [12], Purohit and Senacha [33], Senacha [60], Purohit and Vyas [45] and Purohit *et al.* [48].

Status

Red List: LC (Least Concern in India)

Kelaart's Pipistrelle *Pipistrellus ceylonicus* (Kelaart, 1852)

1852. *Scotophilus ceylonicus* Kelaart, *Prodr. Faun. Zeylanica*, p. 22 (Trincomalee, Sri Lanka)

Taxonomy

Synonyms: *Scotophilus ceylonicus* Kelaart, 1852; *Vesperugo indicus* Dobson, 1878; *Pipistrellus chrysothrix* Wroughton, 1899; *Pipistrellus ceylonicus chrysothrix* (Wroughton, 1899); *Pipistrellus ceylonicus subcanus* Thomas, 1915

Subspecies: One subspecies, namely, *Pipistrellus ceylonicus indicus* (Dobson, 1878), occurs in India [3].

Distribution

India

South Asia: Distributed throughout peninsular India, Bangladesh, Pakistan and Sri Lanka

Rajasthan (Fig. 21.23): *Pipistrellus ceylonicus indicus* (Dobson, 1878) has been reported from Mt. Abu (Sirohi District) [5].

Status

Red List: LC (Least Concern in India)

Genus *Scotozous* Dobson, 1875 (Dormer's Pipistrelle)

Dormer's Pipistrelle *Scotozous dormeri* Dobson, 1875

1875. *Scotozous dormeri* Dobson, *Proc. Zool. Soc. Lond.*, p. 373 (Bellary Hills, Mysore, Karnataka, India)

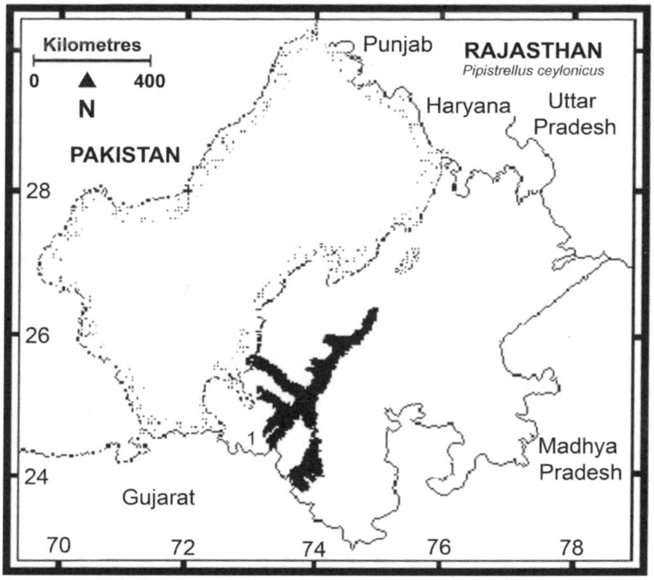

Fig. 21.23 Distribution of *Pipistrellus ceylonicus indicus* in Rajasthan, India. Key: 1—Mt. Abu

Taxonomy

Synonyms: *Pipistrellus dormeri* (Dobson, 1875); *Pipistrellus dormeri dormer* (Dobson, 1875); *Scotozous dormeri caurinus* Thomas, 1915; *Pipistrellus dormeri caurinus* (Thomas, 1915)

Subspecies: No subspecies

Distribution

South Asia: Distributed throughout India, Bangladesh and Pakistan

Rajasthan (Fig. 21.24): *Scotozous dormeri* Dobson, 1875 has been reported from Jodhpur (Jodhpur District), Bharatpur (Bharatpur District), Dungarpur (Dungarpur District), Banswara (Banswara District) [10] and districts of Dungarpur, Banswara, Jhalawar, Bundi, Tonk, Ajmer, Sawai Madhopur, Jodhpur, Kota [18] Alwar and Bharatpur [21]. See Purohit *et al.* [48].

Status

Red List: LC (Least Concern in India)

Genus *Myotis* Kaup, 1829 (Mouse-eared Bats)

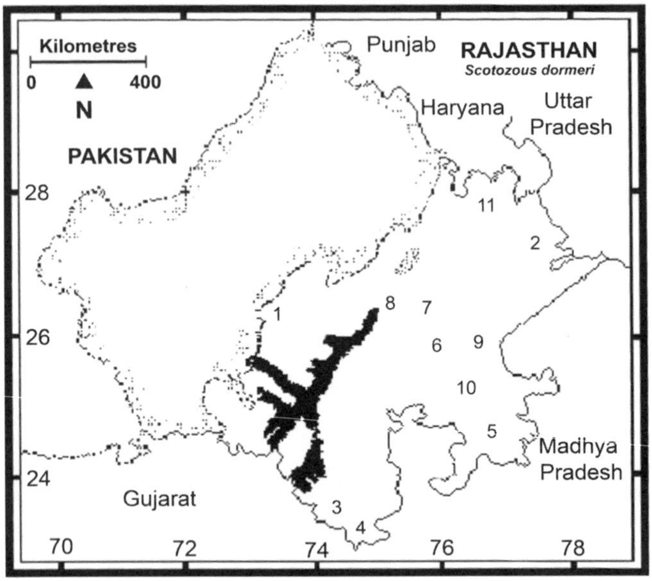

Fig. 21.24 Distribution of *Scotozous dormeri* in Rajasthan, India. Key: 1—Jodhpur, 2—Bharatpur, 3—Dungarpur, 4—Banswara, 5—Jhalawar, 6—Bundi, 7—Tonk, 8—Ajmer, 9—Sawai Madhopur, 10—Kota, 11—Alwar

Lesser Mouse-Eared Myotis *Myotis blythi* (Tomes, 1857)

1857. *Vespertilio blythii* Tomes, *Proc. Zool. Soc. Lond.*, p. 53 (Nasirabad, Rajasthan, India)

Taxonomy

Synonyms: *Vespertilio blythii* Tomes, 1857; *Vespertilio murinoides* Dobson, 1873; =*Vespertilio dobsoni* Trouessart, 1878

Subspecies: One subspecies, namely, *Myotis blythi blythi* (Tomes, 1857), occurs in India [3].

Distribution

South Asia: Distributed in the Himalayan tract of India in Jammu and Kashmir, Himachal Pradesh and Uttaranchal. There is also a historic record from Rajasthan. Other than India, the range of this species also includes Afghanistan, Nepal and Pakistan.

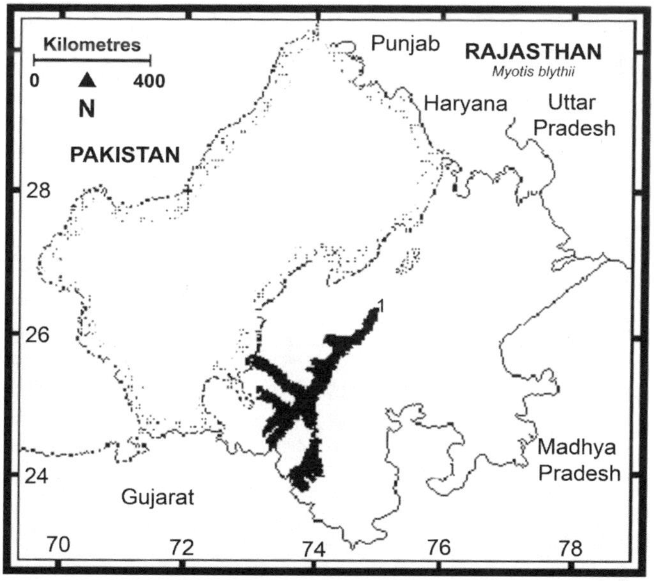

Fig. 21.25 Distribution of *Myotis blythi blythi* in Rajasthan, India. Key: 1—Nasirabad

Rajasthan (Fig. 21.25): *Myotis blythi blythi* (Tomes, 1857) has been reported from Nasirabad (Ajmer District) [10] which is also its type locality, but Topal [51] suggests that the correct locality is Naseerabad, possibly in the Himalayas.

Status

Red List: VU (Vulnerable in India)

Genus *Kerivoula* Gray, 1842 (Woolly Bats)

Painted Woolly Bat *Kerivoula picta* (Pallas, 1767)

1767. *Vespertilio pictus* Pallas, *Spicil. Zool.*, 3: 7 (Ternate Island, Molucca Islands, Indonesia)

Taxonomy

Synonyms: *Vespertilio pictus* Pallas, 1767; *Vespertilio kirivoula* Cuvier, 1832

Subspecies: One subspecies, namely, *Kerivoula picta picta* (Pallas, 1767), occurs in India [3].

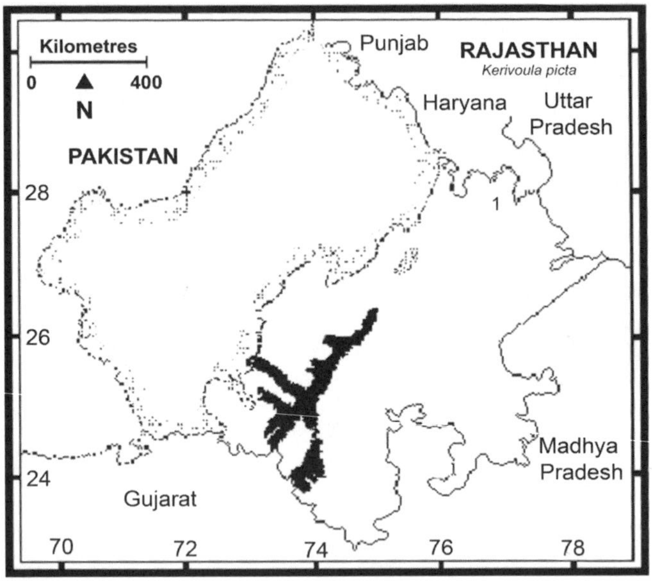

Fig. 21.26 Distribution of *Kerivoula picta picta* in Rajasthan, India. Key: 1—Kishore Pura

Distribution

South Asia: Distributed patchily throughout India, Bangladesh, Nepal and Sri Lanka

Rajasthan (Fig. 21.26): *Kerivoula picta picta* (Pallas, 1767) has been reported from Kishore Pura (Alwar District) [30].

Status

Red List: LC (Least Concern in India)

Zoogeography

Of the 25 species of bats recorded so far from Rajasthan, some species are very restricted in distribution, while others are widespread. As has been mentioned earlier, the unique biogeographic features of Rajasthan play an important role in governing distribution of bats. Unlike the common belief that bats being aerial are widely distributed, they are, in fact, restricted to areas where suitable habitats and food resources exist. In the earlier times, some of the bat species that are presently

Table 21.2 Distribution of bat species in three geographic regions of Rajasthan, India

S. No.	Species	Desert	Non-desert	Aravallis
1.	*Pteropus giganteus*	+	+	+
2.	*Rousettus leschenaultii*	–	+	+
3.	*Cynopterus sphinx*	+	+	–
4.	*Taphozous perforatus*	+	+	–
5.	*Taphozous longimanus*	–	+	+
6.	*Taphozous nudiventris*	+	+	+
7.	*Taphozous melanopogon*	–	–	+
8.	*Rhinopoma hardwickii*	+	+	–
9.	*Rhinopoma microphyllum*	+	+	+
10.	*Megaderma lyra*	+	+	+
11.	*Rhinolophus lepidus*	+	+	–
12.	*Hipposideros fulvus*	+	+	–
13.	*Hipposideros lankadiva*	+	–	–
14.	*Tadarida aegyptiaca*	+	+	+
15.	*Chaerephon plicata*	–	–	+
16.	*Eptesicus serotinus*	–	–	+
17.	*Hesperoptenus tickelii*	–	–	+
18.	*Barbastella darjelingensis*	–	–	+
19.	*Scotophilus kuhlii*	+	+	–
20.	*Scotophilus heathii*	+	+	+
21.	*Pipistrellus tenuis*	+	+	+
22.	*Pipistrellus ceylonicus*	–	–	+
23.	*Scotozous dormeri*	+	+	–
24.	*Myotis blythi*	–	–	+
25.	*Kerivoula picta*	–	+	–

known from the Himalayas have been collected from Rajasthan, indicating that the Aravallis in the past had, at least during certain parts of the year, a climate akin to that of the Himalayas. These records could be due to the fact that at that time specimen collection was allowed and even promoted and so we have authentic specimen records. Secondly, British were very meticulous in keeping records. Now the situation is changed. Specimen collection is generally not allowed, and recorders are also not in the field. Within 150 years, the condition in the Thar Desert has changed so much that the Himalayan elements have disappeared. Probably meticulous researchers have disappeared!

The desert area of Rajasthan was home to two species of bats until the early 1980s, and with the construction of the Indira Gandhi Canal, the general habitat has changed drastically at least in the districts of Sri Ganganagar, Hanumangarh, Bikaner, Churu and parts of Jodhpur and Jaisalmer [50]. Sinha [9] opined that the desert environment was not suited for the fruit bats and reported no sightings of the three species of fruit bats from any location in the desert. However, in the recent times, two species, namely, the Indian Flying Fox and Greater Short-nosed Fruit Bat, have been reported from different localities of the desert [10, 33, 38–41, 46].

A summary of the bat species distribution in different geographic regions of Rajasthan is given in Table 21.2. As many as 15 species have been reported from the desert region, 17 species from the non-desert region and 16 species from the Aravalli Hills. Seven species, namely, Indian Flying Fox, Naked-rumped Tomb Bat, Greater Mouse-tailed Bat, Greater False Vampire Bat, Egyptian Free-tailed bat, Greater Asiatic Yellow House Bat and Least Pipistrelle, have been reported from all the three geographic regions of Rajasthan. Seven species, namely, Greater Short-nosed Fruit Bat, Egyptian Tomb Bat, Lesser Mouse-tailed Bat, Blyth's Horseshoe Bat, fulvous Leaf-nosed Bat, Lesser Asiatic Yellow House Bat and Dormer's Pipistrelle, have been reported from both the desert and non-desert regions of Rajasthan. Two species, namely, Leschenault's Rousette and Long-winged Tomb Bat, have been reported from the non-desert and the Aravalli Hills geographic regions of Rajasthan. Nine species have been reported from only one geographic region, of which one species each, namely, Indian Leaf-nosed Bat and Painted Bat, has been reported from the desert and the non-desert regions, respectively, while the rest have been reported from the Aravalli Hills. An interesting point to be noted is that among these nine species excepting the painted bat, no other species have been collected during the recent times.

Conclusion and Recommendations

Although considerable amount of work has been conducted on bats of Rajasthan, there are clearly certain issues that need to be addressed if at all a true picture of the effect of habitat mosaics and distribution pattern needs to be understood and also to mitigate the problems related to the local extinctions of bat species due to anthropogenic actions. A ban on disturbing bat roosts in old *havelis* and ruins needs to be implemented to protect bats. The Department of Tourism of Rajasthan Government should take stringent measures to promote bat roosting sites as tourist spots and encourage locals to consider them as source of income. Conservation education activities should be taken up to dispel myths and superstitions pertaining to bats among the general public through print and electronic media. More research initiatives need to be taken up to study bat diversity in areas that had not been surveyed in the past, and researchers should continue to report in detail all sightings of bat species.

Acknowledgements We would like to thank Director, Zoological Survey of India, Kolkata; Dr. Ramakrishna and Officer-in-Charge, Mammals Section, Zoological Survey of India, Kolkata for facilities. The first two authors would like to thank the Head, Department of Zoology, Osmania University, Hyderabad, for encouragement and facilities; Vice Chancellor, Maharana Pratap University of Agriculture and Technology, Udaipur for sponsoring Biodiversity Assessment Survey of Hadoti Region; Dean, College of Horticulture and Forestry, Jhalawar and Dr. Parmeshwar S. Chauhan of CH&F, Jhalawar for extending necessary facilities and hospitality. We would also like to thank Ms. Harpreet Kaur, Department of Zoology, Osmania University, Hyderabad and Mast. C. Aditya for help during manuscript preparation. The first two authors acknowledge individual research grants from Council for Scientific and Industrial Research, New Delhi and DBT funding to work on molecular phylogeny of bats.

Appendix 1: Gazetteer of locations from where bats have been recorded in Rajasthan, India (localities whose coordinates are not available are not listed)

Locality name	Coordinates	District
Ajmer	26°29′N, 74°40′E	Ajmer
Aligarh	25°58′N, 76°09′E	Tonk
Alwar	27°32′N, 76°35′E	Alwar
Bagher	23°53′N, 73°48′E	Jhalawar
Balsamand	26°25′N, 73°07′E	Jodhpur
Banswara	23°32′N, 74°28′E	Banswara
Barmer	25°43′N, 71°25′E	Barmer
Berah	25°46′N, 73°26′E	Pali
Bharatpur	27°14′N, 77°28′E	Bharatpur
Bhinmal	25°00′N, 72°26′E	Jalore
Bhim Bharak	26°09′N, 73°08′E	Jodhpur
Bikaner	28°01′N, 73°22′E	Bikaner
Bundi	25°28′N, 75°42′E	Bundi
Darrah	24°48′N, 75°59′E	Kota
Dungarpur	23°53′N, 73°48′E	Dungarpur
Gagron Ka Kila	24°37′N, 76°11′E	Jhalawar
Jaipur	26°53′N, 75°50′E	Jaipur
Jaisalmer	26°52′N, 70°55′E	Jaisalmer
Jaswantgarh	27°39′N, 74°27′E	Udaipur
Jhalrapatan	24°35′N, 76°12′E	Jhalawar
Jhalawar	24°32′N, 76°12′E	Jhalawar
Jhunjhunu	28°05′N, 75°30′E	Jhunjhunu
Jodhpur	26°18′N, 73°08′E	Jodhpur
Kota	25°11′N, 75°58′E	Kota
Lathi	27°03′N, 71°51′E	Jaisalmer
Mandu Fort	22°22′N, 75°24′E	Jodhpur
Marot	27°05′N, 75°05′E	Nagour
Mt. Abu	24°41′N, 72°50′E	Sirohi
Naga	27°50′N, 77°06′E	Jhunjhunu
Nagour	27°12′N, 73°48′E	Nagour
Nangal	27°22′N, 76°13′E	Jhunjhunu
Nasirabad	26°16′N, 74°42′E	Ajmer
Pali	25°46′N, 73°26′E	Pali
Parsad	24°11′N, 73°42′E	Udaipur
Rajgad	26°18′N, 74°37′E	Ajmer
Ranthambhore	26°04′N, 76°32′E	Sawai Madhopur
Salawas	26°07′N, 72°59′E	Sawai Madhopur
Sawai Madhopur	26°00′N, 76°28′E	Sawai Madhopur

(continued)

Locality name	Coordinates	District
Sikar	27°33′N, 75°12′E	Sikar
Sikar Burj	25°23′N, 75°47′E	Bundi
Sirohi	24°53′N, 72°58′E	Sirohi
Tinwari	26°46′N, 73°12′E	Jodhpur
Tonk	25°52′N, 75°50′E	Tonk
Udaipur	27°40′N, 75°32′E	Udaipur

References

1. Mickleburgh SP, Hutson AM, Racey PA (2002) A review of the global conservation status of bats. Oryx 36(1):18–34
2. Simmons NB (2005) Order Chiroptera. In: Wilson DE, Reeder DM (eds) Mammal species of the world: a taxonomic and geographic reference, vol 1, 3rd edn. The John Hopkins University Press, Baltimore, MD, pp 312–529
3. Srinivasulu C, Srinivasulu B (2012) South Asian Mammals—Their Diversity, Distribution, and Status. Springer, New York, p 478
4. Blanford WT (1891) The fauna of British India, Mammalia. Part 2. Taylor & Francis, London
5. Ryley K (1914) Bombay Natural History Society's Mammal Survey of India. No. 12. Palanpur and Mt. Abu. J Bomb Nat Hist Soc 22(4):684–699
6. Garg BL (1955) Significance of the orbitotemporal region in the skulls of the small bats of Ajmer. Curr Sci 24:55–56
7. Prakash I (1961) Die Saugetiere der Rajasthan—Wustein Indien. Natur Volk 91:443–448
8. Prakash I (1963) Taxonomic and biological observations on the bats of Rajasthan Desert. Rec Indian Mus 59(1&2):149–170
9. Sinha YP (1979) Studies on taxonomy, distribution, zoogeography, osteology and field ecology of bats of Rajasthan, especially the Indian Desert. Ph.D. dissertation, Jodhpur University, Jodhpur, Rajasthan, India, p 204
10. Sinha YP (1980) The bats of Rajasthan: taxonomy and zoogeography. Rec Zool Surv India 76(1–4):7–63
11. Sinha YP (1981) New record of Black-bearded tomb bat, *Taphozous melanopogon melanopogon* Temminck from Rajasthan. Geobios, Jodhpur 8(5):225–226
12. Gaur BS (1981) Ecology of bats of the Indian Desert. Unpublished Ph.D. thesis, Jodhpur University, Jodhpur, India, p 110
13. Sinha YP (1996) Bats in Indian Thar Desert, p 349–352. In: Ghosh AK, Baqri QH, Prakash I (eds) Faunal diversity in the Thar Desert: gaps in research. Scientific Publication, Jodhpur, p 140
14. Advani R, Vazirani TG (1981) Studies on ectoparasites of bats of Rajasthan and Gujarat (India). Rec Zool Surv India, Misc. Pub., Occ. Pap. No. 22; p 1–155
15. Advani R (1980) Observations on feeding ecology and behaviour of the Kutch sheath tailed bat, *Taphozous kachhensis kachhensis* in Rajasthan. Zeitschrift Angew Zool 67(3):279–285
16. Advani R (1981) Seasonal fluctuation in the feeding ecology of the Indian false vampire, *Megaderma lyra lyra* (Chiroptera: Megadermatidae) in Rajasthan. Zeits für Säug 46(2): 90–93
17. Advani R (1981) Some observations on the feeding behaviour of the Indian Pygmy Pipistrelle, *Pipistrellus mimus mimus*, Wroughton, 1899 (Mammalia: Chiroptera: Vespertilionidae) in Rajasthan Desert. Säug Mitt 29(4):10–12
18. Advani R (1981) Feeding ecology and behaviour of the Dormer's bat, *Pipistrellus dormeri dormeri* (Dobson, 1875) in the India Desert. Säug Mitt 29(4):56–58

19. Advani R (1982) Feeding, foraging and roosting behaviour of the fruit-eating bats and damage to fruit crops in Rajasthan and Gujarat. Säug Mitt 30(1):46–48
20. Advani R (1982) Feeding ecology of *Tadarida aegyptiaca thomasi* in the Indian desert. Zeits für Säug 47:18–22
21. Advani R (1982) Distribution and status of chiroptera species in Rajasthan, India. Säug Mitt 30(1):49–52
22. Ramaswami LS, Anand Kumar TC (1963) Differential implantation of twin blastocysts in *Megaderma* (Microchiroptera). Cell Mol Life Sci 19(12):641–642
23. Anand Kumar TC (1963) Reproduction in the rat-tailed bat *Rhinopoma kinneari*. J Zool Lond 147:137–155
24. Wason A (1978) Observations on homing ability of some insectivorous bats. Zeits für Säug 44(5):305–306
25. Agarwal A, Gupta BB (1982) The gastric morphology and histology of an insectivorous bat—*Rhinopoma kinneari*. Lynx 21:5–14
26. Lall SB (1985) Folliculogenesis in *Rhinopoma kinneari* Wroughton (Microchiroptera: Mammalia). Myotis 23–24:37–44
27. Bhupathy S (1987) Occurrence of the bicoloured leaf-nosed bat (*Hipposideros fulvus*) in Rajasthan. J Bomb Nat Hist Soc 84(1):199–200
28. Gupta SK, Trivedi KK (1989) Nematode parasites of vertebrates. On two new species of the genus *Litomosoides* Chandler, 1931 (Family: Dipetalonematidae Wehr, 1935) from microbats of Udaipur, Rajasthan, India. Ind J Helminth 41(Suppl):152–161
29. Trivedi S, Lall SB (1989) Histological and histochemical alterations in the ovary of nulliparous and parous bat: *Megaderma lyra*, exhibiting absolute sinistral dominance of genital tract, p 153–160. In: Hanak V, Horacek I, Gaisler J (eds) European bat research. Charles University Press, Praha, p 720
30. Sharma SK (1986) Painted bats and nests of Baya Weaver Bird. J Bomb Nat Hist Soc 81:196
31. Agarwal RL, Agarwal J, Nagar CK, Bhasin S (1981) Perforating injury of cornea by flying bat. Ind J Ophthalmol 29:39–40
32. Trivedi S (1991) Seasonal histochemical and histoenzymological alterations in the ovary and uterus of certain chiroptera (Mammalia) of Rajasthan. Ph.D. thesis, M.L. Sukhadia University, Udaipur, India, p 227
33. Purohit AK, Senacha KR (2002) A review of microchepteran eco-status in Mandore garden, Jodhpur. J Nat Cons 14:251–262
34. Senacha KR (2003) Eco-status and demographic changes among the chiropterans of the Thar desert with special reference to Jodhpur. Ph.D. thesis, JNV University, Jodhpur, India, p 175
35. Senacha KR (2006) Opportunistic survey of Indian Flying Fox *Pteropus giganteus* (Brunnich, 1782). BatNet–CCINSA Newslett 7(1–2):27–29
36. Tak JR, Dookia S (2003) Population fluctuation and ecology of Indian flying fox (Pteropus giganteus giganteus, Brunnich, 1782) in Jodhpur, Rajasthan. Cheetal 42(1–2):47–50
37. Trivedi S, Naruka K, Singh P, Dabi I, Rathore S, Rathore A (2003) Differential leukocyte profile of *Rhinopoma microphyllum kinneari*. Vespertilio 7:169–176
38. Dookia S (2004) Occurrence of the Short-nosed Fruit Bat (*Cynopterus sphinx* Vahl, 1797) in Thar Desert of Rajasthan. Zoos Print J 19(9):1629
39. Dookia S, Tak JR (2004) Status and distribution of Indian Flying Fox (*Pteropus giganteus* Brunnich) in Thar Desert of Rajasthan. Bat Net—CCINSA Newslett 5(1):7–8
40. Purohit AK, Senacha KR (2004) Distribution of bats in and around Jaisalmer of the Great Indian Desert, India. Vespertilio 8:99–104
41. Purohit AK, Senacha KR (2004) Demographic changes among bats in and around Jaisalmer of Great Indian Desert. Cheetal 42(1–2):25–34
42. Senacha KR, Purohit AK (2004) Possible twin birth in the Indian flying fox *Pteropus giganteus*. Bat Res News 45(4):199
43. Trivedi S, Lall SB (2004) Ovarian dehydrogenases of the non-pregnant, pregnant and lactating *Rhinopoma microphyllum kinneari* (Chiroptera: Rhinopomatidae). Vespertilio 8:105–112

44. Trivedi S, Lall SB (2006) Histochemical pattern of ovarian 5′-nucleotidase in *Rhinopoma microphyllum* during different reproductive states. Vespertilio 9–10:175–181
45. Purohit AK, Vyas KB (2006) Review of sex ratio in several bat species inhabiting the Great Indian Desert. Vespertilio 9–10:233–235
46. Senacha KR, Vyas KB, Purohit AK (2006) New records of Short-nosed Fruit Bat *Cynopterus sphinx* (Vahl, 1797) from Thar Desert, Rajasthan. Zoos' Print J 21(10):2419–2420
47. Srinivasulu C, Srinivasulu B (2006) Biodiversity Assessment Survey of Hadoti Region, Rajasthan. Part I. Birds and other wildlife of Jhalawar and Kota districts. Unpublished Report submitted to Maharana Pratap University of Agriculture and Technology, Udaipur, p 24
48. Purohit AK, Vyas KB, Senacha KR (2006) Population dynamics of bats in and around Jodhpur of Great Indian Desert. Tiger Pap 33(3):15–22
49. Gaur BS (1979) Vanishing of Indian False Vampire *Megaderma lyra lyra* (Geoffroy) from Jodhpur. Comp Physiol Ecol 4(4):260
50. Rahmani AR (1997) The effect of Indira Gandhi Nahar Project on the avifauna of the Thar Desert. J Bomb Nat Hist Soc 94(2):233–266
51. Topal G (1971) The taxonomic position of *Myotis dobsoni* (Trouessart, 1879) and some statistical data to the sub-specific examination of *Myotis blythi* (Tomes, 857). Ann Hist Nat Mus Natn Hungar 63:383–400
52. Molur S, Marimuthu G, Srinivasulu C, Mistry S, Hutson AM, Bates PJJ, Walker S, Padma Priya K, Binu Priya AR (eds) (2002) Status of South Asian Chiroptera—Conservation Assessment and Management Plan (CAMP), Workshop Report 2002. Zoo Outreach Organization, Coimbatore, India, CBSG—South Asia and WILD, p 141
53. Prakash I (1960) Breeding mammals in Rajasthan Desert, India. J Mamm 41(3):386–389
54. Purohit AK, Senacha KR (2003) Electrocution of Indian Flying Foxes, *Pteropus giganteus giganteus* in Jodhpur, India. Bat Res News 44(2):63–64
55. Andersen K (1912) Catalogue of Chiroptera in the collection of the British Museum. Vol 1: Megachiroptera. British Museum of Natural History, London, p 854
56. Wroughton RC (1918) Summary of the results from the Indian Mammal Survey of Bombay Natural History Society. Primates and Chiropetra (Part I). J Bomb Nat Hist Soc 25:547–598
57. Bates PJJ, Harrison DL, Muni M (1994) The bats of Western India revisited. Part 1. J Bomb Nat Hist Soc 91(1):1–15
58. Bates PJJ, Harrison DL (1997) Bats of the Indian Subcontinent. Harrison Zoological Museum, Sevenoaks, p 258
59. Purohit AK, Senacha KR (2003) Distribution of chiropteran fauna in and around the Bikaner of Great Indian Desert. Tiger Pap 30(4):3–8
60. Senacha KR (2002) A note on *Prosopis juliflora* (*sic*) *julifera*: emerging threat for the microchiropteran of Thar Desert. BatNet—CCINSA Newslett 3(2):7–8
61. Senacha KR, Purohit AK (2005) Partial albinism in the lesser mouse-tailed bat, *Rhinopoma hardwickii*. Bat Res News 46(3):75–76
62. Advani R (1981) Food and feeding ecology of the rat tailed bat in the Rajasthan Desert. Acta Theriol 26(8–15):269–272
63. Purohit AK, Kaluram (2001) Present population status of *Rhinopoma microphyllum kinneari* at J.N.V. University, Jodhpur. BatNet—CCINSA Newslett 2(1):16
64. Purohit AK, Gaur BS, Senacha KR, Vijayakrishna VK, Chhangani AK (2002) Rare observation of puppies preying upon a greater mouse-tailed bat: *Rhinopoma microphyllum kinneari*. Flora Fauna 8(2):115–116
65. Bates PJJ, Harrison DL, Muni M (1994) The Bats of Western India revisited. Part 3. J Bomb Nat Hist Soc 91(3):360–380
66. Ellerman JR, Morrison-Scott TCS (1951) Checklist of Palearctic and Indian Mammals—1758 to 1946. British Museum of Natural History, London, p 810

Chapter 22
Non-Volant Small Mammals of Rajasthan

Partap Singh, R.S. Tripathi, and B.K. Sharma

Abstract This chapter contains data from small mammal surveys carried out in the entire state by Central Arid Zone Research Institute (CAZRI), Jodhpur, and later by Zoological Survey of India (ZSI), Jodhpur, Rajasthan, India. Non-volant small mammals belong to orders Scandentia, Rodentia, Soricomorpha, and Erinaceomorpha whereas volant small mammals fall under order Chiroptera. Tree shrew belonging to order Scandentia is absent from Rajasthan. The other three orders are well represented in the state, and as many as 26 species of non-volant small mammals have been reported. Order Rodentia with 22 species is the largest order of small mammals, while orders Soricomorpha and Erinaceomorpha are represented by two species each. Rodents of Rajasthan have been well studied because of their pest status since inception of CAZRI. All aspects of rodent pests including physiology, ecology, taxonomy, zoogeography, ethology, and toxicology have been thoroughly worked out. Certain other species of rodents like *Petaurista philippensis*, *Vandeleuria oleracea*, 06 species of genus *Mus*, *Nesokia indica*, and certain desert-dwelling gerbils are poorly studied. Insectivorous orders, Soricomorpha, and Erinaceomorpha, too, have received little attention of ecologists. This chapter includes species composition in various zones, habitat preference, food and feeding ecology, and reproductive biology of lesser known small mammals of Rajasthan.

P. Singh (✉)
Department of Zoology, Government Dungar College, Bikaner, Rajasthan, India
e-mail: partapsk@gmail.com

R.S. Tripathi
Central Arid Zone Research Institute, Jodhpur, Rajasthan, India
e-mail: rstripathi@cazri.res.in

B.K. Sharma
Department of Zoology, R.L. Saharia Government P.G. College,
Kaladera (Jaipur), Rajasthan, India
e-mail: drbksharma@hotmail.com

B.K. Sharma et al. (eds.), *Faunal Heritage of Rajasthan, India: General Background and Ecology of Vertebrates*, DOI 10.1007/978-1-4614-0800-0_22,

Introduction

Owing to its geomorphologic configuration, the state of Rajasthan exhibits spectacular biodiversity of deserticolous elements in the Thar, relict fauna in the south and southeastern zone, and relatively geographically recent biota in its extreme northeast [1]. Moghe [2] has divided the state into nine Agro-climatic zones. The first small mammal survey of the state was carried out by Bombay Natural History Society, and Ryley [3] presented first exhaustive account of small mammals based on the collection of Mr. C. A. Crump. Small mammals have been categorized into two groups, volant small mammals which include order Chiroptera and the non-volant small mammals which include orders Scandentia, Rodentia, Soricomorpha and Erinaceomorpha. Non-volant small mammals in Rajasthan are represented by Rodentia, Soricomorpha, and Erinaceomorpha. Order Scandentia, which includes tree shrews, is not represented in Rajasthan. Systematic surveys of non-volant small mammals have been carried out by scientists of Central Arid Zone Research Institute (CAZRI) and Zoological Survey of India (ZSI), of which we had been the members.

About 417 species of mammals belonging to 48 families had been reported from India [4]. Small mammals, both volant and non-volant, constitute about 60% of total species found in India. Non-volant small mammals constitute 33% of total Indian mammalian fauna and are, thus, the largest group among mammals. Unfortunately, this largest group has remained a neglected lot as far as ecological and ethological studies are concerned. Only those species which have pest status had been worked out, as behavioral understanding makes it easy to take up control operations. Most of non-volant small mammals, especially rodents, are poorly studied. Very little is known about the distribution and other aspects of many Indian small mammals. These “Data Deficient” species and many monotypic species need urgent efforts for protection and conservation.

Food and Foraging

One of the factors behind success of small mammals is the plasticity in their food and feeding behavior. Small mammals of the Thar Desert have little options within their home range and consume whatsoever is abundantly available. Rodents are mainly herbivores and feed on almost all parts of plants and seeds. Some, like *Tatera indica* and *Meriones hurrianae*, feed on insects during acute summer. Depending upon the availability, the gerbils, *M. hurrianae* and *T. indica*, adopt seasonal cyclicity in their food items. For example, they depend mostly on leaves and flowers in monsoon and post-monsoon season, seeds during winters and rhizomes, and stems during summers as food [1, 5, 6]. Other regions of Rajasthan (Aravallis, southeastern Rajasthan and northern Rajasthan) have no dearth of vegetation, and small mammals may have the choice for the food. Natural food of many small mammals has been studied through stomach content analysis. Important species whose food in nature has been studied are *Hemiechinus collaris* [7], *Suncus murinus* [8], *Tatera indica* and *Meriones hurrianae* [9, 47], and *Cremnomys cutchicus* [10].

Advani and Mathur [11] carried out a systematic study on damages caused by rodents to both *kharif* (monsoon Crop) and *rabi* (post-monsoon crop) near Jodhpur. Rodents' damage to crops varied from 4% to 20% (average 8.7%). Pearl Millet (*Pennisetum typhoides*), an important staple food of Rajasthan, suffers lot of yield (107.7 kg/ha) due to rodents in western Rajasthan. The arid pulses viz., moong (*Vigna radiata*) and moth (*Vigna aconitifolia*) can experience 3% pod damage [13]. The desert rodents, especially *T. indica* and *Millardia meltada*, cause 18–21% damage to standing wheat crop [14]. Groundnut cultivation has suffered a major setback due to rodent devastations (30–50%) during last 4–5 years. In irrigated crop system of wheat–mustard–cumin, 5–11% damage by rodents has been recorded [6]. Vegetables are reported to experience 4–30% damage by desert rodents [11]. Palm Squirrels (*Funambulus pennantii*) are a serious menace in fruit orchards [15]. Afforestation plantations are devoured to the tune of 4–10% by slicing and debarking activities of fossorial rodents such as *Nesokia indica*, *Meriones hurrianae*, and *Tatera indica*. Rodent problem in agriculture at preharvest stage is, therefore, very serious as compared to other states like Gujarat (10%) and Tamil Nadu (5–10%) [12].

One more reason which makes rodents destructive is their hoarding behavior. This canny group of small mammals hoard large amount of food material to be consumed during lean period. *Cynodon dactylon* twigs, *Capparis decidua* seeds, rice and wheatears, groundnut, maize, pulses, and many other cereals have been excavated from the burrows of various species. The bandicoots, among rodents, are big hoarders and up to 6 kg of food material has been recovered from their burrows [16]. Sheikher and Malhi [17] reported average hoarding of 390 g of wheat per burrow. Some non-hoarder species, such as Desert Gerbil (*Meriones hurrianae*) has shown changed behavior and is reported to hoard food material in irrigated cropping systems in Shekhawti region of the Thar Desert [1]. Hoarding behavior of rodents makes them more destructive as they store large amount of food in their burrow system as food reserve and nesting material.

Order Erinaceomorpha is represented by two species (*Hemiechinus collaris* and *Paraechinus micropus*) in Rajasthan. Food of *Hemiechinus* has been studied through stomach contents which indicate the type of diet consumed by the animal. Though, many orders of Phylum Arthropoda constituted major chunk of food, it also included scorpion, *Uromastyx*, *Mabuya*, egg shells, and even bones and skin of toads. During winter, hedgehog passes a torpid period of about 2–3 months.

Suncus murinus and *S. stoliczkanus* are two Soricomorphs reported from the state. The knowledge about *S. stoliczkanus*, one of the smallest mammals of world, is limited mostly to distribution records. *Suncus murinus* has widely been studied for food and feeding behavior, and species has been found to feed upon leeches, crickets, cockroaches [18], molluscs [19], fishes [20], frogs [21], geckos [22], rats [23], and even snakes [24]. In the Aravalli ecosystem, 82% of its menu constituted of insects and mammals. Plant material constituted only 8% of total food amount, and remaining 10% was constituted of fish and bird species [8].

All these studies on food and foraging behavior of small mammals indicate that they are opportunistic in the feeding behavior. Rodents though widely considered granivorous consume insects or even a fish [25]. On the other hand, *Suncus murinus* largely considered as an insectivorous also feeds upon the leaves of a shrub *Mimosa hamata*, constituting 91% of annual food [26]. Tree Shrews belonging to order Scandentia are not distributed in the Rajasthan state.

Species Composition and Distribution Patterns

Systematic small mammal capturing has been carried out in the Thar, Aravallis, southeast Rajasthan, and northeast Rajasthan by ZSI and CAZRI. The capture data indicates that the Thar is the most species-rich ecosystem with 22 species as far as non-volant small mammals are concerned (Table 22.1). The Aravalli Mountain Range and well-wooded southeast regions support 17 species each. Northeast Rajasthan is poorly occupied by small mammals and only nine species inhabit this region.

The trapping data of Prakash and coworkers [27] in the Thar Desert indicate *Tatera indica* as the most abundant small mammal (26% of total small mammal population) followed by *Meriones* hurrianae (25%), *Millardia meltada* (14%), *Cremnomys cutchicus* (11%), and *Gerbillus gleadowi* (10%). Fourth relative abundance of Cutch Rock-rat, *Cremnomys cutchicus*, in the desert ecosystem might surprise many. All the rocky outcrops in the Thar are abundantly occupied by this Rock-rat, and it is conjectured that this species was distributed on the hilly outcrops before the desert conditions set in [25], and this species adapted itself according to changing climatic conditions. Based on the abundance of desert rodents in the arid zone of Rajasthan, Prakash *et al.* [27] and Tripathi *et al.* [28] have grouped them in following habitats:

- Sandy habitat: *Meriones hurrianae*>*Tatera indica*>*Gerbillus gleadowi*> *Millardia meltada*
- Gravel habitat: *Meriones hurrianae*>*Tatera indica*>*Mus platythrix*
- Rocky habitat: *Cremnomys cutchicus*>*Mus platythrix*>*Mus cervicolor*> *Funambulus pennantii*
- Ruderal habitat: *Tatera indica*>*Millardia meltada*>*Meriones hurrianae*>*Gerbillus gleadowi*>*Funambulus pennantii*

Seventeen species of small mammals had been reported from the Aravalli region (Table 22.2). The trapping data of Aravallis indicate that Cutch Rock-rat, *C. cutchicus* (36% of total captured small mammals), is the most abundant small mammal followed by *S. murinus* (21%), *T. indica* (10%), *M. meltada* (7%), *Golunda ellioti* (6%), and *Bandicota bengalensis* (5%). All other small mammals were <5% in relative abundance. The abundance of Rock-rat is obvious due to kind of habitat present in the Aravallis. Small crevices present among rock boulders provide ideal habitat to this rock-dwelling species. Moreover, the species has adapted to live in relatively high numbers in the loosely piled stonewalls erected around crop fields to protect the

Table 22.1 Distribution of small mammals in Rajasthan

	Common name	Zoological name	Distribution	Endemicity
Order: Rodentia				
1	Five-striped Palm Squirrel	*Funambulus pennantii*	A, T, SE, NE	
2	Large Brown Flying Squirrel	*Petaurista philippensis*	SE	
3	Indian Gerbil	*Tatera indica*	A, T, SE, NE	
4	Asiatic Long-tailed Climbing Mouse	*Vandeleuria oleracea*	A, SE	
5	House Rat	*Rattus rattus*	A, T, SE, NE	
6	Cutch Rock-rat	*Cremnomys cutchicus*	A, T, SE, NE	EN
7	Soft-furred Metad	*Millardia meltada*	A, T, SE	EN
8	Wroughton's Small Spiny Mouse	*Mus phillipsi*	A, T, SE, NE	EN
9	Brown Spiny Mouse	*Mus platythrix*	A, T, SE	EN
10	Saxicolous Mouse	*Mus saxicola*	A, T, SE	EN
11	Earth-colored Mouse	*Mus terricolor*	A, SE	
12	Indian Bush-rat	*Golunda ellioti*	A, T, SE, NE	
13	Lesser Bandicoot Rat	*Bandicota bengalensis*	A, T, SE, NE	
14	House Mouse	*Mus musculus*	A, T, SE	
15	Dwarf Gerbil	*Gerbillus nanus*	T	
16	Little Hairy-footed Gerbil	*Gerbillus gleadowi*	T	EN
17	Indian Desert Gerbil	*Meriones hurrianae*	T	
18	Sand-colored Metad	*Millardia gleadowi*	T	EN
19	Little Indian Field Mouse	*Mus booduga*	T, SE	
20	Short-tailed Bandicoot Rat	*Nesokia indica*	T	
21	Indian Crested Porcupine	*Hystrix indica*	T, A, NE	
Order: Soricomorpha				
22	House Shrew	*Suncus murinus*	A, T, SE, NE	
23	Anderson's Shrew	*Suncus stoliczkanus*	A, T	
Order: Erinaceomorpha				
24	Indian Long-eared Hedgehog	*Hemiechinus collaris*	A, T, SE	
25	Indian Hedgehog	*Paraechinus micropus*	T	

A = Aravallis, T = Thar, SE = southeast, NE = northeast, EN = endemic

crops from larger animals [29]. These crop fields provide nutritious food all the year round, making it more abundant and thus most successful small mammal of the habitat. Abundance of insectivorous *Suncus murinus* indicates that the region is quite rich in insect diversity and the other food material. Another reason for its abundance may be due to low predator pressure. Snakes and mongooses, though abundant, rarely feed on the species because of foul-smelling musk gland it possesses.

The south and southeast Rajasthan is a stony plateau composed of Vindhyan and Deccan systems. The region is well drained by rivers and witnesses highest rainfall in the state (650–950 mm). The region is well wooded and is abode to 17 non-volant small mammal species, making it second most species-rich region along with the Aravallis. In this region, *Cremnomys cutchicus* (38%) is the most abundant small mammal followed by *Suncus murinus* and *Tatera indica* (13% each) (Table 22.2). The region also holds good proportion of *Mus* species, and six species had been reported from here [30]. Small mammal composition of southeast Rajasthan

Table 22.2 Habitat preference of small mammals in various zones of Rajasthan

		Thar					SE Rajasthan						Aravallis					NE Rajasthan			
	Species	A	B	C	D	E	C	E	F	G	H	I	C	E	H	J	K	C	E	G	H
1	*Funambulus pennantii*	1.1		6.1	4.3	*2.7*		6.3			5.3		6.2	0.8	7.5	5.9		Present			
2	*Petaurista philippensis*						Present														
3	*Tatera indica*	19.0	30.5	4.9	39.0	*43.2*		2.6			25.6	7.7		4.0	37.7		41.7		50.0		33.3
4	*Vandeleuria oleracea*							1.1			0.8			0.8							
5	*Rattus rattus*	Present					7.4	11.6			18.8	30.8		7.9	9.4						66.6
6	*Cremnomys cutchicus*			66.6			22.2		12.5	33.3			67.0	15.1	20.8	88.2		62.5		100.0	
7	*Millardia meltada*	13.0	2.7	3.7	20.1	*21.6*		8.4			0.8			15.1	1.9						
8	*Mus phillipsi*			6.1			37.0	5.3	62.5	20.0	1.5		12.4	0.8	1.9	5.9		25.0	7.1		
9	*Mus platythrix*	1.7	5.4	9.8	1.2		3.7	2.6			0.8			0.8							
10	*Mus saxicola*			2.4				1.1		13.3			Present								
11	*Mus terricolor*									6.7	0.8						8.3				
12	*Golunda ellioti*	1.1	2.7		3.0			12.6			8.3		8.2	4.8	7.5				21.4		
13	*Bandicota bengalensis*					*8.1*	3.7	4.7			0.8			11.9	1.9			Present			
14	*Mus musculus*				0.6	*8.1*		1.1		6.7			Present								
15	*Gerbillus nanus*	4.1																			
16	*Gerbillus gleadowi*	16.6			13.4																
17	*Meriones hurrianae*	41.6	55.5		16.4	*16.2*															
18	*Millardia gleadowi*	1.1	2.7																		
19	*Mus booduga*				1.8			0.5													
20	*Nesokia indica*	Present																			
21	*Hystrix indica*	Present											Present					Present			
22	*Suncus murinus*						25.9	42.6	25.0	20.0	36.8	61.5	6.2	38.1	11.3		50.0	12.5	21.4		
23	*Suncus stoliczkanus*												Present								
24	*Hemiechinus collaris*	Present					Present						Present								
25	*Paraechinus micropus*	Present																			
Total species		09	06	07	09	*06*	06	13	03	06	11	03	05	11	09	03	03	03	04	04	02

A = sandy, B = gravel, C = rocky, D = ruderal, E = crop field, F = hilltop, G = gully, H = scrubland, I = lake bank, J = runnel, K = riverbank

indicates that the insectivorous Asian Musk shrew, *Suncus murinus*, is the most preponderant species (39%) followed by *Rattus rattus* (14%), *Tatera indica* (10%), and *Golunda ellioti* (9%). This region is quite rich in *Mus* species with six species. If we compare the region with Aravallis, *Mus musculus* and *Mus booduga* are two additional species which indicate a higher magnitude of land use transformation from the forest to irrigated cropping. Though the region holds many protected areas and sanctuaries, the richness of rodent species indicates the deterioration of the habitat. Escalating anthropogenic activities are exerting pressure on larger mammals and other wild fauna of the region, though small mammals are finding a good abode in the changed conditions.

Sariska Tiger Reserve is poorly occupied by small mammals, and only nine species had been observed in the region. The arboreal *Funambulus pennantii* and the burrows of *Bandicota bengalensis* with characteristic large mass of excavated soil outside burrow openings were also recorded. *Tatera indica* and *Cremnomys cutchicus* (28% each of total catch). *Suncus murinus* (14%), *Golunda ellioti*, and *Mus phillipsi* were the predominant species. The poor presence of small mammals in the region may be due to developing industries and absence of pastures, scrubs, and wastelands.

Habitat Preference

For successful survival animals have to make four fundamental decisions—where to live, how to gather food, how to avoid predators, and what tactics to use to reproduce [31]. First and foremost, requirement of any animal is to find a suitable habitat/place to live where there is plenty of food, danger of predator is less, and it can successfully mate to raise offsprings. Animals spend considerable time and energy to locate a safe habitat. Small mammals, though have restricted moving ability, are very exploratory and keep on moving small distances until they find a suitable habitat for themselves. Small mammals are rarely territorial and prefer to live in large groups which give them additional benefit of avoiding predators.

Non-volant small mammals have adapted themselves to survive in every kind of condition. There is probably no part of earth which is unoccupied by them except Arctic and Antarctic [32]. They have eventually evolved mechanism to survive in the close proximity of humans and in all other habitats. Trapping data and sighting records of small mammals are summarized in Table 22.2. It clearly indicates that *Gerbillus nanus*, *G. gleadowi*, *Meriones hurrianae*, and *Millardia gleadowi* are species confined to arid zone of Rajasthan and all other species have wide distribution. *Funambulus pennantii*, *Tatera indica*, *Rattus rattus*, *Cremnomys cutchicus*, *Millardia meltada*, *Golunda ellioti*, *Mus phillipsi*, *Mus platythrix*, and *Suncus murinus* are the species which are well distributed all over the Rajasthan state.

Gerbillus nanus and *Millardia gleadowi* are confined to sandy areas in low rainfall zone, while *Gerbillus gleadowi* is fast adapting to survive in the ruderal habitats also (Table 22.3). The diurnal gerbil *Meriones hurrianae* is the most abundant

Table 22.3 Habitat preference in relation to exclusive and habitat sharing species

Habitats	No. of species	Name of species
(a) Exclusive to habitat		
Sandy	One	*G. nanus indus*
Rocky	Two	*C. cutchicus* and *M. cervicolor phillipsi*
Ruderal	Three	*M. musculus*, *M. booduga*, and *B. bengalensis*
(b) Habitat sharing		
Rocky and ruderal	Four	*F. pennanti*, *T. indica*, *M. meltada pallidior*, and *M. platythrix sadhu*
Sandy and gravel	Six	*T. indica*, *M. hurrianae*, *M. meltada pallidior*, *M. platythrix sadhu*, *M. gleadowi*, and *G. ellioti gujerati*
Gravel and ruderal	Five	*T. indica*, *M. hurrianae*, *M. meltada pallidior*, *M. platythrix sadhu*, and *G. ellioti gujerati*
Rocky and ruderal	Four	*F. pennanti*, *T. indica*, *M. meltada pallidior*, and *M. platythrix sadhu*
Sandy and ruderal	Seven	*F. pennanti*, *G. gleadowi*, *T. indica*, *M. meltada pallidior*, *M. platythrix sadhu*, *M. platythrix sadhu*, and *N. indica*

rodent of almost all the habitats in the Thar and can be seen peeping out of its burrow. Work of Late Prof. Prakash has made it one of the most studied mammal species of India [9, 33–39]. *Cremnomys cutchicus* is the most common rodent of rocky areas throughout Rajasthan. This endemic species has been extensively studied [25] and is fast adapting to survive near human habitation. Calorific requirement of the species indicates that this species is expanding its niche and is becoming a peri-commensal species [40].

The Indian gerbil *Tatera indica* is also widely distributed throughout Rajasthan. In Bikaner city, the species has become commensal and is living in human houses along with House Rat, *Rattus rattus*. It can be a serious human health hazard because of sharing its niche with house rat. The diurnal Bush-rat, *Golunda ellioti*, remains under hedges and bushes to avoid diurnal predators. The species though prefers scrubland and crop fields has also been reported from rocky, ruderal, and sandy areas with good bush cover. *Bandicota bengalensis* is a mesic species with Indo-Malayan origin [41] and is distributed in India west of Indus River. This species is very aggressive and is invading the Thar. Because of incoming of Indira Gandhi canal, the species has reached to Bikaner region. It has recently invaded urban areas of Jodhpur [42], and it is conjectured that it will soon spread to the other areas where canal irrigation is being practiced. *Suncus murinus* is most common small

Table 22.4 Peak breeding season and litter size of desert rodents

Rodent species	Peak breeding season	Litter size
H. indica	Monsoon and December	1–8
F. pennanti	March–April and July–September	1–5
G. gleadowi	May, June, and October–January	2–5 (summer) and 5–6 (winter)
T. indica	Monsoon (in arid regions)	1–9
M. hurrianae	(i) February–March	1–9
	(ii) July and September–November	2–7
G. ellioti	March–August	5–10
M. meltada	Spring and Monsoon (in Rajasthan)	3–9
M. musculus	All the year round	1–8
N. indica	January–March and August–October	2–5
B. bengalensis	All the year round	4–12

mammal of southeast Rajasthan and second most common small mammal of Aravalli. Though commonly known as House Shrew, its more relative abundance in the crop field habitat indicates a case of behavioral atavism. Large Brown Flying Squirrel, *Petaurista philippensis* is an arboreal small mammal which prefers to live on the tall trees. The species is a Deccan element and good news is that it is extending its range northwards in the south Rajasthan [43].

Present and past scenario of small mammal distribution in Rajasthan indicates that small mammals of Peninsula and Gangetic Plains are extending their range into the Thar Desert due to the canal irrigation [44]. These observations on changes in species composition and abundance of small mammals should be regularly monitored to prevent the extinction of native and resident species.

The non-volant small mammals, particularly rodents are prolific breeders (litter size of up to 12); therefore, exhibit a very high reproductive potential. Such a high fecundity of rodents is countered by several biotic and abiotic factors operating in nature. Rodents sometimes regulate their population by feeding on their own young ones (cannibalism) during stress periods depending upon the carrying capacity of their habitat. In drought years, the rodent population maintains a low profile due to scarcity of natural food; however, the population explodes in a good rainfall year succeeding any drought year [45]. Breeding season of desert rodents as depicted in Table 22.4 indicates that most rodents breed from March to September, although a few breed all through the year. Minimum births occur during the extreme winter and summer seasons, when the conditions are largely unfavorable in desert. However, both the *Gerbillus* species have shown peak breeding in the month of June also. The major breeding peak during the monsoon season corresponds with the availability of green food in plenty in the desert which might be an important factor in acceleration of their breeding activity.

Endangered Small Mammals and Conservation Status

Small mammal diversity of India is remarkable; 100 species of rodents, three species of Erinaceomorphs, 32 species of Soricomorphs, and three species of Scandents have been reported from here [4]. Every bioclimatic zone of India has some typical small mammal taxa: Marmots, Pikas and Voles in the Himalayas; Bamboo Rats in the northeast; *Bandicota*, *Rattus*, and *Mus* in the plains; Flying Squirrels in Deccan; Porcupines in the rocky regions; and Gerbils and Jerboas in the northwest desert. State of Rajasthan is abode to 25 species of small mammals, and out of these, seven are endemic to Indian subcontinent.

A lot of ecological and ethological studies have been carried on small mammals of Rajasthan since inception of CAZRI, but these studies are confined to only those species which have pest status. As far as IUCN status of small mammals of the state is concerned, most of them are categorized as "Least Concerned" or "Near Threatened" (Table 22.5). General perception among public about small mammals, especially rodents, is that they are vermin and should be exterminated at the very first sight. Howsoever, very few people know that this largest group of mammals is one of the most threatened groups.

Some of the non-volant small mammals, such as, *Funambulus pennantii*, *Tatera indica*, *Rattus rattus*, *Cremnomys cutchicus*, *Golunda ellioti*, *Meriones hurrianae*, *Bandicota bengalensis*, *Mus musculus*, *Mus phillipsi*, and *Suncus murinus* are widely distributed and abundant in number and presently have no serious threats. Some rodents of Thar such as *Gerbillus nanus*, *G. gleadowi*, and *Millardia gleadowi* have restricted distribution to this region, and very little is known about them. Indian Flying Squirrel, *Petaurista philippensis*, is a peninsular element and Rajasthan is its northern limit of distribution. This herbivorous rodent prefers moist deciduous and evergreen forests. This species though widely distributed is in vulnerable category. Anderson's Shrew, *Suncus stoliczkanus*, is another small mammal distributed in central India, Gujarat, Rajasthan, and Sind and Punjab of Pakistan and needs urgent conservation attention since its number is small.

Major threats to the small mammals of Rajasthan are habitat loss due to fragmentation, rampant increase in the use of pesticides during last few years, and road accidents. Slow-moving hedgehogs, particularly, are the main victims of the road accidents which get trampled by fast moving vehicles. These become quite active during breeding season and come out on roads in search of mates and become easy victims of speeding vehicles. The fact is that this group of small mammals is still poorly studied in India and lot is yet to be explored about these secluded animals. Please see Chap. 2 for relevant pictures.

Table 22.5 IUCN status of small mammals of Rajasthan

	Zoological name	Family	IUCN (2012) status	Total population in India	Threats
1	*Funambulus pennantii*	Sciuridae	LRLC	Unknown	T
2	*Petaurista philippensis*	Sciuridae	LRNT	Unknown	I, H, Hf, L, Lf, T
3	*Tatera indica*	Muridae	LRLC	Many	Ps
4	*Vandeleuria oleracea*	Muridae	LRLC	Many	No
5	*Rattus rattus*	Muridae	LRLC	Many	I
6	*Cremnomys cutchicus*	Muridae	LRLC	Unknown	No
7	*Millardia meltada*	Muridae	LRLC	Unknown	Unk
8	*Mus phillipsi*	Muridae	LRLC	Unknown	Dr, Lf, Po
9	*Mus platythrix*	Muridae	LRLC	Unknown	No
10	*Mus saxicola*	Muridae	LRLC	Many	Dr, Po
11	*Mus terricolor*	Muridae	LC		??
12	*Golunda ellioti*	Muridae	LRLC	Unknown	No
13	*Bandicota bengalensis*	Muridae	LRLC	Unknown	No
14	*Mus musculus*	Muridae	LRLC	Many	Unk
15	*Gerbillus nanus*	Muridae	LRNT	Unknown	L
16	*Gerbillus gleadowi*	Muridae	LRLC	Many	I, L
17	*Meriones hurrianae*	Muridae	LRLC	Unknown	L, Ps
18	*Millardia gleadowi*	Muridae	LRNT	Unknown	I, Ps, Po
19	*Mus booduga*	Muridae	LRLC	Many	Dr, Ps, Po
20	*Nesokia indica*	Muridae	LRLC	Unknown	Dr
21	*Hystrix indica*	Hystricidae	LRLC	Many	Hf, Tp, T
22	*Suncus murinus*	Soricidae	LRLC	Many	No
23	*Suncus stoliczkanus*	Soricidae	LRLC	Many	Unk
24	*Hemiechinus collaris*	Erinaceidae	LRLC	Unknown	No
25	*Paraechinus micropus*	Erinaceidae	LC	Unknown	H, Tp, T

LR-LC = Low Risk- Least Concern, LR-NT = Low Risk - Near Threatened, VU = Vulnerable, ?? = not mentioned [46]

L = loss of habitat, Lf = loss of habitat due to fragmentation, H = harvest, Hf = harvest for food, P = predation, Ps = pesticides, T = trade, Tp = trade of parts, I = human interference, Po = pollution, Dr = diseases?, Unk = unknown, No = no threats

References

1. Prakash I, Singh P (2005) Ecology of small mammals of desert and montane ecosystems. Scientific Publishers, Jodhpur, pp 1–153
2. Moghe VB (1994) Resource atlas of Rajasthan. Raja Offset Printers, Jodhpur, pp 1–231
3. Ryley KV (1913) Bombay Natural History Society's mammal survey of India, Burma and Ceylon. Report No. 12. Palanpur and Mount Abu. J Bomb Nat Hist Soc 22:684–699
4. Nameer PO (2008) A note on the checklist of Indian mammals, revised and updated. Zoo's Print 23(8):1–12
5. Prakash I (1959) Food of Indian desert mammals. J Biol Sci 2(2):100–109
6. Prakash I, Ghosh PK (eds) (1992) Rodents in Indian agriculture, Vol 1. Scientific Publishers, Jodhpur, pp 1–707 (State of Art)
7. Krishna D, Prakash I (1955) Hedgehogs of the desert of Rajasthan. Pt I. Distribution and fossorial habits. J Bomb Nat Hist Soc 53(1):38–43
8. Prakash I, Singh H (1999) Food of shrew *Suncus murinus* inhabiting hilly tracts of south and southeastern Rajasthan. Proc Natl Acad Sci USA 698:245–250
9. Prakash I (1962) Taxonomical and ecological account of the mammals of Rajasthan desert. Ann Arid Zone 1(2):142–163
10. Singh P, Prakash I (1997) Food preference and calorific requirement of the Cutch rock-rat *Cremnomys cutchicus medius*. Ann Arid Zone 36:65–72
11. Advani R, Mathur RP (1982) Experimental reduction of rodents damage to vegetable crops in Indian villages. Agro Ecosyst 8:39–46
12. Hopf HS, Morley CFJ, Humphries JR (1976) Rodent damage to growing crops and to farm and village storage in tropical and sub-tropical regions. Centre for Overseas Pest Research and Tropical Product Institute Publication, London, pp 1–115
13. Tripathi RS, Chaudhary V, Idris M (2004) Incidence of rodent pests and their management in pulse crops under arid agro-ecosystem. Pestology 28(6):67–70
14. Advani R, Prakash I, Mathur RP (1982) Assessment of rodent damage and yield losses in standing wheat crop in a desert village complex. Z Angew Zool 99:257–266
15. Prakash I, Kashyap N, Mathur M (1992) The five-striped or Northern Palm Squirrel. In: Prakash I, Ghosh PK (eds) Rodents in Indian agriculture, vol I. Scientific Publishers, Jodhpur, pp 17–24
16. Kamath MK (1961) Studies on Bombay rats, burrows and ectoparasites in Mahim area. M.S. dissertation, Haffkine Institute, Bombay (Unpublished), pp 1–333
17. Sheikher C, Malhi CS (1983) Territorial and hoarding behavior in Bandicota sp. and Mus sp. of Garhwal Himalayas. Proc Ind Natl Acad Sci (B) 49:332–335
18. Roberts TJ (1977) Mammals of Pakistan. Ernest Benn Ltd, London, pp 1–361
19. Lim BL (1966) Land mollusks as food of Malayan rodents and insectivores. J Zool 148:554–560
20. Tiwari JK (1994) Unusual feeding behavior of the grey musk shrew, *Suncus murinus* Lin. J Bomb Nat Hist Soc 91:305
21. Dharamkumarsinhji KS (1946) Musk shrew *Suncus murinus* attacking bull frog (*Rana tigrina*). J Bomb Nat Hist Soc 46:180
22. Sharma SK (1995) Instance of grey musk shrew (*Suncus murinus*) attacking a fat-tailed gecko (*Eublepharis macularius*). J Bomb Nat Hist Soc 92:411
23. Saini MS, Prashad VR (1994) Do shrews prey upon rats? J Bomb Nat Hist Soc 96:446
24. Behura BK (1958) A musk shrew attacking a snake. J Bomb Nat Hist Soc 55:3
25. Singh P (1995) Ecology, population structure and behavior of cutch rock-rat *Cremnomys cutchicus* in the Aravallis. Ph.D. thesis, Jai Narain Vyas Univ., Jodhpur (Unpublished), pp 1–178
26. Advani R, Rana BD (1981) Food of the house shrew, *Suncus murinus sidensis* in the Indian desert. Acta Theriol 26:133–134
27. Prakash I, Gupta RK, Jain AP, Rana BD, Dutta BK (1971) Ecological evaluation of rodent populations in the desert biome of Rajasthan. Mammalia 35(3):384–407

28. Tripathi RS, Jain AP, Kashyap N, Rana BD, Prakash I (1992) North-western desert: special problem areas. In: Prakash I, Ghosh PK (eds) Rodents in Indian agriculture, vol I. Scientific Publishers, Jodhpur, pp 357–396
29. Prakash I, Singh P, Sarvanan A (1995) Niche alteration by the Cutch rock-rat *Cremnomys cutchicus* in the Aravallis. J Bomb Nat Hist Soc 92(2):259
30. Prakash I, Baqri QH (1999) Ecology of small mammals in the hilly tracts of Aravalli and Vindhyan rock systems in southeastern Rajasthan. Final report submitted to Department of Science and Technology, New Delhi, pp 1–78
31. Alcock J (1989) Animal behavior: an evolutionary approach. Sinauer Associates Inc. Publishers, Sunderland, pp 1–596
32. Prakash I, Singh P Ecology and behaviour of murids of South Asia. In: Johnsingh AJT, Manjrekar N (eds) Mammals of South Asia Vol II. Orient Blackswan, Hyderabad (in press)
33. Prakash I (1964) Ecology of Indian desert gerbil *Meriones hurrianae* Jerdon. In: Proceedings of the Symposium on Problems of Indian Arid Zone, Jodhpur, India. 23rd Nov. to 2nd Dec. 1964. Publication with UNESCO collaboration/sponsorship, pp 305–310
34. Prakash I (1969) Ecotoxicology and control of Indian Desert Gerbil *Meriones hurrianae* Jerdon. V. Food preference in the field during monsoon, Jodhpur. J Bomb Nat Hist Soc 65(3): 581–589
35. Prakash I (1971) Breeding season and litter size of Indian desert rodents. Z Angew Zool 58(4):441–454
36. Prakash I (1981) Ecology of the Indian desert gerbil *Meriones hurrianae*. Monograph No. 10. CAZRI Publication, Jodhpur, pp 1–87
37. Prakash I (1988) Rodent pest management. CRC Press, Boca Raton, FL, pp 1–480
38. Prakash I, Idris MD, Soni GR (1987) Scent marking behavior of wild and laboratory bred Indian desert gerbil, *Meriones hurrianae*. Proc Ind Natl Sci Acad (B) 54(1):31–34
39. Prakash I, Jain AP (1971) Bait shyness of two gerbils *Tatera indica* and *Meriones hurrianae*. Ann Appl Biol 69:169–172
40. Singh P, Prakash I (1997) Habitat preference of Indian Bush rat *Gollunda elleoti* Gujerati in the Aravalli mountain ecosystem. J Bomb Nat Hist Soc 94(3):559–561
41. Ellerman JR (1961) The fauna of India including Pakistan Burma and Ceylon, Mammalia (Rodentia), vol 3 (Part II). Govt. of India, Delhi, pp 1–484
42. Chaudhary V, Tripathi RS, Sankhla A (2005) Incidence of *Bandocota bengalensis* in urban locales of Jodhpur city. Rodent Newsletter (ICAR), Jodhpur 29(1–4):9–11
43. Tehsin R (1980) Occurrence of large brown flying squirrel and mouse deer near Udaipur, Rajasthan. J Bomb Nat Hist Soc 77:498
44. Sharma R (2005) Impact of canal irrigation on the ecology of the arid zone of Rajasthan. M.Sc. thesis, Sikkim Manipal University (Unpublished), pp 1–93
45. Tripathi RS (2005) Reproduction in desert rodents. In: Tyagi BK, Baqri QH (eds) Changing faunal ecology in the Thar Desert. Scientific Publishers, Jodhpur, pp 289–304
46. Anonymus (1998) Biodiversity Conservation Prioritization Project (BCPP) India, Endangered Species Project. CAMP Workshop Report, pp 1–174
47. Prakash I (1962) Taxonomical and ecological account of the mammals of Rajasthan desert. Ann Arid Zone 2(2):150–161

Further Reading

Prakash I, Singh P (2005) Small mammals of desert and montane ecosystems. Scientific Publishers, Jodhpur, pp 1–153

Roberts TJ (1977) Mammals of Pakistan. Ernest Benn Ltd, London, pp 1–361

Prakash I (1988) Rodent pest management. CRC Press, Boca Raton, FL, pp 1–480

Chapter 23
Squirrels of Rajasthan with special reference to Elliot's Giant Flying Squirrel *Petaurista petaurista philippensis*

Satish Kumar Sharma and B.K. Sharma

Abstract The chapter is based on a detailed study of the squirrels of Rajasthan. Three species of squirrels are confined to Rajasthan, namely, Common Palm Squirrel and Five-striped Palm Squirrel which are arboreal and terrestrial in nature and the Elliot's Giant Flying Squirrel or Large Brown Flying Squirrel which is an important nocturnal arboreal rodent. Elliot's Giant Flying Squirrel has been recorded from dense forests of Phulwari, Sitamata, and Kumbhalgarh Wildlife Sanctuaries and avoids agricultural fields, grasslands, and human settlements. It is the next giant rodent after Porcupine in the state which was initially believed to be confined to Sitamata Wildlife Sanctuary only, but during recent years, this species has been reported from the southern part of the state. The practice of keeping dried body, bones, and hair of these squirrels by the tribal community known as *bhils* finds a mention in this chapter. In addition, another belief of the tribal that tying pieces of squirrel bones around the neck of underweight infants and fumigating with burning hairs of squirrel will bring good health to them has also been discussed. Protection of old *Mahuwa* (*Madhuca indica*) groves and planting of new *Mahuwa* trees in the distribution range of flying squirrel and awareness programs in tribal zones have been suggested to protect this species. "Arampura Flying Squirrel Watching Point" of Sitamata Wildlife Sanctuary in Udaipur district and "Thala Flying Squirrel Watching Point" of Phulwari Wildlife Sanctuary are regarded as major prospective tourist centers.

S.K. Sharma (✉)
Sajjangarh Wildlife Sanctuary, Department of Forests and Wildlife,
Government of Rajasthan, Udaipur, Rajasthan, India
e-mail: sksharma56@gmail.com

B.K. Sharma
Department of Zoology, R.L. Saharia Government P.G. College,
Kaladera (Jaipur), Rajasthan, India
e-mail: drbksharma@hotmail.com

B.K. Sharma et al. (eds.), *Faunal Heritage of Rajasthan, India: General Background and Ecology of Vertebrates*, DOI 10.1007/978-1-4614-0800-0_23,

Introduction

Order Rodentia is the largest group of animals found in Rajasthan which is represented by the rat, mice, porcupine, striped squirrels, flying squirrel and gerbil. The squirrels, being the harmless and little cuddly creatures, are the most interesting ones. Three species of squirrels are confined to Rajasthan. Two species of striped squirrels, namely, Northern Palm squirrel or Five-striped Palm Squirrel *Funambulus pennantii* and Southern Palm Squirrel or Common Palm Squirrel *F. palmarum*, are arboreal and terrestrial in nature. The third species, Elliot's Flying Squirrel or Large Brown Flying Squirrel *Petaurista petaurista philippensis*, is an important nocturnal arboreal rodent of Rajasthan.

The Five-striped Palm Squirrel is found commonly in almost every part of Rajasthan (Fig. 23.1a–c). It can be seen in the forests, agricultural fields, and human settlements. Despite being a mammalian species, it fabricates nest on a tree or a tall bush for littering. In southern Rajasthan, the nests are seen on clumps of *Dendrocalamus strictus* also. The nest is constructed on *D. strictus* extremities of culms. Fibers of gunny bags, pieces of cloths, and threads are the favorite nesting material. Fibers are transported by holding them in the mouth. Sometimes nests are placed in the holes on the walls.

The Southern Palm Squirrel or Common Palm Squirrel (Fig. 23.1 d-f) is also present in many parts of Rajasthan. Recently, this species has been recorded from Phulwari, Sita Mata, and Kumbhalgarh Wildlife Sanctuaries. In Rajasthan, it is confined to dense forests only and avoids agricultural fields, grasslands, and human settlements. The Common Palm Squirrel is present in all the dense forest areas of Southern Rajasthan. A few confirmed sites of its occurrence are given in Table 23.1.

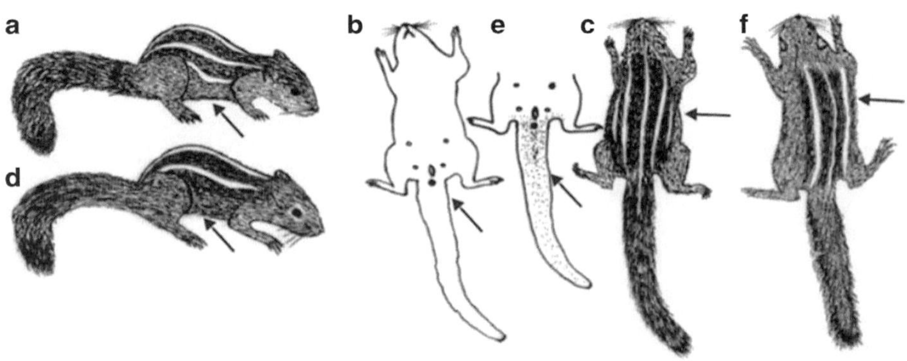

Fig. 23.1 Stripped squirrel (**a–c**) *Funambulus pennantii*, (**a**) lateral view, (**b**) ventral view (*female*), (**c**) dorsal view; (**d–f**) *Funambulus palmarum*, (**d**) lateral view, (**e**) ventral view (*female*), (**f**) dorsal view (*Arrows* show distinguishing features of the two species)

Table 23.1 Confirm localities of occurrence of *F. palmarum* in Rajasthan

S. No.	Locality	District	Status of habitat
1	Phulwari Wildlife Sanctuary • Katawali Jer (Daiya forest block) • Bhildi Mata (Phulwari forest block) • Kanchan (Phulwari forest block) • Luhari Ghati (Phulwari forest block) • Luhari village (Phulwari forest block) • Bhader Baosi (Devli forest block) • Bhader Baosi (Harwa forest block) • Kaduwa Mahuda (Dhedmariya forest block) • Dhovaniya Kund (Dhedmariya forest block) • Ghodapadiya (Dhedmariya forest block) • Amba Village (Ada Haldu forest block)	Udaipur	Sanctuary, dense forest
2	Ram Kunada Temple (Ram Kunda forest block)	Udaipur	Reserve dense forest
3	Nal Mokhi	Udaipur	Reserve dense forest
4	Jargaji Temple Sacred grove (*Naya* Jarga, presenton western aspect of Jarga Hill)	Udaipur	Reserve dense forest
5	Jargaji Temple Sacred grove (*Juna* Jarga, present on eastern aspect of Jarga Hill)	Udaipur	Reserve dense forest
6	Kumbhalgarh Sanctuary, southern forest area toward Jarga side and Chitrawas village	Udaipur	Sanctuary, dense forest
7	Kamalnath Temple sacred grove	Udaipur	Reserve dense forest
8	Valmiki Ashram, Sita Mata Wildlife Sanctuary (range Badi Sadri)	Chittourgarh	Sanctuary, dense forest
9	Polo forest (Adjoining of Daiya forest of Phulwari Wildlife Sanctuary)	Sabarkantha (Gujarat)	Reserve dense forest

Elliot's Giant Flying Squirrel or Large Brown Flying Squirrel

Elliot's Giant Flying Squirrel is known as Large Brown Flying Squirrel (*Petaurista petaurista philippensis*). Corbett and Hill [1] reviewed and revised *P. petaurista* form of this species and identified subspecies as *P. petusauria philippensis*. Elliot's Giant Flying Squirrel has been studied in Rajasthan by Sharma [2] (Fig. 23.2).

Flying Squirrel is known by many names in local dialect in southern Aravalli as described in Table 23.2.

Distribution of Flying Squirrels in Rajasthan

The Elliot's Giant Flying Squirrel is one of the most interesting and important rodent species in Rajasthan. A few decades back, it was believed to be confined to Sita Mata Wildlife Sanctuary only, but during recent years, it has been found widely distributed in the southern part of the state, and its population is also fairly good. A few localities where the presence of Flying Squirrel has been confirmed are given in Table 23.3.

Fig. 23.2 Elliot's Flying Squirrel *Petaurista petaurista philippensis*

Table 23.2 Names of Flying Squirrel in local dialect

Locality	Tribe	Names of Flying Squirrel in local dialect
Phulwari Wildlife Sanctuary (Kotra and Jhadol Tehsils)	*Bhil*	Billari, Gulrawan, Hulrawan, Moor-Chitri, Mrig Chitri, Khank Bola, Jog Hulrawan, Rawai, Ravaya, Khank Balla, Udni-Minki
	Kathodi	Pankha
Kushalgarh (Banswara)	*Bhil*	Ravi Devi, Rawai Devi
Pratapgarh (Chittourgarh)	*Bhil*	Udan Pankhi, Kali Minki

Home Site of Flying Squirrel

Flying Squirrel live in hollows of high, huge, and old trees (Fig. 23.3). *Bhil* tribals residing near Ghodapadiya in Phulwari Wildlife Sanctuary opine that Flying Squirrels feed on termites present on tree trunks. Tribals of Thala of the same sanctuary present the view that Flying Squirrel drinks water during nights from any easily approachable ground water source. The hollows of old trees are used as day roosts by the Flying Squirrel (Fig. 23.3). They emerge 20–45 min after the sunset. After emerging from day roosts, they climb on tree top and jump into the air and glide to reach the nearby tree. If the target tree is closer, they reach directly on the crown and glide to the stem of a far-off tree. If the target tree is far away, they alight

Table 23.3 Occurrence of *Petaurista philippensis* in southern Aravallis

S. No.	Locality	District	Habitat
1	Phulwari Wildlife Sanctuary		
	• Amba village (Ada Haldu block)	Udaipur	*Mahuwa* grove
	• Umariya village (Umariya block)	Udaipur	*Mahuwa* grove
	• Ambasa village (Ambasa block)	Udaipur	*Mahuwa* grove
	• Phuldariya village (Dhedmariya block)	Udaipur	*Mahuwa* grove
	• Kotra	Udaipur	*Mahuwa* grove
	• Dhedmariya village (Dhedmariya block)	Udaipur	*Mahuwa* grove
	• Pipli Nal (Pipaliya Nal) near Dungariya village	Udaipur	*Mahuwa* grove
	• Bhaga Amba Nullah, Champa Khet, and Thala village	Udaipur	*Mahuwa* grove
	• Bujha village near Daiya	Udaipur	*Mahuwa* grove in agricultural field
	• Sarli-ki-Nal	Udaipur	*Mahuwa* grove in agricultural field
	• Dhovaniya Kund and Pipal Dara anicut (Dhedmariya block)	Udaipur	*Mahuwa* grove in agricultural field
	• Bedadhar	Udaipur	*Mahuwa* grove
	• Dhiya	Udaipur	*Mahuwa* grove
	• Chak Kaduwa Mahuda	Udaipur	*Mahuwa* grove
	• Katawali Jer	Udaipur	*Mahuwa* grove
	• Semla Pani, Bhagagarh, Kawel	Udaipur	*Mahuwa* grove
	• Kani Hall (Ambasa block)	Udaipur	*Mahuwa* grove
	• Ashawara (Ashawara block)	Udaipur	*Mahuwa* grove
	• God-Kaucha village (near temple of Ashapura Mata)	Udaipur	*Mahuwa*, *Miliusa tomentosa*, and *Adina cordifolia* grove
2	• Jhameshwar Mahadeo temple grove near Jhamar Kotra	Udaipur	*Mahuwa* and *Terminalia arjuna* grove
3	Sita Mata Wildlife Sanctuary		
	• Arampura Naka	Chittourgarh	*Mahuwa* and Sadar grove
	• Jakham dam road	Chittourgarh	*Mahuwa* and Sadar grove
4	Forests of Jhadol Tehsil		
	• Aamjhad Mahadeo	Udaipur	*Mahuwa* grove
	• Eastern foothill of Ramkuda forest block near Galdhar village	Udaipur	*Mahuwa* grove
	• Nala village	Udaipur	*Mahuwa* grove
	• Khardiya village	Udaipur	*Mahuwa* grove
5	Forests of Gogunda Tehsil		
	• Gogunda-Mokhi road	Udaipur	*Mahuwa* grove (1 km south of Gogunda along the road)
	• Chatiya Khedi	Udaipur	*Mahuwa* grove
	• Padrada (Pala)	Udaipur	*Mahuwa* grove

(continued)

Table 23.3 (continued)

S. No.	Locality	District	Habitat
6	Forests of Dungarpur district		*Mahuwa* grove
	• Near Vaid *Naka*	Dungarpur	*Mahuwa* grove
	• Ratapani block	Dungarpur	*Mahuwa* grove
	• Rani Jhula block	Dungarpur	*Mahuwa* grove
	• Sabla (southern outskirts of village near road)	Dungarpur	*Mahuwa* grove
7	Forests of Banswara district		*Mahuwa* grove
	• Village Hilaj (Ghatol Range)	Banswara	*Mahuwa* grove
	• Village Khamera	Banswara	*Mahuwa* grove
	• Huda Bavji forest block	Banswara	*Mahuwa* grove
	• Sagbari Bhut Khera forest block (Near Borapada village)	Banswara	*Mahuwa* grove
	• Kushalgarh	Banswara	*Mahuwa* grove
8	Forests of Salumber Tehsil		*Mahuwa* grove
	• Manpur near Salumber	Udaipur	*Mahuwa* grove (seen in 1985 by Sh. P.S. Chundawat)
9	Forests of Chittourgarh district		
	• Parsoli village(Pratapgarh forest division)	Chittourgarh	*Mahuwa* grove
	• Nakor Khedi (Pratapgarh forest division)	Chittourgarh	*Mahuwa* grove
10	Gujarat forests at Phulwari border		*Mahuwa* grove
	• Polo forest (Range Dholwani (Ashram)	Sabarkantha	*Mahuwa* grove
	• Near Vanaz forest house (Range Vijai Nagar)	Sabarkantha	*Mahuwa* grove

on the basal zone or on the ground and then start climbing (Fig. 23.3). Their droppings can be seen lying below their holes.

Flying Squirrel spend their daytime sleeping in the hollows of trees. Littering is also done in these hollows. Like other squirrels of the area, they do not fabricate nests on branches. Other Giant Squirrels like *Ratufa indica* also make their nests on the branches of trees and not in hollows. In this regard, flying squirrel is dissimilar with many other Indian squirrels. Both, breeding and nonbreeding periods of their life are spent in hollows. A few trees inhabited by Flying Squirrel for day roost in the Phulwari Wildlife Sanctuary are enlisted in Table 23.4.

Feeding Behavior of Flying Squirrel

Flying Squirrel is a cryptic and nocturnal animal. Nothing is known about food habits of this animal in Rajasthan. The study about food habits of Flying Squirrel in Phulwari Wildlife Sanctuary reveals that it is much dependent on *Mahua*

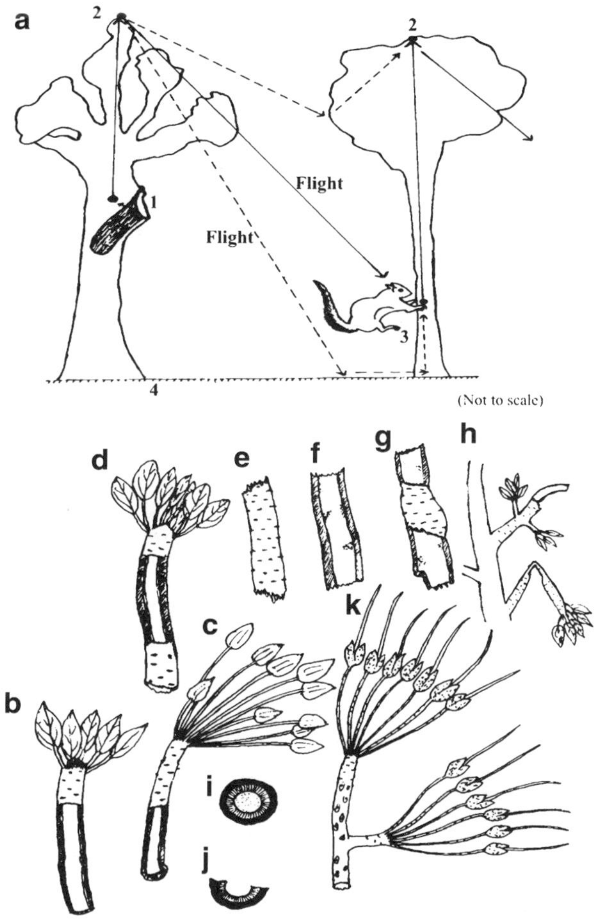

Fig. 23.3 Moving and feeding patterns of *Petaurista petaurista philippensis* (**a**) 1-4 moving pattern: 1-Hollow, the living place, 2-Tree top (from where gliding is started), 3-Squirrel alighting on lower part of the stem of target tree, 4-Ground surface, (**b-k**) feeding pattern on Mahua tree (*Madhuca indica*): (**b**) Unipolar gnawing during non-flowering season, (**c**) Unipolar gnawing during Flowering, (**d**) Intermediate gnawing, (**e**) Bipolar cut twig, (**f**) Bipolar gnawing, (**g**) Bipolar gnawing, (**h**) Gnawed intact twings, (**i**) T.S. of ungnawed portion of the twig, (**j**) T.S. of gnawed portion of the twig, (**k**) Petalless bunch of flowers cut thrown by the squirrel (Petals have been eaten away; Pith eating is avoided at this stage)

(*Madhuca indica*), *Vanda tessellata*, *Terminalia arjuna*, *T. bellerica*, *T. tomentosa*, *Anogeissus latifolia*, *Soymida febrifuga*, *Syzygium heynianum*, *Albizia odoratissim*, and *Ficus racemosa*. The pith of extremities of *Madhuca indica* and many other tree species is a favorite food of the Flying Squirrel. After plucking the twig, squirrel sits on the branch while hanging the tail down to maintain the balance. Then, the animal holds the twig in forehands and starts gnawing it to eat the soft pith. Ripe fruits and kernels are also liked by this rodent.

Mahuwa *Twig-Gnawing Pattern*

Extremities of branches of *Mahuwa* tree is the main food of flying squirrel. Pith gnawing is very common phenomenon of the Flying Squirrel (Fig. 23.3). Extremities, situated toward periphery of the crown are selected for food (Fig. 23.4). Both cut

Table 23.4 Trees providing home site to Flying Squirrel in Phulwari Ki Nal

Roost/host tree	Size of host trees	Locality	Habitat	No. of animal residing
Madhuca indica	Huge	Dhovaniya Kund (Dhedmariya block)	Dense forest	2
Madhuca indica	Huge	Dhovaniya Kund (Dhedmariya block)	Dense forest	1
Miliusa tomentosa	Medium	Temple of Ashapura Mataji God-Kaucha village (western outskirts of Dhedmariya block)	Dense forest	2
Adina cordifolia	Huge	Temple of Ashapura Mataji God-Kaucha village (western outskirts of Dhedmariya block)	Dense forest	2
Syzygium heynianum	Medium	Bhaga Amba Nallah	Reparian forest	1
Madhuca indica	Huge	Bhaga Amba Nallah	Reparian forest	1
Madhuca indica	Huge	Thala	Tree is present in an agricultural field at the border of dense forest	1
Madhuca indica	Huge	Bedadhar	Host tree is present in a grove in an agricultural field. Dense forest is present in the close vicinity	1

and intact twigs are used for devouring the soft, sappy, and nutritious pith (Fig. 23.3b, c). There are two main twig-gnawing patterns adopted by the Flying Squirrel: (1) bark gnawing (Fig. 23.5c) and (2) pith gnawing (Figs. 23.3 and 23.4). Bark of the thick intact branch, situated toward top extremities, is gnawed. Fine incisor marks can be seen on the wood of gnawed twigs (Fig. 23.5c). Debarking phenomenon was seen near Katawali Jer *Mahuwa* grove in Phulwari Wildlife Sanctuary of southern Rajasthan. Gnawed twigs can be seen below the crown of the tree during the morning. Grazing cattle usually consumes gnawed twigs during morning. Sometimes intact stumps of plucked twigs present on the tree itself are also gnawed. The fresh intact twigs may also be gnawed from one side, and such twigs become dry later on (Fig. 23.3h).

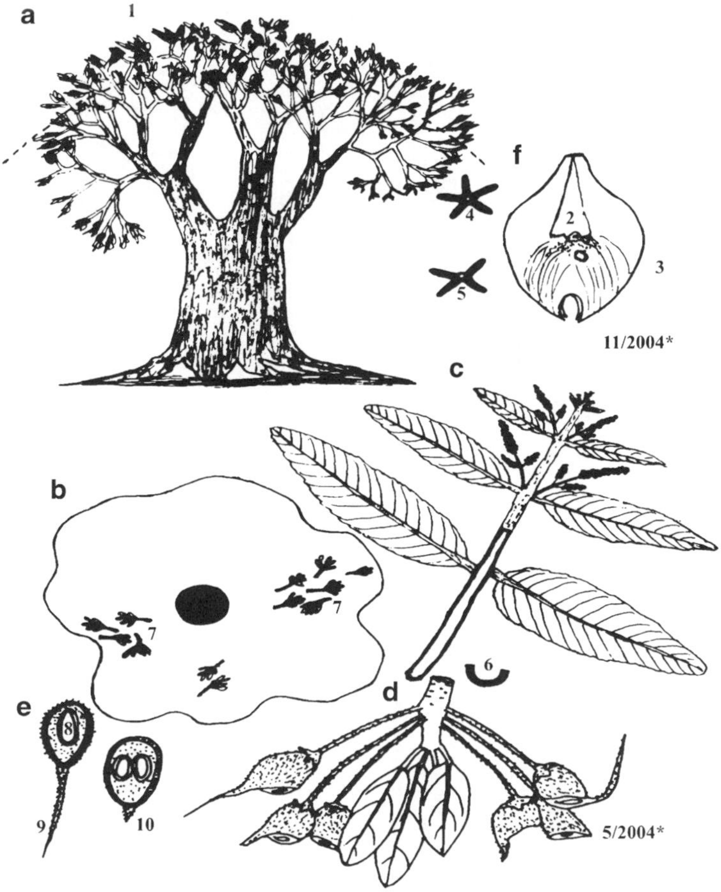

Fig. 23.4 Food and feeding behavior of *Petaurista petaurista philippensis*. (**a**) Huge sized trees are selected for feeding as they provide efficient hiding conditions, (**b**) Discarded material seen below the crown in the morning (the undulating circular line indicates outline of the crown), (**c**) Pith of twig of *Terminalia tomentosa* is liked as food, (**d**) Unripe fruit of *Mahuwa* (*Madhuca indica*) gnawed during the month of May (pith eating avoided), (**e**) Gnawed fruit of *Mahuwa* (*Madhuca indica*), (**f**) Feeding on seed of *Terminalia tomentosa* (Central small circle is the location of seed in the winged fruit): 1-Upper extremities are liked for food, 2-Gnawed one wing of *Terminalia tomentosa fruit*, 3-Ungnawed wing of *Terminalia tomentosa fruit*, 4-T.S. of ungnawed *Terminalia tomentosa fruit*, 5-T.S. of gnawed *Terminalia tomentosa fruit*, 6-T.S. of gnawed twig of *Terminalia tomentosa*, 7-Gnawed twigs lying on the ground, 8-Empty cell of gnawed *Mahuwa* fruit (Seed eatan away), 9-Intact style on gnawed fruit of *Mahuwa*, 10-Damaged style on gnawed fruit of *Mahuwa*. * Denotes month and year of the event

Conservation Problems

The Flying Squirrel is killed by gun or by disturbing the hollows by the tribals. Dried body, bones, and hair are kept in houses by the *Bhils*. Traditionally, pieces of bones are tied around the neck of underweight human infants by the *Bhil* and *Garasia*. Hair is also used to fumigate the underweight infants.

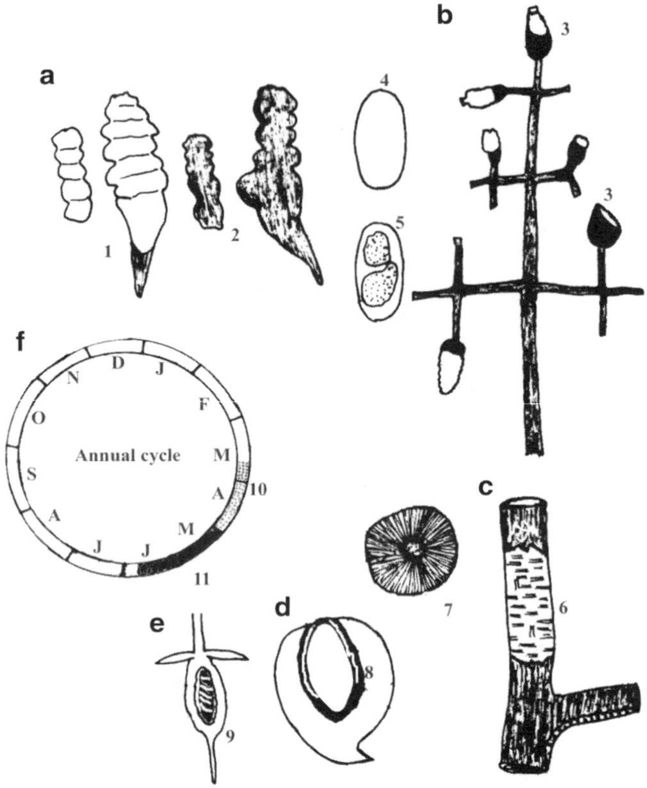

Fig. 23.5 Feeding behavior of *Petaurista petaurista philippensis*. (**a**) Peelings removed from fruit of *Mahuwa* for feeding the pulp, (**b**) Feeding on fruits of *Syzygium heyneanum*, (**c**) Debarking on *Mahuwa* twig, (**d**) Flying fox feeds on *Mangifera indica* fruits (to compare with flying squirrel feeding pattern), (**e**) Flying squirrel likes to feed on immature *Mahuwa* fruit, (**f**) Phenology of *Mahuwa* tree: 1-Ventral surface of peel of *Mahuwa* fruit, 2-Dorsal surface of peel of *Mahuwa* fruit, 3-Remnants of gnawed unripe fruits of *Syzygium heyneanum*, 4-An undamaged kernal of *Syzygium heyneanum fruit*, 5-Gnawed upper surface of a kernal of *Syzygium heyneanum* fruit, 6-Incisor marks on a *Mahuwa* twig, 7-T.S. of gnawed *Mahuwa* twig (bark eaten away, only xylum is visible), 8-Rind and pulp eaten away, seed is intact, 9-Rind peeled off, immature seed eaten away, 10-Flowring, 11-Fruiting

References

1. Corbett GB, Hill JE (1992) The mammals of Indo-malaya region: a systematic review. Natural History Museum Publications. Oxford University Press, Mumbai, p 488
2. Sharma SK (2007) Study of biodiversity and ethnobiology of Phulwari Wildlife Sanctuary, Udaipur (Rajasthan). PhD thesis. MLSU, Udaipur

Chapter 24
Wild Ungulates in Rajasthan

Sumit Dookia, Mamta Rawat, and G.R. Jakher

Abstract This chapter portrays Rajasthan as home to nine species of wild ungulates, including the four of antelopes and five species of deer. Deer and antelopes are protected by the Indian Wildlife (Protection) Amendment Act, 2006 under Schedule I and Schedule III. Out of a total of six species of antelopes found in India, the four are present in Rajasthan, namely, Nilgai *Boselaphus tragocamelus*, Four-horned Antelope *Tetracerus quadricornis*, Indian Gazelle or Chinkara *Gazella bennettii*, and Blackbuck *Antilope cervicapra*. Distribution range and pattern of each species has been described in detail. The significance of these beautiful creatures as herbivores and the major prey base for wild carnivores has been highlighted in addition to their aesthetic value. Nilgai has come in confrontation with locals as crop pest in the recent years, whereas in some areas, Blackbuck also cause damage. Indian Gazelle, however, is in complete harmony with locals. The chapter also discusses the magnificent deer of the state which are distributed largely in protected areas of northeastern Rajasthan and the Aravallis. Sambar *Rusa unicolor* and Chital *Axis axis* are quite common in Rajasthan, while the Hog Deer *Axis porcinus* and Southern Red Muntjac or Barking Deer *Muntiacus muntjak* are confined to Keoladeo National Park. Mouse Deer

S. Dookia (✉)
School of Environment Management, Guru Gobind Singh Indraprastha University, Sector 16-C, Dwarka, New Delhi, India
e-mail:sdookia_gazelle@rediffmail.com

M. Rawat
Ecology and Rural Development Society, 1-A-43, Kudi Housing Board, Jodhpur, Rajasthan, India
e-mail: rawatscorner@gmail.com

G.R. Jakher
Maharaja Ganga Singh University, Bikaner, Rajasthan, India
e-mail: jakhergr@rediffmail.com

B.K. Sharma et al. (eds.), *Faunal Heritage of Rajasthan, India: General Background and Ecology of Vertebrates*, DOI 10.1007/978-1-4614-0800-0_24,

are limited to only one district in southern Rajasthan. The need for further detailed study regarding their population estimates and conservation has been taken into consideration in this chapter.

Introduction

The even-toed ungulate is the most successful group of large herbivores on the earth. Artiodactyls are found in every zoogeographic region except Antarctica and Australia, but many species have been introduced into areas which are present outside their natural range, including Australia, New Guinea, and the Islands of Oceania. Approximately, 240 ungulate members of this order show incredible diversity in size, form, dietary preferences, and climatic tolerance. This order contains the majority of domestic mammalian species including cattle, reindeer, camels, pigs, goats, and sheep.

Recent molecular evidence has radically altered the classification of this order, notably proving that whales and dolphins (order Cetacea) belong to this order. To accommodate the merging of these orders, a hybridized name "Cetartiodactyla" has been created [1, 2] which is still debatable.

Rajasthan harbors a large number of faunal elements. It is home to nine species of wild ungulates, out of which four belong to antelopes and five are deer species. The antelopes found in the Thar Desert are Nilgai *Boselaphus tragocamelus*, Chousingha or Four-horned Antelope *Tetracerus quadricornis*, Indian Gazelle or Chinkara *Gazella bennettii*, and Blackbuck *Antilope cervicapra*. Nilgai and Four-horned Antelope belong to subfamily Bovinae, whereas Blackbuck and Chinkara belong to subfamily Antilopinae. The genus *Gazella* has spread from the Sahara to the Thar, whereas endemic to the oriental region are the Blackbuck, Nilgai, and Four-horned Antelope. Their western limit is the Thar Desert.

Sambar (*Rusa unicolor*) is the largest Indian deer found in wooded forest, including forests of Rajasthan. Chital (*Axis axis*), the most common deer of India, is found in foothills of Himalayas and throughout peninsula including the forests of Rajasthan. Hog Deer (*Axis porcinus*) smaller than Chital is largely found in alluvial grass plains of north India, from Punjab to Assam. The Southern Red Muntjac or Barking Deer (*Muntiacus muntjak*) is a small deer, with a dog-bark-like call, found in thickly wooded hills and forest areas. In Rajasthan, it has been only reported from Keoladeo NP [3], in southern Aravalli Hills in Udaipur and Chittourgarh area [4]. The smallest deer White-spotted Chevrotain or Mouse Deer (*Tragulus meminna*) is very distinct in size and body characters and is included in family Tragulidae. Mouse deer is a tiny little creature with very slender limbs and high hindquarters found in the grass-covered rocky hillsides and in the forests, distributed in peninsular India up to 24° latitude as northwardly limit [5]. In Rajasthan, Mouse Deer is reported from the southern Rajasthan forests of Udaipur and Pratapgarh areas [4].

Status and Distribution of Wild Ungulates

Family Cervidae: Deer Family

Sambar *Rusa unicolor* (Kerr, 1792)

Sambar, the largest deer among all deer, is included as Vulnerable in IUCN Red List (2012) and Schedule III of the Wildlife (Protection) Amendment Act, 2006 of India, whereas it is not listed in CITES.

Distribution

In Rajasthan, distribution of Sambar is restricted to southeast part of the state. It can be easily seen in the wooded areas of Ranthambhore and Sariska forests.

Threats

The major threats to Sambar are poaching, habitat loss, and fragmentation of the surrounding forest area outside of the PAs.

Conservation Importance

Sambar, being the largest deer of the Indian forest, constitutes the major part of the diet of large carnivores such as tiger, leopard, and dhole. In Rajasthan, the forest of Sariska Tiger Reserve supports highest sambar density in comparison to all other tiger reserves of the country [6]. They are just next to chital, numerically the second most important prey species of the large carnivores of India. Taking into consideration its preference for cover and avoidance of disturbance, the abundance of Sambar would be reliable indication of the health of any particular forest area as its potential to support adequate carnivore numbers.

Chital or Spotted Deer *Axis axis* (Erxleben, 1777)

Chital, the third largest deer inhabiting the plains and undulating terrain of India, is included under Schedule III of the Wildlife (Protection) Amendment Act, 2006 of India, whereas it is not listed in CITES and kept under Least Concern category of IUCN Redlist of Threatened species (2012).

Distribution

Chital inhabits almost all forest types in India, from dry deciduous to moist deciduous, thorn, and mangrove forests, whereas introduced population in Andaman Islands is found in evergreen forests. In Rajasthan, distribution of Chital is restricted to southern Aravalli Hills and PAs of southeastern part of the state. It can be easily seen in the wooded areas of Ranthambhore and Sariska forests.

Threats

The main conservation threats to Chital are predation by domestic dogs, diseases, and road accidents near the protected areas.

Conservation Importance

The Chital forms one of the important prey species of major carnivores in the forest area of India. Their large number and big herds make it an easily available prey for the tiger and the leopard. Chital is a species that is most amendable to wildlife management practices, and just a little effort and care is required to increase the number of this prolific breeder, in addition to maintaining the grassland–woodland edge habitat so essential for the survival of the species. The high number of Chital population in any PA is an excellent indicator of the health of prey base, which can support good population of large carnivores.

Hog Deer *Axis porcinus* (Zimmerman, 1780)

The Asian subspecies of the Hog Deer (*Axis porcinus porcinus*) is included under Appendix I in CITES, Endangered of IUCN Red List of Threatened Species (2012), and Schedule III of the Wildlife (Protection) Amendment Act, 2006 of India.

It is a shy animal and restricted to moist alluvial and tall grasslands of south and southeast Asia [7]. The native range of distribution of Hog Deer was extended from Pakistan in the west to the southern Thailand and Vietnam in the east, through northern India, Nepal, Bhutan, Bangladesh, and Myanmar [8, 9]. In Rajasthan, this deer was reported from the Keoladeo National Park, Bharatpur [4, 8].

This animal is facing the fear of extinction throughout its distribution range. Altered land use, change in flooding regime, weed invasion, woodland succession, agricultural conversions for sugarcane, paddy cultivation, and growing habitations are a few factors behind its declining population. Beside habitat loss, habitat fragmentation and indiscriminate hunting of this species throughout its range also seem to be a crucial factor affecting its distribution and population [10].

Barking Deer or Muntjac *Muntiacus muntjak* (Zimmerman, 1780)

The Barking Deer or Muntjak is included under Least Concern Category of in IUCN Red List (2012) and Schedule III of the Wildlife (Protection) Amendment Act, 2006 of India, whereas it is not listed in any appendix of CITES.

It is a shy animal and found over most of the oriental biogeographic region of the world [11]. It is expected to be found in a variety of forested habitats and climatic conditions. It is a solitary deer species and pairs are found only in the breeding season. Direct sighting of such a cryptic animal is rare, whereas its loud alarm "bark" call (like dog bark, hence named as Barking Deer), fecal pellets, and hair in the predator's scat give its presence in particular forest patch. In Rajasthan, it was reported from Bharatpur [3, 4] and in the bordering forest of Udaipur, Sirohi, and Chittourgarh districts adjacent to Gujarat [4]. At present, their status is yet to be ascertained in the Rajasthan. Muntjac largely depends on the good-quality forest habitat, both for food and cover. Due to this, its habitat suffers from forest degradation resulting from the cattle grazing, woodcutting, or fodder collection.

Mouse Deer or White-spotted Chevrotain *Moschiola meminna* (Erxleben, 1777)

Mouse deer, the smallest deer species, is included under Least Concern Category of IUCN Redlist of Threatened Species (2012) and Schedule I of the Wildlife (Protection) Amendment Act, 2006 of India, whereas it is not listed in CITES.

It is small, compactly built, shy, and mostly nocturnal in habit [11]. This deer belongs to an ancient mammalian lineage that was probably abundant and distributed worldwide during the Oligocene and Miocene, from 3.5 to 5 million years ago [12]. Prater [5] reports the northern limit up to 24°N in the Indian Peninsula. In Rajasthan, it is reported to be occurring in the southern Rajasthan [4]. They are reported to favor rocky habitats, grass-covered rocky hillsides and forest, seldom far from water and often occur along forest streams and rivers [13].

The main threat for the Mouse Deer is hunting by local communities and forest fire, whereas they are sensitive to habitat loss, alteration, and degradation. This species is least studied and requires a field research to understand its ecology and habitat requirements.

Family Bovidae: Antelope and Gazelle Family

Nilgai *Boselaphus tragocamelus* (Pallas, 1766)

The Nilgai or Bluebull, the largest antelope of Asia, is included under Least Concern in IUCN Red List (2012) and Schedule III of the Wildlife (Protection) Amendment Act, 2006 of India, whereas it is not included in CITES.

Distribution

Nilgai is the species of peninsular India, including Pakistan and Nepal. The present distribution of Nilgai ranges from Himalayan foothills situated southwards through central India, down to the southern part of Andhra Pradesh and to semiarid part in the west. It is absent in the northeast India and the southernmost part of the peninsular region. It is found in crop fields as they are like pseudo-grasslands for this open-country animal, and most importantly, it is tolerated by human beings due to religious reasons. Nilgai naturally prefers broken countryside with scattered trees and patches of grasslands. Generally found in variety of habitats, ranging from plains, undulating hills, and scattered trees to the cultivated plains, it avoids dense forest and steep hilly terrain [14]. In many parts of India, they enjoy complete protection, being regarded as a relative of cow, close to Lord Shiva and hence considered sacred by various communities. Northern states, especially Haryana, Madhya Pradesh, Rajasthan, and Uttar Pradesh, account for a large population (about 60%) of the total Nilgai population in the country [15].

Threats

In Rajasthan, Nilgai is found largely outside the PAs, where no natural predator is found. However, it is regularly killed by village dogs.

Conservation Importance

Its overall range has shrunk, particularly in the margin areas of its natural distribution, but in some areas, it is abundant. The constant degradation of the natural forests, the absence of large predators, and the increased agricultural activities have offered unfavorable habitat conditions. Invariably, this situation forced Nilgai to become serious pest as crop raider and a major issue of man–animal conflict. One of the possible solutions is to include a selective culling program linked to licensed hunting permit under the Wildlife (Protection) Amendment Act, 2006. Despite of the severity of the problem, very few come up to get license for culling. Relocation of problematic animal, for the time being, seems the safest solution, but the problem spreads with the animal invading the newer places too.

Four-horned Antelope or Chousingha *Tetracerus quadricornis* (Blainville, 1816)

The Four-horned Antelope, a rare antelope with four horns, is included as Vulnerable in IUCN Red List and Schedule I of the Wildlife (Protection) Ammendment Act, 2006 of India, whereas it is not listed in CITES.

The Four-horned Antelope is a monotypic species of its genus and endemic to India. Four-horned Antelope and nilgai belong to Boselaphini, the last survivors of a form very similar to that of the ancestors of the entire subfamily Bovinae. Both of these antelopes have relatively primitive anatomical and behavioral characteristics, i.e., absence of rings on horn, keeled in front part from true antelopes [16].

Distribution

The Four-horned Antelope is distributed from Uttar Pradesh to Tamil Nadu, from Orissa in the east to Gujarat in the west [17, 18], mainly the forest-dweller species found in the modest population in some national parks in its entire distributional range. In Rajasthan, its small population is found in Kumbhalgarh Wildlife Sanctuary in the Aravalli Range [17]. It prefers to live in the dry deciduous forest patches, with moderate undergrowth. Within dry deciduous forest, this antelope inhabits from open savannah to dense miscellaneous patches [16].

Threats

The biggest threat to its population is from habitat destruction or habitat alteration. It needs a mixture of different habitats, including grasslands, thick under cover, and closed canopy forest.

Conservation Importance

The current status and trends of the overall population of Four-horned Antelope is poorly known. It has a very selective choice of food and shelter because of its specific metabolic, thermoregulatory, and antipredatory requirements [16]. It needs prefect equilibrium of the habitat since increase in certain herbivore populations may lead to decline in their stronghold as a result of increased competition, whereas improving certain habitat without keeping this animal in mind may cause further decline in its distribution.

Indian Antelope or Blackbuck *Antilope cervicapra* (Linnaeus, 1758)

The Blackbuck is included under Near Threatened Category of IUCN Red List (2012) and Schedule I of the Wildlife (Protection) Amendment Act, 2006 of India, whereas it is listed in Appendix III of CITES. The Blackbuck, the state animal of Andhra Pradesh, Punjab, and Haryana states in India, is the only representative of the genus *Antilope* in India and is endemic to Indian subcontinent. There are total four subspecies of genus Antilope, i.e., *A. c. centralis*, *A. c. cervicapra*, *A. c. rajputanae*, and *A. c. rupicapra*, distributed in the wide range of habitats, from semiarid grasslands and scrub to open forest [19–21]. It belongs to subfamily Antilopinae and their closest relatives are the *Gazelle* [22, 23].

Distribution

The Blackbuck is found in wide range of habitats, but highest densities are found in semiarid, open, and short-grass plains. The present distribution of the Blackbuck is much reduced; their populations are now small, scattered, fragmented, and relatively isolated. The most recently countrywide survey of its populations from 1981 to 1988 estimated to be around 29,000–38,000 individuals [23, 24]. The highest population was recorded from Rajasthan, Punjab, Gujarat, and Maharashtra states [25]. In Rajasthan, it is solely found in the semiarid region in the northwestern side of Aravalli Hill Range. This is largely protected in the sites where local communities prevent hunting influenced by religious sentiments. Some large populations in Rajasthan are found in the Taal Chapper Blackbuck Sanctuary (Churu); Guda Bishnoian area; Dhawa-Doli area (Jodhpur); Jaroda, Rotu, and Gachipura areas (Nagour); and Gajner Sanctuary (Bikaner), and smaller ones are found around Anupgarh (Sri Ganganagar) in small groups.

Threats

The main reasons of population decline are hunting and loss of suitable habitat. The areas where Blackbuck thrives have been converted into agricultural lands.

Conservation Importance

The Blackbuck thrives, when given proper protection, as a result of their high fecundity and ability to survive well. This tendency often brings them into conflict with local communities as they became serious crop pest. The current status and trends of the overall population of Blackbuck is poorly known. It largely depends on the forage quality and grass. In Rajasthan, this antelope is facing serious problem by conversion of grasslands into dense thorny scrubland habitat due to exotic *Prosopis juliflora*. This is a fairly well-studied animal and proper management of this wild ungulate can save this species from decline.

Chinkara or Indian Gazelle *Gazella bennettii* (Sykes, 1831)

The Chinkara is included under Least Concern in IUCN (2012) and Schedule I of the Wildlife (Protection) Amendment Act, 2006 of India, whereas it is not listed under any category of CITES.

It is the state animal of Rajasthan. Groves [26–28] classified this species into six subspecies based on the differences in the coat color, body skeleton, and measurements: *G. b. bennettii*, *G. b. christyi*, *G. b. salinarum*, and *G. b. fuscifrons* found in India, whereas *G. b. shikarii* and *G. b. karamii* restricted to Iran.

Distribution

The Chinkara is found in almost every type of ecosystem in the western Rajasthan. Unlike the blackbuck, the chinkara is not restricted to only a few pockets but is much more scattered and is sometimes found far away from human settlement and water holes. Interestingly, the chinkara rarely, if ever, shares the same habitat as the blackbuck. The chinkara is a familiar faunal element of shifting dune countries and the interdunal valleys in the extreme arid part of western Rajasthan (rainfall 100–150 mm). The occurrence of this animal in a hilly and humid area like Jaisamand (Udaipur), located in southern Rajasthan, points to its versatility of niche utilization.

Threats

The Chinkara is largely found in the western Rajasthan. There has been a tremendous pressure on its habitat, as increase in agriculture activities in hitherto uncultivated land of the region. The local communities like *Bishnoi, Jat, Jain, Rajpurohit,* and *Meghwal*, through generations, have been traditionally holding these wild ungulates with great reverence and have been protecting them from natural calamities, predators, and poachers [29].

Conservation Importance

Thar Desert in Rajasthan holds 89% of Chinkara population in the state [30]; major proportion is ironically found outside rather than inside the PAs. This antelope is very selective in food plants and any little changes in its habitat can restrict its distribution [31]. Please see Chap. 2 and Chaps. 1 and 18 of from Faunal Heritage of Rajasthan: Conservation and Management of Vertebrates Vol. 2 Sharma, B.K. *et al.* (eds.) 2013, Springer. (ISBN 978-3-319-01344-2) for pictures.

Acknowledgements The authors are thankful to the volunteers and members of Ecology and Rural Development Society (ERDS) for their constant support in assessing the status of blackbuck, chinkara, and nilgai in western Rajasthan. The first author (SD) is highly thankful to the Rufford Small Grants Foundation for supporting the "Chinkara Conservation Project" since last 4 years.

References

1. Hernandez-Fernandez M, Vrba ES (2005) A complete estimate of the phylogenetic relationships in Ruminata: a dated species-level super tree of the extant ruminants. Biol Rev 80:269–302
2. Nikaido M, Rooney AP, Okada N (1999) Phylogenetic relationships among Cetartiodactyls based on insertions of short and long interspersed elements: hippopotamuses are the closest extant relatives of whales. Proc Natl Acad Sci USA 96:10261–10266
3. Nowak RM (1991) Walker's mammals of the world, 5th edn. Johns Hopkins University Press, Baltimore, MD, pp 1936

4. Sharma S, Sharma SK, Sharma S (2003) Notes on mammalian fauna of Rajasthan. Zoos' Print J 18(4):1085–1088
5. Prater SH (1971) The book of Indian animals. Bombay Natural History Society and Oxford University Press, Bombay, pp 483
6. Sankar K, Acharya B (2004) Sambar (*Cervus unicolor* Kerr, 1792). In: Sankar K, Goyal SP (eds) Ungulates of India. ENVIS bulletin: wildlife and protected areas, vol 7, no 1. Wildlife Institute of India, Dehradun, India, p 163–170
7. Biswas T (2004) Hog Deer (*Axis porcinus* Zimmerman, 1780). In: Sankar K, Goyal SP (eds) Ungulates of India. ENVIS bulletin: wildlife and protected areas, vol 7, no 1. Wildlife Institute of India, Dehradun, India, p 61–78
8. Dhungel SK, O'Gara WB (1991) Ecology of Hog deer in Royal Chitwan National Park, Nepal. Wild Mono 119:1–40
9. Biswas T, Mathur VB (2000) A review of the present conservation scenario of hog deer (*Axis porcinus*) in its native range. Indian Forester 126(10):1068–1084
10. Biswas T, Mathur VB, Sawarkar VB (2002) Status of hog deer (*Axis porcinus*) in India. Report submitted to Wildlife Institute of India, Dehradun
11. Price SA, Bininda-Emonds ORP, Gittleman JL (2005) A complete phylogeny of the whales, dolphins and even-toed hoofed mammals (Cetartiodactyla). Biol Rev 80:445–473
12. Dubost G (2001) Chevrotains. In: MacDonald D, Norris S (eds) The new encyclopedia of mammals. Oxford University Press, Oxford, pp 500–501
13. Raman TRS (2004) Mouse Deer (*Moschiola meminna*, Eaxleben, 1777). In: Sankar K, Goyal SP (eds) Ungulates of India. ENVIS bulletin: wildlife and protected areas, vol 7, no 1. Wildlife Institute of India, Dehradun, India, p 131–140
14. Blanford WT (1888–1991) The fauna of British India, including Ceylon and Burma. Mammalia. Taylor and Francis, London, p 617
15. Sankar K, Johnsingh AJT, Acharya B (2004) Blue Bull or Nilgai (*Boselaphus tragocamelus*, Pallas, 1766). In: Sankar K, Goyal SP (eds) Ungulates of India. ENVIS bulletin: wildlife and protected areas, vol 71. Wildlife Institute of India, Dehradun, India, p 120–143
16. Sharma K, Rahmani AR (2004) Four-horned Antelope or Chowsingha (*Tetracerus quadricornis* Blainville, 1816). In: Sankar K, Goyal SP (eds) Ungulates of India. ENVIS bulletin: wildlife and protected areas, vol 7, no 1. Wildlife Institute of India, Dehradun, India, p 53–60
17. Krishna YC, Clyne PJ, Krishnaswamy J, Kumar NS (2009) Distributional and ecological review of the four horned antelope, *Tetracerus quadricornis*. Mammalia 73:1–6
18. Rice C (1991) The status of four-horned antelope *Tetracerus quadricornis*. J Bomb Nat Hist Soc 88:63–66
19. Schaller GB (ed) (1967) The deer and the tiger: a study of wildlife in India. University of Chicago Press, Chicago, IL, p 370
20. Mungall EC, Patel BH, Prasad NLNS, Doherty SE (1981) Conservation and management of the Indian blackbuck antelope (*Antilope cervicapra*)—a final report US Fish and Wildlife Services, September 1981, Unpublished
21. Prasad NLNS (1981) Home range, dispersal, and movement of blackbuck (*Antilope cervicapra*) population in relation to seasonal change in Mudmal and environs. Ph.D. dissertation, Dept. of Biosciences, Saurashtra University, Rajkot, India
22. Effron M (1976) Deciphering gazelle taxonomy with chromosomes. Zoo Nooz 49:12–14
23. Rahmani AR (1991) Present distribution of the blackbuck (*Antilope cervicapra* Linn.) in India, with special emphasis on the lesser known populations. J Bomb Nat Hist Soc 88:35–46
24. Rebholz W, Harley E (1999) Phylogenetic relationships in the bovid subfamily Antilopinae based on mitochondrial DNA sequences. Mol Phylogenet Evol 12:87–94
25. Isvaran K (2004) Indian Antelope or Blackbuck (*Antilope cervicapra* Linn. 1758). In: Sankar K, Goyal SP (eds) Ungulates of India. ENVIS bulletin: wildlife and protected areas, vol 7, no 1. Wildlife Institute of India, Dehradun, India, p 79–90
26. Groves CP (1985) An introduction to the gazelles. Chinkara 1:4–16

27. Groves CP (1993) The Chinkara (*Gazella bennettii*) in Iran, with the description of two new subspecies. J Sci I R Iran 4(3):166–178
28. Groves CP (2003) Taxonomy of ungulates of the Indian subcontinent. J Bomb Nat Hist Soc 100(2&3):341–362
29. Dookia S (2007) Participation of local villagers in conservation of Indian Gazelle or Chinkara (*Gazella bennettii*) in Thar Desert of Rajasthan, India. Report submitted to Rufford Small Grant Foundation, UK, p 32
30. Rahmani AR, Sankaran R (1991) Blackbuck and Chinkara in the Thar: a changing scenario. J Arid Environ 20:379–391
31. Dookia S (2002) Habitat preference, abundance and group size of Indian Gazelle *Gazella bennetti* in semi-arid region of Rajasthan. Ph.D. thesis, JNV University, Jodhpur, Rajasthan, India, p 179

Appendices

Appendix 1: Forest Statistics of the State of Rajasthan

Table 1a Forest scenario at a glance

S. No.	Item	Unit	Rajasthan	India
1.	Total geographical area	km^2	342,239	3,287,263
2.	Forest area (as on 31 March, 2007)	km^2	32,639	765,253
3.	% of forest area with reference to Geographical Area	km^2	9.54	23.28
4.	Forest cover as per SFR (2003 Report)	km^2	15,826	678,333
5.	Actual per Capita Forest Cover	ha	0.02	0.06
6.	Waste land	Million ha	18	93.7
7.	Net area sown	Thousand ha	15,509	142,819
8.	National parks	Nos.	0.2*	87
9.	Area of national parks	km^2	421.23	40,631.64
10.	Wildlife sanctuaries	Nos.	25	485
11.	Area of wildlife sanctuaries	km^2	8,739.98	115,374.42
12.	Tiger projects	Nos.	2	25
13.	Population of tiger as of the year 2010	Nos.	70	3,435

Source: Rajasthan Forest Statistics 2007, Department of Forests, Government of Rajasthan, India. Website: http://rajforest.nic.in/ and Ministry of Environment and Forests (MoEF), Government of India, 2011

*The newly formed Mukundra Hills National Park is not yet included in this table on the official website: http://rajforest.nic.in/ of Department of Forest, Government of Rajasthan, India. Downloaded on Nov 22, 2012

B.K. Sharma et al. (eds.), *Faunal Heritage of Rajasthan, India: General Background and Ecology of Vertebrates*, DOI 10.1007/978-1-4614-0800-0,

Table 1b Total forest area by legal status of Rajasthan

Legal status	Area (km^2)
Reserved forest area	12453.92
Protected forest area	17415.96
Unclassified forest area	2768.86
Total forest area	32638.64

Source: Rajasthan Forest Statistics 2007, Department of Forests, Government of Rajasthan, India. Downloaded from the website http://rajforest.nic.in/ on Nov. 24, 2012

Table 1c District-wise forest statistics as per legal status (area in km^2)

District	Reserved forest	Protected forest	Unclassified forest	Total
Ajmer	194.99	418.09	0.02	613.10
Alwar	1,006.06	636.83	141.25	1,784.14
Banswara	0.00	1,236.67	0.00	1,236.67
Baran	0.00	2,226.74	12.58	2,239.32
Barmer	0.00	568.33	44.77	609.10
Bharatpur[a]	0.00	369.57	12.82	382.39
Bhilwara	437.80	289.62	66.77	794.18
Bikaner	0.00	234.29	1,014.45	1,248.73
Bundi	837.29	706.65	16.04	1,559.98
Chittourgarh	1,584.70	1,181.36	0.56	2,766.62
Churu	7.20	10.84	53.18	71.22
Dausa	133.37	148.69	0.57	282.63
Dholpur	7.92	597.78	32.75	638.45
Dungarpur	251.29	433.25	8.71	693.25
Ganganagar	0.00	50.65	582.79	633.44
Hanumangarh	0.00	113.25	126.21	239.46
Jaipur	679.34	263.10	5.63	948.68
Jaisalmer	0.00	199.77	383.52	581.29
Jalore	122.24	298.05	30.40	450.68
Jhalawar	413.45	930.62	5.73	1,349.79
Jhunjhunu	6.02	392.57	6.77	405.36
Jodhpur	4.68	175.52	62.70	242.89
Karouli	62.99	1,675.55	64.27	1,802.81
Kota	874.83	412.58	22.63	1,310.04
Nagour	0.80	206.23	33.89	240.93
Pali	819.45	141.62	2.51	963.58
Rajsamand	277.44	119.14	0.000	396.58
Sawai Madhopur	792.88	154.16	6.67	953.71
Sikar	9.92	619.18	8.59	637.68
Sirohi	866.60	749.75	22.30	1,638.65
Tonk	101.42	230.75	3.80	335.97
Udaipur	2,961.25	1,626.17	0.00	4,587.42

Source: Working Plan and Forest Settlement Cell Department of Forests, Government of Rajasthan, India (Till 31.3.2007). Downloaded from the website: http://rajforest.nic.in/ on Nov 22, 2012

[a]A total of 4,757.66 ha land is being converted as Reserved Forest area in District Bharatpur (Annual report, 2009–2010 of Forest Department, Govt. of Rajasthan, India)

Table 1d District-wise percent forest area with reference to geographic area and per capita forest area in Rajasthan (as on March, 2007)

S. No.	District	Geographic area (km^2)	Forest area (km^2)	Percent forest area with reference to geographic area	Population (Census, 2007)	Per capita forest area (ha)
1.	Ajmer	8,481	613.10	7.23	2,181,670	0.03
2.	Alwar	8,380	1,784.14	21.29	2,992,592	0.06
3.	Banswara	5,037	1,236.67	24.55	1,501,589	0.08
4.	Baran	6,955	2,239.32	32.20	1,021,653	0.22
5.	Barmer	28,387	609.10	2.15	1,964,835	0.03
6.	Bharatpur	5,066	382.39	7.55	2,101,142	0.02
7.	Bhilwara	10,455	794.18	7.60	2,013,789	0.04
8.	Bikaner	27,244	1,248.73	4.58	1,674,271	0.07
9.	Bundi	5,550	1,559.98	28.11	962,620	0.16
10.	Chittoragarh	10,856	2,766.62	25.48	1,803,524	0.15
11.	Churu	16,830	71.22	0.42	1,923,878	0.00
12.	Dausa	2,950	282.63	9.58	1,317,063	0.02
13.	Dholpur	3,034	638.45	21.04	983,258	0.06
14.	Dungarpur	3,770	693.25	18.39	1,107,643	0.06
15.	Ganganager	7,944	633.44	7.97	1,789,423	0.04
16.	Hanumangarh	12,690	239.46	1.89	1,518,005	0.02
17.	Jaipur	11,588	948.68	8.19	5,251,071	0.02
18.	Jaisalmer	38,401	581.29	1.51	508,247	0.11
19.	Jalore	10,640	450.68	4.24	1,448,940	0.03
20.	Jhalawar	6,219	1,349.79	21.70	1,180,323	0.11
21.	Jhunjhunu	5,928	405.36	6.84	1,913,689	0.02
22.	Jodhpur	22,850	242.89	1.06	2,886,505	0.01
23.	Karouli	5,052	1,802.81	35.69	1,209,665	0.15
24.	Kota	5,481	1,310.04	23.90	1,568,525	0.08
25.	Nagour	17,718	240.93	1.36	2,775,058	0.01
26.	Pali	12,387	963.58	7.78	1,820,251	0.05
27.	Rajsamand	4,768	396.58	8.32	987,024	0.04
28.	Sawai Madhopur	5,005	953.71	19.06	1,117,057	0.09
29.	Sikar	7,732	637.68	8.25	2,287,788	0.03
30.	Sirohi	5,136	1,638.65	31.91	851,107	0.19
31.	Tonk	7,194	335.97	4.67	1,211,671	0.03
32.	Udaipur	12,511	4,587.42	36.67	2,633,312	0.17
	Total	342,239	32,638.74	9.54	56,507,188	0.06

Source: Working Plan & Forest Settlement Cell, Department of Forest, Government of Rajasthan, India. Downloaded from the website: http://rajforest.nic.in/ on Nov 22, 2012

Appendix 2: Ravine Area in Rajasthan

Table 2a District-wise ravine area

District	Estimated area (ha)
Alwar	10,000
Banswara	500
Bharatpur	53,000
Bhilwara	3,300
Bundi	86,000
Dungarpur	800
Jaipur	20,000
Jhalawar	6,900
Kota	132,600
Sawai Madhopur	130,000
Sirohi	1,200
Tonk	4,400
Total	451,700

Source: Rajasthan Forest Statistics, 2007. Department of Forest, Government of Rajasthan, India. Downloaded from the website: http://rajforest.nic.in/ on Nov 22, 2012

Table 2b River-wise ravine area

Name of river	Area (ha)	District covered
Chambal	130,000	Kota, Bundi, Sawai Madhpur, Dholpur
Kalisindh	40,000	Kota
Parvati	40,000	Kota
Parwan	20,000	Kota
Alniya	5,000	Kota
Talera	14,000	Bundi
Ghora Pachhar	14,000	Bundi
Mangli	20,000	Bundi
Moj	30,000	Bundi
Morel	10,000	Sawai Madhpur
Banas	40,000	Sawai Madhopur, Tonk, Bhilwara
Banganga	10,000	Jaipur
Mahi	700	Banswara, Dungarpur
Sahibi	10,000	Alwar
Gambhiri	10,000	Sawai Madhopur
Sanwan	10,000	Jaipur
Parvati (Bharatpur)	15,000	Bharatpur
Saraswati	3,000	Ajmer
Panchana	30,000	Sawai Madhopur
Total	451,700	

Source: Rajasthan Forest Statistics 2007. Department of Forest, Government of Rajasthan, India. Downloaded from the website: http://rajforest.nic.in/ on Nov 22, 2012

Appendix 3: Checklists of Vertebrate Fauna of Rajasthan

Table 3a Anurans

S. No.	Scientific name	Common name
Family Dicroglossidae		
1.	*Euphlyctis cyanophlyctis*	Indian Skipping Frog (LC)
2.	*Euphlyctis hexadactylus*	Indian Green Frog (LC)
3.	*Hoplobatrachus tigerinus*	Indian Bullfrog (LC)
4.	*Fejervarya limnocharis*	Asian Grass Frog (LC)
5.	*Sphaerotheca breviceps*	Short-headed Burrowing Frog (LC)
6.	*Sphaerotheca rolandae*	Roland's Burrowing Frog (LC)
Family Bufonidae		
7.	*Duttaphrynus melanostictus*	Black-spectacled Toad (LC)
8.	*Duttaphrynus stomaticus*	Marbled Toad (LC)
9.	*Pseudepidalea viridis*	Green Toad (LC)
Family Microhylidae		
10.	*Microhyla ornata*	Ornamented Pygmy Frog (LC)
11.	*Uperodon systoma*	Marbled Balloon Frog (LC)
12.	*Kaloula taprobanica*	Sri Lankan Bullfrog (LC)
Family Rhacophoridae		
13.	*Polypedates maculatus himalayensis*	Himalayan Tree Frog (LC)

IUCN Red List of Threatened Species. Version 2013.1. www.iucnredlist.org

Table 3b Reptiles

S. No.	Order	Family	Zoological name	Distribution	IUCN Status (ver 2013.1)/WPA
1.	I. Crocodylia (Gharials and Crocodiles)	I. Crocodylidae	*Crocodylus palustris*	Chambal River,Dhaulpur	CR/Sch-I
2.		II. Gavialidae	*Gavialis gangeticus*	Chamal River,Dhaulpur	CR/Sch-I
3.	II. Testudines Tortoises (land-dwelling) and turtles (aquatic forms)	III. Testudinidea (tortoise)	*Geochelone elegans*	W.Raj	LC/LR
4.		IV. Geoemydidae (fresh water turtles)	*Batagur kachuga*		CR/Sch-I
5.			*Pangshura tentoria*		LR/LC
6.			*Pangshura tecta*		LC/Sch-I
7.			*Batagur dhongoka*		EN
8.			*Geoclemys hemiltonii*		VU/Sch-I
9.			*Hardella thurjii*		VU
10.		V. Trionychidea (fresh water turtles)	*Chitra indica*		VU/Sch-IV
11.			*Nilssonia hurum*		VU/Sch-I
12.			*N. gangeticus*		NE/Sch-0
13.			*Lissemys punctata*		LC/Sch-I
14.	III. Squamata Lacertilia (Lizards)	VI. Gekkonidae	*Crossobamon orientalis*	W.Raj.	–
15.			*Bunopus tuberculatus*	W.Raj.	–
16.			*Cyrtodactylus scabar*	W. and N. Raj.	–
17.			*Hemidactylus flaviviridis*	Whole Raj.	–
18.			*H. brookii*	E. and S. Raj.	–
19.			*H. triedrus*	Ajmer (C Raj.)	–
20.			*H. leschenaulti*	Jaisalmer (W. Raj)	–
21.		VII. Eublepharidae	*Eublepharius macularius*	E.S. and W. Raj.	–
22.		VIII. Agamadae	*Calotes versicolor*	Whole Raj.	–
23.			*Phrynocephalus laungwalansis*	W. Raj.	Endemic to Thar Desert
24.			*Trapelus agilis*	W. And N. Raj.	–
25.			*Sitana ponticeriana*	S. Raj.	LC

26.			*Uromastyx hardwickii*	W. And E. Raj.	Sch-II
27.		IX.Chamaeleonidea	*Chamaeleon zeylanicus*	S. Raj.	Sch-I
28.		X. Scincidae	*Eutropis carinata*	Whole Raj.	–
29.			*M. disssimilis*	Ajmer (C. Raj.)	–
30.			*Lygosoma guentheri*	E. and S. Raj.	
31.			*Eurylepis taeniolatus*	W. Raj.	(LC)
32.			*Ophiomorus raithmai*	W.Raj.	
33.			*Ophiomorus tridactylus*	W. and E. Raj.	
34.			*Ablepharus grayanus*	Mt. Abu. (S. Raj.)	
35.		XI. Lacertidae	*Opisops jerdoni*	W. and E. Raj.	
36.			*O. microlepis*	Ajmer (C. Raj)	
37.			*Eramias guttulata watsonana*	Jaisalmer. (W. Raj.)	
38.			*Acanthodactylus cantoris cantoris*	W. Raj.	
39.		XII. Varanidae	*Varanus bengalensis*	Whole Raj.	LC/Sch-II
40.			*Varanus griseus*	W. Raj.	Sch-II
41.	IV. Serpentes	XIII. Typholpidae	*Typhlops porrectus*		Sch-IV
42.			*Ramphotyphlops braminus*		Sch-IV
43.			*Rhinotyphlops acutus*		Sch-IV
44.		XIV.Pythonidae	*Python molurus molurus*		Sch-I
45.		XV. Boidae	*Gongylophis conicus*		Sch-IV
46.			*Eryx johnli*		Sch-IV
47.		XVI. Colubridae	*Coelognathus helena helena*		Sch-IV
48.			*Ptyas mucosa*		Sch-II
49.			*Platyceps ventromaculatus*		Sch-IV
50.			*Coluber gracilis*		Sch-IV
51.			*Argyrogena fasciolata*		Sch-IV
52.			*Spalerosophis diadema diadema*		Sch-IV
53.			*S. diadema atriceps*		Sch-II
54.			*S. arenarius*		Sch-IV

(continued)

Table 3b (continued)

S. No.	Order	Family	Zoological name	Distribution	IUCN Status (ver 2013.1)/WPA
55.			*Lytorhyncus paradoxus*		Sch-IV
56.			*Oligodon arnensis*		Sch-IV
57.			*Dendrelaphis tristis*		Sch-IV
58.			*O. taeniolatus*		Sch-II
59.			*Lycodon aulicus*		Sch-IV
60.			*Sibynophis subpunctatus*		Sch-IV
61.			*L. striatus*		Sch-IV
62.			*Xenochrophis piscator*		Sch-II
63.			*Amphiesma stolatum*		Sch-IV
64.			*Macropisthodon plumbicolor*		Sch-II
65.			*Boiga trigonata*		Sch-IV
66.			*Boiga forsteni*		Sch-IV
67.			*Psammophis schokari*		Sch-IV
68.			*P. leithii*		Sch-IV
69.			*Ahaetulle nastua*		Sch-II
70.		XVII. Elapidae	*Bungarus caeruleus*		Sch-IV
71.			*B. sindanus sindanus*		Sch-IV
72.			*Naja naja*		Sch-II
73.			*Naja oxiana*		Sch-II
74.		XVIII. Viperidae	*Daboia russelii*		DD/Sch-II
75.			*Echis carinatus*		LC/Sch-IV

Source: Baqri, Q.H. (2004), Studies of faunal diversity in the Thar Desert of Rajasthan. Desert Regional Station, Zoological Survey of India. Submitted to The Ministry of Environment and Forest, New Delhi

Gupta R K and Prakash I (Eds.) (1975). Environmental analysis of the Thar Desert. English Book Depot, Dehradun. pp.484

Sharma S.K. (1995), An overview of amphibians and reptilian fauna of Rajasthan. Flora and Fauna. 1(1):47–48

Whitaker, R. and Captain A (2004). Snakes of India: the field guide. Draco Books Chennai, India

IUCN Red List of Threatened Species. Version (2013.1). www.iucnredlist.org

Table 3c Aves

S. No.	Common/English name	Scientific name (IUCN Status ver 2013.1)
1.	Black-necked Grebe	*Podiceps nigricollis* (LC)
2.	Great Crested Grebe	*Podiceps cristatus* (LC)
3.	Little Grebe	*Tachybaptus ruficollis* (LC)
4.	Spot-billed Pelican	*Pelecanus philippensis* (NT)
5.	Dalmatian Pelican	*Pelecanus crispus* (VU)
6.	Great White Pelican	*Pelecanus onocrotalus* (LC)
7.	Indian Cormorant	*Phalacrocorax fuscicollis* (LC)
8.	Great Cormorant	*Phalacrocorax carbo* (LC)
9.	Little Cormorant	*Phalacrocorax niger* (LC)
10.	Oriental Darter	*Anhinga melanogaster* (NT)
11.	Western Reef-egret	*Egretta gularis* (LC)
12.	Little Egret	*Egretta garzetta* (LC)
13.	Great Egret	*Casmerodius albus* (LC)
14.	Intermediate Egret	*Mesophoyx intermedia* (LC)
15.	Eastern Cattle Egret (Cattle Egret)	*Bubulcus ibis* (LC)
16.	Grey Heron	*Ardea cinerea* (LC)
17.	Purple Heron	*Ardea purpurea* (LC)
18.	Indian Pond-heron	*Ardeola grayii* (LC)
19.	Black-crowned Night-heron	*Nycticorax nycticorax* (LC)
20.	Striated Heron	*Butorides striatus* (LC)
21.	Black Bittern	*Ixobrychus flavicollis* (LC)
22.	Cinnamon Bittern	*Ixobrychus cinnamomeus* (LC)
23.	Yellow Bittern	*Ixobrychus sinensis* (LC)
24.	Little Bittern	*Ixobrychus minutus* (LC)
25.	Great Bittern	*Botaurus stellaris* (LC)
26.	Wooly-necked Stork	*Ciconia episcopus* (LC)
27.	Asian Openbill	*Anastomus oscitans* (LC)
28.	White Stork	*Ciconia ciconia* (LC)
29.	Painted Stork	*Mycteria leucocephala* (NT)
30.	Black Stork	*Ciconia nigra* (NT)
31.	Black-necked Stork	*Ephippiorhynchus asiaticus* (NT)
32.	Glossy Ibis	*Plegadis falcinellus* (NT)
33.	Black-headed Ibis	*Threskiornis melanocephalus* (LC)
34.	Eurasian Spoonbill	*Platalea leucorodia* (LC)
35.	Red-naped Ibis	*Pseudibis papillosa* (LC)
36.	Lesser Adjutant	*Leptoptilos javanicus* (VU)
37.	Greater Adjutant	*Leptoptilos dubius* (EN)
38.	Lesser Flamingo	*Phoeniconaias minor* (NT)
39.	Greater Flamingo	*Phoenicopterus roseus* (LC)
40.	Bar-headed Goose	*Anser indicus* (LC)
41.	Greylag Goose	*Anser anser* (LC)
42.	Lesser Whistling-duck	*Dendrocygna javanica* (LC)
43.	Common Shelduck	*Tadorna tadorna* (LC)
44.	Comb Duck	*Sarkidiornis melanotos* (LC)
45.	Common Teal	*Anas crecca* (LC)

(continued)

Table 3c (continued)

S. No.	Common/English name	Scientific name (IUCN Status ver 2013.1)
46.	Garganey	*Anas querquedula* (LC)
47.	Falcated Duck	*Anas falcata* (NT)
48.	Gadwall	*Anas strepera* (LC)
49.	Eurasian Wigeon	*Anas penelope* (LC)
50.	Northern Shoveler	*Anas clypeata* (LC)
51.	Northern Pintail	*Anas acuta* (LC)
52.	Western Spot-billed Duck	*Anas poecilorhyncha* (LC)
53.	Mallard	*Anas platyrhynchos* (LC)
54.	White-headed Duck	*Oxyura leucocephala* (EN)
55.	Greater Scaup	*Aythya marila* (LC)
56.	Tufted Duck	*Aythya fuligula* (LC)
57.	Ferruginous Duck	*Aythya nyroca* (NT)
58.	Common Pochard	*Aythya ferina* (LC)
59.	Red-crested Pochard	*Netta rufina* (LC)
60.	Cotton Pygmy-goose	*Nettapus coromandelianus* (LC)
61.	Black-winged Kite	*Elanus caeruleus* (LC)
62.	Brahminy Kite	*Haliastur indus* (LC)
63.	Black Kite	*Milvus migrans migrans/govinda* (LC)
64.	Besra	*Accipiter virgatus* (LC)
65.	Shikra	*Accipiter badius* (LC)
66.	Northern Goshawk	*Accipiter gentilis* (LC)
67.	Eurasian Sparrowhawk	*Accipiter nisus* (LC)
68.	Common Buzzard	*Buteo buteo* (LC)
69.	Long-legged Buzzard	*Buteo rufinus* (LC)
70.	Oriental Honey-buzzard	*Pernis ptilorhyncus* (LC)
71.	White-eyed Buzzard	*Butastur teesa* (LC)
72.	Crested Serpent-eagle	*Spilornis cheela* (LC)
73.	Short-toed Snake-Eagle	*Circaetus gallicus* (LC)
74.	Booted Eagle	*Hieraaetus pennatus* (LC)
75.	Bonelli's Eagle	*Aquila fasciatus* (LC)
76.	Changeable Hawk-eagle	*Nisaetus cirrhatus* (LC)
77.	Black Eagle	*Ictinaetus malayensis* (LC)
78.	Lesser Spotted Eagle	*Aquila pomarina* (LC)
79.	Greater Spotted Eagle	*Aquila clanga* (VUL)
80.	Steppe Eagle	*Aquila nipalensis* (LC)
81.	Tawny Eagle	*Aquila rapax* (LC)
82.	Golden Eagle	*Aquila chrysaetos* (LC)
83.	Eastern Imperial Eagle (Imperial Eagle)	*Aquila heliaca* (VU)
84.	Osprey	*Pandion haliaetus* (LC)
85.	Grey-headed Fish-eagle	*Ichthyophaga icthyaetus* (NT)
86.	White-bellied Sea-eagle	*Haliaeetus leucogaster* (LC)
87.	Pallas's Fish-eagle	*Haliaeetus leucoryphus* (VU)
88.	White-tailed Eagle	*Haliaeetus albicilla*
89.	White-rumped Vulture	*Gyps bengalensis* (CR)

(continued)

Table 3c (continued)

S. No.	Common/English name	Scientific name (IUCN Status ver 2013.1)
90.	Indian Vulture	*Gyps indicus* (CR)
91.	Griffon Vulture	*Gyps fulvus* (LC)
92.	Himalayan Vulture	*Gyps himalayensis* (LC)
93.	Egyptian Vulture	*Neophron percnopterus* (EN)
94.	Red-headed Vulture	*Sarcogyps calvus* (CR)
95.	Cinereous Vulture	*Aegypius monachus* (NT)
96.	Montagu's Harrier	*Circus pygargus* (LC)
97.	Pallid Harrier	*Circus macrourus* (NT)
98.	Pied Harrier	*Circus melanoleucos* (LC)
99.	Northern Harrier	*Circus cyaneus* (LC)
100.	Western Marsh Harrier	*Circus aeruginosus* (LC)
101.	Merlin	*Falco columbarius* (LC)
102.	Lesser Kestrel	*Falco naumanni* (LC)
103.	Common Kestrel	*Falco tinnunculus* (LC)
104.	Oriental Hobby	*Falco severus* (LC)
105.	Eurasian Hobby	*Falco subbuteo* (LC)
106.	Red-headed Falcon	*Falco chicquera* (LC)
107.	Peregrine Falcon	*Falco peregrinus* (LC)
108.	Laggar Falcon	*Falco jugger* (NT)
109.	Saker Falcon	*Falco cherrug* (EN)
110.	Grey Francolin	*Francolinus pondicerianus* (LC)
111.	Black Francolin	*Francolinus francolinus* (LC)
112.	Painted Francolin	*Francolinus pictus* (LC)
113.	Common Quail	*Coturnix coturnix* (LC)
114.	Rain Quail	*Coturnix coromandelica* (LC)
115.	Blue Quail	*Coturnix chinensis* (LC)
116.	Jungle Bush-quail	*Perdicula asiatica* (LC)
117.	Rock Bush-quail	*Perdicula argoondah* (LC)
118.	Barred Buttonquail	*Turnix suscitator* (LC)
119.	Yellow-legged Buttonquail	*Turnix tanki* (LC)
120.	Small Buttonquail	*Turnix sylvaticus* (LC)
121.	Red Spurfowl	*Galloperdix spadicea* (LC)
122.	Painted Spurfowl	*Galloperdix lunulata* (LC)
123.	Grey Junglefowl	*Gallus sonneratii* (LC)
124.	Indian Peafowl	*Pavo cristatus* (LC)
125.	Common Crane	*Grus grus* (LC)
126.	Demoiselle Crane	*Anthropoides virgo* (LC)
127.	Sarus Crane	*Grus antigone* (VU)
128.	Siberian Crane	*Leucogeranus leucogeranus* (CR)
129.	Lesser Florican	*Sypheotides indicus* (EN)
130.	Houbara Bustard	*Chlamydotis undulata* (VU)
131.	Great Indian Bustard	*Ardeotis nigriceps (CR)*
132.	Slaty-breasted Rail	*Gallirallus striatus* (LC)
133.	Water Rail	*Rallus aquaticus* (LC)
134.	Baillon's Crake	*Porzana pusilla* (LC)

(continued)

Table 3c (continued)

S. No.	Common/English name	Scientific name (IUCN Status ver 2013.1)
135.	Spotted Crake	*Porzana porzana* (LC)
136.	Ruddy-breasted Crake	*Porzana fusca* (LC)
137.	Brown Crake	*Amaurornis akool* (LC)
138.	White-breasted Waterhen	*Amaurornis phoenicurus* (LC)
139.	Watercock	*Gallicrex cinerea* (LC)
140.	Purple Swamphen	*Porphyrio porphyrio* (LC)
141.	Common Moorhen	*Gallinula chloropus* (LC)
142.	Common Coot	*Fulica atra* (LC)
143.	Greater Painted-snipe	*Rostratula benghalensis* (LC)
144.	Pied Avocet	*Recurvirostra avosetta* (LC)
145.	Black-winged Stilt	*Himantopus himantopus* (LC)
146.	Bronze-winged Jacana	*Metopidius indicus* (LC)
147.	Pheasant-tailed Jacana	*Hydrophasianus chirurgus* (LC)
148.	Oriental Pratincole	*Glareola maldivarum* (LC)
149.	Small Pratincole	*Glareola lactea* (LC)
150.	Cream-coloured Courser	*Cursorius cursor* (LC)
151.	Indian Courser	*Cursorius coromandelicus* (LC)
152.	Great Thick-knee	*Esacus recurvirostris* (LC)
153.	Eurasian Thick-knee	*Burhinus oedicnemus* (LC)
154.	Red-wattled Lapwing	*Vanellus indicus* (LC)
155.	Yellow-wattled Lapwing	*Vanellus malarbaricus* (LC)
156.	River Lapwing	*Vanellus duvaucelii* (NT)
157.	Grey-headed Lapwing	*Vanellus cinereus* (LC)
158.	Sociable Lapwing	*Vanellus gregarius (CR)*
159.	White-tailed Lapwing	*Vanellus leucurus* (LC)
160.	Northern Lapwing	*Vanellus vanellus* (LC)
161.	Grey Plover	*Pluvialis squatarola* (LC)
162.	Pacific Golden Plover	*Pluvialis fulva* (LC)
163.	Greater Sand Plover	*Charadrius leschenaultii* (LC)
164.	Lesser Sand Plover	*Charadrius mongolus* (LC)
165.	Common Ringed Plover	*Charadrius hiaticula* (LC)
166.	Little Ringed Plover	*Charadrius dubius* (LC)
167.	Kentish Plover	*Charadrius alexandrinus* (LC)
168.	Eurasian Curlew	*Numenius arquata* (NT)
169.	Black-tailed Godwit	*Limosa limosa* (NT)
170.	Bar-tailed Godwit	*Limosa lapponica* (LC)
171.	Terek Sandpiper	*Xenus cinereus* (LC)
172.	Common Greenshank	*Tringa nebularia* (LC)
173.	Wood Sandpiper	*Tringa glareola* (LC)
174.	Green Sandpiper	*Tringa ochropus* (LC)
175.	Common Sandpiper	*Actitis hypoleucos* (LC)
176.	Marsh Sandpiper	*Tringa stagnatilis* (LC)
177.	Common Redshank	*Tringa totanus* (LC)
178.	Spotted Redshank	*Tringa erythropus* (LC)
179.	Ruddy Turnstone	*Arenaria interpres* (LC)

(continued)

Table 3c (continued)

S. No.	Common/English name	Scientific name (IUCN Status ver 2013.1)
180.	Ruff	*Philomachus pugnax* (LC)
181.	Curlew Sandpiper	*Calidris ferruginea* (LC)
182.	Dunlin	*Calidris alpina* (LC)
183.	Temminck's Stint	*Calidris temminckii* (LC)
184.	Little Stint	*Calidris minuta* (LC)
185.	Red-necked Phalarope	*Phalaropus lobatus* (LC)
186.	Jack Snipe	*Lymnocryptes minimus* (LC)
187.	Pintail Snipe	*Gallinago stenura* (LC)
188.	Eurasian Woodcock	*Scolopax rusticola* (LC)
189.	Caspian Gull	*Larus cachinnans* (LC)
190.	Pallas's Gull	*Larus ichthyaetus* (LC)
191.	Brown-headed Gull	*Larus brunnicephalus* (LC)
192.	Black-headed Gull	*Larus ridibundus* (LC)
193.	Little Tern	*Sterna albifrons* (LC)
194.	Common Tern	*Sterna hirundo* (LC)
195.	Black-bellied Tern	*Sterna acuticauda* (LC)
196.	Gull-billed Tern	*Sterna nilotica* (LC)
197.	River Tern	*Sterna aurantia* (NT)
198.	Whiskered Tern	*Chlidonias hybridus* (LC)
199.	White-winged Tern	*Chlidonias leucopterus* (LC)
200.	Indian Skimmer	*Rynchops albicollis* (VU)
201.	Black-bellied Sandgrouse	*Pterocles orientalis* (LC)
202.	Pin-tailed Sandgrouse	*Pterocles alchata* (LC)
203.	Chestnut-bellied Sandgrouse	*Pterocles exustus* (LC)
204.	Spotted Sandgrouse	*Pterocles senegallus* (LC)
205.	Painted Sandgrouse	*Pterocles indicus* (LC)
206.	Rock Pigeon	*Columba livia* (LC)
207.	Pale-backed Pigeon	*Columba eversmanni* (VU)
208.	Oriental Turtle-dove	*Streptopelia orientalis* (LC)
209.	Laughing Dove	*Streptopelia senegalensis* (LC)
210.	Red Collared-dove	*Streptopelia tranquebarica* (LC)
211.	Spotted Dove	*Stigmatopelia chinensis* (LC)
212.	Eurasian Collared-dove	*Streptopelia decaocto* (LC)
213.	Yellow-footed Green-pigeon	*Treron phoenicoptera* (LC)
214.	Plum-headed Parakeet	*Psittacula cyanocephala* (LC)
215.	Rose-ringed Parakeet	*Psittacula krameri* (LC)
216.	Alexandrine Parakeet	*Psittacula eupatria* (LC)
217.	Grey-bellied Cuckoo	*Cacomantis passerinus* (LC)
218.	Pied Cuckoo	*Clamator jacobinus* (LC)
219.	Asian Koel	*Eudynamys scolopacea* (LC)
220.	Indian Cuckoo	*Cuculus micropterus* (LC)
221.	Common Cuckoo	*Cuculus canorus* (LC)
222.	Himalayan Cuckoo	*Cuculus saturatus* (LC)
223.	Common Hawk-Cuckoo	*Hierococcyx varius* (LC)
224.	Sirkeer Malkoha	*Phaenicophaeus leschenaultii* (LC)

(continued)

Table 3c (continued)

S. No.	Common/English name	Scientific name (IUCN Status ver 2013.1)
225.	Greater Coucal	*Centropus sinensis* (LC)
226.	Brown Hawk-Owl	*Ninox scutulata* (LC)
227.	Barn-Owl	*Tyto alba* (LC)
228.	Brown Fish-Owl	*Ketupa zeylonensis* (LC)
229.	Dusky Eagle-Owl	*Bubo coromandus* (LC)
230.	Eurasian Eagle-Owl	*Bubo bubo* (LC)
231.	Long-eared Owl	*Asio otus* (LC)
232.	Short-eared Owl	*Asio flammeus* (LC)
233.	Mottled Wood-Owl	*Strix ocellata* (LC)
234.	Collared Scops-Owl	*Otus bakkamoena* (LC)
235.	Oriental Scops-Owl	*Otus sunia* (LC)
236.	Spotted Owlet	*Athene brama* (LC)
237.	Jungle Owlet	*Glaucidium radiatum* (LC)
238.	Sykes's Nightjar	*Caprimulgus mahrattensis* (LC)
239.	Eurasian Nightjar	*Caprimulgus europaeus* (LC)
240.	Grey Nightjar	*Caprimulgus indicus* (LC)
241.	Indian Nightjar	*Caprimulgus asiaticus* (LC)
242.	Large-tailed Nightjar	*Caprimulgus macrurus* (LC)
243.	Savanna Nightjar	*Caprimulgus affinis* (LC)
244.	Crested Treeswift	*Hemiprocne coronata* (LC)
245.	Asian Palm-swift	*Cypsiurus balasiensis* (LC)
246.	Little Swift	*Apus affinis* (LC)
247.	Fork-tailed Swift	*Apus pacificus* (LC)
248.	Alpine Swift	*Tachymarptis melba* (LC)
249.	European Roller	*Coracias garrulus* (NT)
250.	Indian Roller	*Coracias benghalensis* (LC)
251.	Common Hoopoe	*Upupa epops* (LC)
252.	Stork-billed Kingfisher	*Halcyon capensis* (LC)
253.	Black-capped Kingfisher	*Halcyon pileata* (LC)
254.	White-throated Kingfisher	*Halcyon smyrnensis* (LC)
255.	Pied Kingfisher	*Ceryle rudis* (LC)
256.	Common Kingfisher	*Alcedo atthis* (LC)
257.	European Bee-Eater	*Merops apiaster* (LC)
258.	Blue-cheeked Bee-Eater	*Merops persicus* (LC)
259.	Blue-tailed Bee-Eater	*Merops philippinus* (LC)
260.	Little Green Bee-Eater	*Merops orientalis* (LC)
261.	Indian Grey Hornbill	*Ocyceros birostris* (LC)
262.	Coppersmith Barbet	*Megalaima haemacephala* (LC)
263.	Brown-headed Barbet	*Megalaima zeylanica* (LC)
264.	Eurasian Wryneck	*Jynx torquilla* (LC)
265.	Brown-capped Woodpecker	*Dendrocopos nanus* (LC)
266.	Yellow-crowned Woodpecker	*Dendrocopos mahrattensis* (LC)
267.	Streak-throated Woodpecker	*Picus xanthopygaeus* (LC)
268.	Rufous Woodpecker	*Celeus brachyurus* (LC)
269.	Black-rumped Flameback	*Dinopium benghalense* (LC)
270.	White-naped Woodpecker	*Chrysocolaptes festivus* (LC)

(continued)

Table 3c (continued)

S. No.	Common/English name	Scientific name (IUCN Status ver 2013.1)
271.	Indian Pitta	*Pitta brachyura* (LC)
272.	Singing Bushlark	*Mirafra cantillans* (LC)
273.	Indian Lark	*Mirafra erythroptera* (LC)
274.	Rufous-winged Lark	*Mirafra assamica* (LC)
275.	Eurasian Skylark	*Alauda arvensis* (LC)
276.	Oriental Skylark	*Alauda gulgula* (LC)
277.	Crested Lark	*Galerida cristata* (LC)
278.	Tawny Lark	*Galerida deva* (LC)
279.	Ashy-crowned Sparrow-Lark	*Eremopterix griseus* (LC)
280.	Black-crowned Sparrow-Lark	*Eremopterix nigriceps* (LC)
281.	Bimaculated Lark	*Melanocorypha bimaculata* (LC)
282.	Greater Hoopoe Lark	*Alaemon alaudipes* (LC)
283.	Rufous-tailed Lark	*Ammomanes phoenicura* (LC)
284.	Desert Lark	*Ammomanes deserti* (LC)
285.	Greater Short-toed Lark	*Calandrella brachydactyla longipennis* (LC)
286.	Hume's Short-toed Lark	*Calandrella acutirostris* (LC)
287.	Indian Short-toed Lark	*Calandrella raytal* (LC)
288.	Plain Martin	*Riparia paludicola* (LC)
289.	Sand-Martin	*Riparia riparia* (LC)
290.	Dusky Crag-Martin	*Hirundo concolor* (LC)
291.	Eurasian Crag-Martin	*Hirundo rupestris* (LC)
292.	Streak-throated Swallow	*Hirundo fluvicola* (LC)
293.	Barn Swallow	*Hirundo rustica* (LC)
294.	Red-rumped Swallow	*Hirundo daurica* (LC)
295.	Wire-tailed Swallow	*Hirundo smithii* (LC)
296.	Yellow Wagtail	*Motacilla flava leucocephala* (LC)
297.	Grey Wagtail	*Motacilla cinerea* (LC)
298.	Citrine Wagtail	*Motacilla citreola citreola* (LC)
299.	Forest Wagtail	*Dendronanthus indicus* (LC)
300.	White Wagtail	*Motacilla alba dukhunensis* (LC)
301.	White-browed Wagtail	*Motacilla madaraspatensis* (LC)
302.	Olive-backed Pipit	*Anthus hodgsoni hodgsoni* (LC)
303.	Tree Pipit	*Anthus trivialis trivialis* (LC)
304.	Red-throated Pipit	*Anthus cervinus* (LC)
305.	Rosy Pipit	*Anthus roseatus* (LC)
306.	American Pipit	*Anthus rubescens* (LC)
307.	Water Pipit	*Anthus spinoletta* (LC)
308.	Richard's Pipit	*Anthus richardi* (LC)
309.	Paddyfield Pipit	*Anthus rufulus* (LC)
310.	Tawny Pipit	*Anthus campestris* (LC)
311.	Blyth's Pipit	*Anthus godlewskii* (LC)
312.	Long-billed Pipit	*Anthus similis* (LC)
313.	White-breasted Woodswallow	*Artamus leucorynchus* (LC)
314.	Common Woodshrike	*Tephrodornis pondicerianus* (LC)
315.	Black-winged Cuckooshrike	*Coracina melaschistos* (LC)

(continued)

Table 3c (continued)

S. No.	Common/English name	Scientific name (IUCN Status ver 2013.1)
316.	Black-headed Cuckooshrike	*Coracina melanoptera* (LC)
317.	Large Cuckooshrike	*Coracina macei* (LC)
318.	Long-tailed Minivet	*Pericrocotus ethologus* (LC)
319.	Scarlet Minivet	*Pericrocotus flammeus* (LC)
320.	White-bellied Minivet	*Pericrocotus erythropygius* (LC)
321.	Small Minivet	*Pericrocotus cinnamomeus* (LC)
322.	Red-vented Bulbul	*Pycnonotus cafer* (LC)
323.	Red-whiskered Bulbul	*Pycnonotus jocosus* (LC)
324.	White-eared Bulbul	*Pycnonotus leucotis* (LC)
325.	White-browed Bulbul	*Pycnonotus luteolus* (LC)
326.	Common Iora	*Aegithina tiphia* (LC)
327.	White-tailed Iora	*Aegithina nigrolutea* (LC)
328.	Blue-winged Leafbird	*Chloropsis cochinchinensis* (LC)
329.	Black-headed Long-tailed Shrike	*Lanius schach tricolor* (LC)
330.	Rufous-backed Long-tailed Shrike	*Lanius schach erythronotus* (LC)
331.	Great Grey Shrike	*Lanius excubitor* (LC)
332.	Grey-backed Shrike	*Lanius tephronotus* (LC)
333.	Brown Shrike	*Lanius cristatus cristatus* (LC)
334.	Rufous-tailed Shrike	*Lanius isabellinus* (LC)
335.	Bay-backed Shrike	*Lanius vittatus* (LC)
336.	Red-backed Shrike	*Lanius collurio* (LC)
337.	Black-naped Monarch	*Hypothymis azurea* (LC)
338.	Asian Paradise Flycatcher	*Terpsiphone paradisi* (LC)
339.	White-browed Fantail	*Rhipidura aureola* (LC)
340.	White-throated Fantail	*Rhipidura albicollis* (LC)
341.	Blue Rock-Thrush	*Monticola solitarius* (LC)
342.	Blue-capped Rock-Thrush	*Monticola cinclorhynchus* (LC)
343.	Orange-headed Thrush	*Zoothera citrina* (LC)
344.	Malabar Whistling-Thrush	*Myiophonus horsfieldii* (LC)
345.	Black-throated Thrush (Dark-throated Thrush)	*Turdus ruficollis atrogularis* (LC)
346.	Red-throated Thrush	*Turdus ruficollis ruficollis* (LC)
347.	Tickell's Thrush	*Turdus unicolor* (LC)
348.	Eurasian Blackbird	*Turdus merula* (LC)
349.	Grey-winged Blackbird	*Turdus boulboul* (LC)
350.	Eurasian Scaly Thrush	*Zoothera dauma* (LC)
351.	Bluethroat	*Luscinia svecica svecica* (LC)
352.	White-tailed Rubythroat	*Luscinia pectoralis* (LC)
353.	Siberian Rubythroat	*Luscinia calliope* (LC)
354.	Rufous-tailed Scrub-Robin	*Cercotrichas galactotes* (LC)
355.	Oriental Magpie-Robin	*Copsychus saularis* (LC)
356.	Indian Robin	*Saxicoloides fulicatus* (LC)
357.	Indian Chat	*Cercomela fusca* (LC)
358.	Black Redstart	*Phoenicurus ochruros rufiventris* (LC)
359.	Blue-fronted Redstart	*Phoenicurus frontalis* (LC)
360.	Isabelline Wheatear	*Oenanthe isabellina* (LC)

(continued)

Table 3c (continued)

S. No.	Common/English name	Scientific name (IUCN Status ver 2013.1)
361.	Northern Wheatear	*Oenanthe oenanthe* (LC)
362.	Desert Wheatear	*Oenanthe deserti deserti* (LC)
363.	Variable Wheatear	*Oenanthe picata capistrata* (LC)
364.	Kurdish Wheatear	*Oenanthe xanthoprymna* (LC)
365.	Pied Bushchat	*Saxicola caprata* (LC)
366.	Common Stonechat	*Saxicola torquata* (LC)
367.	White-browed Bushchat	*Saxicola macrorhynchus* (VU)
368.	Grey Bushchat	*Saxicola ferreus* (LC)
369.	Spotted Flycatcher	*Muscicapa striata* (LC)
370.	Asian Brown Flycatcher	*Muscicapa dauurica* (LC)
371.	Rusty-tailed Flycatcher	*Muscicapa ruficauda* (LC)
372.	Ultramarine Flycatcher	*Ficedula superciliaris* (LC)
373.	Red-breasted Flycatcher	*Ficedula parva* (LC)
374.	Slaty-blue Flycatcher	*Ficedula tricolor* (LC)
375.	Tickell's Blue Flycatcher	*Cyornis tickelliae* (LC)
376.	Verditer Flycatcher	*Eumyias thalassinus* (LC)
377.	Yellow-eyed Babbler	*Chrysomma sinense* (LC)
378.	Tawny-bellied Babbler	*Dumetia hyperythra* (LC)
379.	Common Babbler	*Turdoides caudata* (LC)
380.	Striated Babbler	*Turdoides earlei* (LC)
381.	Large Grey Babbler	*Turdoides malcolmi* (LC)
382.	Jungle Babbler	*Turdoides striata* (LC)
383.	Indian Scimitar-Babbler	*Pomatorhinus horsfieldii* (LC)
384.	Puff-throated Babbler	*Pellorneum ruficeps* (LC)
385.	Zitting Cisticola	*Cisticola juncidis* (LC)
386.	Lanceolated Warbler	*Locustella lanceolata* (LC)
387.	Common Grasshopper Warbler	*Locustella naevia* (LC)
388.	Ashy Prinia	*Prinia socialis* (LC)
389.	Grey-breasted Prinia	*Prinia hodgsonii* (LC)
390.	Jungle Prinia	*Prinia sylvatica* (LC)
391.	Plain Prinia	*Prinia inornata* (LC)
392.	Rufous-fronted Prinia	*Prinia buchanani* (LC)
393.	Rufous-vented Prinia	*Prinia burnesii* (NT)
394.	Graceful Prinia	*Prinia gracilis* (LC)
395.	Cetti's Warbler	*Cettia cetti* (LC)
396.	Moustached Warbler	*Acrocephalus melanopogon* (LC)
397.	Clamorous Reed-Warbler	*Acrocephalus stentoreus brunnescens* (LC)
398.	Thick-billed Warbler	*Acrocephalus aedon* (LC)
399.	Blyth's Reed-warbler	*Acrocephalus dumetorum* (LC)
400.	Paddyfield Warbler	*Acrocephalus agricola* (LC)
401.	Blunt-winged Warbler	*Acrocephalus concinens stevensi* (LC)
402.	Booted Warbler	*Hippolais caligata* (LC)
403.	Common Tailorbird	*Orthotomus sutorius* (LC)
404.	Green-crowned Warbler	*Seicercus burkii* (LC)
405.	Grey-headed Canary-flycatcher	*Culicicapa ceylonensis* (LC)
406.	Common Chiffchaff	*Phylloscopus collybita tristis* (LC)
407.	Mountain Chiffchaff	*Phylloscopus sindianus* (LC)

(continued)

Table 3c (continued)

S. No.	Common/English name	Scientific name (IUCN Status ver 2013.1)
408.	Plain Leaf-Warbler	*Phylloscopus neglectus* (LC)
409.	Dusky Warbler	*Phylloscopus fuscatus fuscatus* (LC)
410.	Smoky Warbler	*Phylloscopus fuligiventer* (LC)
411.	Sulphur-bellied Warbler	*Phylloscopus griseolus* (LC)
412.	Tickell's Leaf-Warbler	*Phylloscopus affinis* (LC)
413.	Greenish Warbler	*Phylloscopus trochiloides* (LC)
414.	Large-billed Leaf-Warbler	*Phylloscopus magnirostris* (LC)
415.	Western Crowned Warbler	*Phylloscopus occipitalis* (LC)
416.	Hume's Leaf-Warbler	*Phylloscopus humei* (LC)
417.	Brooks's Leaf-Warbler	*Phylloscopus subviridis* (LC)
418.	Orphean Warbler	*Sylvia hortensis* (LC)
419.	Common Whitethroat	*Sylvia communis* (LC)
420.	Desert Warbler	*Sylvia nana* (LC)
421.	Lesser Whitethroat	*Sylvia curruca halimodendri* (LC)
422.	White-naped Tit	*Parus nuchalis* (VU)
423.	Great Tit	*Parus major* (LC)
424.	Black-lored Tit	*Parus xanthogenys* (LC)
425.	Spotted Creeper	*Salpornis spilonotus* (LC)
426.	Fire-capped Tit	*Cephalopyrus flammiceps* (LC)
427.	Chestnut-bellied Nuthatch	*Sitta castanea* (LC)
428.	Pale-billed Flowerpecker	*Dicaeum erythrorhynchos* (LC)
429.	Thick-billed Flowerpecker	*Dicaeum agile* (LC)
430.	Oriental White-eye	*Zosterops palpebrosus* (LC)
431.	Purple-rumped Sunbird	*Nectarinia zeylonica* (LC)
432.	Purple Sunbird	*Nectarinia asiatica* (LC)
433.	Gould's Sunbird	*Aethopyga gouldiae* (LC)
434.	Crested Bunting	*Melophus lathami* (LC)
435.	Thick-billed Reed Bunting	*Emberiza schoeniclus pyrrhuloides* (LC)
436.	Reed Bunting	*Emberiza schoeniclus pallidior* (LC)
437.	Chestnut-breasted Bunting	*Emberiza stewarti* (LC)
438.	House Bunting	*Emberiza striolata* (LC)
439.	Black-headed Bunting	*Emberiza melanocephala* (LC)
440.	Red-headed Bunting	*Emberiza bruniceps* (LC)
441.	Grey-necked Bunting	*Emberiza buchanani* (LC)
442.	Ortolan Bunting	*Emberiza hortulana* (LC)
443.	Trumpeter Finch	*Bucanetes githagineus* (LC)
444.	Mongolian Finch	*Eremopsaltria mongolicus* (LC)
445.	Common Rosefinch	*Carpodacus erythrinus roseatus* (LC)
446.	Green Avadavat	*Amandava formosa* (VU)
447.	Red Avadavat	*Amandava amandava* (LC)
448.	Tricoloured Munia	*Lonchura malacca* (LC)
449.	White-throated Munia	*Lonchura malabarica* (LC)
450.	White-rumped Munia	*Lonchura striata* (LC)
451.	Scaly-breasted Munia	*Lonchura punctulata* (LC)
452.	House Sparrow	*Passer domesticus* (LC)
453.	Eurasian Tree Sparrow	*Passer montanus* (LC)
454.	Spanish Sparrow	*Passer hispaniolensis* (LC)

(continued)

Table 3c (continued)

S. No.	Common/English name	Scientific name (IUCN Status ver 2013.1)
455.	Chestnut-shouldered Petronia	*Petronia xanthocollis* (LC)
456.	Baya Weaver	*Ploceus philippinus philippinus* (LC)
457.	Streaked Weaver	*Ploceus manyar* (LC)
458.	Black-breasted Weaver	*Ploceus benghalensis* (LC)
459.	Eurasian Golden Oriole	*Oriolus oriolus* (LC)
460.	Black-hooded Oriole	*Oriolus xanthornus* (LC)
461.	Black Drongo	*Dicrurus macrocercus* (LC)
462.	Ashy Drongo	*Dicrurus leucophaeus* (LC)
463.	White-bellied Drongo	*Dicrurus caerulescens* (LC)
464.	Greater Racket-tailed Drongo	*Dicrurus paradiseus* (LC)
465.	Hair-crested Drongo	*Dicrurus hottentottus* (LC)
466.	Common Starling	*Sturnus vulgaris* (LC)
467.	Brahminy Starling	*Sturnus pagodarum* (LC)
468.	Rosy Starling	*Sturnus roseus* (LC)
469.	Chestnut-tailed Starling	*Sturnus malabaricus* (LC)
470.	Asian Pied Starling	*Sturnus contra* (LC)
471.	Bank Myna	*Acridotheres ginginianus* (LC)
472.	Common Myna	*Acridotheres tristis* (LC)
473.	Jungle Myna	*Acridotheres fuscus* (LC)
474.	House Crow	*Corvus splendens* (LC)
475.	Large-billed Crow	*Corvus macrorhynchos culminatus* (LC)
476.	Common Raven	*Corvus corax tibetanus* (LC)
477.	Rufous Treepie	*Dendrocitta vagabunda* (LC)

Source: Bombay Natural History Society (BNHS), Mumbai, India, 2012: BirdLife International 2012. IUCN Red List of Threatened Species. Version 2013.1. www.iucnredlist.org

Table 3d Mammals

S. No.	Scientific name	Common name (IUCN Status ver 2013.1)
Order Primates		
Family Cercopithecidae (Monkeys)		
1.	*Macaca mulatta*	Rhesus Monkey (LC)
2.	*Semnopithecus entellus*	Northern Plains Gray Langur (LC)
Order Artiodactyla		
Family Cervidae (Deers)		
3.	*Rusa unicolor*	Sambar (VU)
	Muntiacus muntjak	Southern Red Muntjac (LC)
4.	*Axis axis*	Chital (LC)
Family Tragulidae		
5.	*Moschiola meminna*	White-spotted Chevrotain (LC)
Family Bovidae (Cattle, Antelope, Goat, Sheep)		
6.	*Boselaphus tragocamelus*	Nilgai (LC)
7.	*Antilope cervicapra*	Blackbuck (NT)
8.	*Tetracerus quadricornis*	Four-horned Antelope (VUL)
9.	*Gazella bennettii*	Chinkara (LC)

(continued)

Table 3d (continued)

S. No.	Scientific name	Common name (IUCN Status ver 2013.1)
Family Suidae (Pigs)		
10.	*Sus scrofa*	Wild Boar (LC)
Order Perissodactyla		
Family Equidae (Wild Asses)		
11.	*Equus hemionus*	Asiatic Wild Ass (EN)
Order Carnivora		
Family Ursidae (Bears)		
12.	*Melursus ursinus*	Sloth Bear (VU)
Family Canidae (Dogs)		
13.	*Canis aureus*	Golden Jackal (LC)
14.	*Canis lupus*	Gray Wolf (LC)
Past	*Cuon alpinus*	Dhole (EN)
15.	*Vulpes bengalensis*	Bengal Fox (LC)
16.	*Vulpes vulpes*	Red Fox (LC)
Family Hyaenidae (Hyaena)		
17.	*Hyaena hyaena*	Striped Hyaena (NT)
Family Felidae (Cats)		
18.	*Panthera tigris*	Bengal Tiger (EN)
19.	*Panthera pardus*	Leopard (NT)
20.	*Felis silvestris*	Wild Cat (LC)
21.	*Caracal caracal*	Caracal (LC)
22.	*Felis chaus*	Jungle Cat (LC)
23.	*Prionailurus rubiginosus*	Rusty-spotted Cat (VU)
24.	*Prionailurus viverrinus*	Fishing Cat (EN)
Family Mustelidae (Mustelids)		
25.	*Mellivora capensis*	Honey Badger (LC)
26.	*Lutrogale perspicillata*	Smooth-Coated Otter (VU)
Family Viveridae (Civets)		
27.	*Viverricula indica*	Small Indian Civet (LC)
28.	*Paradoxurus hermaphroditus*	Common Palm Civet (LC)
Family Herpestidae (Mongooses)		
29.	*Herpestes edwardsii*	Indian Grey Mongoose (LC)
30.	*Herpestes smithii*	Ruddy Mongoose (LC)
Order Pholidota		
Family Manidae (Pangolins)		
31.	*Manis crassicaudata*	Thick-tailed Pangolin (NT)
Order Logomorpha		
Family Leporidae (Hares)		
32.	*Lepus nigricollis*	Indian Hare (LC)
Order Insectivora		
Family Erinaceidae (Hedgehogs)		
33.	*Paraechinus micropus*	Indian Hedgehog (LC)
34.	*Paraechinus collaris*	Indian Long-eared Hedgehog (LC)
Family Soricidae (Shrews)		
35.	*Suncus murinus*	House Shrew (LC)

(continued)

Table 3d (continued)

S. No.	Scientific name	Common name (IUCN Status ver 2013.1)
Order Rodentia		
Family Sciuridae (Squirrels)		
36.	*Funambulus palmarum*	Common Palm Squirrel (LC)
37.	*Funambulus pennantii*	Five-striped Palm Squirrel (LC)
38.	*Petaurista philippensis*	Large Brown Flying Squirrel (LC)
Family Muridae (Rats, Mice)		
39.	*Bandicota indica*	Greater Bandicoot Rat (LC)
40.	*Bandicota bengalensis*	Lesser Bandicoot Rat (LC)
41.	*Tatera indica*	Indian Gerbil (LC)
42.	*Meriones hurrianae*	Indian Desert Gerbil (LC)
43.	*Gerbillus nanus*	Dwarf Gerbil (LC)
44.	*Gerbillus gleadowi*	Little Hairy-footed Gerbil (LC)
45.	*Rattus rattus*	House Rat (LC)
46.	*Millardia meltada*	Soft-furred Metad (LC)
47.	*Millardia gleadowi*	Sand-coloured Metad (LC)
48.	*Vandeleuria oleracea*	Asiatic Long-tailed Climbing Mouse (LC)
49.	*Mus musculus*	House Mouse (LC)
50.	*Mus booduga*	Little Indian Field Mouse (LC)
Family Hystricidae (Porcupines)		
51.	*Hystrix indica*	Indian Crested Porcupine (LC)
Order Chiroptera		
Family Pteropodidae (Fruit Bats)		
52.	*Pteropus giganteus*	Indian Flying Fox (LC)
53.	*Pteropus leschenaulti*	Fulvous Fruit Bat (LC)
54.	*Cynopteris brachyotis*	Lesser Dog-faced Fruit Bat (LC)
55.	*Cynopteris sphinx*	Greater Short-nosed Fruit Bat (LC)
Family Rhinopomatidae (Mouse-tailed Bats)		
56.	*Rhinopoma microphyllum*	Greater Mouse-tailed Bat (LC)
57.	*Rhinopoma hardwickii*	Lesser Mouse-tailed Bat (LC)
Family Emballonuridae (Tomb Bats)		
58.	*Taphozous nudiventris*	Naked-rumped Tomb Bat (LC)
59.	*Taphozous perforatus*	Egyptian Tomb Bat (LC)
Family Rhinolophidae (Horse-shoe Bats)		
60.	*Rhinolophus lepidus*	Blyth's Horseshoe Bat (LC)
Family Hipposideridae (Leaf-nosed Bats)		
61.	*Hipposideros lankadiva*	Indian Leaf-nosed Bat (LC)
Family Megadermatidae (False Vampires)		
62.	*Megaderma lyra*	Greater False Vampire (LC)
Family Molossidae (Free-tailed Bats)		
63.	*Tadarida aegyptiaca*	Egyptian Free-tailed Bat (LC)
Family Vespertilionidae (Evening Bats)		
64.	*Scotophilus heathii*	Greater Asiatic Yellow House Bat (LC)
65.	*Pipistrellus tenuis*	Least Pipistrelle (LC)
Order Cetaceans		
Family Platanistidae (Dolphins)		
66.	*Platanista gangetica*	Ganges River Dolphin (EN)

Source: IUCN Red List of Threatened Species. Version 2013.1. www.iucnredlist.org

Appendix 4

Appendix 4 Names of selected faunal species of Rajasthan in local/regional (Rajasthani) dialect along with common, Hindi and Zoological names

S. No.	Zoological name	Common name	Hindi name	Name in Rajasthani/local dialect
1.	*Aorichthys seenghala*	Giant River Catfish	Seenghri	Seenghri
2.	*Catla catla*	Catla	Catla	Catla
3.	*Channa marulius*	Giant Snake-headed Greyei	Sal/Pumuri/Bohr	–
4.	*Channa punctatus*	Spotted Snakehead	Savank	–
5.	*Cirhinus mrigala*	Mrigal	Mrigal	–
6.	*Labeo rohita*	Rohu	Rohu/Bita	Rohu
7.	*Tortor*	Mahaseer	Mahaseer	–
8.	*Wallago attu*	Boal	Lachi	Lachi
9.	*Euphlyctis hexadactylus*	Indian Pond Frog	Mendhak	Meendhko/Dadhdu
10.	*Geochelone elegans (Schoepff)*	Indian Star Tortoise	Kachhua	Kachhuo
11.	*Calotes versicolor*	Garden Lizard	Girgit/Kirkantia	Girgitan
12.	*Hemidactylus sp.*	Common House Lizard	Chhipkali/Kirkant	Bichhmara
13.	*Uromastix hardwickii*	Spiny-tailed Lizard	Sanda	Sanda
14.	*Varanus.griseus*	Desert Monitor	Goh	Gohera
15.	*V. bengalensis*	Bengal Monitor	Goh Pada, Goh	Gohera
16.	*Ahaetulla nasuta*	Common Vine Snake	Hara Saanp	Haryo Shaanp
17.	*Atretium schistosum*	Olive Keel Back	Pani Ka Saanp	Dendu
18.	*Daboia russelli*	Russell's Viper	Daboia	Chital Saanp
19.	*Denrelaphis tristis*	Common Bronze Back	–	Udni
20.	*Echis carnatus*	Saw-scaled Viper	Afai	Phoopsya Kakargada
21.	*Eryx johnii*	Red Sand Boa	Dumbhi	Syanwad
22.	*Mabuya carinata*	Common Skink	Bamni	Bamni
23.	*Mabuya macularia*	Bronze Grass Skink	Bamni	Bamni
24.	*Naja naja*	Indian Cobra	Cobra	Kaldar Sarp/Kaalo Naag
25.	*Opisops jerdoni*	Jerdon's Snake-eye	Bamni	Bamni

26.	*Python molurus*	Indian Rock Python	Ajgar	Ajgar
27.	*Lygosoma guentheri*	Snake Skink	Bamni	Bamni
28.	*Sanochophis piscator*	Checkered Keelback	–	Dendu
29.	*Crocodylus palustris*	Indian Crocodile	Magarmachh	Magar
30.	*Accipiter badius*	Shikra	Shikra	Sikra
31.	*Acridotheres tristis*	Common Myna	Mayna	Gudsali
32.	*Aegypus monachus*	Cinereous Vulture	Kala Giddh	Kanwla
33.	*Aeordotes nigriceps*	Great Indian Bustard	Godavan	Godavan
34.	*Bubo bubo*	Eurasian Eagle-Owl	Ghugghu	Ghugghu
35.	*Carvus splendenus*	House Crow	Kowa	Kaglo
36.	*Casmerodeus albus*	Large Egret	Bada Bagula	Bado Buglo
37.	*Columbo livia*	Rock Pigeon	Kabutar	Parewta
38.	*Coturnix coturnix*	Common Quail	Bater	–
39.	*Dendrocopos maharattensis*	Yellow-fronted Woodpecker	Kathhfodwa	Kathfodwo/Khatichida
40.	*Egretta garzetta*	Cattle Egret	Chhota Bagula	Chhoto Buglo
41.	*Eudynamys scopacea*	Asian Koel	Koyal	Koyal
42.	*Francolinus pondicerianus*	Grey Francolin	Safed Teetar	Teetar
43.	*Gallinulla chloropus*	Common Moorhen	Jalmurgi	Jalmurgi
44.	*Grus antigone*	Sarus Crane	Saras	Saras
45.	*Grus virgo*	Demoisselle Crane	Kurja	Kurjan
46.	*Halcyon smyrnensis*	White-breasted Kingfisher	Neelkanth	Neelkanth
47.	*Milvus migrans*	Black Kite	Cheel	Cheeldo
48.	*Orthotamus sutorius*	Common Tailor Bird	Darzee Chidiya	Phutki
49.	*Parus nuchulis*	Pied Tit	–	Ramcharkali
50.	*Passer domesticus*	House Sparrow	Chidiya	Chidakali/Chidi
51.	*Pavo crystatus*	Indian Peafowl	Mor	Moryo
52.	*Phoenicopterus roseus*	Greater Flamingo	Rajhans	Hans
53.	*Psittacula sp.*	Parakeet	Tota	Sua

(continued)

Appendix 4 (continued)

S. No.	Zoological name	Common name	Hindi name	Name in Rajasthani/local dialect
54.	*Pycnonotus cafer*	Red-vented Bulbul	Bulbul	Bulbul
55.	*Sarcogypus calvus*	Red-headed Vulture	Raj-Giddh	Kanwla
56.	*Streptopelia sp.*	Dove	Fakhta	Kamedi
57.	*Turdoides striatus*	Jungle Babbler	Satbhai	Dumrhee
58.	*Tyto alba*	Barn Owl	Chugad	Kochar
59.	*Venellus indicus*	Red-wattled Lapwing	Tithari	Titodi
60.	*Hemiechinus auritus*	Long-eared Hedgehog	Jhau Chuha	Jhabulya/Jahulya
61.	*Suncus murinus*	Grey Musk-Shrew	Chhachhundar	Chakchundri
62.	*Pteropus giganteus*	Flying Fox	Badi Chamgadad	Bagal/Chamched
63.	*Rhinolophus sp.*	Small-sized Bat	Chamgadad	Chamched
64.	*Rattus rattus*	Black Rat	Kala Chuha	Undra
65.	*Mus musculus*	Mouse	Chuhiya	Undri
66.	*Hystrix indica*	Indian Porcupine	Sehi	Sevdi
67.	*Funambulus pennantii*	Five-striped Palm Squirrel	Gilhari	Gilgawri/Gilgotri
68.	*Lepus nigricollis*	Indian Hare	Khargosh	Sushlyo/Susso
69.	*Herpestes edwardsi*	Common Mongoose	Newla	Nolyo
70.	*Hyaena hyaena*	Stripped Hyaena	Lakadbaggha	Jarakh
71.	*Vulpes bengalensis*	Indian Fox	Lomdi	Lukati/Lunkadi/Lungti
72.	*Cuon alpinus*	Wild Dog	Dhole	Dhole
73.	*Canis aureus*	Jackal	Siyar	Gathdo/Gaadra
74.	*Melursus ursinus*	Sloth Bear	Bhalu	Reechh
75.	*Viverricula indica*	Small Indian Civet	Bijju	Bijuda

76.	*Macaca mulatta*	Rhesus Macaque	Bandar	Lal Muh Walo Bandro/ Molyo
77.	*Presbytis entellus*	Common Langur	Langur	Kala Muh Walo Bandro/ Molyo
78.	*Felis leo*	Lion	Babbar Sher	Nahar/Naar
79.	*Panthera tigris tigris*	Bengal Tiger	Bagh	Bagh
80.	*Panthera pardus fusca*	Leopard, Panther	Bhagera	Baghero
81.	*Felis caracal*	Caracal	Siyahgosh	Mormari Bagheri
82.	*Felis domesticus*	Common Cat	Billi	Bilai/Mazri/Minhki
83.	*Boselphus tragocamelus*	Blue Antelope	Neel-gaya	Roz/Rozda
84.	*Cervus sp.*	Sambar/All Deer	Hiran	Hiran/Hirnyo
85.	*Capra hircus*	Goat	Bakri	Tat/Chhyali/Chhedi
86.	*Canis famaliaris*	Dog	Kutta	Gandkda/Kookra
87.	*Cameleus sp.*	Camel	Uunt	Syand
88.	*Ovis aries*	Sheep	Bhed	Larhdi/Laldi
89.	*Bos bubalus*	Buffalo	Bhains	Pada
90.	*Bos indian*	Ox	Bail	Saand/Naardo
91.	*Equus asinus*	Donkey/Ass	Gadha	Gadho
92.	*Sus scrofa*	Pig	Suar	Soorda

Glossary

Acharya A head teacher (synonymous to a Professor of today)

Advertisement Calls Sexually mature males produce advertisement calls to attract females

Aggregation A group of individuals of the same species, comprised of more than just a mated pair or a family, gathered in the same place but not internally organized

Agonistic Referring to any activity related to fighting, whether aggression or conciliation

Ahimsa Nonviolence

Ahoi Ashthmi A festival falling on the eighth day before Diwali (The festival of lights) where the Sehi or porcupine is worshipped and prayers are offered for the well-being and long-life of children

Allo-grooming Grooming directed at another individual, as opposed to self-grooming which is directed at one's own body

Alpha Referring to the highest-ranking individual within a dominance hierarchy

Altruism Self-destructive behavior performed for the benefit of others

Arbuda To boil

Avatar Incarnation of God

Bagh A garden

Band The term applied to groups of certain social mammals, including coatis and human beings

Bani-Thani The famous ancient portrait of a queen from Rajasthan painted in Kishangarh style. This particular painting has been sold in millions of copies both on paper and cloth

Baori A traditional step-well linked with a natural water source

Bargad The tree *Ficus bengalensis*

Beed/Charagah Community-protected grassland

B.K. Sharma et al. (eds.), *Faunal Heritage of Rajasthan, India: General Background and Ecology of Vertebrates*, DOI 10.1007/978-1-4614-0800-0,

Ber *Ziziphus* sp.

Bhagvad Gita Popularly called Geeta. This holy book of Hindus is based on Lord Krishna's preaching to Arjuna in the battle field of Kurukshetra when he refused to fight the mega war of Mahabharata after witnessing his teacher, relatives, and kins as enemies. Arjuna was one of the five Pandavas and a childhood pal of Lord Krishna.

Bhakar The western-most ridges

Bhoori Pahadi The brown hill

Bishnoi The conservationist community of Rajasthan who are known to sacrifice their lives for protecting plants and animals

Bund/Bandh/Bandha A check dam.

Chappan Fifty Six

Chhapania Kaal The ill-famous famine of 1956 Vikram Samvat

Churel *Holoptela integrifolia*

Cooperative breeding When more-or-less closely related relatives assist in raising young one

Dadupanth An important sect formed by the followers of Saint Dadudayal

Dal-bati-Churma A typical Rajasthani Cuisine. Dal is made of lentils and Pulses in the form of thick curry, *bati* is a solid ball of wheat flour, grilled and mixed with clarified butter or ghee and *churma* is a sweet dish made by mashing up *batis* with sugar and clarified butter or ghee.

Danda A wooden stick.

Demography The rate of growth and the age structure of populations and the processes that determine these properties.

Dhok/Dhawla *Angiossus pendula.*

Dhoran Sand dunes.

Dominance (Social dominance) Social relationship which addresses the management of social conflict including the allocation of limited resources, through the exertion of control and influence in a way that minimizes the risk of overt aggression by way of the use of conventionalized ritual display behaviors and involves a cost–benefit evaluation of the benefits of seeking to win a particular social conflict versus the likely associated cost

Dominance hierarchy The physical domination of some members of a group by other members, in relatively orderly and long-lasting patterns. Except for the highest- and lowest-ranking individuals, a given member dominates one or more of its companions and is dominated in turn by one or more of the others. The hierarchy is initiated and sustained by hostile behavior, albeit sometimes of a subtle and indirect nature

Drive An internal factor that determines how likely an animal is to perform a behavior

Dussehra A major Indian festival which marks the victory of good over bad and celebrated by burning the effigies of Ravana and his brothers. Lord Rama fought with the demon King Ravana who had abducted his wife. The famous epic Ramayana based on the life of Lord Rama aptly describes the fight

Fitness The reproductive success of an organism

Frequency domain In electronics, control systems engineering and statistics, frequency domain is a term used to describe the domain for analysis of mathematical functions or signals with respect to frequency rather than time

Gadia A bullock-cart

Garh A fort

Garud Puran An epic narrating onward journey of the human soul following death

Gauchar A common waste land in a village used for fuel and fodder where agriculture is generally prohibited

Ghana Thicket of the forest, an older name for the world famous Keoladeo National Park, located at Bharatpur

Giri A hill

Girwa A girdle of the hills

Gogamedhi Mela The annual fair held at Gogamedhi in the Churu district of Rajasthan to mark the birthday of the local deity Gogaji

Golden Sal *Shorea robusta*

Gram Panchayat Village Development Block

Grooming The cleaning of the body surface by licking, nibbling, picking with the fingers, or other kinds of manipulation. When the action is directed toward one's own body, it is called self-grooming; when directed at another individual, it is referred to as allo-grooming

Guada A grazing camp in a village

Guar or Gwar *Cyamopsis tetragonoloba* plant, the pods of which are used for making a delicious vegetable in Rajasthan

Gurjan *Dipterocarpus turbinatus*

Guru Spiritual Teacher

Habitat preference Preference for a particular habitat in the ecosystem where chances of survival are maximum

Hadouti Land of Hada Kings, now called Hadouti region

Harem A group of females guarded by a male who prevents other males from mating with them

Havelies Traditionally decorated palatial villas owned by the rich of older times

Hierarchy In general, a system of two or more levels of units, the higher levels controlling at least to some extent the activities of the lower levels in order to integrate the group as a whole. In dominance systems within societies, a hierarchy is the sequence of dominant and dominated individuals

Home range The area that an animal learns thoroughly and patrols regularly. The home range may or may not be defended; those portions that are defended constitute the territory

Icchadhari sarp Some snakes are believed to have the power to fulfill desires and transform in any form including a human being

Imli Tamarind *Tamarindus indicus*

Imprinting Capacity to learn specific types of information at certain critical periods in development

Inbreeding Preferential mating between relatives, mating between sibs, half-sibs, and parent offspring

Jagir Estate

Jagirdar The owner of a large estate

Jamun *Syzygium cumini*

Janamasthmi Birthday of Lord Krishna usually falling on Ashthmi or the eighth day of the Hindi Shrawan or rainy month

Jila Prishad District administrative block

Johad A structure consisting of semicircular mud walls built on small streams, the primary function of which is to recharge ground water

Johar The act of Rajput women to enter into the funeral pyre of their husbands in a bid to save their grace after the men were killed in the battle against Moughals

Kadamb *Mitragyna parvi flora*

Kair Sangri A tasty dry vegetable made of the pods and fruits of xeric trees like *Prosopis cineraria* and *Capparis decidua* . The dish is a popular delicacy which emanated from the villages of Rajasthan

Karaya *Tragacanth* sp.

Kartika A holy month as per Hindu calendar generally falling in October or November each year

Kavad The portable shrine with multiple folding doors, each of which is painted with representations of epics and myths

Khejadli The well-known village in the Jodhpur district of Rajasthan inhabited mainly by the Bishnoi people

Khoh Inclined beds of rock in the Aravallis forming continuous strike ridges and long and narrow valleys, considered an important tiger niche

Kin selection The selection of genes due to one or more individuals favoring or disfavoring the survival and reproduction of relatives (other than offspring) who possess the same genes by common descent. It is one of the extreme forms of group selection contrary to interdemic kins as enemies

Kinship Possession of a common ancestor in the not too distant past. Kinship is measured precisely by the coefficient of kinship and coefficient of relationship (q.v.)

Lahnga and odhni The typical attire of Rajasthani women. Lahnga is a long skirt and odhni is like a stole made of colorful flowing cloth

Lasadia A dissected plateau situated in the east of Jaisamand (325–650 m).

Maa Durga The fearless Goddess who rides a Lion.

Machan A temporary tree house typically used by hunters to keep track of wild animals

Magra A hill

Mahabharat The famous Hindu mythological epic depicting the famous war fought between the *Kaurvas* (hundred sons of a blind king Dhritrashtha, the ruler of Hastinapur) and the *Pandavas* (five brothers, sons of king Pandu, brother of Dhritrashtha) for the throne of Hastinapur

Maharana The king of a big state

Mahua *Madhuca longifolia*

Mal Plains
Mandir A temple
Maru Desert
Marusthali Desert Region
Marwar The region of low-lying plains in the western Rajasthan
Matrilineal Passed from the mother to her offspring, for example, access to a territory or status within a dominance system
Mawa Boiled and condensed milk used for making traditional Indian sweet dishes
Megadiversity An estimate of the total number of all the organisms in an ecosystem and means that a place has a larger percentage of living species in its territory than what would correspond to it if that percentage were proportional to its surface
Meghadoot An epic written by the famous Sanskrit poet Kalidasa during the Gupta Period in India
Mela The socioreligious gathering during festivals
Meru A mountain
Mewar A region having high tableland in the eastern side of Rajasthan
Naal Ravine area
Naga Snake
Nagapanchami An important festival dedicated to snakes and celebrated on the fifth day of the Craven where snakes are worshipped and offered milk to gain knowledge, wealth, and fame
Nali A narrow river
Narsimha Incarnation of Lord Vishnu who appeared in the form of half man and half lion to save and bless his greatest follower Prahlada—a 12-year-old prince—and killed his father, the demon King Hiranyakashyapa
Neem Margosa *Azadirachta indica*
Nullah Outlet for city waste
Oran A common property reserve in a typical Rajasthani village
Outbreeding Preferential mating between nonrelatives
Palash *Butea monosperma*
Pals Clusters of detached huts among the hills where each hut stands on a small mound in the midst of its path of cultivated land. The settlement or the pal is divided into a number of paras or phalas which afford cover and protection in case of attack
Panchayat Samiti Administrative office of the village council
Panchayati Raj Administration consisting of the village council or *Panchayat*
Paraat A large round vessel made of steel or other metal used to knead dough
Parental investment Behavior towards offspring that increases the chances of the offspring's survival at the cost of the parent's ability to invest in other offspring
Patal Loka According to Hindu mythology, the universe has three big spaces or worlds or Lokas—the Earth (*Mrityu loka*) where humans live, the sky (*Dev Loka*) where gods live and *Patal Loka* (in the depth of the earth's crust) where the snakes live
Pathar The rocky area

Phylogeny The evolutionary history of a particular group of organisms; also, the diagram of the "family tree" that shows which species (or groups of species) gave rise to others (contrast with ontogeny)

Proximate explanation Explanation those appeal to motivational variables, experience, and genotype as the cause of behavior

Puran A long ancient story based on a definite subject which is a genre of important Hindu, Jain, and Buddhist religious texts, notably consisting of narratives of the history of the universe from creation to destruction, genealogies of kings, heroes, sages, and demigods, and descriptions of Hindu cosmology, philosophy, and geography

Rahant A typical Persian wheel used to lift water from a well for irrigation

Rajput The warrior clan from Rajasthan

Rajputana A term given to Rajasthan in the pre-Independence era which meant "the land of Rajputs"

Ramayana The famous Hindu mythological Epic written by Rishi Valmiki narrating the interesting story of Lord Rama's life

Rana King of a small state

Reciprocal altruism The trading of altruistic acts by individuals at different times For example, one person saves a drowning person in exchange for the promise (or at least the expectation) that his altruistic act will be repaid if the circumstances are reversed at some time in the future

Riyasat Erstwhile princely state

Sagar A large fresh water wetland

Saka The tradition of sacrificing life to protect trees, also known as Khadana

Salar *Boswellia serrata*

Samadhi A grave

Sand Fish *Ophiomorus tridactylus*

Satya Truth

Sevan *Lasiurus sindicus* grass

Shabadvani Teachings of the founder of Bishnoi religion Guru Jambeshwar or Jambhoji originally scripted in a poetic form

Shikargah Hunting lodge

Shikar-haudi A hunting tower made by the erstwhile rulers

Shikari A hunter

Shikarkhana Department A special department which maintained hunting records

Shikhar Top of the Mountain

Socio-biology Interdisciplinary field of knowledge that explores the biological basis of social behavior including morality

Sperm competition A form of post-copulatory competition which occurs when ejaculates from more than one male might be in female's reproductive tract

Tadka Cumin (Jeera) seeds and/or onion and garlic pieces with red chilli powder mixed in butter or *Ghee* usually added to cooked pulses (Dal) for enhancing the taste

Talaab A pond
Teak *Tectona grandis*
Tehsil Administrative office of a township
Tendu *Diospyros melanoxylon* Roxb
Time–energy budget The amounts of time and energy allotted by animals to various activities
Upanishad Hindu Epics consisting of philosophical texts and considered as sources of knowledge about the Hindu religion. More than 200 are known, of which the first dozen or so, the oldest and most important, are variously referred to as the principal, main, or old Upanishads
Uparmal Plateau
Vahana Mounts or vehicle of gods, goddesses, and deities
Vallabhacharya Sect An important sect formed by the followers of Saint Vallabhacharya
Vedas Indian epics having a large body of texts originated in ancient India. Composed in Vedic Sanskrit, the texts were written in the oldest layer of Sanskrit literature. They are the oldest scriptures of Hinduism based on mantras which are used to worship nature Gods and Goddesses
Vidis Grasslands near villages.
Vrikshmitra Friend of trees
Zoogeography The branch of science which deals with the geographic distribution of animal species and their attributes

Further Readings

Ali S (2002) The book of Indian birds, 13th edn. BNHS-Oxford University Press, Mumbai, India, p 326

Ali S, Ripley RD (1983) Handbook of the birds of India and Pakistan. Oxford University Press, Delhi, India, p 268

Anonymous (1993) Directory of Indian wetlands. Compiled by WWF - India, New Delhi and Asian Wetland Bureau (AWB), Kualalampur, p 263

Anonymous (1995) Wasteland Atlas of India, vol. II. National Remote Sensing Service Agency, Hyderabad, India, p 4706

Anonymous (1998) Thar Desert biosphere reserve. Ministry of Environment & Forests, Government of India, New Delhi, India. Project Document No. 7, p 100

Batten Peter (1976) Living trophies. Crowell, New York, USA, p 246

Bhardwaj GS (2008) Tracking tigers in Ranthambhore. Aureole Publishing, New Delhi, India, p 156

Bird Life International (2008) Critically endangered birds: a global audit. Cambridge, UK, p 15

Blanford WT (1888–1891) The fauna of British India including Ceylon and Burma. Mammalia. Taylor and Francis, London, England, p. 617.

Bond Ruskin (2007) The tribes on my frontier. Penguin Books India Pvt. Ltd., Delhi, India, p 184

Daniel JC (2002) The book of Indian reptiles and amphibians. Bombay Natural History Society and Oxford University Press, Mumbai, India, p 252

Das I (1995) Turtles and tortoises of India. Oxford University Press, Mumbai, India, p 179

Divyabhanusinh (2006) End of the trail: the cheetah in India, 3rd ed. Oxford University Press, New Delhi, India, p 302

Divyabhanusinh (2005) The story of Asia's lions. Marg Publications, Mumbai, India, p 260

Divyabhanusinh (2008) The lions of India. Black Kite Publication, Ranikhet, India, p 267

Ellerman J, Morrison-Scott T (1951) Checklist of Palaearctic and Indian mammals (1758 to 1946). British Museum of Natural History, London, UK, p 810

Grewal Bikram, Sehgal Bittu (2006) The Bharatpur inheritance. Eastern Books Corporation, New Delhi, India, p 168

Grimmett R, Inskipp C, Inskipp T (2006) Pocket guide to the birds of Indian subcontinent. Oxford University Press, New Delhi, India, p 384

Islam MZ, Rahmani AR (2004) Threatened birds of India. Bombay Natural History Society, Mumbai, India, p 102

Linzey Andrew (2001) Animal theology. Kyobunkan Publishing, Tokyo, p 224

Linzey Andrew, Regan Tom (2010) Other nations: animals in modern literature. Baylor University Press, Waco, Texas, USA, p 214

B.K. Sharma et al. (eds.), *Faunal Heritage of Rajasthan, India: General Background and Ecology of Vertebrates*, DOI 10.1007/978-1-4614-0800-0,

McKenna CM, Bell SK (2002) Classification of mammals: above the species level. Eastern Book Corporation, New Delhi, India, p 631
Menon Vivek (2003) Field guide to Indian mammals. Eastern Book Corporation, New Delhi, India, p 200
Murthy TSN (2010) Reptile fauna of India. BR Publishing Corporation, Delhi, India, p 331
Naoroji RK (2006) Birds of prey of the Indian Subcontinent. Christopher Helm, London, UK., p 480
Nigam Venulata (2004) Bibliography of Rajasthan Fauna. Rec. Zool. Surv. India. Occasional Paper No. 224. Zoological Survey of India Publication, Kolkata, India, p 352
Pande Satish, Tambe Saleel, Francis Clement, Sant Niranjan (2009) Birds of Western Ghat and grasslands. Bombay Natural History Society (BNHS) Publication, Mumbai, India, p 377
Prakash I (1994) Mammals of Thar Desert. Scientific Publishers, Jodhpur, India, p 230
Prater SH (1971) The book of Indian animals. Bombay Natural History Society, Mumbai, India, 11th edn. Oxford University Press, Chicago, Illinois, USA, p 384
Rahmani AR, Gayatri U (2005) Birds of wetland and grasslands. BNHS Publications, Mumbai, India, p 226
Rahmani Asad R (1997) Wildlife in the Thar. World Wide Fund for Nature-India, New Delhi, India, p 100
Rahmani AR (2005) Common birds of India. Publication Division, Ministry of Information and Broadcasting, Govt. of India, New Delhi, p 106
Rahmani Asad R, Islam MZ (2009) Ducks, geese and swans of India (their status and distribution). BNHS - Oxford University Press Publication, Mumbai, India, p 374
Rangarajan Mahesh (2001) India's wildlife history: an introduction. Permanent Black, Ranikhet, Uttrakhand, India, p 135
Roonwal ML, Mohnot SM (1977) Primates of South Asia: ecology, sociobiology and behaviour. Harvard University Press, Cambridge, MA, p 421
Sankar K, Goyal SP (2007) Ungulates of India. ENVIS Bulletin: wildlife and protected areas, 7(1). Wildlife institute of India, Dehradun, India. p 91–102
Sankhla Kailash (1977) The story of Indian tiger. Simon & Schuster, New York, USA, p 220
Sharma RC (2008) The wildlife memoirs: a forester recollects. Concept Publishing Co., New Delhi, India, p 192
Sharma VD, Sankhala K (1984) Vanishing cats of Rajasthan. In: Jackson P (ed) The plight of cats, Bougly-Villars, Switzerland, p 117–135. Proceedings of the meeting and workshop of the IUCN/SSC Cat Specialist Group, Kanha National Park, Madhya Pradesh, India, p 228
Sharma VD, Rajpal S (1998) Wild wonders of Rajasthan. Prakash, New Delhi, India, p 189
Singh Arjan Billie (1998) Tiger haven. In: John Moorehead (ed). OUP, p 237
Snodgrass Jeffrey G (2006) Casting kings: bards and Indian modernity. Oxford University Press, Oxford, UK, p 256
Thapar Valmik (2006) The last tiger: struggling for survival. Oxford University Press, New Delhi, India, p 274
Thapar Valmik (2010) The tiger: soul of India. Oxford University Press, New Delhi, India, p 168
Tikadar BK (1983) Threatened animals of India. Zoological survey of India. Govt. of India Press, Calcutta, India, p 308

Index

B.K. Sharma et al. (eds.), *Faunal Heritage of Rajasthan, India: General Background and Ecology of Vertebrates*, DOI 10.1007/978-1-4614-0800-0,

C

H

The Book and Its Audience

This is the first ever scientific documentation of the faunal wealth of the Indian Desert state—Rajasthan, covering the species diversity, distribution and conservation status A scholarly contribution to the field of knowledge, this monumental work provides novel and vital information on the vertebrate faunal heritage of India's largest state.

Broadly falling under the Indo-Malaya Ecozone, the three major biomes of Rajasthan include Deserts and Xeric Shrublands; Tropical and Sub-tropical Dry Broadleaf Forests and Tropical and Sub-tropical Moist Broadleaf Forests. The ecoregions thus covered are North Western Thorn Scrub Forests and the Thar Desert; Khathiar-Gir Dry Deciduous Forests and the Upper Gangtic Plains Moist Deciduous Forests, respectively.

Contrary to popular belief, the well-known Thar or Great Indian Desert occupies only a part of the state. In fact, for the convenience of study, Rajasthan can be seen as diagonally divided by Aravalli mountain ranges into arid and semi-arid regions. The latter has a spectacular variety of highly diversified and unique yet fragile ecosystems comprising lush green fields, marshes, grasslands, rocky patches and hilly terrains, dense forests, the southern plateau, fresh water wetlands and salt lakes.

Apart from the floral richness, the faunal abundance from fishes to mammals including the flagship and threatened species namely Tiger, Leopard, Great Indian Bustard and Lesser Florican, White-naped Tit, Raptors, Demoiselle and Sarus Crane, Chelones, Bats, Wild Ungulates, Small Cats, Bear, Wolf, Smooth-coated Indian Otter, Spiny-tailed Lizard, Giant Flying Squirrel, Gharial and Gangetic River Dolphin described in the 24 chapters penned by top notch wildlife experts research scholars and academics make this volume more palatable and wholesome.

Chapters covering Fossil Records; Conservation of Biodiversity via the age old Public Science of the Desert; Anthropological Account of Communities and Tribes; Socio-cultural, Mythological and Historical aspects of Faunal Conservation and the Fauna in Retrospect (covered in Vol-1 titled *Faunal Heritage of Rajasthan: General Background and Ecology of Vertebrates*); Wildlife Trade; Ecotourism; Climate and other environmental factors like Indira Gandhi Nahar Pariyojna (IGNP) which are believed to have changed the ecological face of Western Rajasthan; Protected Area

B.K. Sharma et al. (eds.), *Faunal Heritage of Rajasthan, India: General Background and Ecology of Vertebrates*, DOI 10.1007/978-1-4614-0800-0,

Network; Tiger Re-introduction Experiment; and Community Conservation are key attractions. The world famous Heronry, Tiger Reserves, Wildlife Sanctuaries and some threat-ridden biodiversity rich areas shall certainly draw the attention of readers from around the world.

The last chapter highlighting issues and insights on conservation and management, initiatives and gaps in research (covered in Vol-2 titled *Faunal Heritage of Rajasthan: Conservation and Management of Vertebrates*) would help researchers from India and abroad to identify potential areas of future collaborative work. The strategies suggested herein can be a powerful tool for international conservational advocacy.

The elaborate content supported by rare photographs and paintings has implications for faunal ecology in similar habitats elsewhere on the Earth. Through these mammoth volumes, the editors have dearly embraced the state of Rajasthan as a whole with particular emphasis on the vertebrate faunal diversity and aspects of its conservation management.

The original manuscript initially conceived and titled as *Faunal Heritage of Rajasthan, India: Ecology and Conservation of Vertebrates* was a bit too large to be presented as a single volume and so the same was split in to two separate books/ volumes entitled *Faunal Heritage of Rajasthan, India: General Background and Ecology of Vertebrates* and *Faunal Heritage of Rajasthan, India: Conservation and Management of Vertebrates*. It is strongly advised that the two books are read and consulted in conjunction with one another rather as a set of two closely related books to have an overall picture of the vertebrate faunal abundance of Rajasthan and its conservation management.

Type of Work

Text and Reference Book

Audience/Written For

Teachers, researchers, amateur and advanced students of Zoology, Environmental Science, Willdife and Conservation Biology, Animal Behaviour, wildlife organizations, freelancers, nature lovers, wildlife photographers, policy makers, and citizens in general.

Keywords

Indian Desert State; Rajasthan; Faunal Heritage; Tiger Reserve; Fauna in Retrospect; Arid Ecosystem; Thar Desert; Bishnoi Community; UNESCO-World Heritage Site; Aravalli Hills; Vertebrate Ecology; Conservation Management; History of Faunal Conservation; Anthropology; Mythology; Hunting Tribes; Tiger Reintroduction Programme; Great Indian Bustard; *Panthera tigris,* Nature Reverence.

About the Editors

Dr. B.K. Sharma (Chief Editor) is Associate Professor and Head, Department of Zoology at R.L. Saharia Government P.G. College, Kaladera (Jaipur), Rajasthan, India. Dr. Sharma is a recipient of coveted national and international fellowships, awards and recognitions namely, Doctoral and Post-doctoral research fellowships of the Indian Council of Medical Research (ICMR), Council of Scientific and Industrial Research (CSIR) and Department of Science and Technology (DST), Government of India at Maulana Azad Medical College, New Delhi; Commonwealth Academic Staff Fellowship at University College London (UCL), UK; Visiting Fellowship under Indian National Science Academy (INSA) and German Research Foundation's (DFG) Bilateral Exchange Programme at University of Heidelberg, Germany; Royal Society's Incoming Fellowship at University of Nottingham, UK; Visiting Scientist under DST's Indo-Slovenian Joint Scientific and Technological Cooperation Programme (2011–2014); Country Advisor for India under "Development Partnerships in Higher Education (DelPHE) Programme" of the British Council, UK and Humane Education Award-2009 by the International Network for Humane Education (InterNICHE, UK) and Proefdiervrij—the Dutch Society for Replacement of Animal Testing. During 2010–2011, he was invited as a key member of the "Core Expert Committee to Consider Discontinuation of Dissection of Animals in Zoology/Life Science in Indian Universities and Colleges". The committee was constituted by the Ministry of Human Resource Development (MHRD), Government of India and the University Grants Commission (UGC)—apex body governing the Indian higher education system. A member of several national and international academic, scientific and professional bodies including the Indian Science Congress Association, American Association for Advancement of Science (AAAS) and International Union for Conservation of

B.K. Sharma et al. (eds.), *Faunal Heritage of Rajasthan, India: General Background and Ecology of Vertebrates*, DOI 10.1007/978-1-4614-0800-0,

Nature (IUCN), Species Survival Commission (SSC), Flamingo Specialist Group (FSG), he has published research papers in national and international peer-reviewed scientific journals apart from organizing a national conference and a workshop cum symposium. Areas of his research interests include "Ecology", "Wildlife and Conservation Biology", "Directed Differentiation of Embryonic Stem Cells into Hepatocytes for use in Bio-artificial liver (BAL) Device and other potential Tissue Engineering Applications" and "Development of Novel Software Technologies, Artificial Intelligence Devices and other superior Educational Paradigms like simulated dissections and Computer Assisted Learning (CAL) methods as Better Pedagogical Tools for Life Science (Zoology) Curriculum with a Perspective on Humane Education". He is currently editing a manual entitled "Humane Alternatives to Dissection and Animal Experimentation in Life Science Education & Training: A Manual for Universities and Colleges" for Cambridge Scholars Publishing (CSP), UK in addition to translating in Hindi the world famous book titled "Animal Liberation" written by Peter Singer- the Australian Philosopher Professor currently at Princeton University, USA for Prakrit Bharati Academy, Jaipur, India.

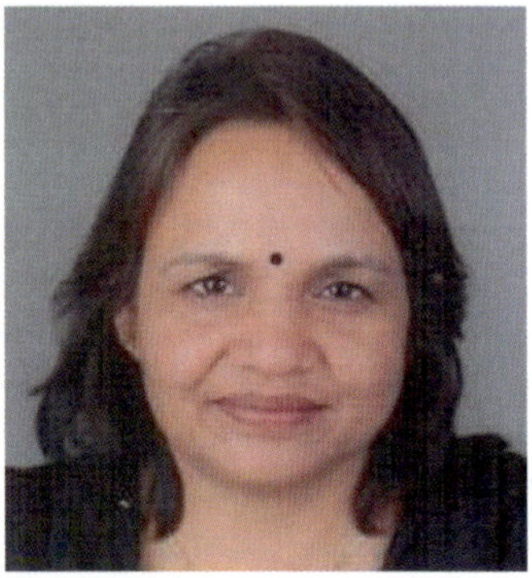

Dr. Seema Kulshreshtha (Co-editor) Associate Professor and Head of Zoology Department at Government Shakambhar PG College, Sambhar Lake (Jaipur), Rajasthan, India, she received her Ph.D. degree from University of Rajasthan, Jaipur in 1989. She has qualified University Grants Commission's National Eligibility Test (NET) in addition to availing fellowships from Indian Council of Medical Research and Indian Council of Agriculture Research. She has worked as a Post-doctoral fellow in a National Fellow Project during 1992–1996 at G.B. Pant University of Agriculture and Technology, Pant Nagar, Uttaranchal, India to develop an indigenous Dipstick Veterinary Pregnancy Diagnostic kit for farmers. She has devoted over 20 years in research and teaching Zoology and contributed papers in national and international journals. She has completed a research project on the study of wetland birds of Sambhar Salt lake Rajasthan. The areas of her research interests include Immunology, Reproductive Physiology and Biodiversity conservation and management.

Dr. Asad R. Rahmani (Co-editor) An eminent ornithologist of international repute, after his PhD on fish, he joined the Bombay Natural History Society where he was Senior Scientist (BNHS) and Project Scientist of Endangered Species Project, Principal Investigator of Grassland Ecology Project and Stork Ecology Project. In 1992, he joined Centre of Wildlife and Ornithology, Aligarh Muslim University, Aligarh where he became Chairman. In 1997, he rejoined BNHS as Director. He has published more than 160 research papers and 250 popular articles, authored nine books and has submitted more than 30 project technical reports to the Government of India and funding agencies. Currently, he is a Global Council member of Birdlife International, UK, and Chairman of BirdLife Asia Council. He serves in many national and state wildlife committees, such as the National Board for Wildlife, Wetland Committee, Bustard Task Force of the Ministry of Environment and Forests. He has been a research guide to ten Ph.D. students and Executive Editor of *Journal of Bombay Natural History Society*, and two popular magazines of BNHS, *Mistnet* and *Hornbill*. Presently, he is working as the Director, BNHS-India, which is a leading institution of India known the world over for high quality of research on wildlife and environment.

Printed by Printforce, the Netherlands